Advances in Convex Analysis and Global Optimization

# Nonconvex Optimization and Its Applications

Volume 54



***The titles published in this series are listed at the end of this volume.***

# Advances in Convex Analysis and Global Optimization

## Honoring the Memory of C. Caratheodory (1873–1950)

Edited by

Nicolas Hadjisavvas
*University of the Aegean*

and

Panos M. Pardalos
*University of Florida*

KLUWER ACADEMIC PUBLISHERS
DORDRECHT / BOSTON / LONDON

Library of Congress Cataloging-in-Publication Data

Advances in convex analysis and global optimization : honoring the memory of C. Caratheodory (1873-1950) / edited by Nicolas Hadjisavvas, Panos M. Pardalos.
p. cm. -- (Nonconvex optimization and its applications ; v. 54)
ISBN 0-7923-6942-4 (alk. paper)
1. Convex functions--Congresses. 2. Mathematical optimization--Congresses. 3. Nonlinear programming--Congresses. I. Carathéodory, Constantin, 1873-1950. II. Hadjisavvas, Nicolas, 1953- III. Pardalos, P. M. (Panos M.), 1954- IV. Series.

QA331.5 .A328 2001
515'.8--dc21

2001029504

ISBN 0-7923-6942-4

---

Published by Kluwer Academic Publishers,
P.O. Box 17, 3300 AA Dordrecht, The Netherlands.

Sold and distributed in North, Central and South America
by Kluwer Academic Publishers,
101 Philip Drive, Norwell, MA 02061, U.S.A.

In all other countries, sold and distributed
by Kluwer Academic Publishers,
P.O. Box 322, 3300 AH Dordrecht, The Netherlands.

*Printed on acid-free paper*



Printed in the Netherlands.

*" As you set out for Ithaka*
*hope your road is a long one,*
*full of adventure, full of discovery.*
*Laistrygonians, Cyclops,*
*angry Poseidon-dont'be afraid of them:*
*you'll never find the things like that on your way*
*as long as you keep thoughts raised high,*
*as long as a rare excitement*
*stirs your spirit and your body.*
*Laistrygonians, Cyclops,*
*wild Poseidon-you won't encounter them*
*unless you bring them along inside your soul,*
*unless your soul sets them up in front of you. "*

- Constantine P. Cavafy

# Contents

**Preface** .................... xi

**Constantin Carathéodory: His Life and Work** .................... xiii
*C. Phili*

1. **Inner Approximation of State-constrained Optimal Control Problems** .................... 1
*F. H. Clarke and R. J. Stern*

2. **Nonsmooth Problems in Mathematical Diagnostics** .................... 11
*V. F. Demyanov, A. Astorino and M. Gaudioso*

3. **Deterministic Global Optimization for Protein Structure Prediction** .................... 31
*J. L. Klepeis and C. A. Floudas*

4. **Some Remarks on Minimum Principles** .................... 75
*F. Giannessi*

5. **Transversal Hypergraphs and Families of Polyhedral Cones** .................... 105
*L. Khachiyan*

6. **SDP Relaxations in Combinatorial Optimization from a Lagrangian Viewpoint** .................... 119
*C. Lemaréchal, F. Oustry*

7. **Convex Analysis in the Calculus of Variations** .................... 135
*R. T. Rockafellar*

8. **Global Minimization and Parameter Estimation in Computational Biology** .................... 153
*J.B. Rosen, A.T. Phillips, S.Y. Oh and K.A. Dill*

9. **Lagrangian Quadratic Bounds in Polynomial Nonconvex and Boolean Models with Superfluous Constraints** .................... 181
*N. Z. Shor*

10. **Generalized Duality in Variational Analysis** .................... 205
*S. M. Robinson*

**11. Clustering via D. C. Optimization** ........ **221**
*H. Tuy, A.M. Bagirov and A.M. Rubinov*

**12. Algorithms and Merit Functions for the Principal Eigenvalue** ........ **235**
*G. Auchmuty*

**13. Modified Versions of the Cutting Angle Method** ........ **245**
*A.M. Bagirov and A.M. Rubinov*

**14. Theoretical and Computational Results for a Linear Bilevel Problem** ........ **269**
*M. Campêlo and S. Scheimberg*

**15. The Lagrangian Search Method** ........ **283**
*P. S. Efraimidis and P. G. Spirakis*

**16. An $\varepsilon$–maximum Principle for Generalized Control Systems** ........ **295**
*A. H. Hamel*

**17. D.C. Optimization Approaches via Markov Models for Restoration of Signal (1-D) and (2-D)** ........ **303**
*Le Thi Hoai An and Pham Dinh Tao*

**18. New Positive Semidefinite Relaxations for Nonconvex Quadratic Programs** ........ **319**
*J. B. Lasserre*

**19. Interval Analysis Applied to Global Minimization** ........ **333**
*C. Lavor and N. Maculan*

**20. Approximate Analytic Center Quadratic Cut Method for Strongly Monotone Variational Inequalities** ........ **345**
*H. J. Lüthi and B. Büeler*

**21. Generating Convex Functions** ........ **361**
*P. Maréchal*

**22. The Method of Moments for Nonconvex Variational Problems** ........ **371**
*R. Meziat, J. J. Egozcue and P. Pedregal*

**23. A Pivoting-based Heuristic for the Maximum Clique Problem** .......... **383**
*A. Massaro and M. Pelillo*

**24. An Analytic Center Self Concordant Cut Method for the Convex Feasibility Problem** .......... **395**
*F. S. Mokhtarian and J. L. Goffin*

**25. Strengthened Semidefinite Programming Relaxations for the Max-Cut Problem** .......... **409**
*M. F. Anjos and H. Wolkowicz*

**26. Supervised Training Using Global Search Methods** .......... **421**
*V. P. Plagianakos, G. D. Magoulas and M. N. Vrahatis*

**27. Learning Rate Adaptation in Stochastic Gradient Descent** .......... **433**
*V. P. Plagianakos, G. D. Magoulas and M. N. Vrahatis*

**28. Improving the Particle Swarm Optimizer by Function "Stretching"** .......... **445**
*K. E. Parsopoulos, V. P. Plagianakos, G. D. Magoulas and M. N. Vrahatis*

**29. Some Convergence Properties of the Steepest Descent Algorithm Revealed by Renormalisation** .......... **459**
*L. Pronzato, H. P. Wynn and A. A. Zhigljavsky*

**30. Interior–Point Algorithm for Dantzig and Wolfe Decomposition Principle** .......... **473**
*Marcos Augusto dos Santos and Paulo Roberto Oliveira*

**31. Stochastic Perturbation Methods for Affine Restrictions** **487**
*M. Bouhadi, R. Ellaia and J. E. Souza de Cursi*

**32. Directed Derivatives of Convex Compact-Valued Mappings** .......... **501**
*R. Baier and E. M. Farkhi*

**33. A Perturbed Auxiliary Problem Method for Paramonotone Multivalued Mappings** .......... **515**
*G. Salmon, J.J. Strodiot and V.H. Nguyen*

**34. A Note on Random Variational Inequalities and Simple Random Unilateral Boundary Value Problems** .............. **531**
*J. Gwinner*

**35. A Comparison Principle and the Lipschitz Continuity for Minimizers** ................................................ **545**
*C. Mariconda and G. Treu*

**36. Tunneling and Genetic Algorithms for Global Optimization** ............................................ **553**
*S. Gómez, J. Solano, L. Castellanos and M. I. Quintana*

**37. Convexity and Monotonicity in Global Optimization** ... **569**
*H. Tuy*

# Preface

There has been much recent progress in global optimization algorithms for nonconvex continuous and discrete problems from both a theoretical and a practical perspective. Convex analysis plays a fundamental role in the analysis and development of global optimization algorithms. This is due essentially to the fact that virtually all nonconvex optimization problems can be described using differences of convex functions and differences of convex sets.

A conference on *Convex Analysis and Global Optimization* was held during June 5 – 9, 2000 at Pythagorion, Samos, Greece. The conference was honoring the memory of C. Caratheodory (1873-1950) and was endorsed by the Mathematical Programming Society (MPS) and by the Society for Industrial and Applied Mathematics (SIAM) Activity Group in Optimization. The conference was sponsored by the European Union (through the EPEAEK program), the Department of Mathematics of the Aegean University and the Center for Applied Optimization of the University of Florida, by the General Secretariat of Research and Technology of Greece, by the Ministry of Education of Greece, and several local Greek government agencies and companies. This volume contains a selective collection of refereed papers based on invited and contributing talks presented at this conference. The two themes of convexity and global optimization pervade this book. The conference provided a forum for researchers working on different aspects of convexity and global optimization to present their recent discoveries, and to interact with people working on complementary aspects of mathematical programming.

We would like to take the opportunity to thank the sponsors, the anonymous referees, Kluwer Academic Publishers, and the authors of the papers, for making the conference successful and the publication of this volume possible. We hope that this book will motivate further research on global optimization and will be a useful reference for students and researchers in the field.

N. Hadjisavvas and P. M. Pardalos
University of the Aegean and University of Florida
January 2001

# CONSTANTIN CARATHÉODORY: HIS LIFE AND WORK

Christine Phili
*Department of Mathematics*
*Faculty of Applied Sciences*
*National Technical University of Athens*
*Zografou Campus*
*15780 Athens*
*Greece*
xfili@math.ntua.gr

## I. AN INTERNATIONAL JOURNEY

A Greek of the diaspora - descendant from a family of erudites, physicians and diplomats, well-educated, with many intellectual interests - Constantin Carathéodory was born on 13 September of 1873 in Berlin where his father Stephanos represented the Ottoman Empire in Berlin's Conference (the Conference treated the arrangement of the Empire's frontiers). Two years later S. C. was appointed plenipotentiary ambassador in Brussels, a capital where he lived even after his retirement (1902) until the end of his life (1907).

C. C. lost his mother when he was 6 years old. He was brought up by his grandmother Euthalia Petrokokinnou (descendant of a noble family of Chio's island) in a house frequented by diplomats (Graf Brandenburg, Lord Vivian), by musicians (Massenet, Paderewski), by historians of music (Gevaert, Bourgault - Ducoudray), by painters (Paul Meyerheim, Emile Wanters, Constantin Meunier), and also by professors of international Law (Aphonse Rivier, F. Martens). Perfectly bilingual in Greek and French (later he learnt German, Italian, English, Turkish and Dutch), he started his education in the winter of 1881. He went to the private primary school of Vanderstock, and later for reasons of health he attended a primary school in the Italian Riviera (1883 - 85, in Bordighera and San Remo). In the classical high school in Brussels he was especially attracted to geometry. In 1886 he studied in the gymnasium Athénée Royal d'Ixelles, from where he graduated in 1891. Although he was qualified for classical studies (later when he was professor in Göttingen, he read many Greek and Latin authors) he was attracted by the quality of the teaching of mathematics by Professor Angenot. For two successive years, he won the first prize in mathematics in the Concours Généraux.

The famous Ecole Polytechnique in Paris, created by the French Revolution and (together with the Ecole Normale Supérieure) center of the higher mathematical education in France, became a model for establishing other analogous institutions in Europe. Following the traces of the distinguished school, the French officer J. Chapelié (1792-1864) established in 1834 in Belgium the Military School of Brussels and became its first commander. The presence of the geometers Dandelin (1794-1847) and Quetelet (1796-1874) stressed the primary role of mathematics.

So when in 1891 C. C. entered this Institution, mathematics constituted an important part of his formation. As he remarked in his Autobiographical Notes, he studied Infinitesimal Calculus from the classical book of Ch. Sturm, while the teaching of the descriptive geometry by Chomé, - who followed Mongian style - familiarized him with geometry. The exemplary teaching of Ch. Lagrange in probabilities, astronomy and geodesy completed his mathematical education.

In 1895 the young engineer paid a visit to the commander of Crete Alexander S. Carathéodory (his relative who was to become his father in law), where he met for the first time Eleftherios Venizelos (1864 - 1936), the most important statesman of Greece in the XX*th* century. During his stay he established a strong friendship with Venizelos and probably the discussions with the expert diplomat and amateur of mathematics A. Carathéodory modified the plans of the young engineer.

It seems that the events of 1895 forced him to abandon Crete and to take refuge in Mytilene, where his cousin Ioannis Aristarchis constructed the road network of the island. Constantin helped him to plan the roads of the island of Samos, but the Greek-Turkish war of 1897 prevented its realization.

After short stays in London and Paris, in the winter of 1858 he started to work as assistant engineer in the construction of the barrages of Assouan and Assiout. There, after working hard all day (as mentioned in his Autobiographical Notes), he studied pure mathematics from the books of C. Jordan, Cours d'Analyse (2nd ed.) and the Analytic Geometry of Salmon-Fiendler.

In the beginning of 1900, he left Egypt and abandoned his career as engineer. His book on Egypt (Athens 1901) constitutes his farewell gift to this fascinating country. He took the decision, against the objections of his family and those of his friends, to devote himself to mathematics. Hesitating between France and Germany he chose Germany and in the first days of May, he was registered in the University of Berlin, where the brilliant triad of Kronecker, Kummer and Weierstrass was succeeded by the promising one of Fuchs, Schwarz and Frobenius.

He attended their lectures with enthusiasm, but he befriended only H. A. Schwarz, who was professor of his uncle Telemachos Carathodory in the Polytechnical School of Zurich, from whom he learnt the foundations of Projective Geometry and theory of Functions. There he made two other faithful friends: E. Schmidt and L. Fejér.

When in 1901 Schmidt continued his studies in Göttingen, Carathéodory following his friend's advise moved to Göttingen too, where F. Klein and D. Hilbert taught. With Klein - whom he later succeeded - he discussed on mathematics, on the French geometrical school of Monge (a subject that Carathéodory knew well from the period of Belgian Military School) and the reform of curriculum in high schools, while attending the course of the History of Mathematics which Klein taught in the University. Later Klein continued to teach it privately (at home) to the young gifted mathematician.

When in 1903 Hans Hahn came from Vienna to Göttingen, he gave a lecture in the Mathematical Society of Göttingen on the Calculus of Variations, especially on the theory of Escherich of two variations in Lagrange's problem. This lecture opened new horizons to the young student who, wishing to construct a geometrical example, conceived the subject of his Thesis. His collaboration with Schwarz in Berlin on E-Weierstrassian functions, led to the final form of his work "On the discontinuous solutions in the Calculus of Variations". He gave it first to Minkowski (because he felt closer to him than to Hilbert and Klein), and defended it in July of 1904. A few weeks later, attending the 3rd International Congress of Mathematicians in Heidelberg, he had the opportunity to meet some great mathematicians such as Painlevé, Lorentz and Lindolöf.

Although his thesis offered many possibilities for a distinguished academic career, it could not open the gates of the University of Athens or those of the Military School as in this period no position was available. So following Hilbert's advice he started to work on his *Habilitationsschrift* "Strong maxima and minima in simple integrals", which he presented in 1905. With it he was nominated Privatdozent in the University of Göttingen.

As Privatdozent of Göttingen University (a post he occupied until 1908) he started to teach kinematics. At the end of the semester, P. Bourtroux arrived in Göttingen. Carathéodory's discussions with him on Borel's proof of Picard's Theorem, led to a generalization of the theorem that appeared in his communication in the French Academy of Sciences (1905).

In 1908 he was elected professor in Hannover's Technical University. This same year he presented in Mathematische Annalen his paper Research on the Foundation of Thermodynamics, which, being published in a mathematical journal, was initially unnoticed by physicists. As late as 1921, after Max Born referred to it, C.C. won the interest and the praise of Max Planck and A. Sommerfeld. In this paper he exposes an alternative logical structure on the foundation of Thermodynamics (the so-called Carathéodory's Principle). The next year he became full professor in the newly established Technical University of Breslau. At this time the young Cretan lawyer Venizelos, already elevated to Greece's leadership and desiring to organize the University Community, invited him to participate in the evaluating committee for professorships in Athens University.

The next years C. C. succeeded Klein in the University of Göttingen, (1913) while in 1918 he became Frobenius' successor in the University of Berlin, a position he maintained for two almost years, until the Greek Prime Minister Eleftherios Venizelos invited him to participate actively to an enterprising and ambitious plan: the Foundation of the University of Smyrni.

## II. THE UNIVERSITY OF SMYRNI - LIGHT FROM THE EAST

With his book on the Theory of Functions (Vorlesungen ber reelle Funktionen, Teubner 1918) C.C. established his international reputation. Next year 1919 Max Planck, the father of Quantum theory, received him with an inspiring speech in the Prussian Academy of Sciences.

The reestablishment of the Ionian coasts as a beacon of Hellenism constitutes one of the basic axes of Venizelos' policy. From the beginning of 1919 the idea to create a University dominated his thoughts, and Carathéodory adopted it and participated in its realization. On the 20th of October 1919, he submitted his proposal on the University's foundation in Smyrni (liberated from the Turks on 2.5.1919) having as a model the Anglo-Saxon Universities. In 1920 he abandoned an international career to support the dream of his old friend. Two months later on the cruiser Elli in the bay of Smyrni, Venizelos, Carathéodory and some others made plans for the new University. Carathéodory wished to bring together all Greek personalities to reach their target. So he invited the professor of the University of Berlin G. Ioakimoglou - born in Ionia- to assist in its organization. Ioakimoglou adopting Venizelos' exhortation, that Greece must civilize Minor Asia, accepted the invitation.

By a decree of the High Committee, the University of Smyrni was founded, having Greek as its official language and comprising the following departments:

1. School of agriculture (to teach botanologists and zoologists, and to provide training to farmers) and of sciences (to teach civil engineers, architects, chemists, geologists etc).
2. School of eastern languages (Turkish, Arabic, Persian, Armenian, ancient and new Hebrew).
3. School of State administration.
4. Commercial school.
5. School of foremen.
6. High Islamic theological seminar.
7. Institute of sanitation (with free tests, preparation of vaccines etc).
8. Library science.

Carathéodory tried to create a rich library unique in the whole of the East and also to provide instruments for the microbiological laboratory (with the assistance of Ioacimoglou). He engaged Dr. Auserer for the library (a member of the National Library of Berlin who spoke Turkish fluently) and created an annex where all information concerning the sale of private libraries and the catalogues of double reprints in public German libraries was gathered. So with the assistance of the Greek state Carathéodory was able to buy these books from the University of Louvain (Germany in Versailles agreed to replace the library of the University, which was burnt in August 1914). The books were packed in 36 large boxes and sent to Smyrni. To those an important number of rare books were added, collected by the Austrian archaeologists working in Ephessos. Carathéodory, being highly cultivated in art and aesthetics (he was a renowned collector), took an interest at the University's furniture and obtained it from the well-known Berliner House Zelder und Platen.

As for the faculty, Carathéodory proposed (besides Prof. Ioakimoglou) F. Theodorides (a graduate of the Polytechnic School of Zurich and student of Pierre Weiss) for the chair of Physics, and P. Karopoulos (assistant of Tamann) for the chair of chemistry. For the department of agriculture he indicated Th. Kessissoglou, a graduate of the agricultural School of Gembloux (Belgium) and organizer of some agricultural schools in China, Colombia and Uruguay. Also following Carathéodory's proposition, Paschevitz was appointed machinist of the University; its secretary was Nicolaos Kriticos (who later became Professor of the National Technical University of Athens) who in this period served in the army division of Cydonies (Asia Minor).

In October of 1921 Carathéodory returned to Smyrni and worked with enthusiasm to organize the University, whose emblem was defined by him: Light from the East, believing that the new Institution will be the most perfect University of the East. The lectures never started as the tragic events in August 1922 swept everything. His daughter Mrs Despina Rodopoulou-Carathéodory referred to this period: "He stayed to save anything he could: books, instruments etc. These were put in different ships hoping that one day they will arrive in Athens. My father stayed there until the very last moment". George Horton, consul of U.S.A. in Smyrni wrote a book, *All about Turkey*, which was translated in Greek. In this book Horton notes: "One of the last Greeks I saw on the streets of Smyrna before the entry of the Turks was Professor Carathéodory, president of the doomed University. With him departed the incarnation of Greek genius of culture and civilization in the Orient".

A refugee in Athens too, he was nominated Professor of Mathematical Analysis in the University of Athens (1922) and Professor of Mechanics in the National Technical University of Athens (1923). Nevertheless, in these difficult years the horizon of his scientific activity remains limited. "*They make great efforts to keep me here* – he wrote to his friend I. Kalitsounakis in 24.2.1924 – *and of course I would wish to stay if I was sure that I could serve my country. But given the situation in our Department (as it continues to be for many years) this seems to me to be very difficult*". So when in this same year (1924) the famous physicist A. Sommerfeld invited him to succeed F. Lindeman in the University of München, he accepted this invitation and he stayed there for the rest of his life (1950).

## III. THE FERTILE YEARS

In 1928, invited by the American Mathematical Society he visited the U.S.A., as guest professor for one semester in Harvard, giving lectures in various Universities (Austin, Berkley etc). In a letter to his relative Penelopi Delta (distinguished Greek writer, a kind of Greek Madame de Sevigny), he defended the decision for his settlement abroad. "*I assure you that in the few months that I am here I achieved more than in two years in Athens. My activity in Smyrni could be fruitful, nevertheless Greece did not lose anything from my departure, on the contrary! Rather than waste my time there, I can be here more useful through my propaganda for Greece*".

The victory of the Liberal Party (August 1928), gave the opportunity to its founder to become concerned again with the educational problem of the country. For the reform of the University education Venizelos asked for Carathéodory's contribution, which he accepted, hoping to convince the Government to take effective measures.

So Venizelos recalled Carathéodory to reform the University of Athens and to establish the University of Thessaloniki, an idea that he had conceived at the end of the First World War. In the session of the Greek Parliament on 17 December of 1929, the Prime Minister E. Venizelos declared: "*I remind you that after the end of the Great War, Venizelos' Government at that time decided not only to establish a second University but a third one Wishing that the new Universities will be established as perfectly as possible, it recalled Professor Carathéodory, who is not only a great mathematical genius but he is preeminently distinguished organizer of Institutions, since the Great Germany charged him with the organization of the Polytechnic School of Breslau. So, we invited Mr. Carathéodory and we told him 'We need to establish these two Universities, we consider you as the best one to organize them. We think that it is better to start with the University of Smyrni and after that of Thessaloniki*' ".

In 1930 he was appointed Governmental commissioner (a post which he kept until 1933) in the Universities of Athens and Thessaloniki. In the sessions of the University Senate (according to the law 5143 of 10.7.1931 he was Rector) he gave the directions for the creation of internal regulations, he planned the formation of the veterinarian school (unique in Greece which starts functioning only in 1950). "*The Reform of the University of Athens*" Athens (1930) is his report which he submitted to the Government about the restructuring of the National and Kapodistrian University, where he proposes measures for the improvement of the financial situation of the institution, the reduction of the number of students, the reform of the exams, the procedure for choosing professors etc. The State Law no. 5343/1932, regulating all University matters until 1982, was based on that report.

Wishing to contribute to the Greek mathematical activities, he participated in 1934 in the First Interbalcanic Mathematical Conference with his paper "*On the Equations of Mechanics*" which was presented by his old colleague in Berlin University, Prof. Nicolaos I. Hadjidakis.

In 1936, visiting again United States, he was impressed by the progress of Greek scientists, as he wrote in a letter addressed to the Rector of the University of Athens, Professor Panagiotis Zervos: "*What made a great impression on me in America is the truly great progress of Greek scientists not only the number of Greeks who we met as professors in the Universities is astonishingly great, but some among them occupy quite exceptional positions*".

This same year, he was member of the Committee for the Fields Medal with Cartan, Birchoff, Tagaki and president F. Severi. The Medal was awarded, during the 10th International Congress in Oslo, to Jesse Douglas and Lars Ahlfors.

In 1937 he returned once more to Greece and gave lectures in the National Technical University. He also participated in the festivities for the centenary of the Archeological Society with his paper "*On the curves of the pedestal of the Parthenon and on the distance of its columns*", where he rejected the conception of Penrose and Stevens, who asserted that the curves of the Parthenon are parabolas. Carathéodory remarks that Iktinos constructed the only curves known in his period, circles of great diameters, since the concept of conic sections came after the 5th century.

## IV. HIS CONTRIBUTION TO MATHEMATICS

For many years Carathéodory was the editor of Mathematische Annalen and member of the editorial board of the Rendiconti del Circolo Matematico di Palermo [earlier, another distinguished Greek mathematician was on the editorial board of this same journal, Kyparissos Stephanos (1857 - 1917)]. He was elected member of many Academies (Berlin, München, Göttingen, Bologna, Lincei, Athens), while noteworthy is his election in the Papal Academy.

"*Everything which causes our admiration, always according to Euclid's recollection, is the unified perfection of two abilities, which are rarely found together in the human intellect. That of the imaginative perception of space together with the deepest power of the abstract calculus. This combination resulting from the inherited treasure of distant ancestors constitutes the philosophical stone of Carathéodory's creation*". With these words, E. Schmidt describes the background of the man who could be compared with the ancient Greeks, in his post mortem tribute to Carathéodory (Ges. W. Bd. V p. 412).

His research covers many fields of mathematics, pure and applied. His entry by chance (as that of the young Lagrange few centuries ago) into the temple of mathematics is marked by his interest in the Calculus of Variations, on which he continued to work for a long time. Carathéodory in his Thesis (1904) demonstrated that the theory of smooth functions could be extended to non-differentiable curves. Besides this, as he knew quite well the history of this branch [later he wrote the "*Introduction into Euler's publications on the Calculus of Variations*" in L. Euler Opera Omnia (1) 24 1952 p. viii-lxii] he generalized an idea of Johann I. Bernoulli for the resolution of the brachystochrone problem which became a turning point of the theory.

The theory of functions constitutes Carathéodory's second main field of research, which he extended in many areas. His first communication "*On some generalizations of M. Picard's theorem*" in the French Academy of Sciences in 1905, led him to many generalizations as for example his work with E. Landau "*Contribution on convergence of functions sequences*" (1911) or "*About a generalization of Picard's theorems*" (1920), both of them published in the Proceedings of the Prussian Academy of Sciences.

Another area comprises coefficient problems in power series expansions, which he starts to work in 1907 (see "*On the domain of variation of power series coefficients which do not accept a given value*" in Mathematische Annalen; also for Fourier's coefficients see "*On the constant Fourier's coefficients of monotone functions*" (1920) in the Proceedings of the Prussian Academy of Sciences).

Problems arising from Schwarz's lemma (which took this name from his proposition) compose an important part of his research; see for example "*A generalization of Schwarz's lemma*" Bull. Amer. Math. Soc. (1937).

He makes another significant advance in the theory of several variables in "*A theorem for analytic functions of several variables analogous to Vitali's theorem*" published in Crelle's Journal (1931).

Nevertheless his most remarkable contribution is focused on the field of conformal representation on which he started to work in 1912-1913: see "*Research on conformal representation in constant and variable regions*". (Math. Ann. 1912) or "*About the reciprocal correspondence of boundary in conformal representation of the interior of a Jordan's curve*" (Math. An. 1913).

After Riemann's first germ of conformal representation in 1851, the main theorem of conformal representation of simply connected regions on the circle of unit radius was demonstrated (in a greatly simplified proof) by Carathéodory (see *Elementary Proof of the fundamental theorem of conformal representation* published in the Math. Abh. for H. A. Schwarz's 50th anniversary of his thesis (1914)). Nevertheless his main contribution is his theory of boundary correspondence in which he introduced the geometrical set theoretic properties of those boundaries (theory of Primenden).

His third main field comprises the theory of real functions. The profound study of C. Jordan's book Cours d' Analyse [2nd ed. (1893) where the author presented the new analysis as it was developed in the works of Weierstrass, Dedekind, Cantor and Jordan himself] offers him the rigorous background for his research, which took its definite form in his book *Lectures on Real Functions* (1918) where he completes Borel and Lebesgue and makes his first step of axiomatization. O. Perron in his tribute for Carathéodory, stresses that with this book Carathéodory gave a wonderful gift to the mathematicians. He also worked on the measure and integration theory. His paper *On the linear measure of point sets, a generalization of length's concept* in the Scientific Society of Göttingen (1914) is characteristic. Carathéodory returned to it, in a more generalized and abstract form, in *Algebrization's sketch of Intergral's concept* which he presented in 1938 in the Proceedings of the Bavarian Academy of Sciences. The first paper of N. Bourbaki (published in 1935 in the Proceedings of the French Academy of Sciences) was related to the theory of integration and was based on one of his theorems.

Wishing to reform his classic book on real functions, he decided that its 3rd edition would be divided to two volumes. The first one comprises numbers, point-sets, functions and in the second volume, as his student Nicolaos Criticos (1894-1985) pointed out in his obituary, would comprise the new Carathéodory's theory concerning our objects of thought which the great mathematician names them *Soma* in German (from the Greek word $\sigma\tilde{\omega}\mu\alpha$=body). These *Somen* are defined by simple postulates and so the well-known concepts of Lebesgue's and Stieltjes integrals result as an application of his theory. Probably Hilbert's Grundlagen der Geometrie (1899) and his conviction that the objects of mathematical thought are the symbols themselves (Math. An. 1926) inspired him to this attempt of abstractness. The form of his theory appeared in his posthumous book *Measure and Integral and its algebrization* (1956).

Ending our tribute to C.C., we wish to stress his interest in applied mathematics. His papers on thermodynamics are noteworthy. The first one *Research on the Foundations of Thermodynamics* (Math. An. 1909) was completed in 1925 by another important paper: *On the determination of the energy and the absolute temperature with irreversible processes*, which was published in the Proceedings of the Berlin Academy of Sciences. His interest in the theory of relativity was expressed in 1924 in his paper: *On the axiomatization of the special theory of relativity* presented in the Proceedings of the Prussian Academy of Sciences where he developed an axiomatic foundation of Einstein's special theory of relativity and presented general transformation equations, partial case of which are the Lorentz - Minkowski's transformations.

As Perron pointed out, C.C. was very interested in the History of Mathematics. His knowledge of ancient Greek mathematics was remarkable. His friend, professor I. Kallitsounakis in his posthumous tribute to Carathéodory in the Proceedings of the Academy of Athens (session of 23rd February 1950), reveals that in the last years of his life he encouraged him to work on Plato's mathematics. But Carathéodory considered that his age constituted an obstacle to working in this vast area of research.

Nevertheless his knowledge of the history of mathematics was not restricted to ancient Greek mathematics. He knew quite well Euler's work as well as that of his contemporaries. He wished to understand the mathematical methods of every period and, using the precious ideas of the past, to arrive to solutions of present problems. So apart from his contribution on Euler's Opera Omnia he presented various papers as for example *The beginning of research in the calculus of variations*, Osiris (1937) or *The Significance of Erlanger Program* (1919) or his contribution on *Mathematics* (with W. van Dyck) for the 50 years of German Science on the occasion of the 70th anniversary of Friedrich Schmidt - Ott.

"*All publications of Carathéodory* – wrote Nicolaos Criticos – *have something plastic and all are well crafted, even the less important, with the love and persistence and conscientiousness of a good craftsman, thus giving to the lecture an analogous aesthetic pleasure, like the well-conceived and presented with perfection work of art*".

## V. BIBLIOGRAGHY

Carathéodory's articles were collected in Cesammelte Mathematische Schriften in 5 Vols (Munich 1954 - 1957) (already in 1944 the Bavarian Academy of Sciences decided to edit Carathéodory's Opera Omnia).

Carathéodory's books are:

*Vorlesungen uber reelle Funktionen*, Teubner, Leipzig - Berlin 1918; 2nd edition 1928).

*Conformal Representation*, Cambridge Tracts in Mathematics and Mathematical Physics No 28 (Cambridge, 1932).

*Variationsrechnung und partielle Differentialgleichungen erster Ordnung*, Teubner, Leipzig - Berlin 1935; 2nd ed. 1956.

# 1 INTRODUCTION

We consider a standard control system of the form

$$\dot{x}(t) = f(x(t), u(t)) \quad , u(t) \in \mathcal{U}, t \in [0,T] \text{ a.e.}, \tag{1.1}$$

and the following optimal control problem $P$ associated with it: To minimize $\ell(x(T))$ over the state trajectories $x$ satisfying (1.1) as well as the constraints

$$x(0) = x_0, x(t) \in S \text{ a.e.}$$

The underlying time interval $[0,T]$, the sets $\mathcal{U}$ and $S \subset \mathbf{R}^n$, the point $x_0 \in S$ and the functions $\ell : \mathbf{R}^n \to \mathbf{R}$ and $f : \mathbf{R}^n \times \mathcal{U} \to \mathbf{R}^n$ are all given. The term *state trajectory* refers to an absolutely continuous solution $x$ of (1.1) for some measurable control function $u$ with values in $\mathcal{U}$. Under our standard hypotheses (see below), this is equivalent to being a solution of the differential inclusion

$$\dot{x}(t) \in F(x(t)) \text{ a.e.}, \tag{1.2}$$

where the multifunction $F$ is given by

$$F(x) := \{f(x,u) : u \in \mathcal{U}\}.$$

The unilateral state constraint $x(t) \in S$ is the main focus of this article. It makes the problem $P$ significantly harder than its counterpart which does not incorporate that constraint. For example, the necessary conditions of maximum principle type are much more complex in the presence of state constraints, and analytical solutions correspondingly harder to identify.

A well-known approach to approximating the state-constrained problem by one free of such unilateral constraints involves replacing the constraint by a penalty term. A well-known instance of this involves the study of the problem $P_k$ of minimizing

$$\ell(x(T)) + k \int_0^T d_S(x(t))\, dt$$

over the state trajectories $x$ satisfying $x(0) = x_0$, where $d_S$ signifies the (Euclidean) distance function associated with $S$ and $k$ is a positive parameter (see for example [5]). If $V(P_k)$ is the value of the problem $P_k$ ((i.e., its infimum) and similarly for $V(P)$, then clearly we have $V(P_k) \leq V(P)$. Under reasonable hypotheses on the data, one can show that $V(P_k)$ converges to $V(P)$ as we make the penalization increasingly severe by letting $k$ tend to infinity.

We think of $P_k$ as being an *outer* approximation of $P$, since it is a "bigger" problem (admits more feasible trajectories) and gives rise to a lower approximation of $V(P)$, one for which the state constraint is absent. Our purpose is to develop a useful *inner* approximation to $P$.

A simple way to obtain an inner approximation to $P$ is to replace $S$ by a subset $S'$ of $S$. Denoting the new problem by $P'$, we have evidently $V(P') \geq V(P)$, and $P'$ is a smaller problem than $P$ (it admits fewer feasible trajectories). But this approach fails in one important respect: the new problem still incorporates a state constraint, and is no easier than the original. We shall nonetheless study such approximations, but only as a stepping stone to the ones that truly achieve our goal of eliminating the state constraint. This will be achieved through a certain redefinition of the multifunction $F$ near the boundary.

## 2 HYPOTHESES

The following standing hypotheses are made on the data of the problem.

F1 The function $\ell$ is locally Lipschitz; f is continuous and locally Lipschitz in the state variable $x$ uniformly for $u \in \mathcal{U}$; that is, for each bounded subset $\Gamma$ of $\mathbf{R}^n$, there exists $K_\Gamma > 0$ such that

$$\|f(x,u) - f(y,u)\| \leq K_\Gamma \|x - y\|$$

whenever $(x,u)$ and $(y,u)$ lie in $\Gamma \times \mathcal{U}$.

F2 The function $f$ exhibits *linear growth* in the state variable: There exist positive numbers $\gamma_1, \gamma_2$ such that

$$\|f(x,u)\| \leq \gamma_1 \|x\| + \gamma_2 \quad \forall (x,u) \in \mathbf{R}^n \times \mathcal{U}.$$

F3 The velocity set $F(x) := \{f(x,u) : u \in \mathcal{U}\}$ is closed and convex for each $x$.

**Remark**

1. These standard assumptions guarantee that for each control function $u$ on $[0,T]$, there corresponds via (1.1) and the initial condition a unique trajectory $x$.

2. Hypothesis $[F3]$ amounts to saying that we are considering *relaxed systems* (see for example [1]).

A natural question that arises in connection with the problem $P$ is whether it is feasible to find trajectories that remain in $S$. The set $S$

is said to be *weakly invariant* (or *viable*) if, whenever $x_0$ belongs to $S$, then there is at least one trajectory beginning at $x_0$ which remains in $S$. A characterization of this important property is available through the *lower hamiltonian* of the system, the function $h$ defined by

$$h(x,\zeta) := \min_{u \in \mathcal{U}} \langle \zeta, f(x,u) \rangle.$$

If $S$ is closed (we shall require rather more below), then it is known that $S$ is weakly invariant if and only if

$$h(x,\zeta) \leq 0 \ \ \forall \zeta \in N_S^P(x), \ \ \forall x \in S,$$

where $N_S^P(x)$ signifies the *proximal normal cone* to $S$ at $x$. (For details of this and other concepts of nonsmooth analysis we refer to [4].) If $S$ is closed and weakly invariant, then there exists at least one state trajectory that begins at $x_0$ and remains in $S$ thereafter, and standard existence theorems proven via "sequential compactness of trajectories" imply that a solution to problem $P$ exists (see [4]).

The following hypotheses are made in connection with the set $S$ and the behavior of the dynamics near its boundary:

S1 The set $S$ is compact and *wedged*: for each $x \in S$, the (Clarke) tangent cone $T_S^C(x)$ has nonempty interior.

S2 The *strict inwardness* condition:

$$h(x,\zeta) < 0 \ \ \forall 0 \neq \zeta \in N_S^C(x), \ \ \forall x \in S,$$

where $N_S^C(x)$ signifies the (Clarke) normal cone to $S$ at $x$.

**Remark**

1. Since $N_S^C(x) \supset N_S^P(x)$, hypothesis [S2] implies that $S$ is weakly invariant, as we would wish in order for the problem to be well-posed. But more than that, it requires that for each point $x$ on the boundary of $S$, and for each nonzero outward normal vector $\zeta$ at $x$, there exists an admissible velocity vector $v \in F(x)$ which makes a strictly negative inner product with $\zeta$. Thus the dynamics retain the possibility to leave the boundary of $S$. It can be shown (see [6]) that hypotheses [S1] and [S2] imply the existence of $\delta > 0$ such that each point $x \in \operatorname{bdry} S$ admits $u_x \in \mathcal{U}$ so that

$$f(x,u_x) + \delta \bar{B} \subset T_S^C(x). \tag{1.3}$$

2. The class of sets $S$ that are wedged (see [4]) enjoys a number of properties that has made it useful in a variety of other applications;

it includes smooth manifolds with boundary as well as convex sets with nonempty interior; more generally, it is the class of sets which are locally describable as epigraphs of Lipschitz functions. A set that is wedged is the closure of its interior, and its boundary has measure zero.

# 3 INNER SET APPROXIMATIONS

We denote by $\hat{S}$ the closure of the complement of $S$, and (for $r > 0$) by $S_r$ the *r-inner approximation* of $S$ given as follows:

$$S_r := \{x : d_{\hat{S}}(x) \geq r\}.$$

Consider now the version of problem $P$ in which the state constraint set is $S_r$, and with underlying time interval $[\tau, T]$ and initial condition $x(\tau) = \alpha$ (where $\tau \in [0, T]$.) We shall refer to this problem as $P(S_r, F)(\tau, \alpha)$, where our notation explicitly keeps track of both the state constraint set and the dynamics involved in the problem, and where the initial data $(\tau, \alpha)$ is set apart as a parameter to be handled differently. The value associated to the problem $P(S_r, F)(\tau, \alpha)$ is denoted $V(S_r, F)(\tau, \alpha)$. Note that $P(S_r, F)(\tau, \alpha)$ is only defined when $r \leq d_{\hat{S}}(\alpha)$ (so that the initial condition and the state constraint are not in conflict).

The following results are a consequence of a familiar type of "tracking argument"; proofs are given in [6].

**Proposition 1** *$V(S, F)$ is Lipschitz: There is a constant $K_V$ such that*

$$|V(S, F)(\tau, \alpha) - V(S, F)(\tau', \alpha')| \leq K_V \|(\tau - \tau', \alpha - \alpha')\|$$

*for all $(\tau, \alpha), (\tau', \alpha') \in [0, T] \times S$.*

**Proposition 2** *There exists $r_0 > 0$ and a constant $c_0$ such that for each $r \in [0, r_0]$ and $(\tau, \alpha) \in [0, T) \times \operatorname{int} S$ having $r \leq d_{\hat{S}}(\alpha)$, one has*

$$|V(S_r, F)(\tau, \alpha) - V(S, F)(\tau, \alpha)| \leq c_0 r.$$

This result says that $P(S_r, F)(\tau, \alpha)$ is a good (inner) approximation to $P(S, F)(\tau, \alpha)$ when $r$ is small, but it is not the approximation we seek, since in general its state constraint will be active. The approximation we seek will be obtained by making an appropriate change in the dynamics of the problem near the boundary.

# 4 MODIFYING THE DYNAMICS

It is a consequence of (1.3) that every point $x$ on the boundary of $S$ admits an admissible velocity vector $v(x) \in F(x)$ which points strictly inward; the following establishes that the dependence on $x$ can be made sufficiently regular; the proof appears in [6].

**Proposition 3** *There exists a Lipschitz function $v : S \to \mathbf{R}^n$ which is a selection for $F$ on $S$, and such that*

$$v(x) + \frac{\delta}{2}\bar{B} \subset T_S^C(x) \quad \forall x \in S. \tag{1.4}$$

For each $r > 0$ sufficiently small, we define a new multifunction $F_r$ in the following way: if $x \in S_r$, then $F_r(x) = F(x)$; if $x \in S \backslash S_r$, then $F_r(x)$ is obtained by interpolation between $F(x)$ and $v(x)$ as follows:

$$F_r(x) := \frac{d_{\hat{S}}(x)}{d_{\hat{S}}(x) + d_{S_r}(x)} F(x) + \frac{d_{S_r}(x)}{d_{\hat{S}}(x) + d_{S_r}(x)} v(x).$$

Note that $F_r(x)$ reduces to $\{v(x)\}$ when $x$ lies on the boundary of $S$. Denoting by the same letter $v$ the standard Lipschitz extension of $v$ to all of $\mathbf{R}^n$ (see for example [4, 1.11.6]), we then set $F_r(x) := \{v(x)\}$ for all $x \in \mathbf{R}^n \backslash S$.

The values of $F_r$ outside $S$ are irrelevant, but it is convenient to have $F_r$ globally defined. It is easily seen to be compact and convex valued as well as globally Lipschitz. Note also that $v(x) \in F_r(x)$ $\forall x$. The differential inclusion $\dot{x} \in F_r(x)$ is thus well-posed, and the fact that $F_r(x)$ reduces to $\{v(x)\}$ on the boundary of $S$, where $v(x)$ is a vector pointing strictly inwards, implies (unsurprisingly) that solutions of the differential inclusion that begin in $S$ must immediately enter its interior, and remain in $S$ thereafter. Explicitly, we have the following consequence of (1.4) (see [6]).

**Proposition 4** *The constant $r_0$ of Proposition 2 can be chosen small enough so that for some positive constant $c_1$, for any $r \in (0, r_0)$ and for any $x_0 \in S \backslash S_{r_0}$, any solution $x$ of the differential inclusion*

$$\dot{x} \in F_r(x), \; x(0) = x_0$$

*satisfies*

$$d_{\hat{S}}(x(t)) - d_{\hat{S}}(x_0) \geq c_1 t$$

*at least as long as $t$ still satisfies $x(t) \notin S_r$.*

We now examine the optimal control problem $P(S, F_r)(0, x_0)$. Note that it constitutes an inner approximation to the original problem $P(S, F)$ $(0, x_0)$, since $F_r(x) \subset F(x)$. Thus $V(S, F_r)(0, x_0) \geq V(S, F)(0, x_0)$. Also, the problem $P(S, F_r)(0, x_0)$ does not really have a state constraint, since its trajectories all remain in $S$ even when the state constraint is deleted (in light of the preceding proposition). We now prove that $P(S, F_r)(0, x_0)$ is a good approximation to $P(S, F)(0, x_0)$ when $r$ is small.

**Theorem 1** *For any $\epsilon > 0$, for all $r > 0$ sufficiently small and for any $(\tau, \alpha) \in [0, T] \times S$ we have*

$$|V(S, F_r)(\tau, \alpha) - V(S, F)(\tau, \alpha)| \leq \epsilon.$$

Proof: We shall make explicit how small $r$ must be. Let $K_\ell$ be a Lipschitz constant for $\ell$ on $S$, and let $M_F$ be a bound on the values of $F(x)$, uniform for $x \in S$. We shall prove the assertion of the proposition for any $r$ such that

$$r < r_0, r(c_0 + K_V(1 + M_F)/c_1) < \epsilon, 2rK_\ell M_F/c_1 < \epsilon.$$

Consider first the case in which $\tau \in [T - r/c_1, T]$. Let $x_1$ solve the problem $P(S, F_r)(\tau, \alpha)$ and $x_2$ solve $P(S, F)(\tau, \alpha)$. Since $\|\dot{x}_i\| \leq M_F$, we have

$$\begin{aligned} 0 &\leq V(S, F_r)(\tau, \alpha) - V(S, F)(\tau, \alpha) = \ell(x_1(T)) - \ell(x_2(T)) \\ &= \ell(x_1(T)) - \ell(\alpha) + \ell(\alpha) - \ell(x_2(T)) \\ &\leq K_\ell\|x_1(T) - \alpha\| + K_\ell\|x_2(T) - \alpha\| \\ &\leq 2K_\ell M_F(T - \tau) \\ &\leq 2K_\ell M_F r/c_1 < \epsilon, \end{aligned}$$

which gives the required inequality.

Consider now the case in which $\tau \in [0, T - r/c_1)$, and the subcase in which $\alpha \in S_r$. Then

$$V(S, F_r)(\tau, \alpha) \leq V(S_r, F)(\tau, \alpha) \leq c_0 r + V(S, F)(\tau, \alpha),$$

in view of Proposition (2). Once again the required inequality follows.

There remains only the case in which $\tau \in [0, T - r/c_1)$, and $\alpha \in S \backslash S_r$. Let $y$ be the solution of the Cauchy problem

$$\dot{y} = v(y), y(\tau) = \alpha,$$

where $v$ is the function of Proposition 3. Note that $y$ is a trajectory for $F_r$. It is a consequence of Proposition 4 that for some positive $t$ satisfying $\tau < t \leq \tau + r/c_1$, we have $y(t) \in S_r$. An optimal path for $P(S_r, F_r)(t, y(t))$ (or equivalently, $P(S_r, F)(t, y(t))$) can then be followed, giving rise to the conclusion

$$V(S, F_r)(\tau, \alpha) \leq V(S_r, F)(t, y(t)).$$

Noting that $\|y(t) - \alpha\| \le M_F(t - \tau)$, we derive

$$\begin{aligned} V(S, F_r)(\tau, \alpha) &\le V(S_r, F)(t, y(t)) \\ &\le V(S, F)(t, y(t)) + c_0 r \end{aligned}$$

(in view of Proposition 2)

$$\le V(S, F)(\tau, \alpha) + K_V \|(t - \tau, y(t) - \alpha)\| + c_0 r$$

(in view of Proposition 1)

$$\begin{aligned} &\le V(S, F)(\tau, \alpha) + r(c_0 + K_V(1 + M_F)/c_1) \\ &< V(S, F)(\tau, \alpha) + \epsilon, \end{aligned}$$

which again implies the required assertion. □

**Remark**

1. Notice that the theorem establishes that we have an approximation that is uniform over the initial data $(\tau, \alpha) \in [0, T] \times S$. Although this uniformity may not appear to be important in the present context, it is in fact crucial for developing optimal state-constrained *feedback* , as shown in [6]. The approach taken there, roughly speaking, involves the construction of a feedback $\hat{u}(t, x)$ based upon the function $V(S, F_r)(t, x)$ viewed as a Lyapunov function, and through the technique called *proximal aiming* (see [3], [2]).

2. We remark finally that the theorem leads to the following new route for deriving *necessary conditions* for state-constrained optimal control problems from the ones for the unconstrained case: let $\bar{x}$ be the solution of the problem $P(S, F)(0, x_0)$, and let $x_i$ be (for each positive integer $i$) a solution of $P(S, F_{r_i})(0, x_0)$, where $r_i$ decreases to 0 and where $x_i$ converges uniformly to $\bar{x}$. (It is easy to arrange to have such a sequence.) Then necessary conditions of maximum principle type can be written for the solution $x_i$ of $P(S, F_{r_i})(0, x_0)$, since the state constraint here is superfluous. Now study the convergence of these necessary conditions as $i \to \infty$. The resulting necessary conditions will involve measures in general; see [7].

## References

[1] F.H. Clarke. *Optimization and Nonsmooth Analysis.* Wiley-Interscience, New York, 1983. Republished as vol.5 of *Classics in Applied Mathematics*, SIAM, 1990.

[2] F.H. Clarke, Yu.S. Ledyaev, L.Rifford, and R.J. Stern. Feedback stabilization and Lyapunov functions. *SIAM J. Control and Optim.* To appear.

[3] F.H. Clarke, Yu.S. Ledyaev, R.J. Stern, and P.R. Wolenski. Qualitative properties of trajectories of control systems: a survey. *J. Dyn. Control Sys.*, 1:1–48, 1995.

[4] F.H. Clarke, Yu.S. Ledyaev, R.J. Stern, and P.R. Wolenski. *Nonsmooth Analysis and Control Theory.* Graduate Texts in Mathematics, vol. 178. Springer-Verlag, New York, 1998.

[5] F.H. Clarke and P.D. Loewen. State constraints in optimal control: a case study in proximal normal analysis. *SIAM J. Control and Optim.*, 25:1440–1456, 1987.

[6] F.H. Clarke, L. Rifford, and R.J. Stern. Feedback in state constrained optimal control. To appear.

[7] R.B.Vinter. *Optimal Control.* Birkhäuser, 2000.

Chapter 2

# NONSMOOTH PROBLEMS IN MATHEMATICAL DIAGNOSTICS

V.F.Demyanov*

*Applied Mathematics Dept.*

*St.Petersburg State University*

*Staryi Peterhof, St.Petersburg*

*198904 RUSSIA*

A.Astorino

*Dip. di Elettronica Informatica e Sistemistica*

*Università della Calabria*

*Via Pietro Bucci, Cubo 41C*

*87030 Rende (CS)*

*ITALY*

M.Gaudioso

*Dip. di Elettronica Informatica e Sistemistica*

*Università della Calabria*

*Via Pietro Bucci, Cubo 41C*

*87030 Rende (CS)*

*ITALY*

**Abstract** Different problems of practical importance, such as problems of pattern recognition, medical and technical diagnostics, identification, study of experimental data etc., can be described by mathematical models which require the solution of the problem of separation of two or more sets. In many cases the sets mentioned are inseparable, and the problem arises

*The research was supported by the Russian Foundation for Fundamental Studies (grant RFFI No. 00-01-00454)



*N. Hadjisavvas and P.M. Pardalos (eds.), Advances in Convex Analysis and Global Optimization,* 11–30.

to "separate" them in the "best" possible way (e.g., to identify as many points as possible or to minimize the number of unidentified points).

In the paper the identification problems are treated as optimization problems. More sophisticated models are described by nonsmooth optimization problems. An algorithm is suggested allowing to construct a sequence of linear criterion functions which can be used for the identification of the points of the sets.

**Keywords:** Separation Theorem, Exact Penalty Function, Criterion Function, Mathematical Diagnostics, Diagnostic problem, Identification Problem, Quasidifferentiable Function, Necessary Optimality Conditions, Nonsmooth Analysis.

# 1 INTRODUCTION

Different problems of practical importance, such as problems of pattern recognition, medical and technical diagnostics, identification, study of experimental data etc., can be described by mathematical models which require the solution of the problem of separation of two or more sets. In this paper the case of two sets is discussed. If the convex hulls of these sets do not intersect, the separation theorem can be employed. However, in many cases the sets mentioned are inseparable, and the problem arises to "separate" them in the "best" possible way (for example, to identify as many points as possible or to minimize the number of unidentified points).

If these sets, $A$ and $B$, intersect, then one can try to "isolate" the intersection of these sets. If one of the sets, say $A$, is "inside" the convex hull of the set $B$, one can try to find, e.g., a polyhedral set $\Omega$ containing (in some sense) the set $A$ and to minimize the number of "displaced" points of both sets (for example, the sum of the number of points of $B$, *belonging* to the set $\Omega$, and the number of points of $A$, *not belonging* to the set $\Omega$).

The idea of "isolation" was implemented in [1].

In many cases the identification problems can be treated as optimization problems. More sophisticated models are described by nonsmooth optimization problems. In the paper an algorithm is suggested allowing to construct a sequence of linear criterion functions which can be used for the identification of the points of the sets. The results and tools of Quasidifferential Calculus are employed to solve the related optimization problems.

The paper is organized as follows. In Section 2 the problem is formulated. In Section 3 the convex case is considered. The discrete case is discussed in Section 4. Optimality conditions are formulated in Section 5. An example illustrating the algorithm is presented in Section 6.

## 2 STATEMENT OF THE PROBLEM

Let $\mathcal{A} \subset I\!R^n$ and $\mathcal{B} \subset I\!R^n$ be some closed bounded sets. The problem is to get a rather simple criterion for distinguishing between elements of the sets $\mathcal{A}$ and $\mathcal{B}$. A point $c \in \mathcal{A} \cup \mathcal{B}$ is identified if it is possible to indicate that either $c \in \mathcal{A} \setminus \mathcal{B}$, or $c \in \mathcal{B} \setminus \mathcal{A}$. In the case $c \in \mathcal{A} \cap \mathcal{B}$ the point $c$ is *essentially unidentifiable.*

We discuss two cases: the first one is the case where both sets $\mathcal{A}$ and $\mathcal{B}$ are convex or one of them is convex. If each of the sets $\mathcal{A}$ and $\mathcal{B}$ is convex we can try to "isolate" the intersection of the sets $\mathcal{A}$ and $\mathcal{B}$. If only one of the sets $\mathcal{A}$ and $\mathcal{B}$ is convex, we can try to "isolate" only the set $\mathcal{A}$ or only the set $\mathcal{B}$.

The second case is the case where the sets $\mathcal{A}$ and $\mathcal{B}$ are *a priori* unknown, however, two sets of points

$$A = \{a_i \mid i \in I = 1 : N_1\} \subset \mathcal{A}, \quad B = \{b_j \mid j \in J = 1 : N_2\} \subset \mathcal{B}$$

are given. The problem is again to get a criterion for distinguishing between elements of the sets $\mathcal{A}$ and $\mathcal{B}$. Since $\mathcal{A}$ and $\mathcal{B}$ are unknown, we can use the elements of $A$ and $B$ only. Naturally, it is assumed that $A$ and $B$ are representative samplings of the sets $\mathcal{A}$ and $\mathcal{B}$, respectively, therefore we shall manipulate with the sets $A$ and $B$.

In the present paper we use only linear functions to perform the desired "isolation". Namely, we are looking for a sequence of linear functions (in Section 3) or a family of polyhedral sets (Section 4) which can be used to identify points belonging to $\mathcal{A} \cup \mathcal{B}$.

## 3 THE CONVEX CASE

First let us consider the case where each of the sets $\mathcal{A}$ and $\mathcal{B}$ is convex. Different dispositions of $\mathcal{A}$ and $\mathcal{B}$ are possible. Let us consider only some of them.

**1)** $\mathcal{A} \cap \mathcal{B} = \emptyset$ (see Fig. 2.1);

**2)** $\mathcal{A} \cap \mathcal{B} \neq \emptyset$, $\mathcal{A} \not\subset \mathcal{B}$, $\mathcal{B} \not\subset \mathcal{A}$ (see Figs. 2.2, 2.3);

**3)** $\mathcal{A} \subset \mathcal{B}$ (or $\mathcal{B} \subset \mathcal{A}$) (see Fig. 2.3).

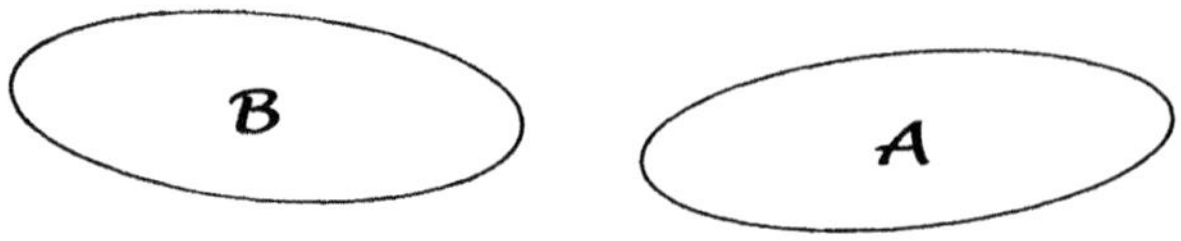


*Figure 2.1*

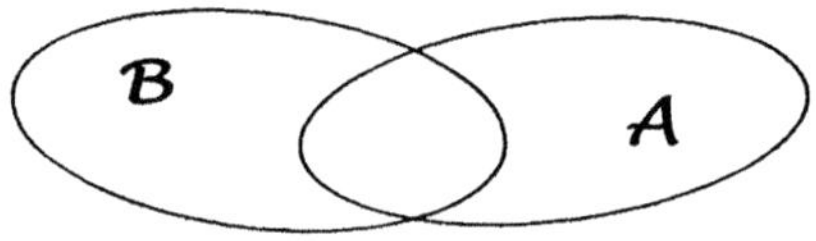


*Figure 2.2*

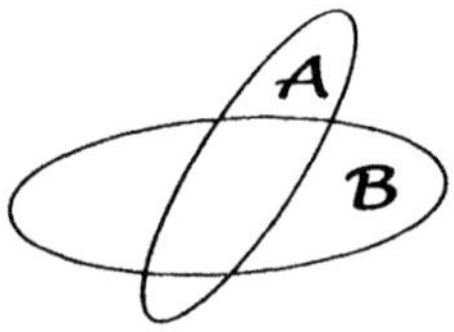


*Figure 2.3*

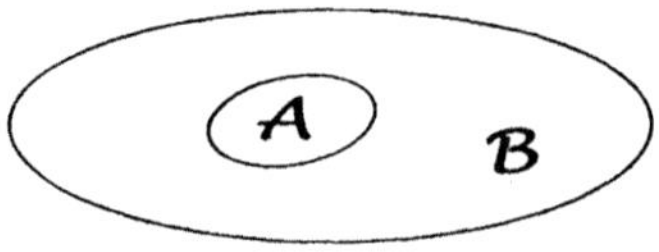


*Figure 2.4*

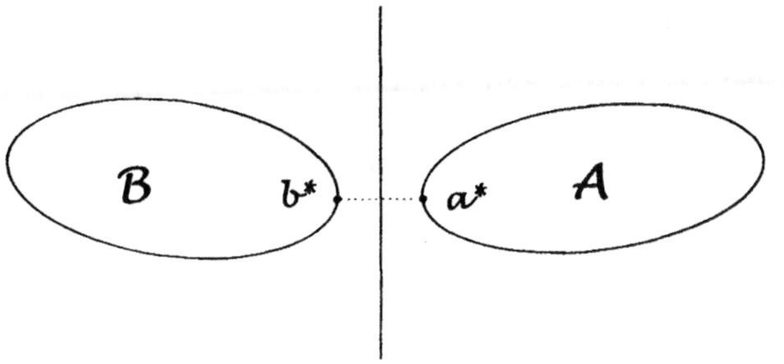


*Figure 2.5*

## 3.1 IDENTIFICATION BY MEANS OF ONE HYPERPLANE. THE CASE $A \cap B = \emptyset$.

The simplest criterion function is a linear one. In the case 1) (see Fig. 2.1) by the separation theorem (see [15, 8]) there exist a vector $y \in \mathbb{R}^n$ and a real $d \in \mathbb{R}$ such that

$$(x, y) + d > 0 \quad \forall x \in \mathcal{A}, \tag{2.1}$$

$$(x, y) + d \leq 0 \quad \forall x \in \mathcal{B}. \tag{2.2}$$

To find $\overline{y} = [y, d] \in R^{n+1}$, satisfying (2.1), (2.2), it is sufficient to find

$$f(\mathcal{A}, \mathcal{B}) = \min_{a \in \mathcal{A}, b \in \mathcal{B}} ||a - b||^2 = ||a^* - b^*||^2. \tag{2.3}$$

Since $\mathcal{A} \cap \mathcal{B} = \emptyset$ then $f(\mathcal{A}, \mathcal{B}) > 0$. Put

$$y^* = \frac{a^* - b^*}{||a^* - b^*||}, \quad d^* = (a^* + b^*, y^*) = \frac{||a^*||^2 - ||b^*||^2}{||a^* - b^*||}.$$

By the necessary condition for a minimum (see, e.g., [8]) the pair $[y^*, d^*]$ satisfies the inequalities (see Fig. 2.5)

$$(x, y^*) + d^* > \frac{1}{2}||a^* - b^*|| > 0 \quad \forall x \in \mathcal{A}, \tag{2.4}$$

$$(x, y^*) + d^* \leq -\frac{1}{2}||a^* - b^*|| < 0 \quad \forall x \in \mathcal{B}. \tag{2.5}$$

The function

$$r(x, \overline{y}^*) = (x, y^*) + d^* \tag{2.6}$$

can be used as a criterion one: Let it be known that $c \in \mathcal{A} \cup \mathcal{B}$.

Inequalities (2.4) and (2.5) imply that

if

$$r(c, \overline{y}^*) > 0 \quad then \quad c \in \mathcal{A}, \tag{2.7}$$

if

$$r(c, \overline{y}^*) < 0 \quad then \quad c \in \mathcal{B}. \tag{2.8}$$

## 3.2 THE CASE $\mathcal{A} \cap \mathcal{B} \neq \emptyset$.

Now consider the case $\mathcal{A} \cap \mathcal{B} \neq \emptyset$. Let us construct the functions

$$\overline{f}_{\mathcal{A}}(y) = \max_{a \in \mathcal{A}}(a, y), \quad \overline{f}_{\mathcal{B}}(y) = \max_{b \in \mathcal{B}}(b, y), \tag{2.9}$$

$$\underline{f}_{\mathcal{A}}(y) = \min_{a \in \mathcal{A}}(a, y), \quad \underline{f}_{\mathcal{B}}(y) = \min_{b \in \mathcal{B}}(b, y), \tag{2.10}$$

$$F_{\mathcal{AB}}(y) = |\overline{f}_{\mathcal{A}}(y) - \overline{f}_{\mathcal{B}}(y)| + |\underline{f}_{\mathcal{A}}(y) - \underline{f}_{\mathcal{B}}(y)|. \tag{2.11}$$

Since $\mathcal{A} \cap \mathcal{B} \neq \emptyset$ then

$$\overline{f}_{\mathcal{A}}(y) \geq \underline{f}_{\mathcal{B}}(y), \ \overline{f}_{\mathcal{B}}(y) \geq \underline{f}_{\mathcal{A}}(y) \quad \forall y \in I\!R^n. \tag{2.12}$$

Note that $F_{\mathcal{AB}}(y) = F_{\mathcal{AB}}(-y)$.

The function $r(x, y) = (x, y)$, for any $y \neq 0_n$, can be taken as a criterion function:

Let $c \in \mathcal{A} \cup \mathcal{B}$. If

$$r(c, y) > \min\{\overline{f}_{\mathcal{A}}(y), \overline{f}_{\mathcal{B}}(y)\} = \overline{f}_{\mathcal{AB}}(y) \tag{2.13}$$

or

$$r(c, y) < \max\{\underline{f}_{\mathcal{A}}(y), \underline{f}_{\mathcal{B}}(y)\} = \underline{f}_{\mathcal{AB}}(y), \tag{2.14}$$

then the point $c$ can be identified (see Figs. 2.6-2.9):

If $r(c, y) > \overline{f}_{\mathcal{AB}}(y)$ then

$$c \in \begin{cases} \mathcal{A} & if \ \overline{f}_{\mathcal{A}}(y) > \overline{f}_{\mathcal{B}}(y), \\ \mathcal{B} & if \ \overline{f}_{\mathcal{A}}(y) < \overline{f}_{\mathcal{B}}(y), \end{cases} \tag{2.15}$$

if $r(c, y) < \underline{f}_{\mathcal{AB}}(y)$ then

$$c \in \begin{cases} \mathcal{A} & if \ \underline{f}_{\mathcal{A}}(y) < \underline{f}_{\mathcal{B}}(y), \\ \mathcal{B} & if \ \underline{f}_{\mathcal{A}}(y) > \underline{f}_{\mathcal{B}}(y). \end{cases} \tag{2.16}$$

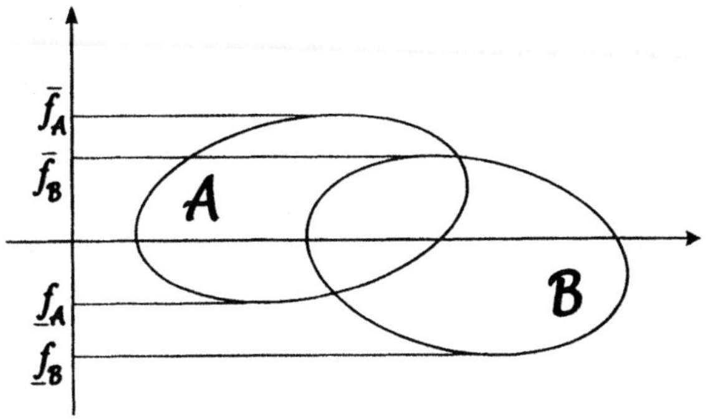


*Figure 2.6*

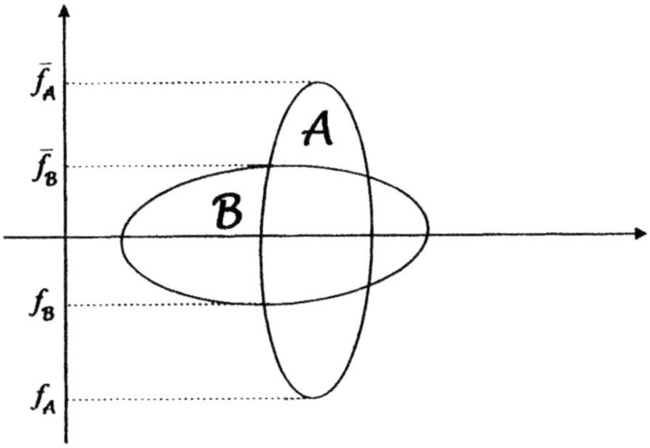


*Figure 2.7*

In the case

$$r(c,y) \in [\underline{f}_{\mathcal{AB}}(y)\}, \overline{f}_{\mathcal{AB}}(y)] \tag{2.17}$$

we are unable to identify the point $c$. If

$$F_{\mathcal{AB}}(y) = 0 \quad \forall y \neq 0_n$$

then $\mathcal{A} = \mathcal{B}$.

Let $h_1(y)$ be the width of the set $\mathcal{A} \cap \mathcal{B}$ in the direction $y$. It is clear that if $||y|| = 1$ then $h_1(y) \leq \overline{f}_{\mathcal{AB}}(y) - \underline{f}_{\mathcal{AB}}(y)$.

Observe that

$$h_2(y) \;=\; \max\{\overline{f}_{\mathcal{A}}(y), \overline{f}_{\mathcal{B}}(y)\} \;-\; \min\{\underline{f}_{\mathcal{A}}(y), \underline{f}_{\mathcal{B}}(y)\} \tag{2.18}$$

is the width of the set $\mathcal{A} \cup \mathcal{B}$ in the direction $y$ (see Figs. 2.6-2.9). Note that any point $c \in \mathcal{A} \cup \mathcal{B}, c \notin \mathcal{A} \cap \mathcal{B}$, can be identified by the function

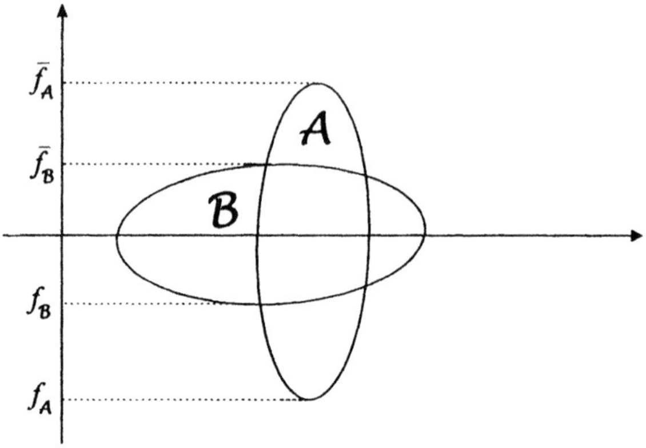


*Figure 2.8*

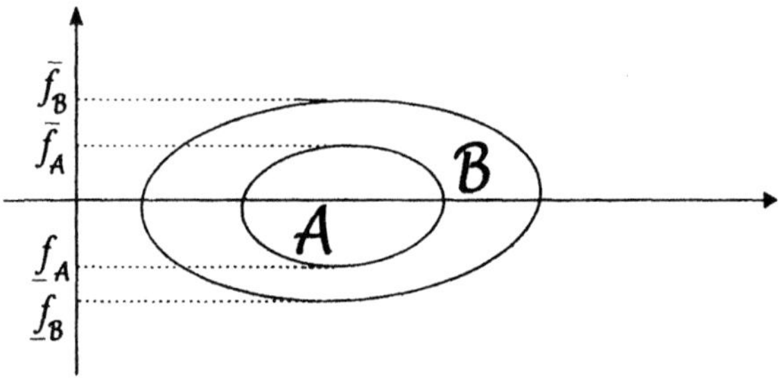


*Figure 2.9*

$r(x, y)$ for some $y$. The set $\mathcal{A} \cap \mathcal{B}$ is the set of points which can not be identified (there exists no $y$ such that $c$ can be identified by means of the function $r(x, y)$). The set $\mathcal{A} \cap \mathcal{B}$ is the set of *essentially unidentifiable points.*

## 3.3 AN ALGORITHM IN THE CASE $\mathcal{A} \cap \mathcal{B} \neq \emptyset$.

From the preceding considerations one can use the following identification procedure. Put

$$\mathcal{A}_1 = \mathcal{A}, \quad \mathcal{B}_1 = \mathcal{B}, \quad F_1(y) = F_{\mathcal{A}_1\mathcal{B}_1}(y).$$

Find

$$\max_{\|y\|=1} F_1(y) \quad = \quad F_1(y_1^*). \tag{2.19}$$

Let us construct the set

$$\mathcal{D}_1 \quad = \quad \left\{ c \in \mathcal{A}_1 \cup \mathcal{B}_1 \;\middle|\; r(c, y_1^*) \notin [\underline{f}_{\mathcal{A}_1\mathcal{B}_1}(y_1^*), \overline{f}_{\mathcal{A}_1\mathcal{B}_1}(y_1^*)] \right\}. \tag{2.20}$$

The points $c \in \mathcal{D}_1$ can be identified by the function $r(x, y_1^*)$ as described in the previous subsection.

Construct the sets

$$\mathcal{A}_2 = \mathcal{A}_1 \setminus \mathcal{D}_1, \;\; \mathcal{B}_2 = \mathcal{B}_1 \setminus \mathcal{D}_1. \tag{2.21}$$

These sets are convex and closed.

Now put $F_2(y) \;=\; F_{\mathcal{A}_2\mathcal{B}_2}(y)$ and find

$$\max_{\|y\|=1} F_2(y) \quad = \quad F_2(y_2^*). \tag{2.22}$$

Let us construct the set

$$\mathcal{D}_2 \quad = \quad \left\{ c \in \mathcal{A}_2 \cup \mathcal{B}_2 \;\middle|\; r(c, y_2^*) \notin [\underline{f}_{\mathcal{A}_2\mathcal{B}_2}(y_2^*), \overline{f}_{\mathcal{A}_2\mathcal{B}_2}(y_2^*)] \right\}. \tag{2.23}$$

The points $c \in \mathcal{D}_2$ can be identified by the function $r(x, y_2^*)$.

Continuing in the same manner we construct the sequence of hyperplanes $\{l_s\}$ where $l_s = \{x \in \mathbb{R}^n \mid r(x, y_s^*) = 0\}$ and the sequences of sets $\{\mathcal{A}_s\}$ and $\{\mathcal{B}_s\}$ (see (2.21)). If for some $s$ we get $F_s(y_s^*) = 0$, it means that $\mathcal{A}_s \;=\; \mathcal{B}_s \;=\; \mathcal{A} \cap \mathcal{B}$, and the process terminates.

For practical and computational reasons we have to terminate this procedure after some $k$ steps. As a result, we get $k$ hyperplanes $l_1^*, ..., l_k^*$. The vectors $y_1^*, ..., y_k^*$ can be used for the identification of a given point $c \in \mathcal{A} \cup \mathcal{B}$. First, we check whether one of the inequalities

$$r(c, y_1^*) > \overline{f}_{\mathcal{A}_1\mathcal{B}_1}(y_1^*)\}, \;\; r(c, y_1^*) < \underline{f}_{\mathcal{A}_1\mathcal{B}_1}(y_1^*)\} \tag{2.24}$$

holds. If yes, the point $c$ is identified as described above. If not, we check whether one of the inequalities

$$r(c, y_2^*) > \overline{f}_{\mathcal{A}_2\mathcal{B}_2}(y_2^*)\}, \quad r(c, y_2^*) < \underline{f}_{\mathcal{A}_2\mathcal{B}_2}(y_2^*)\} \tag{2.25}$$

holds. If yes, the point $c$ is identified. If not, we take $y_3^*$ and continue as above.

It is possible to show that for every point $c \in \mathcal{A} \cup \mathcal{B}, c \notin \mathcal{A} \cap \mathcal{B}$, there exists $k(c)$ such that $c$ can be identified by the vectors $y_1^*, ..., y_{k(c)}^*$.

## 4 THE DISCRETE CASE

In practice it is natural to assume that $\mathcal{A}$ and $\mathcal{B}$ are convex hulls of the sets $A$ and $B$:

$$\mathcal{A} = co\,A = co\,\{a_i \mid i \in I = 1 : N_1\},\ \mathcal{B} = co\,B = co\,\{b_j \mid j \in J = 1 : N_2\}. \tag{2.26}$$

Then the above described algorithm becomes implementable since the sets $\mathcal{D}_s$ (see (2.20), (2.23)), $\mathcal{A}_s$ and $\mathcal{B}_s$ are easily constructed by deleting some $a_i$'s and $b_j$'s and inserting some new points.

In this case it is also possible to introduce instead of $F_{\mathcal{AB}}(y)$ (see (2.11)) other performance functionals. Let us consider in detail the case (see Fig. 3) $co\ A \subset co\ B$ (now it is not required that $co\ A = \mathcal{A}$, $co\ B = \mathcal{B}$). Let us try to "isolate" only the set $A$.

Choose the vectors $\overline{y}_1, ..., \overline{y}_m$ such that

$$\overline{y}_k = [y_k, d_k],\ y_k \in I\!R^n,\ d_k \in I\!R,\ ||y_k|| = 1 \quad \forall k \in 1 : m \tag{2.27}$$

and consider the functions $h_k(x, \overline{y}_k) = (x, y_k) + d_k$ and the hyperplanes

$$l_k(\overline{y}_k) = \{x \in I\!R^n \mid h_k(x, \overline{y}_k) = 0\},\ k \in 1 : m.$$

Put

$$\overline{Y} = [\overline{y}_1, ..., \overline{y}_m] \in I\!R^{(n+1)m}, \quad h(x, \overline{Y}) = \max_{k \in 1:m} h_k(x, \overline{y}_k)$$

and construct the set

$$\Omega(\overline{Y}) = \{x \in I\!R^n \mid h_k(x, \overline{y}_k) \leq 0 \quad \forall k \in 1 : m\} = \{x \in I\!R^n \mid h(x, \overline{Y}) \leq 0\}. \tag{2.28}$$

Assume that

$$\Omega(\overline{Y}) \neq \emptyset. \tag{2.29}$$

The set $\Omega(\overline{Y}) \subset I\!R^n$ is convex and closed but may be unbounded (it is a polyhedral set)(see Fig. 2.10). Let

$\alpha_A(\overline{Y})$ be the number of points of $A$ belonging to $\Omega(\overline{Y})$,

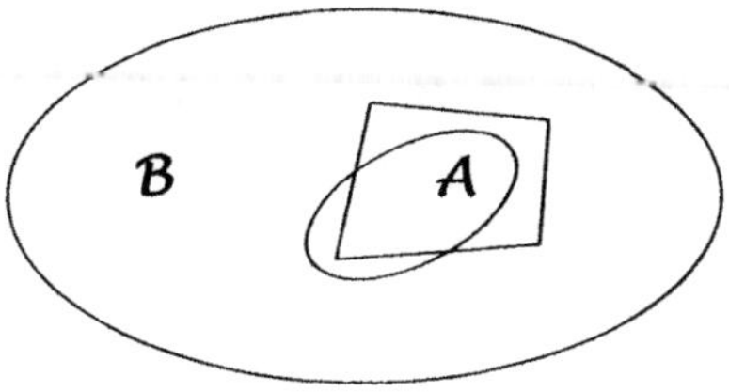


*Figure 2.10*

$\beta_A(\overline{Y})$ be the number of points of $A$ *not* belonging to $\Omega(\overline{Y})$,
$\alpha_B(\overline{Y})$ be the number of points of $B$ belonging to $\Omega(\overline{Y})$,
$\beta_B(\overline{Y})$ be the number of points of $A$ *not* belonging to $\Omega(\overline{Y})$,
$\alpha_B^0(\overline{Y})$ be the number of points of $B$ belonging to $\text{int}\,\Omega(\overline{Y})$.

The ratio $p_{A1}(\overline{Y}) = \frac{\alpha_A(\overline{Y})}{N_1}$ is the part of points of the set $A$ belonging to $\Omega(\overline{Y})$;

the ratio $p_{A2}(\overline{Y}) = \frac{\beta_A(\overline{Y})}{N_1}$ is the part of points of the set $A$ *not* belonging to $\Omega(\overline{Y})$;

the ratio $p_{B1}(\overline{Y}) = \frac{\alpha_B(\overline{Y})}{N_2}$ is the part of points of the set $B$ belonging to $\Omega(\overline{Y})$,

the ratio $p_{B2}(\overline{Y}) = \frac{\beta_B(\overline{Y})}{N_2}$ is the part of points of the set $A$ *not* belonging to $\Omega(\overline{Y})$.

Note that

$$p_{A1}(\overline{Y}) + p_{A2}(\overline{Y}) = 1, \;\; p_{B1}(\overline{Y}) + p_{B2}(\overline{Y}) = 1. \tag{2.30}$$

We will interpret the above defined ratios as probabilities.

The function $h(x, \overline{Y})$ may serve as a criterion function in the following sense:

Let $c \in A \cup B$.

If $h(c, \overline{Y}) \leq 0$ then $c \in \Omega(\overline{Y})$ and the probability $p_{\alpha 1}(\overline{Y}) = \frac{\alpha_A(\overline{Y})}{\alpha_A(\overline{Y})+\alpha_B(\overline{Y})}$ is that $c \in A$, while the probability $p_{\alpha 2}(\overline{Y}) = \frac{\alpha_B(\overline{Y})}{\alpha_A(\overline{Y})+\alpha_B(\overline{Y})}$ is that $c \in B$.

If $h(c, \overline{Y}) > 0$ then $c \notin \Omega(\overline{Y})$ and the probability $p_{\beta 1}(\overline{Y}) = \frac{\beta_A(\overline{Y})}{\beta_A(\overline{Y})+\beta_B(\overline{Y})}$ is that $c \in A$, while the probability $p_{\beta 2}(\overline{Y}) = \frac{\beta_B(\overline{Y})}{\beta_A(\overline{Y})+\beta_B(\overline{Y})}$ is that $c \in B$.

Assuming that $A$ and $B$ are representative samplings of $\mathcal{A}$ and $\mathcal{B}$, respectively, we may expect that the above result holds in the case $c \in \mathcal{A} \cup \mathcal{B}$ as well, i.e.

*If $h(c,\overline{Y}) \leq 0$ then the probability $p_{\alpha 1}(\overline{Y})$ is that $c \in \mathcal{A}$, and the probability $p_{\alpha 2}(\overline{Y})$ is that $c \in \mathcal{B}$. If $h(c,\overline{Y}) > 0$ then the probability $p_{\beta 1}(\overline{Y})$ is that $c \in \mathcal{A}$, while the probability $p_{\beta 2}(\overline{Y})$ is that $c \in \mathcal{B}$.*

Now let us construct proper performance functionals. Choose and fix any $\varepsilon > 0$. Let us introduce the functions

$$f_{11}(\overline{Y}) = \frac{1}{N_1}\sum_{i\in I}\max\{0, h(a_i,\overline{Y})\}, \tag{2.31}$$

$$f_{12}(\overline{Y}) = \frac{1}{N_2}\sum_{j\in J}\max\{0, -h(b_j,\overline{Y})\}, \tag{2.32}$$

$$f_{1\varepsilon}(\overline{Y}) = \sum_{i\in I}\max\{0, \frac{h(a_i,\overline{Y})}{|h(a_i,\overline{Y})|+\varepsilon}\}, \tag{2.33}$$

$$f_{2\varepsilon}(\overline{Y}) = \sum_{j\in J}\max\{0, -\frac{h(b_j,\overline{Y})}{|h(b_j,\overline{Y})|+\varepsilon}\}. \tag{2.34}$$

If $f_{11}(\overline{Y}) = 0$ then $A \subset \Omega(\overline{Y})$. If $f_{12}(\overline{Y}) = 0$ then $B \cap int\ \Omega(\overline{Y}) = \emptyset$. Note that

$$f_{1\varepsilon}(\overline{Y}) \underset{\varepsilon\downarrow 0}{\longrightarrow} \beta_A(\overline{Y}), \tag{2.35}$$

$$f_{2\varepsilon}(\overline{Y}) \underset{\varepsilon\downarrow 0}{\longrightarrow} \alpha_B^0(\overline{Y}), \tag{2.36}$$

If $f_{1\varepsilon}(\overline{Y}) = 0$ then $A \subset \Omega(\overline{Y})$. If $f_{2\varepsilon}(\overline{Y}) = 0$ then $B \cap int\ \Omega(\overline{Y}) = \emptyset$.

From the above considerations it is clear that we are interested to get $\overline{Y}^* \in I\!R^{(n+1)m}$ such that

$$f_{11}(\overline{Y}^*) = 0,\ f_{12}(\overline{Y}^*) = 0 \tag{2.37}$$

(and then $f_{1\varepsilon}(\overline{Y}^*) = f_{2\varepsilon}(\overline{Y}^*) = 0$). However, it may happen that such a $\overline{Y}^*$ does not exist. Therefore the following performance functionals are of interest:

$$F_1(\overline{Y}) = f_{11}(\overline{Y}) + f_{12}(\overline{Y}), \tag{2.38}$$

$$F_2(\overline{Y}) = \max\{f_{11}(\overline{Y}), f_{12}(\overline{Y})\}, \tag{2.39}$$

$$F_3(\overline{Y}) = \frac{1}{N_1}f_{1\varepsilon}(\overline{Y}) + \frac{1}{N_2}f_{2\varepsilon}(\overline{Y}), \tag{2.40}$$

$$F_4(\overline{Y}) \;=\; \max\{\frac{1}{N_1}f_{1\varepsilon}(\overline{Y}),\; \frac{1}{N_2}f_{2\varepsilon}(\overline{Y})\}. \tag{2.41}$$

Now find

$$\min_{\overline{Y}\in G} F_s(\overline{Y}) = \; F_s(\overline{Y}^*_s),\; s \in 1:4, \tag{2.42}$$

where

$$G = \Big\{ \overline{Y} = [\overline{y}_1, ..., \overline{y}_n] \in \mathbb{R}^{(n+1)m} \;\Big|$$

$$\overline{y}_k = [y_k, d_k],\; y_k \in \mathbb{R}^n,\; d_k \in \mathbb{R},\; ||y_k|| = 1 \quad \forall k \in 1:m \Big\}. \tag{2.43}$$

If there exists $\overline{Y}^*$ satisfying (2.37) then

$$F_1(\overline{Y}^*) \;=\; F_2(\overline{Y}^*) \;=\; F_3(\overline{Y}^*) \;=\; F_4(\overline{Y}^*) \;=\; 0 \tag{2.44}$$

(and $\overline{Y}^*_s \;=\; \overline{Y}^* \quad \forall i \in 1:4$).

It may occur that for different practical situations different functionals are preferable.

In the discrete case now under consideration it seems that the performance functionals $F_s$ ($s \in 1:4$) (see (2.38), (2.39),(2.40), (2.41)) are better than the functional (2.11) since they are less dependent on "random" points in the sets $A$ and $B$ (whose existence may be caused, for instance, by experimental errors).

**Remark 4.1** *It has already been observed that $f_{1\varepsilon}(\overline{Y})$ represents (as $\varepsilon \downarrow 0$) the number of "displaced" points of the set $A$ (the number of points of $A$, not belonging to the set $\Omega(\overline{Y})$) while $f_{2\varepsilon}(\overline{Y})$ represents (as $\varepsilon \downarrow 0$) the number of "displaced" points of the set $B$ (the number of points of $A$, belonging to the set $\Omega(\overline{Y})$).*

*An analogous algorithm can be introduced to "isolate" only the set $B$. The idea of "isolation" was first implemented in [1].*

## 5 OPTIMALITY CONDITIONS

Thus, for the solution of the identification problem we have introduced 5 functionals: $F(y)$ (see (2.11)) and $F_s(\overline{Y}), s \in 1:4$ (see (2.38)–(2.41)). In the case $F(y)$ the problem is to find

$$\max_{y\in G_0} F(y) \tag{2.45}$$

where

$$G_0 = \{y \in \mathbb{R}^n \mid ||y|| = 1\}. \tag{2.46}$$

In the case $F_s(\overline{Y})$ the problem is to find

$$\min_{\overline{Y}\in G} F_s(\overline{Y}) \tag{2.47}$$

where $G \in \mathbb{R}^{(n+1)m}$ is given by (2.43).

The problems (2.45) and (2.47) are problems of Nonsmooth Optimization (to be more precise, those of Quasidifferentiable Optimization). In the sequel we shall discuss only problem (2.45).

Thus, let $\mathcal{A} \cap \mathcal{B} \neq \emptyset$. Consider the problem of maximizing the functional $F(y)$ defined by (2.11):

$$F(y) = |\max_{a\in\mathcal{A}}(a,y) - \max_{b\in\mathcal{B}}(b,y)| + |\min_{a\in\mathcal{A}}(a,y) - \min_{b\in\mathcal{B}}(b,y)| \quad (2.48)$$

on the set

$$G_0 = \{y \in \mathbb{R}^n \mid ||y|| = 1\}. \quad (2.49)$$

The set $G_0$ can be described in the equivalent form:

$$G_0 = \{y \in \mathbb{R}^n \mid \varphi(y) = 0\} \quad (2.50)$$

where

$$\varphi(y) = \Big| \; ||y|| = 1 \Big|. \quad (2.51)$$

Clearly,

$$\varphi(y) \geq 0 \quad \forall y \in \mathbb{R}^n. \quad (2.52)$$

It follows from the theory of Exact Penalization (see, e.g., [11, 16, 10, 3, 5, 4]) that there exists a threshold value $\lambda^*$ such that for any $\lambda > \lambda^*$ the problem of maximizing the functional $F(y)$ (see (2.48)) on the set $G_0$ described by (2.50) is equivalent to the problem of minimizing the functional

$$\Phi_\lambda(y) = -F(y) + \lambda\varphi(y) \quad (2.53)$$

on the space $\mathbb{R}^n$ (this problem is an *unconstrained* minimization problem).

Note in passing that the problem of minimizing the functional $F(y)$ on the set $G_0$ is equivalent to the problem of minimizing the functional

$$\Phi_{1\lambda}(y) = F(y) + \lambda\varphi(y) \quad (2.54)$$

on $\mathbb{R}^n$.

The functions $F(y)$ and $\varphi(y)$ are quasidifferentiable (see [8, 7]). Recall that a function $f : \mathbb{R}^n \to \mathbb{R}$ is called quasidifferentiable at a point $x \in \mathbb{R}^n$ if $f$ is directionally differentiable at $x$ and if there exist compact convex sets $\underline{\partial} f(x) \subset \mathbb{R}^n$ and $\overline{\partial} f(x) \subset \mathbb{R}^n$ such that

$$f'(x,g) = \lim_{\alpha\downarrow 0} \frac{f(x+\alpha g) - f(x)}{\alpha} = \max_{v\in\underline{\partial} f(x)}(v,g) + \min_{w\in\overline{\partial} f(x)}(w,g) \quad \forall g \in \mathbb{R}^n. \quad (2.55)$$

The pair $Df(x) = [\underline{\partial} f(x), \overline{\partial} f(x)]$ is called a quasidifferential of $f$ at $x$.

Let us find quasidifferentials of the functions $F$ and $\varphi$ at $y \in G_0$.

The following cases are possible:

a)(See Fig. 2.6).

$$\max_{a\in\mathcal{A}}(a,y) \geq \max_{b\in\mathcal{B}}(b,y), \quad \min_{a\in\mathcal{A}}(a,y) \geq \min_{b\in\mathcal{B}}(b,y), \tag{2.56}$$

b)(See Figs. 2.7 and 2.9).

$$\max_{a\in\mathcal{A}}(a,y) < \max_{b\in\mathcal{B}}(b,y), \quad \min_{a\in\mathcal{A}}(a,y) \geq \min_{b\in\mathcal{B}}(b,y), \tag{2.57}$$

c)(See Fig. 2.8).

$$\max_{a\in\mathcal{A}}(a,y) \geq \max_{b\in\mathcal{B}}(b,y), \quad \min_{a\in\mathcal{A}}(a,y) < \min_{b\in\mathcal{B}}(b,y), \tag{2.58}$$

d)(See Fig. 2.6 where $\mathcal{A}$ and $\mathcal{B}$ are permuted).

$$\max_{a\in\mathcal{A}}(a,y) < \max_{b\in\mathcal{B}}(b,y), \quad \min_{a\in\mathcal{A}}(a,y) < \min_{b\in\mathcal{B}}(b,y). \tag{2.59}$$

Applying the rules of Quasidifferential Calculus (see [8, 7]) one gets:

In the case a):

$$\underline{\partial} F(y) = co\ \{a | a \in R_{\mathcal{A}}(y)\} + co\ \{-b | b \in Q_{\mathcal{B}}(y)\}, \tag{2.60}$$

$$\overline{\partial} F(y) = co\ \{-b | b \in R_{\mathcal{B}}(y)\} + co\ \{a | a \in Q_{\mathcal{A}}(y)\}, \tag{2.61}$$

where

$$R_{\mathcal{A}}(y) = \{a \in \mathcal{A} \mid (a,y) = \max_{a'\in\mathcal{A}}(a',y)\},\ Q_{\mathcal{A}}(y) = \{a \in \mathcal{A} \mid (a,y) = \min_{a'\in\mathcal{A}}(a',y)\}, \tag{2.62}$$

$$R_{\mathcal{B}}(y) = \{b \in \mathcal{B} \mid (b,y) = \max_{b'\in\mathcal{B}}(b',y)\}, \quad Q_{\mathcal{B}}(y) = \{b \in \mathcal{B} \mid (b,y) = \min_{b'\in\mathcal{B}}(b',y)\}. \tag{2.63}$$

In the case b):

$$\underline{\partial} F(y) = co\{b | b \in R_{\mathcal{B}}(y)\} + co\{-b | b \in Q_{\mathcal{B}}(y)\}, \tag{2.64}$$

$$\overline{\partial} F(y) = co\{-a | a \in R_{\mathcal{A}}(y)\} + co\{a | a \in Q_{\mathcal{A}}(y)\}. \tag{2.65}$$

In the case c):

$$\underline{\partial} F(y) = co\{a | a \in R_{\mathcal{A}}(y)\} + co\{-a | a \in Q_{\mathcal{A}}(y)\}, \tag{2.66}$$

$$\overline{\partial}F(y) = co\{-b|b \in R_{\mathcal{B}}(y)\} + co\{b|b \in Q_{\mathcal{B}}(y)\}. \quad (2.67)$$

In the case d):

$$\underline{\partial}F(y) = co\{b|b \in R_{\mathcal{B}}(y)\} + co\{-a|a \in Q_{\mathcal{A}}(y)\}, \quad (2.68)$$

$$\overline{\partial}F(y) = co\{-a|a \in R_{\mathcal{A}}(y)\} + co\{b|b \in Q_{\mathcal{B}}(y)\}. \quad (2.69)$$

It is easy to show that if $||y|| = 1$ then

$$\underline{\partial}\varphi(y) = co\{y, -y\}, \quad \overline{\partial}\varphi(y) = \{0_n\} \quad (2.70)$$

where $0_n = (0, ..., 0) \in I\!R^n$.

It follows from the rules of Quasidifferental Calculus that $\Phi_\lambda(y)$ is quasidifferentiable and

$$\underline{\partial}\Phi_\lambda(y) = -\overline{\partial}F(y) + \lambda\underline{\partial}\varphi(y), \quad \overline{\partial}\Phi_\lambda(y) = -\underline{\partial}F(y) + \lambda\overline{\partial}\varphi(y) = -\underline{\partial}F(y). \quad (2.71)$$

If $y^* \in G_0$ is a maximizer of $F$ on $G_0$ then $y^*$ is an unconstrained minimizer of $\Phi_\lambda(y)$. Then the following necessary condition should hold (see [8, 7]):

$$-\overline{\partial}\Phi_\lambda(y^*) \subset \underline{\partial}\Phi_\lambda(y^*). \quad (2.72)$$

Note that the condition

$$-\overline{\partial}\Phi_\lambda(y^*) \subset int\ \underline{\partial}\Phi_\lambda(y^*) \quad (2.73)$$

is a sufficient condition for $y^*$ to be a local minimizer of $\Phi_\lambda$.

Taking into account (2.71) one gets from (2.72)

$$\underline{\partial}F(y^*) \subset \{-\overline{\partial}F(y^*) + \lambda\underline{\partial}\varphi(y^*)\}. \quad (2.74)$$

A point $y^*$ satisfying condition (2.74) is called a sup-stationary point of $F$ on $G_0$.

Condition (2.74) can be used to check whether a point $y_0$ is sup-stationary or not, to find a direction of steepest ascent and to construct numerical methods for finding sup-stationary points.

## 6 AN EXAMPLE

To illustrate the algorithm discussed in Subsection 3.3, let us consider the following elementary example. Let

$$\mathcal{A} = co\ \{(2,2), (-2,-2)\} \subset I\!R^2, \quad \mathcal{B} = co\ \{(1,-1), (-1,1)\} \subset I\!R^2.$$

Clearly (see Fig. 2.11) $\mathcal{A} \cap \mathcal{B} = \{(0,0)\} \neq \emptyset$. We are looking for a

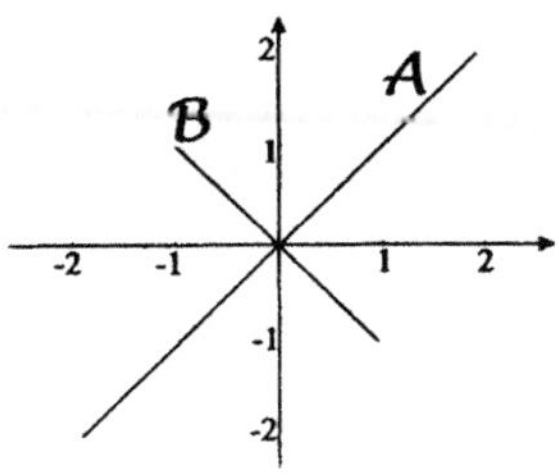


*Figure 2.11*

maximizer of the function

$$F(y) = |\max_{a\in\mathcal{A}}(a,y) - \max_{b\in\mathcal{B}}(b,y)| + |\min_{a\in\mathcal{A}}(a,y) - \min_{b\in\mathcal{B}}(b,y)|$$

on the set $G_0 = \{y \in \mathbb{R}^n \mid ||y|| = 1\}$.

Consider the point $y_0 = (\frac{1}{\sqrt{2}}, \frac{1}{\sqrt{2}})$. Clearly, $y_0 \in G_0,\ f(y_0) = 4\sqrt{2}$. We have

$$R_{\mathcal{A}}(y_0) = \{(2,2)\},\ Q_{\mathcal{A}}(y_0) = \{(-2,-2)\},\ R_{\mathcal{B}}(y_0) = Q_{\mathcal{B}}(y_0) = \mathcal{B}. \tag{2.75}$$

Since

$$\overline{f}_{\mathcal{A}}(y_0) = \max_{a\in\mathcal{A}}(a,y_0) = 2\sqrt{2} > \overline{f}_{\mathcal{B}}(y_0) = \max_{b\in\mathcal{B}}(b,y_0) = 0,$$

$$\underline{f}_{\mathcal{A}}(y_0) = \min_{a\in\mathcal{A}}(a,y_0) = -2\sqrt{2} < \underline{f}_{\mathcal{B}}(y_0) = \min_{b\in\mathcal{B}}(b,y_0) = 0$$

then (see (2.58), (2.66), (2.67), and (2.75))

$$\underline{\partial}F(y_0) = \{(2,2)\} + \{(2,2)\} = \{(4,4)\},$$

$$\overline{\partial}F(y_0) = co\ \{-b|b \in \mathcal{B}\} + co\ \{b|b \in \mathcal{B}\} \ = \ co\ \{(2,-2),(-2,2)\},$$

$$\underline{\partial}\varphi(y_0) \ = \ co\ \{(\frac{1}{\sqrt{2}}, \frac{1}{\sqrt{2}}), (-\frac{1}{\sqrt{2}}, -\frac{1}{\sqrt{2}})\}.$$

Let us check condition (2.74). For $\lambda > 4\sqrt{2}$

$$\underline{\partial}F(y_0) = \{(4,4)\} \ \subset \ -\overline{\partial}F(y_0) + \lambda\underline{\partial}\varphi(y_0) \ =$$

$$= -co\ \{(2,-2),(-2,2)\} \ + \ \lambda\ co\ \{(\frac{1}{\sqrt{2}}, \frac{1}{\sqrt{2}}), (-\frac{1}{\sqrt{2}}, -\frac{1}{\sqrt{2}})\}$$

i.e. the point $y_0 \in G_0$ satisfies the necessary condition for a minimum of the function $\Phi_\lambda$ (and, hence, for a maximum of $F$ on $G_0$). As a matter of fact this point is a global maximizer of $F$ on $G_0$ but our theory allows only to claim that $y_0$ is a sup-stationary point of $F$ on $G_0$.

Take the function $r(x, y_0) = (x, y_0)$. Since

$$\overline{f}_{\mathcal{AB}}(y_0) = \min\{\overline{f}_{\mathcal{A}}(y_0), \overline{f}_{\mathcal{B}}(y_0)\} = 0,$$

$$\underline{f}_{\mathcal{AB}}(y_0) = \max\{\underline{f}_{\mathcal{A}}(y_0), \underline{f}_{\mathcal{B}}(y_0)\} = 0$$

then (see (2.15), (2.16))

$$c \in \mathcal{A} \quad if\ r(c, y_0) > 0,$$

$$c \in \mathcal{A} \quad if\ r(c, y_0) < 0.$$

Thus, all points of $\mathcal{A}$, except for the point $x_0 = (0, 0)$, are identified by the function $r(x, y_0)$.

Now let us construct the set $\mathcal{D}_1$ (see (2.20)):

$$\mathcal{D}_1 = \left\{ c \in \mathcal{A}_1 \cup \mathcal{B}_1 \,\middle|\, r(c, y_1) \neq 0 \right\}.$$

Clearly (see (2.21))

$$\mathcal{A}_2 = \mathcal{A}_1 \setminus \mathcal{D}_1 = \{(0, 0)\}, \ \mathcal{B}_2 = \mathcal{B}_1 \setminus \mathcal{D}_1 = \mathcal{B}.$$

Find (see (2.22))

$$\max_{y \in G_0} F_2(y)$$

where

$$F_2(y) = |\max_{a \in \mathcal{A}_\in}(a, y) - \max_{b \in \mathcal{B}_\in}(b, y)| + |\min_{a \in \mathcal{A}_\in}(a, y) - \min_{b \in \mathcal{B}_\in}(b, y)| =$$

$$= |\max_{b \in \mathcal{B}}(b, y)| + |\min_{b \in \mathcal{B}}(b, y)|.$$

It is obvious that

$$\max_{y \in G_0} F_2(y) = F_2(y_1)$$

where $y_1 = (-\frac{1}{\sqrt{2}}, \frac{1}{\sqrt{2}})$ and $F_2(y_1) = 2\sqrt{2}$.

Take the function $r(x, y_1) = (x, y_1)$. Since

$$\overline{f}_{\mathcal{A}_1}(y_1) = \max_{a \in \mathcal{A}_1}(a, y_1) = 0, \quad \overline{f}_{\mathcal{B}_1}(y_1) = \max_{a \in \mathcal{B}_1}(a, y_1) = \sqrt{2},$$

$$\underline{f}_{\mathcal{A}_1}(y_1) = \min_{a \in \mathcal{A}_1}(a, y_1) = 0, \quad \underline{f}_{\mathcal{B}_1}(y_1) = \min_{a \in \mathcal{B}_1}(a, y_1) = -\sqrt{2},$$

$$\overline{f}_{\mathcal{A}_1\mathcal{B}_1}(y_1) = \min\{\overline{f}_{\mathcal{A}_1}(y_1), \overline{f}_{\mathcal{B}_1}(y_1)\} = 0,$$

$$\underline{f}_{\mathcal{A}_1\mathcal{B}_1}(y_1) = \max\{\underline{f}_{\mathcal{A}_1}(y_1), \underline{f}_{\mathcal{B}_1}(y_1)\} = 0$$

then (see (2.15), (2.16))

$$c \in \mathcal{B}_1 \quad if \ r(c, y_1) > 0,$$

$$c \in \mathcal{B}_1 \quad if \ r(c, y_1) < 0.$$

Thus, all points of $\mathcal{B}_1$ (note that $\mathcal{B}_1 = \mathcal{B}$), except for the point $x_0 = (0, 0)$, are identified by the function $r(x, y_1)$.

Hence, applying the procedure described in Section 3, we are able (using the functions $r(x, y_0)$ and $r(x, y_1)$) to identify all points of $\mathcal{A}$ and $\mathcal{B}$, except for the point $x_0 = \mathcal{A} \cap \mathcal{B}$.

**Remark 6.1** *In the paper only linear hyperplanes were employed. Using different tools (such as ellipsoids, balls, clusters etc.) it is possible to get other performance functionals (see, e.g., [14, 6]). In many cases such functionals are essentially of nonsmooth nature.*

## References

[1] Astorino A., Gaudioso M. (2000), Polyhedral separability through successive LP, submitted to *Journal of Optimization Theory and Applications.*

[2] Boukary D., Fiacco A. V. (1995), Survey of penalty, exact-penalty and multiplier methods from 1968 to 1993. *Optimization,* Vol. 32, No. 4, pp. 301–334.

[3] Demyanov V. F. (1994), Exact penalty functions in nonsmooth optimization problems. *Vestnik of St. Petersburg University,* ser. 1, issue 4 (No. 22), pp. 21-27.

[4] Demyanov V. F. (2000), *Conditions for an extremum and variational problems.* St. Petersburg, St. Petersburg University Press.

[5] Demyanov V. F., Di Pillo G., Facchinei F. (1998), Exact penalization via Dini and Hadamard conditional derivatives. *Optimization Methods and Software,* Vol. 9, pp. 19–36.

[6] Demyanov V. F., Bagirov A. M., Rubinov A. M. (2001), A method of truncated codifferential with application to some problems of cluster analysis. Forthcoming.

[7] Demyanov V. F., Rubinov A. M. (1995), *Constructive Nonsmooth Analysis.* Frankfurt a/M., Peter Lang Verlag.

[8] Demyanov V. F., Vasiliev L. V. (1986), *Nondifferentiable Optimization.* New-York, Springer–Optimization Software.

[9] Di Pillo G., Grippo L. (1988), On the Exactness of a class of Nondifferentiable Penalty functions. *J. Optim. Theory Appl.,* Vol. 57, pp. 397–408.

[10] Di Pillo G., Facchinei F. (1989), Exact penalty functions for nondifferentiable programming problems. In *Nonsmooth Optimization and Related Topics*. Eds. F. H. Clarke, V. F. Demyanov and F. Giannessi, pp. 89-107. New York, Plenum.

[11] Eremin I. I. (1967), The method of penalties in convex programming. *Doklady Acad. Nauk SSSR*, Vol. 173, pp. 748 - 751.

[12] Fletcher R. (1983), Penalty functions. In *Mathematical programming: the state of the art.* (Eds. A. Bachen, M. Grötschel, B. Korte), Springer–Verlag, Berlin, pp. 87–114.

[13] Han S., Mangasarian O. (1979), Exact penalty functions in nonlinear programming. *Mathematical Programming*, Vol.17, pp. 251–269.

[14] Mangasarian O. L. (1997), Mathematical programming in data mining. *Data Mining and Knowledge Discovery*, Vol. 1, pp. 183–201.

[15] Rockafellar R. T.(1970) *Convex Analysis.* Princeton, N.J., Princeton University Press.

[16] Zangwill W. L. (1967), Nonlinear programming via penalty functions. *Management Science*, Vol. 13, pp. 344–358.

# Chapter 3

# DETERMINISTIC GLOBAL OPTIMIZATION FOR PROTEIN STRUCTURE PREDICTION

John L. Klepeis
*Department of Chemical Engineering*
*Princeton University*
*Princeton, NJ 08544*
john@titan.princeton.edu

Christodoulos A. Floudas
*Department of Chemical Engineering*
*Princeton University*
*Princeton, NJ 08544*
floudas@titan.princeton.edu

**Abstract** Deterministic global optimization plays an essential role in the solution of many difficult problems with applications ranging from economics and operations research to computational chemistry and molecular biology. In this chapter we explore the application of deterministic global optimization approaches to problems related to protein structure prediction. Due to the complex nature of protein interactions, energy landscapes which model these systems display huge numbers of local minima often separated by high energy barriers. Since the number of local minima is vast, the corresponding formulation has earned the simple yet suggestive title of "multiple-minima" problem. Based on the complexity of the energy hypersurface, there is an obvious need for the development of effective global optimization techniques. In this work, we have focused on the development of such global optimization methods through the foundations of the $\alpha$BB deterministic global optimization approach.

**Keywords:** Deterministic global optimization, Protein folding, Structure prediction, Energy modeling



*N. Hadjisavvas and P.M. Pardalos (eds.), Advances in Convex Analysis and Global Optimization,* 31–73.

# 1 INTRODUCTION

Proteins are undoubtedly the most complex and vital molecules in nature. This complexity arises from an intricate balance of intra- and inter-molecular interactions which define the native three-dimensional structure of the system, and subsequently its biological functionality. Recent advances in genetic engineering have heightened the interest in research related to understanding the dynamics and predicting the equilibrium native protein folding and docking conformations. The ability to predict these structures is of great theoretical interest, especially in the fields of biophysics and biochemistry. Moreover, the applications of such knowledge also promise to be exciting. For example, the ability to predict these structures would greatly increase our understanding of hereditary and infectious diseases and aid in the interpretation of genome data. Such knowledge would also likely revolutionize the process of de novo drug design.

Anfinsen's thermodynamic hypothesis [9] suggests that this native structure is in a state of thermodynamic equilibrium corresponding to the system with the lowest free energy. Experimental studies have since shown that, under native physiological conditions and after denaturation, globular proteins spontaneously refold to their unique, native structure [34]. Understanding the transition of a protein from a disordered state to its native state defines the protein folding problem.

The use of computational techniques and simulations in addressing the protein folding and peptide docking problems became possible through the introduction of qualitative and quantitative methods for modeling these systems. The development of realistic energy models also established a link to the field of global optimization, where, based on Anfinsen's hypothesis, the quantity to be optimized is the free energy of the system. Because the number of local minima is vast, the corresponding problem formulation has earned the simple yet suggestive title of "multiple-minima" problem. The basis for these difficulties is best summarized by Levinthal's paradox [43]. This paradox suggests a contradiction between the almost infinite number of possible stable states that the system may sample and the relatively short time scale required for actual protein folding. Levinthal's observations suggest that the native state is the lowest kinetically accessible free energy minimum, which may be different from the true global minimum. These principles have been used to develop computational techniques for predicting protein folding pathways [11, 15, 14, 42, 65]. Such techniques attempt to map the shape of the energy hypersurface and determine whether this surface "funnels" a protein towards a dominant conformational basin. By invoking the thermodynamic hypothesis, the overall shape of the energy hypersurface is neglected and the problem can be formulated in terms of global minimization, which requires the use of effective global optimization techniques. If this formulation is to reproduce the behavior of realistic systems, the folding of actual proteins should not be kinetically hindered. This has been verified for various systems by performing denaturation–refolding experiments. In addition, by introducing structural characteristics whose formation may act as kinetic barriers, such as the formation of disulfide bonds, the performance of the thermodynamic equilibrium model should be improved.

Based on the complexity of the energy hypersurface, there is an obvious need for the development of efficient global optimization techniques. Although the energy can be expressed analytically, exhaustive searches are possible for only the smallest of systems. These observations, and the importance of the protein folding and peptide docking problems, have propelled the introduction of new global search strategies specifically designed for these problems.

In the sequel, we first outline the basics of the deterministic global optimization approach, $\alpha$BB, which has been used extensively to study the protein structure prediction. This is followed by a comprehensive study of ab-initio modeling for structure prediction of single chain polypeptides. An extensive comparison of energy modeling, including solvation, entropic effects and free energy calculations, is provided for the oligopeptides. The related problem of restrained structure refinement in the presence of sparse experimentally derived restraints is also discussed.

## 2 THEORY

The generic optimization problem to be addressed has the following form:

$$\begin{array}{rl} \min\limits_{\mathbf{x}} & f(\mathbf{x}) \\ \text{subject to} & \mathbf{g}(\mathbf{x}) \leq 0 \\ & \mathbf{h}(\mathbf{x}) = 0 \\ & \mathbf{x} \in [\mathbf{x}^L, \mathbf{x}^U] \end{array} \tag{3.1}$$

where $\mathbf{x}$ is a vector of $n$ continuous variables, $f(\mathbf{x})$ is the objective function, $\mathbf{g}(\mathbf{x})$ is a vector of inequality constraints, and $\mathbf{h}(\mathbf{x})$ is a vector of equality constraints. Both the objective function and constraint equations are assumed to be twice continuously differentiable. $\mathbf{x}^L$ and $\mathbf{x}^U$ denote the lower and upper bounds on the $\mathbf{x}$ variables, respectively. The constraints define the feasible region for the problem.

Two main classes of global optimization techniques have been developed to address problem (3.1), namely, stochastic and deterministic approaches. Stochastic methods, such as those based on genetic algorithms [25] and simulated annealing [35], can be used to treat unconstrained nonconvex problems. However, the stochastic nature of the search strategy invalidates any claims regarding global optimality since it is impossible to obtain valid bounds on the solution of the problem. The addition of nonconvex constraints further complicates these solution schemes. In contrast, deterministic methods rely on a theoretically-based search of the domain space to guarantee the identification of the global optimum solution.

A common characteristic of deterministic global optimization algorithms is the progressive reduction of the domain space until the global solution has been found with arbitrary accuracy. The solution is approached from above and below by generating converging sequences of upper and lower bounds, and the generation of these bounds on the global optimum solution is an essential part of all deterministic global optimization algorithms [20, 28, 29].

The $\alpha$BB algorithm has been developed to address general twice continuously differentiable models of type 3.1 [5, 3, 4, 2, 7]. The algorithm is built on a branch-and-bound framework and can handle generic nonconvex optimization problems represented by formulation (3.1). $\epsilon$-convergence to the global optimum solution is guaranteed when the functions $f(\mathbf{x})$, $\mathbf{g}(\mathbf{x})$ and $\mathbf{h}(\mathbf{x})$ are twice continuously differentiable. The algorithm has been shown to terminate in a finite number of iterations for this broad class of problems [4, 2, 45, 46].

The $\alpha$BB global optimization approach is based on the convex relaxation of the original nonconvex formulation (3.1). This requires convex lower bounding of all expressions, and these expressions can be classified as : (i) convex terms; (ii) nonconvex terms of special structure; and (iii) nonconvex terms of general structure. Obviously, convex lower bounding functions are not required for original convex expressions (e.g., linear terms). Certain nonconvex terms, including bilinear, trilinear and univariate concave functions, possess special structure that can be exploited in developing lower bounding functions. All other nonconvex terms can be underestimated using a general expression [7].

When applying the $\alpha$BB approach to the protein folding problem, formulation (3.1) involves only nonconvex expressions of general structure. For this reason, the following exposition will briefly cover underestimation for terms of special structure and then focus on the development of a convex lower bounding formulation for global optimization involving generic nonconvex objective and constraint functions.

## 2.1 UNDERESTIMATING TERMS OF SPECIAL STRUCTURE

In the case of a bilinear term $xy$, [6] showed that the tightest convex lower bound over the domain $[x^L, x^U] \times [y^L, y^U]$ is obtained by introducing a new variable $w_B$ which replaces every occurrence of $xy$ in the problem and satisfies the following relationship:

$$w_B = \max\{x^L y + y^L x - x^L y^L;\ x^U y + y^U x - x^U y^U\}. \tag{3.2}$$

This lower bound can be relaxed and included in the minimization problem by adding two linear inequality constraints,

$$\begin{aligned} w_B &\geq x^L y + y^L x - x^L y^L, \\ w_B &\geq x^U y + y^U x - x^U y^U. \end{aligned} \tag{3.3}$$

Moreover, an upper bound can be imposed on $w$ to construct a better approximation of the original problem [48]. This is achieved through the addition of two linear constraints:

$$\begin{aligned} w_B &\leq x^U y + y^L x - x^U y^L, \\ w_B &\leq x^L y + y^U x - x^L y^U. \end{aligned} \tag{3.4}$$

A trilinear term of the form $xyz$ can be underestimated in a similar fashion [47]. A new variable $w_T$ is introduced and bounded by the following eight inequality constraints:

$$\begin{aligned} w_T &\geq xy^L z^L + x^L y z^L + x^L y^L z - 2x^L y^L z^L, \\ w_T &\geq xy^U z^U + x^U y z^L + x^U y^L z - x^U y^L z^L - x^U y^U z^U, \\ w_T &\geq xy^L z^L + x^L y z^U + x^L y^U z - x^L y^U z^U - x^L y^L z^L, \\ w_T &\geq xy^U z^L + x^U y z^U + x^L y^U z - x^L y^U z^L - x^U y^U z^U, \\ w_T &\geq xy^L z^U + x^L y z^L + x^U y^L z - x^U y^L z^U - x^L y^L z^L, \\ w_T &\geq xy^L z^U + x^L y z^U + x^U y^U z - x^L y^L z^U - x^U y^U z^U, \\ w_T &\geq xy^U z^L + x^U y z^L + x^L y^L z - x^U y^U z^L - x^L y^L z^L, \\ w_T &\geq xy^U z^U + x^U y z^U + x^U y^U z - 2x^U y^U z^U. \end{aligned} \tag{3.5}$$

Fractional terms of the form $x/y$ are underestimated by introducing a new variable $w_F$ and two new constraints [47] which depend on the sign of the bounds on $x$.

$$\begin{array}{lll} w_F & \geq & \begin{cases} x^L/y + x/y^U - x^L/y^U & \text{if } x^L \geq 0 \\ x/y^U - x^L y/y^L y^U + x^L/y^L & \text{if } x^L < 0 \end{cases} \\ w_F & \geq & \begin{cases} x^U/y + x/y^L - x^U/y^L & \text{if } x^U \geq 0 \\ x/y^L - x^U y/y^L y^U + x^U/y^U & \text{if } x^U < 0 \end{cases} \end{array} \tag{3.6}$$

For fractional trilinear terms, eight new constraints are required [47]. The fractional trilinear term $xy/z$ is replaced by the variable $w_{FT}$ and the constraints for $x^L, y^L, z^L \geq 0$ are given by

$$\begin{aligned} w_{FT} &\geq xy^L/z^U + x^L y/z^U + x^L y^L/z - 2x^L y^L/z^U, \\ w_{FT} &\geq xy^L/z^U + x^L y/z^L + x^L y^U/z - x^L y^U/z^L - x^L y^L/z^U, \\ w_{FT} &\geq xy^U/z^L + x^U y/z^U + x^U y^L/z - x^U y^L/z^U - x^U y^U/z^L, \\ w_{FT} &\geq xy^U/z^U + x^U y/z^L + x^L y^U/z - x^L y^U/z^U - x^U y^U/z^L, \\ w_{FT} &\geq xy^L/z^U + x^L y/z^L + x^U y^L/z - x^U y^L/z^L - x^L y^L/z^U, \\ w_{FT} &\geq xy^U/z^U + x^U y/z^L + x^L y/z - x^L y^U/z^U - x^U y^U/z^L, \\ w_{FT} &\geq xy^L/z^U + x^L y/z^L + x^U y^L/z - x^U y^L/z^L - x^L y^L/z^U, \\ w_{FT} &\geq xy^U/z^L + x^U y/z^L + x^U y^U/z - 2x^U y^U/z^L. \end{aligned} \tag{3.7}$$

Univariate concave functions are trivially underestimated by their linearization at the lower bound of the variable range. Thus the convex envelope of the concave function $ut(x)$ over $[x^L, x^U]$ is the linear function of $x$:

$$ut(x^L) + \frac{ut(x^U) - ut(x^L)}{x^U - x^L}(x - x^L). \tag{3.8}$$

The generation of the best convex underestimator for a univariate concave function does not require the introduction of additional variables or constraints.

## 2.2 UNDERESTIMATING GENERAL NONCONVEX TERMS

A general nonconvex term $f(\mathbf{x})$ belonging to the class of twice continuously differentiable functions can be underestimated over the entire domain $\mathbf{x} \in [\mathbf{x}^L, \mathbf{x}^U]$ by the function $\hat{f}(\mathbf{x})$ defined as

$$\hat{f}(\mathbf{x}) = f(\mathbf{x}) + \sum_{i=1}^{n} \alpha_i (x_i^L - x_i)(x_i^U - x_i) \tag{3.9}$$

where the $\alpha_i$'s are non negative scalars.

$\hat{f}(\mathbf{x})$ is a guaranteed underestimator of $f(\mathbf{x})$ because the original nonconvex expression is augmented by the addition of separable quadratic functions which are negative over the entire domain $[\mathbf{x}^L, \mathbf{x}^U]$. Furthermore, since the quadratic term is convex, all nonconvexities in the original term $f(\mathbf{x})$ can be overpowered by using sufficiently large values of the $\alpha_i$ parameters.

The convex lower bounding function $\hat{f}(\mathbf{x})$, defined over the rectangular domain of $\mathbf{x}^L \leq \mathbf{x} \leq \mathbf{x}^U$, possesses a number of important properties which guarantee the convergence of the $\alpha$BB algorithm to the global optimum solution :

(i) $\hat{f}(\mathbf{x})$ is a valid underestimator of $f(\mathbf{x})$. That is:

$$\forall\, \mathbf{x} \in \left[\mathbf{x}^L, \mathbf{x}^U\right] \text{ it can be shown that } \hat{f}(\mathbf{x}) \leq f(\mathbf{x}),$$

(ii) $\hat{f}(\mathbf{x})$ matches $f(\mathbf{x})$ at all corner points.

(iii) $\hat{f}(\mathbf{x})$ is convex in $\mathbf{x} \in \left[\mathbf{x}^L, \mathbf{x}^U\right]$.

(iv) The maximum separation between the nonconvex term of generic structure, $f(\mathbf{x})$, and its convex relaxation, $\hat{f}(\mathbf{x})$, is bounded and also proportional to the positive $\alpha$ parameters and to the square of the diagonal of the current box constraints :

$$\max_{\mathbf{x}^L \leq \mathbf{x} \leq \mathbf{x}^U} \left[f(\mathbf{x}) - \hat{f}(\mathbf{x})\right] = \frac{1}{4}\sum_{i}^{n} \alpha_i \left(x_i^U - x_i^L\right)^2 . \quad (3.10)$$

(v) The underestimators constructed over supersets of the current set are always less tight than the underestimator constructed over the current box constraints for every point within the current box constraints.

The key development in the convex lower bounding formulation is the definition of the $\alpha$ parameters. Specifically, the magnitude of the $\alpha$ parameters may be related to the minimum eigenvalue of the Hessian matrix of the nonconvex term $f(\mathbf{x})$.

$$\alpha \geq \max\left\{0, -\frac{1}{2} \min_{i, \mathbf{x}^L \leq \mathbf{x} \leq \mathbf{x}^U} \lambda_i(\mathbf{x})\right\} \quad (3.11)$$

where $\lambda(\mathbf{x})$ represent the eigenvalues of the Hessian matrix ($H_f(\mathbf{x})$) for the nonconvex term. An explicit minimization problem can be written to find the minimum eigenvalue ($\lambda_{min}$):

$$\begin{aligned} \min_{\mathbf{x},\lambda} \quad & \lambda \\ s.t. \quad & \det\left(H_f(\mathbf{x}) - \lambda I\right) = 0 \\ & \mathbf{x} \in [\mathbf{x}^L, \mathbf{x}^U] \end{aligned}$$

The solution of this problem is a non-trivial matter for arbitrary non-convex functions.

One method for the rigorous determination of $\alpha$ parameters for general twice differentiable problems involves interval analysis of Hessian matrices to calculate bounds on the minimum eigenvalue [3, 5]. The difficulties arising from the presence of the variables in the convexity condition can be alleviated through the transformation of the exact $\mathbf{x}$-dependent Hessian matrix to an interval matrix $[H_f]$ such that $H_f(\mathbf{x}) \subseteq [H_f]$, $\forall\, \mathbf{x} \in [\mathbf{x}^L, \mathbf{x}^U]$. The elements of the original Hessian matrix are treated as independent when calculating their natural interval extensions [56, 53]. The interval Hessian matrix family $[H_f]$ is then used to formulate a theorem in which the $\alpha$ calculation problem is relaxed [3]. In other words, a valid lower bound on the minimum eigenvalue can be used to calculate rigorous $\alpha$ values :

$$\alpha \geq \left\{0, -\frac{1}{2}\lambda_{min}\left([H_f]\right)\right\} \tag{3.12}$$

where $\lambda_{min}\left([H_f]\right)$ is the minimum eigenvalue of the interval matrix family $[H_f]$.

An $\mathcal{O}(n^2)$ method to calculate these $\alpha$ values is the straightforward extension of Gerschgorin's theorem [21] to interval matrices. For a real matrix $A = (a_{ij})$, the well-known theorem states that the eigenvalues are bounded below by $\lambda_{min}$ such that

$$\lambda_{min} = \min_i \left( a_{ii} - \sum_{j \neq i} |a_{ij}| \right). \tag{3.13}$$

For an interval matrix $[A] = ([\underline{a}_{ij}, \overline{a}_{ij}])$, a lower bound on the minimum eigenvalue is given by

$$\lambda_{min} \geq \min_i \left[ \underline{a}_{ii} - \sum_{j \neq i} \max\left(|\underline{a}_{ij}|, |\overline{a}_{ij}|\right) \right].$$

This procedure provides a single $\alpha$ value which is valid for all variables.

Non-uniform diagonal shift matrices can be used to calculate a different $\alpha$ value for each variable in order to construct an underestimator of the form shown in Equation (3.9). The non-zero elements of the diagonal shift matrix can no longer be related to the minimum eigenvalue of the interval Hessian matrix $[H_f]$. If all elements of the scaling vector are set to 1, the equation for the $\alpha_i$ values becomes :

$$\alpha_i = \max\left\{0, -\frac{1}{2}\left(\underline{a}_{ii} - \sum_{j \neq i} |a|_{ij}\right)\right\},$$

However, the choice of scaling is arbitrary, and different $\alpha_i$ parameters can be estimated through various scaling techniques.

## 2.3 CONVEXIFICATION OF FEASIBLE REGION

To obtain a valid lower bound on the global solution of the nonconvex problem, the lower bounding problem generated in each domain must have a unique solution. This implies that the formulation include only convex inequality constraints, linear equality constraints and an increased feasible region relative to that of the original nonconvex problem. The left-hand side of any nonconvex inequality constraint, $g(\mathbf{x}) \leq 0$, in the original problem can simply be replaced by its convex underestimator $\hat{g}(\mathbf{x})$, constructed according to Equation (3.9), to yield the relaxed convex inequality $\hat{g}(\mathbf{x}) \leq 0$.

For an equality constraint containing general nonconvex terms, the equation obtained by simple substitution of the appropriate underestimators is also nonlinear. Therefore, the original equality $h(\mathbf{x}) = 0$ must be rewritten as two inequalities of opposite signs,

$$\begin{aligned} h^+(\mathbf{x}) &= \quad h(\mathbf{x}) \quad \leq 0 \\ h^-(\mathbf{x}) &= -h(\mathbf{x}) \quad \leq 0. \end{aligned} \tag{3.14}$$

These two inequalities must then be underestimated independently to give $\hat{h}^+(\mathbf{x})$ and $\hat{h}^-(\mathbf{x})$.

## 2.4 CONVEX LOWER BOUNDING PROBLEM FORMULATION

Summarizing the concepts introduced so far, a convex relaxation for any nonconvex problem of type (3.1) belonging to the broad class of twice continuously differentiable continuous NLPs can be constructed as

$$\begin{array}{ll} \min\limits_{\mathbf{x}} & \hat{f}(\mathbf{x}) \\ \text{subject to} & \hat{\mathbf{g}}(\mathbf{x}) \leq 0 \\ & \hat{\mathbf{h}}^{+}(\mathbf{x}) \leq 0 \\ & \hat{\mathbf{h}}^{-}(\mathbf{x}) \leq 0 \\ & \mathbf{x} \in [\mathbf{x}^L, \mathbf{x}^U] \end{array} \tag{3.15}$$

where ^ denotes the convex underestimator of the specified function over the domain $\mathbf{x} \in [\mathbf{x}^L, \mathbf{x}^U]$. Since the inclusion of convex terms and nonconvex terms of special structure has been neglected, these functions involve only $\alpha$ type underestimating expressions. These underestimators are functions of the size of the domain under consideration, and because the $\alpha$BB algorithm follows a branch-and-bound approach, this domain is systematically reduced at each new node of the tree. Tighter lower bounding functions can therefore be generated by updating the underestimating equations. The lower bounds on the problem form a non-decreasing sequence, and the underestimating strategy is therefore consistent, as required for convergence.

## 2.5 VARIABLE BOUND UPDATES

The quality of the convex lower bounding problem can also be improved by ensuring that the variable bounds are as tight as possible. These variable bound updates can either be performed at the onset of an $\alpha$BB run or at each iteration.

In both cases, the same procedure is followed in order to construct the bound update problem. Given a solution domain, the convex underestimator for every constraint in the original problem is formulated. The bound problem for variable $x_i$ is then expressed as

$$x_i^{L,NEW}/x_i^{U,NEW} = \left\{ \begin{array}{ll} \min\limits_{\mathbf{x}} / \max\limits_{\mathbf{x}} & x_i \\ \text{subject to} & \hat{\mathbf{g}}(\mathbf{x}) \leq 0 \\ & \mathbf{x}^L \leq \mathbf{x} \leq \mathbf{x}^U \end{array} \right. \tag{3.16}$$

where $\hat{\mathbf{g}}(\mathbf{x})$ are the convex underestimators of the constraints, and the bounds on the variables, $\mathbf{x}^L$ and $\mathbf{x}^U$ are the best calculated bounds. Thus, once a new lower bound $x_i^{L,NEW}$ on $x_i$ has been computed via a minimization, this value is used in the formulation of the maximization problem for the generation of an upper bound $x_i^{U,NEW}$.

Because of the computational expense incurred by an update of the bounds on all variables, it is often desirable to define a smaller subset of the variables on which this operation is to be performed. The criterion devised for the selection of the branching variables can be used in this instance, since it provides a measure of the sensitivity of the problem to each variable.

## 2.6 THE $\alpha$BB ALGORITHM

The global optimization method $\alpha$BB deterministically locates the global minimum solution of (3.1) based on the refinement of converging lower and upper bounds. The lower bounds are obtained by the solution of (3.15), which is formulated as a convex programming problem. Upper bounds are based on the solution of (3.1) using local minimization techniques.

As previously mentioned, the maximum separation between the generic nonconvex terms and their respective convex lower bounding representations is proportional to the square of the diagonal of the current rectangular partition. As the size of the rectangular domains approach zero, this separation also become infinitesimally small. That is, as the current box constraints $[\mathbf{x}^L, \mathbf{x}^U]$ collapse to a point, the maximum separation between the original objective function of (3.1) and its convex relaxation in (3.15) becomes zero. This implies that for the positive numbers $\epsilon$ and $\mathbf{x}$ there always exists another positive number $\delta$ which, by reducing the rectangular region $[\mathbf{x}^L, \mathbf{x}^U]$ around $\mathbf{x}$ so that $\|\mathbf{x}^U - \mathbf{x}^L\| \leq \delta$, cause the difference between the feasible region of the original problem (3.1) and its convex relaxation (3.15) to become less than $\epsilon$. Therefore, any feasible point $\mathbf{x}$ of problem (3.15), including the global minimum solution, becomes at least $\epsilon$–feasible for problem (3.1) by sufficiently tightening the bounds on $\mathbf{x}$ around this point.

Once the solutions for the upper and lower bounding problems have been established, the next step is to modify these problems for the next iteration. This accomplished by successively partitioning the initial rectangular region into smaller subregions. The number of variables along which subdivision is required is equal to the number of variables $\mathbf{x}$ participating in at least one nonconvex term of the (3.1) formulation. The default partitioning strategy used in the algorithm involves successive subdivision of the original rectangle into two sub-rectangles by halving on the midpoint of the longest side of the initial rectangle (bisection). Therefore, at each iteration a lower bound of the objective function (3.1) is simply the minimum over all the minima of problem (3.15) in each sub-rectangle of the initial rectangle. In order to ensure lower bound improvement, the sub-rectangle to be bisected is chosen by selecting the sub-rectangle which contains the infimum of the minima of (3.15) over all the sub-rectangles. This procedure guarantees a non–decreasing sequence for the lower bound. A non–increasing sequence for the upper bound is found by solving the original nonconvex problem (3.1) locally and selecting it to be the minimum over all the previously recorded upper bounds. Obviously, if the single minimum of (3.15) for any sub–rectangle is greater than the current upper bound, this sub–rectangle can be discarded because the global minimum cannot be within this subdomain (fathoming step).

Figure 3.1 diagrams an unconstrained one-dimensional example of the approach. The mathematical proof that the $\alpha$BB global optimization algorithm converges to the global optimum solution is presented in [45]. In addition to computational chemistry related problems, the $\alpha$BB approach has been applied to a variety of constrained optimization problems [4, 2, 3, 7].

# 3 STRUCTURE PREDICTION OF OLIGOPEPTIDES

The use of computational techniques and simulations in addressing the protein folding problem became possible through the introduction of qualitative and quantitative methods for modeling these systems. Given a sufficiently accurate description of the intramolecular forces, it is in principle possible to predict the folded conformation through global optimization. In our work, we have focused both on the development of global optimization methods and on the verification of energy modeling techniques.

In the area of energy modeling, our work has involved the investigation of numerous detailed representations of protein systems. In addition to the traditional all-atom potential energy models, our work has explored the effects of solvation contributions. In fact, although the problem of considering solvation effects in global conformational energy searches has been made tractable by the development of implicit solvation models, results for such formulations are essentially nonexistent, and those that have appeared are for limited searches only. In our work, both solvent accessible area and volume effects have been considered in the context of global searches for oligopeptides. In addition, we have examined the effects of several parameterizations for these models, and have been able to identify those that provide the best correspondence between computational and experimental results.

## 3.1 POTENTIAL ENERGY MODELS

There are a number of approaches that may be used to model protein interaction energies. In reality, the dynamics of atoms are governed by the quantum theory of their participating electrons. Using the Born-

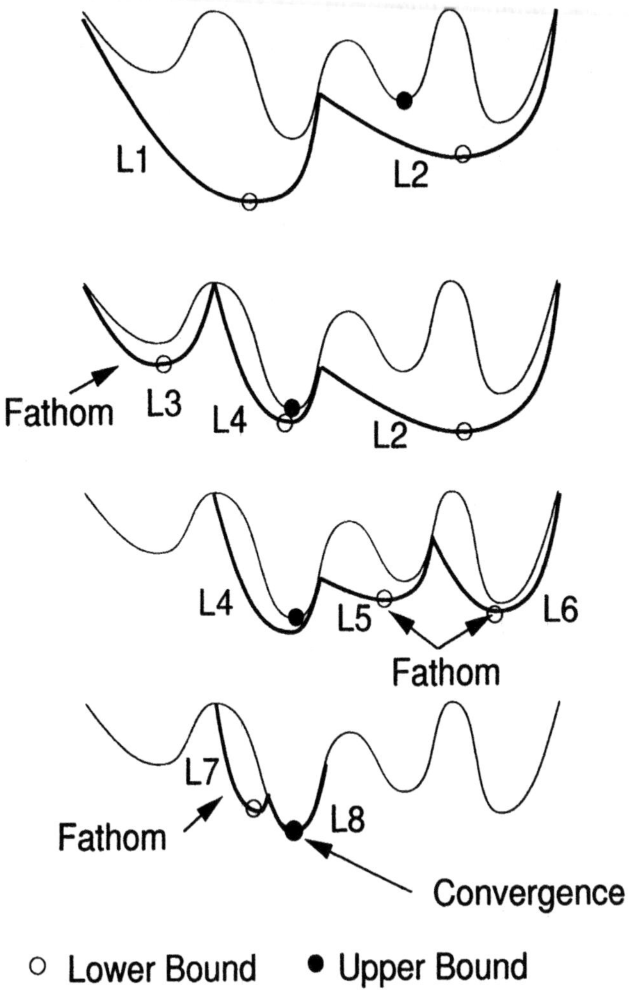


*Figure 3.1* One-dimensional illustrative example of the $\alpha$BB approach. In iteration 1 the overall domain is bisected, the two convex lower bounding functions are created and their unique minima ($L1$ and $L2$) are identified. An upper bound is also identified. Since $L1$ is less than $L2$, the region containing $L1$ is further bisected in iteration 2, while the other region is stored. The minimum of one region ($L3$) is greater than the new upper bound, so this region can be fathomed. The other region is stored. In iteration 3 the region with the next lowest lower bound ($L2$) is bisected and since both new lower bound minima ($L5$ and $L6$) are greater than the current best upper bound, the entire region is fathomed. Finally, by iteration 4, the region containing $L4$ is bisected which results in a region that can be fathomed (containing $L7$) and a convex region whose minimum ($L8$) equals the current upper bound and is the global minimum.

Oppenheimer approximation, one can determine the energy for fixed atomic nuclei from the smallest eigenvalue of the Hamiltonian of the electron system. These approximations and their derivatives are calculated using ab–initio methods. However, due to their computational complexity, such calculations are limited to extremely small molecules. Less detailed, semi-empirical methods are based on all atom representations of the peptide. In general, these models, also known as force fields, are expressed as summations of empirically derived potential functions, with the mathematical form of individual energy terms based on the phenomenological nature of that term. Other simplified models have been used to reduce the degrees of freedom associated with the conformational energy expressions.

For this work the ECEPP/3 (Empirical Conformational Energy Program for Peptides) [52] potential model is utilized. In this force field, it is assumed that the covalent bond lengths and bond angles are fixed at their equilibrium values. Then, the conformation is only a function of the independent torsional angles of the system, also known as dihedral angles. The total conformational energy is calculated as the sum of the electrostatic, nonbonded, hydrogen bonded, and torsional contributions. There is also a pseudo–potential for loop closing if the polypeptide contains two or more sulfur–containing residues. The main energy contributions (electrostatic, nonbonded, hydrogen bonded) are computed as the sum of terms for each atom pair (i,j) whose interatomic distance is a function of at least one dihedral angle. The general potential energy terms of ECEPP/3 are shown in Figure 3.2, while the development of the appropriate parameters is discussed and reported elsewhere [52].

## 3.2 SOLVATION ENERGY MODELS

Solvation contributions are generally believed to be a significant force in stabilizing the native conformations of proteins. Explicit methods can be used to include solvation effects by actually surrounding the polypeptide with solvent molecules and calculating solvent–peptide and solvent–solvent interactions. Although these methods are conceptually simple, explicit inclusion of solvent molecules greatly increases the computational time needed to simulate the polypeptide system. Therefore, most simulations of this type are limited to restricted conformational searches. In addition, it is difficult to quantify the effect of hydrophobic interactions that result from the ordering of water molecules.

Methods for estimating solvent free energies have also been developed using both integral equations and continuum models. Integral equation methods can be used to evaluate solvent structure and thermodynamic

$$
\begin{aligned}
E = & \sum_{(i,j)\in ES} 332.0\frac{q_i q_j}{D r_{ij}} && \text{(Electrostatic)}\\
+ & \sum_{(i,j)\in NB} F\frac{A}{r_{ij}^{12}} - \frac{C}{r_{ij}^{6}} && \text{(Nonbonded)}\\
+ & \sum_{(hx)\in HX} F\frac{A'}{r_{hx}^{12}} - \frac{B}{r_{hx}^{10}} && \text{(Hydrogen bonded)}\\
+ & \sum_{k\in TOR} \left(\frac{E_0}{2}\right)(1 \pm \cos n_k\theta_k) && \text{(Torsional)}\\
+ & \sum_{l\in LOOP} B_L \sum_{il=1}^{il=3} (r_{il} - r_{io})^2 && \text{(Cystine Loop-Closing)}\\
+ & \sum_{l\in LOOP} A_L (r_{4l} - r_{4o})^2 && \text{(Cystine Torsional)}
\end{aligned}
$$

*Figure 3.2* Potential energy terms in ECEPP/3 force field. $r_{ij}$ refers to the interatomic distance of the atomic pair (ij). $Q_i$ and $Q_j$ are dipole parameters for the respective atoms, in which the dielectric constant of 2 has been incorporated. $F_{ij}$ is set equal to 0.5 for 1–4 interactions and 1.0 for 1–5 and higher interactions. $A_{ij}$, $C_{ij}$, $A'_{ij}$ and $B_{ij}$ are nonbonded and hydrogen bonded parameters specific to the atomic pair. $E_{o,k}$ and $E_{o,l}$ are parameters corresponding to torsional barrier energies for a given dihedral angle. $\theta_k$ represents any dihedral angle. $c_k$ takes the values -1,1, and $n_k$ refers to the symmetry type for the particular dihedral angle. The cystine loop–closing term is calculated as a penalty term of three distances involved in loop–closing, where $r_{il}$ represents the actual distance and $r_{io}$ represents the required distance. $B_i$, the penalty parameter, is set equal to 100. Finally, $E_p$ is a fixed internal energy that is added for each proline residue in the protein. Energy units are kcal/mole and distance units are $\AA$

properties. Typically, molecular dynamics or Monte Carlo simulations are used to calculate ensemble averages from which free energy differences can be obtained. A number of methods have been proposed to estimate these solvation free energies from simulations based on molecular dynamics and Monte Carlo averages [17, 41, 63]. The integral equation method has also been used to analyze the solvent structure of a protein system [36]. In contrast, continuum models use a simplified representation of the solvent environment by neglecting the molecular nature of the water molecules. Calculations of solvation free energies using electrostatic continuum models rely on numerical solutions to the Poisson–Boltzmann equation from which dielectric and ionic strength effects are obtained [27]. Other continuum models estimate free energies of solvation as a function of surface areas and volumes.

In our work, solvation contributions are included implicitly using empirical correlations with both surface area [55] and volume [10]. The main assumption of these models is that, for each functional group of the peptide, a hydration free energy can be calculated from an averaged free energy of interaction of the group with a layer of solvent known as the hydration shell. In addition, the total free energy of hydration is expressed as a sum of the free energies of hydration for each of the functional groups of the peptide; that is, an additive relationship is assumed.

Accessible surface area methods assume that the free energy of hydration is proportional to the solvent–accessible surface area of the peptide, as described by the following equation:

$$E_{HYD} = \sum_{i=1}^{N} (A_i)(\sigma_i) \tag{3.17}$$

In Equation (3.17), an additive relationship for N individual functional groups is assumed. ($A_i$) represents the solvent–accessible surface area for the functional group, and ($\sigma_i$) is an empirically derived free energy density parameter. Once the solvent–accessible surface areas have been calculated, these values must be multiplied by the appropriate ($\sigma_i$) parameters as shown in Equation (3.17). A variety of atomic solvation parameter (ASP) sets have been developed to model the transfer of atoms from a gaseous to a hydrated environment. In our work, five ASP sets, namely the WE1 [66], WE2 [66], OONS [54], SCKS [60], and JRF [67] parameters, have been studied.

For volume shell models, the free energy of hydration is assumed to be proportional to the water–accessible volume of a hydration layer surrounding the peptide. This can be represented in the form:

$$E_{HYD} = \sum_{i=1}^{N}(VHS_i)(\delta_i) \tag{3.18}$$

An additive relationship for the $N$ individual atoms of the peptide is assumed, and ($\text{VHS}_i$) represents the solvent–accessible volume of hydration shell for each atom $i$ which is exposed to water. The ($\delta_i$) parameters are empirically determined free energy of hydration densities for these atoms. The hydration shell is defined by the volume inside a sphere of radius $R_i^h$ but outside a sphere of radius $R_i^v$, with both radii centered on atom $i$. The larger radius, $R_i^h$, corresponds to the radius of the first hydration shell of atom $i$, while $R_i^v$ is equal to the van der Waals radius. Free energy density parameters for solvent accessible volumes have been developed for nonionic and charged organic solute molecules [32, 33, 31]. In this work, RRIGS specific ($\delta_i$) parameters, which were developed by a least square fitting of experimental free energy of solvation data for 140 small organic molecules [10], are used.

## 3.3 GLOBAL OPTIMIZATION FRAMEWORK

The energy minimization problem is formulated as a unconstrained nonconvex global optimization problem, which is fashioned after the general formulation given in problem 3.1. Let $i = 1, \ldots, N_{RES}$ be an indexed set describing the sequence of amino acid residues in the peptide chain. There are $\phi_i, \psi_i, \omega_i,\ i = 1, \ldots, N_{RES}$ dihedral angles along the backbone of this peptide. In addition, let $K^i$ denote the number of dihedral angles for the side chain of the $i^{th}$ residue and $J^N$ and $J^C$ denote the number of dihedral angles for the amino and carboxyl end groups, respectively. Using these definitions the optimization problem takes the following form:

$$\min \quad E(\phi_i, \psi_i, \omega_i, \chi_i^k, \theta_j^N, \theta_j^C) \tag{3.19}$$

$$\begin{aligned}
\text{subject to} \quad -\pi &\le \phi_i \le \pi, \ i = 1, \ldots, N_{RES} \\
-\pi &\le \psi_i \le \pi, \ i = 1, \ldots, N_{RES} \\
-\pi &\le \omega_i \le \pi, \ i = 1, \ldots, N_{RES} \\
-\pi &\le \chi_i^k \le \pi, \ i = 1, \ldots, N_{RES}, \ k = 1, \ldots, K^i \\
-\pi &\le \theta_j^N \le \pi, \ j = 1, \ldots, J_N
\end{aligned}$$

$$-\pi \leq \theta_j^C \leq \pi, \quad j = 1, \ldots, J_C$$

In general, $E$ represents the total potential energy function and the free energy of solvation. However, in the case of one ASP set, in particular the JRF ASP [67], the potential energy function is minimized before adding the hydration energy contributions for this ASP set. In other words, gradient contributions from solvation are not considered. This approach is represented by the following equation:

$$E_{JRF}^{Total} = E_{Min}^{Unsol} + E_{JRF}^{Sol} \tag{3.20}$$

Even after reducing this optimization problem to a function of internal variables (dihedral angles), the multidimensional surface that describes the energy function has an astronomically large number of local minima. A large number of techniques have been developed to search this nonconvex conformational space. In general, the major limitation is that these methods depend heavily on the supplied initial conformation. As a result, there is no guarantee for global convergence because large sections of the domain space may be bypassed. To overcome these difficulties, the $\alpha$BB global optimization approach [1, 4, 2, 3, 7] has been extended to identifying global minimum energy conformations of solvated peptides. The $\alpha$BB global optimization algorithm effectively brackets the global minimum solution by developing converging sequences of lower and upper bounds. These bounds are refined by iteratively partitioning the initial domain. Upper bounds on the global minimum are obtained by local minimizations of the original energy function, $E$. Lower bounds belong to the set of solutions of the convex lower bounding functions, which are constructed by augmenting $E$ with the addition of separable quadratic terms.

The determination of the global minimum energy conformation using $\alpha$BB requires the interfacing of a number of programs ($\alpha$BB [1, 4, 2, 3, 7], PACK [59], NPSOL [22] and potential and solvation energy modules). PACK, a peptide generation program, is called once directly by $\alpha$BB in order to initialize the current problem. In subsequent steps PACK is called through NPSOL [22], a local nonlinear optimization solver used to solve both the upper and lower bounding problems. PACK internally transforms to and from Cartesian and internal coordinate systems, and provides potential energy and gradient contributions for the ECEPP/3 potential model at every step of the local minimizations. When considering surface–accessible solvation, surface areas are calculated using MSEED [55]; whereas volumes of hydration shells are determined using the RRIGS module [10]. Finally, an additional module, UBC (Upper Bound Check), is used to verify the quality of the upper bound solutions. The entire suite of programs has been combined to form the

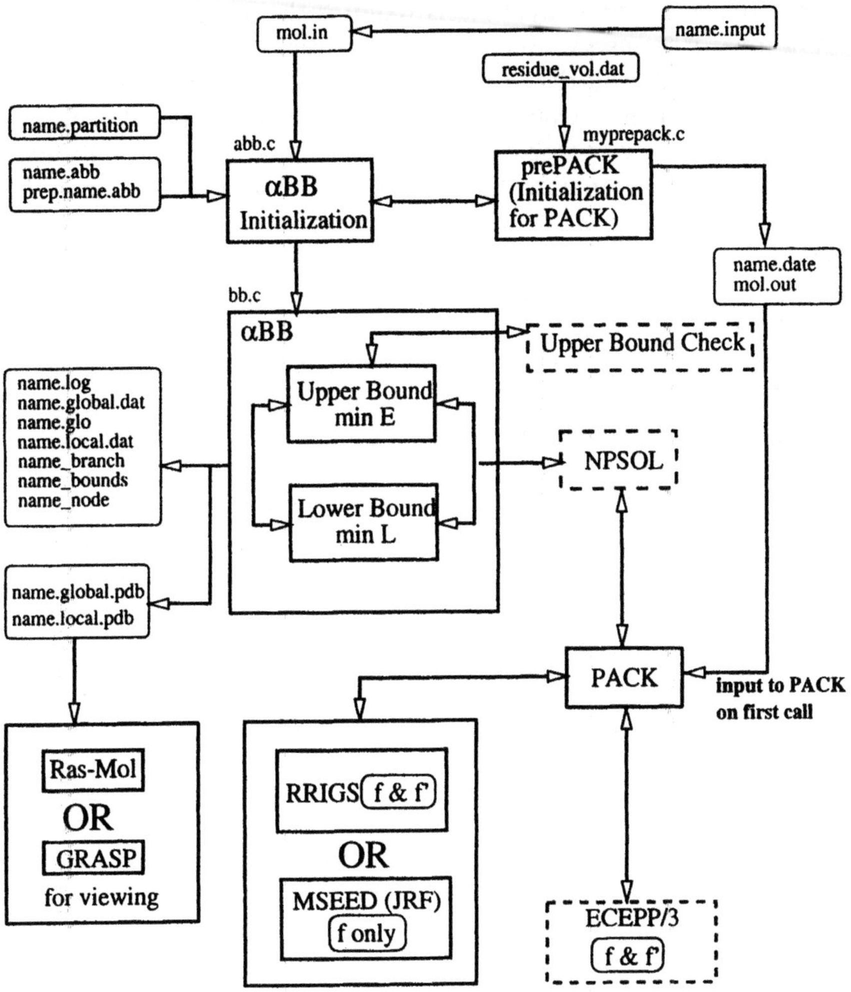


*Figure 3.3* Interface for $\alpha$BB within GLO-FOLD. The arrows indicate the direction of information flow. The names of the input, output, and intermediate files are indicated, in addition to selected source code files. References to "$f$ & $f'$" and "$f$ only" describe whether gradient evaluations or only function evaluations are used in the respective modules.

GLO-FOLD software package for the prediction of protein structure, as shown in Figure 3.3.

## 4 FREE ENERGY MODELING

Locating the global minimum *potential* energy or the global minimum *potential plus solvation* energy conformation is not sufficient because

Anfinsen's thermodynamic hypothesis requires the minimization of the conformational free energy. Specifically, potential energy minimization neglects the entropic contributions to the stability of the molecule. An approximation to these entropic contributions can be developed by using information about low energy conformations. That is, once a sufficient ensemble of low energy minima has been identified, a statistical analysis can be used to estimate the relative entropic contributions, and thus the relative free energy, for conformations in the ensemble.

Therefore, the analysis of the free energy of peptides requires efficient methods for locating not only the global minimum energy structure but also large numbers of low energy conformers. A variety of methods have been used to find such stationary points on potential energy surfaces. For example, periodic quenching during a Monte Carlo or molecular dynamics trajectory can be used to identify local minima [62]. However, a drawback of these approaches is their inherent stochastic nature. In its original form, the $\alpha$BB *deterministic* global optimization algorithm [3, 1, 4, 2, 7] has been shown to be an efficient method for finding the global minimum energy conformation for both unsolvated and solvated peptide systems [8, 37, 38]. Here, novel methods are proposed within the framework of the $\alpha$BB algorithm to optimize the free energy of peptide systems. These modifications facilitate the generation of ensembles of low energy conformers, which can be used to identify the global minimum free energy conformation, as well as perform detailed free energy rankings.

In peptide systems, this entropic contribution arises from fluctuations around a local conformational state. There exists a number of procedures, including both exact and approximate calculations, that can be used to determine the entropic contributions, and thus the free energy, of peptide systems.

First, assume that the full conformational space $R$ can be considered as the union of disjoint basins of attraction, and the conformational space associated with a given basin (denoted by $\gamma$) is defined by $R_\gamma$. The energy, $E$, is a function of the variable set $\theta$, which corresponds to the set of dihedral angles used to describe the conformational state of the system. Each basin of attraction is characterized by a unique local minimum at position $\theta^*_\gamma$, with a corresponding energy $E^*_\gamma$. That is, local minimization starting at any point in $R_\gamma$ will lead to the local minimum at $\theta^*_\gamma$. It should be noted that this approximation of the conformational space excludes all maxima and saddle point conformations.

A rigorous procedure can be envisioned for calculating the exact probability associated with a given basin. First, a sample of conformations must be generated with initial starting energies $E_i$, as defined by the total set $I$. Each structure is minimized to identify its corresponding basin minimum ($\theta_\gamma^*$). These structures define the set $I(\gamma)$ (i.e., those structures associated with basin $\gamma$). As the sampling goes to infinity, the probability associated with basin $\gamma$ can be calculated by the following expression :

$$p_\gamma^{exact} = \frac{\sum_{i(\gamma)\in I(\gamma)} exp(-\beta E_{i(\gamma)})}{\sum_{i\in I} exp(-\beta E_i)} \tag{3.21}$$

Obviously, such a method is intractable for large systems, and this is the impetus for developing approximate methods.

A tractable method for including entropic effects for proteins relies on the concept of the harmonic approximation. Initially, the theoretical development of this approximation for polymer systems generated debate in the literature [23, 24, 19]. In the work of [23] [23] a classical rigid model was used to characterize a partition function based on the fixed bond length and bond angle assumptions. In contrast, [19] [19] derived a different partition function using a classical flexible model. Later analysis by [24] [24] actually showed that the flexible model was also applicable to the fixed bond length and bond angle system (i.e., a peptide described by the internal coordinate system).

An approximate probability associated with a given basin ($\gamma$) can be calculated using the following equation :

$$p_\gamma^{approx} = \frac{\left[[Det\,(H_\gamma)]\right]^{-1/2} \exp\left(-\beta E_\gamma^*\right)}{\sum_{i=1}^N \left[[Det\,(H_i)]\right]^{-1/2} \exp\left(-\beta E_i^*\right)} \tag{3.22}$$

To develop a meaningful comparison of relative free energies, the total partition function (i.e., the denominator of Equation (3.22)) must include an adequate ensemble of low energy local minima, as well as the global minimum energy conformation.

These probabilities can be used to estimate the occupancy of each individual basin, or summed in order to calculate cumulative probabilities for an ensemble of structures exhibiting similar physical or energetic properties. It should be noted that the determination of free energy using the harmonic approximation does not require the explicit inclusion of a contribution based on the density of states. That is, the harmonic approximation decomposes the energetic states within a basin of attraction into one energetic value represented by the local minimizer of the basin. In contrast to counting methods, which estimate probabilities based on the density of states, the contribution of each structure should be accounted for only once. Therefore, using the harmonic approximation requires a structural comparison of all local minimizers.

The probabilities obtained through the harmonic approximation can also be used to calculate thermodynamic quantities. Once the set of unique minimizers has been identified, these structures can be ranked according to their free energy values, and then divided into bins of a specified energy width. Probabilities for each bin can be calculated by summing the individual probabilities (as defined in Equation (3.22)) :

$$P_j^{approx} = \sum_{\gamma=1}^{n_j} p_\gamma^{approx} \tag{3.23}$$

Here $P_j^{approx}$ signifies the probability for energy bin $j$. The summation includes the $n_j$ individual probabilities ($p_\gamma^{approx}$) belonging to bin $j$. Average thermodynamic quantities can now be estimated using the equations with the following form :

$$< E >_T = \sum_j P_j^{approx} < E >_j \tag{3.24}$$

Here the total average energy, $< E >_T$, is calculated by summing the bin probabilities multiplied by the mean energy of bin $j$, $< E >_j$.

## 4.1 FREE ENERGY PROBLEM FORMULATION

As before, the energy minimization problem for proteins is formulated as a nonconvex nonlinear optimization problem. The inclusion of free energy modeling into the protein folding problem does not change the general formulation. However, an additional condition must be satisfied; that is, an ensemble of local minimum low energy conformations must be generated along with the global minimum energy conformation. Once this ensemble has been compiled, a free energy ranking can be performed using the harmonic approximation presented in the previous section.

Several rigorous methods can be envisioned for locating local minimum energy conformations using the $\alpha$BB deterministic global optimization approach. As an introduction to the ideas used here, two rigorous approaches for finding all local minimum energy conformations are discussed.

The first method relies on the introduction of a single inequality constraint to the problem formulation given by (3.19). The new formulation is :

$$\begin{aligned} \min \quad & E(\phi_i, \psi_i, \omega_i, \chi_i^k, \phi_j^N, \phi_j^C) \qquad (3.25) \\ \text{subject to} \quad & (E^* - E) + \epsilon^* < 0 \end{aligned}$$

$$
\begin{aligned}
-\pi &\leq \phi_i \leq \pi, \ i = 1, \ldots, N_{RES} \\
-\pi &\leq \psi_i \leq \pi, \ i = 1, \ldots, N_{RES} \\
-\pi &\leq \omega_i \leq \pi, \ i = 1, \ldots, N_{RES} \\
-\pi &\leq \chi_i^k \leq \pi, \ i = 1, \ldots, N_{RES}, \ k = 1, \ldots, K^i \\
-\pi &\leq \phi_j^N \leq \pi, \ j = 1, \ldots, J^N \\
-\pi &\leq \phi_j^C \leq \pi, \ j = 1, \ldots, J^C
\end{aligned}
$$

The additional constraint requires that the objective function values be larger than the energy value at some local (or global) minimum, as denoted by $E^*$, plus a positive parameter, $\epsilon^*$. When $\epsilon^* = 0$, the solution of the corresponding global optimization problem will give the best local minimum energy conformation with an energy larger than $E^*$. The original formulation given by (3.19) is actually a special case of this problem in which $E^* = -\infty$ and $\epsilon^* = 0$. That is, in (3.19) no bounds are placed on the value of the objective function, $E$. The global minimum energy conformation is only required to take some finite value. In order to locate all local minima, a set of global optimization problems must be solved iteratively with updating of the parameter $E^*$.

The problem of finding all local minimum energy conformations can also be formulated as a single global optimization problem, which can be deterministically solved using the $\alpha$BB algorithm [47]. This method stems from the idea that all stationary points (i.e., minima, maxima and transition states) of the energy hypersurface satisfy the constraint $\nabla E(\theta) = \mathbf{0}$. This can be written as :

$$
\frac{\partial E(\theta)}{\partial \theta_i} = 0 \ , \qquad i = 1, \ldots, N_\theta \tag{3.26}
$$

Here $N_\theta$ represents the total number of dihedral angles defined by the variable set $\theta$. The problem of finding local minima is equivalent to finding all solutions of Equation (3.26) for which the Hessian of $E$ is positive definite.

Both methods for rigorously locating all local minimum energy conformations have some disadvantages. On one hand, the first approach should effectively locate low energy conformers in order of increasing energy. However, locating each minimum requires the solution of a full global optimization problem. The second approach avoids this drawback because it can be solved as a single global optimization problem. However, when dealing with a high dimensional search space, the number of necessary subdivisions may be computationally inhibitive. In addition, this method will potentially locate stationary points other than local minima. Therefore, the development of other methods for locating low energy local minimum energy conformations were pursued.

## 4.2 ENSEMBLE OF LOCAL MINIMUM ENERGY CONFORMATIONS

Since the number of local minima on a given energy hypersurface may become astronomically large (e.g., the number of local minima for met-enkephalin is estimated to be on the order of $10^{11}$ [44]), methods that do not necessarily provide all local minima were developed. Specifically, it was determined that the generation of ensembles of low energy conformers is possible through algorithmic modifications of the general $\alpha$BB procedure. Rigorous implementation of the global optimization algorithm requires the minimization of a *convex* lower bounding function in each domain. The unique solution $\theta$ for each lower bounding minimum can then used as a starting point for the minimization (or function evaluation) of the original energy function in the current domain. In the case of local minimization, each partitioned region provides a single minimum energy conformation as the algorithm proceeds. Using this information, along with the global minimum energy conformation, a list of low energy conformers can be constructed.

A method for increasing the number of local minima produced within each subdomain would involve the selection of multiple random starting points for minimizing the upper bounding function. At first, this approach appears to be equivalent to choosing random points for local minimization. Initially, when the subdomains constitute significant portions of the original domain space, this is the case. However, as the separation between lower and upper bounds decreases, the subdomains are localized in regions of low energy. Therefore, the random point selection is localized in regions which contain low energy local minima.

However, this approach does not take advantage of the information provided by the lower bounding functions. Rigorously, these functions posses a single minimum in each subdomain. Since the choice of $\alpha$ affects the convexity of the lower bounding functions, the $\alpha$ values can be modified to ensure a certain nonconvexity in these functions. In this case, the lower bounding functions possess multiple minima, and these functions can be minimized several times in each domain. In addition, since the lower bounding functions smooth the original energy hypersurface, the location of these multiple minima provide information on the location of low energy minima for the upper bounding function. Therefore, by using the location of the minima of the lower bounding function as starting points for local minimization of the upper bounding function, an improved set of low energy conformations can be identified. As before, these conformations are also localized in those domains with low energy as the subdomains decrease in size.

A second approach incorporates free energy information into the branch and bound algorithm. Specifically, harmonic entropic contributions are calculated and included at each minima of the upper and lower bounding functions. In this way, the progression of lower and upper bounds includes a temperature dependent entropic term. A similar modification to the Monte Carlo minimization method has also been proposed [64], and has been shown to be effective in locating low energy conformers of peptides [49, 50].

The problem formulation is identical to the one given in (3.19). That is, the minimization of $E$ and $L$ are still performed using only potential and solvation energy contributions. However, once local minima have been located, the free energy is calculated by the following expression:

$$G = U_{Min} + \frac{1}{2\beta} \ln \left[ Det \left( H_{Min} \right) \right] \tag{3.27}$$

$U_{Min}$ represents the local minimum energy of $E$ or $L$, and $Det\,(H_{Min})$ is the determinant of the Hessian evaluated at this local minimum. The specification of a thermodynamic temperature ($\beta = 1/k_B T$) is required as an additional input parameter.

A single rigorous application of the $\alpha$BB algorithm to this problem will result in the identification of the global minimum free energy at a given temperature. However, the goal is to identify an ensemble of low energy and, in this case, low free energy conformers so that a free energy ranking and comparison can be made. Therefore, the algorithmic steps for the Free Energy Directed Approach (FEDA) are similar to those for EDA, with the additional evaluation of the free energy ($G$) at each local minima of $E$ and $L$. The thermodynamic temperature used in Equation (3.27) must be specified as an additional input parameter.

## 4.3 FREE ENERGY COMPUTATIONAL STUDIES

The EDA was first applied to the isolated form of met-enkephalin. All 24 dihedral angles were considered variable, with the 10 dihedral angles of the backbone residues acting as global variables (variables on which branching occurs). For both peptides, the EDA algorithm detailed above, was applied 10 times. The input conditions correspond to initial $\alpha$ values of 5 and 10, with a subsequent reduction of these values based on the current level in the branch and bound tree.

Once the ensemble of local minima had been compiled, a set of distinct conformations was identified by checking for repeated and symmetric conformations. In addition, a conformation was only considered unique if at least one dihedral angle differed by at least $50^o$ when comparing each pair of conformations. These conformations were then used to generate results and distributions according to energy and free energy values. Energy bins were used to characterize a group of distinct structures between a range of energy values (every 0.5 kcal/mol) relative to the global minimum energy structure. For example, Bin 1 contains structures that are 0.0-0.5 kcal/mol above the global minimum energy structure, Bin 2 contains structures that are 0.5-1.0 kcal/mol above the global minimum energy structure, etc.

In the case of isolated met-enkephalin, the 10 (EDA) runs generated a total of 83908 distinct local minima. The potential energy global minimum (PEGM) conformation for met-enkephalin possesses an energy of -11.707 kcal/mol. This conformation exhibits a type II' $\beta$–bend along the N–C' peptidic bond of $Gly^3$ and $Phe^4$. Essentially, this structure corresponds to the free energy global minimum (FEGM) conformation for a temperature of 0 K, that is, when entropic contributions are not included. When considering the harmonic free energy, the prediction of the FEGM can be calculated over a range of temperatures. The results suggest that the inclusion of entropic contributions greatly affects the relative stability of individual low energy structures.

The EDA was also applied to the RRIGS solvated form of met-enkephalin using the same protocol and conditions as detailed above. Qualitatively, the PEGM (in this case PEGM refers to potential+solvation) for solvated met-enkephalin exhibits a more extended conformation than that which is observed for the isolated form. As detailed in Table 3.1, the PEGM structure persists as the FEGM at 100 K. However, at each subsequent temperature, the FEGM structure changes, and this change is accompanied by an increase in total energy (potential and solvation). As with isolated met-enkephalin, the difference in total energy between the PEGM and FEGM at 500 K is greater than 5 kcal/mol. This suggests that entropic effects are important in defining the predicted native structure. When considering individual structures, entropic effects tend to produce more extended FEGM conformations at higher temperatures, especially with regard to the placement of the aromatic rings. It is interesting to note that in a previous study the positioning of aromatic rings was found to be a major difference when considering the ability of solvation models to predict extended PEGM conformations for the solvated enkephalin peptides [37]. The sequence of FEGM structures is illustrated in Figure 3.4.

*Table 3.1* Dihedral angle values for PEGM and FEGM structures of solvated met-enkephalin. The temperatures are provided in the first row. The last two rows indicate the harmonic free energy (kcal/mol) and the potential energy value (kcal/mol), respectively.

| Residue | DA | PEGM | 100 K | 200 K | 300 K | 400 K | 500 K |
|---|---|---|---|---|---|---|---|
| $Tyr_1$ | $\phi$ | -168.2 | -168.2 | -170.9 | -168.4 | -168.4 | -152.5 |
| | $\psi$ | -30.9 | -30.9 | -28.5 | -34.3 | -34.3 | 153.2 |
| | $\omega$ | 178.6 | 178.6 | 177.5 | -178.9 | -178.9 | 178.5 |
| | $\chi_1$ | -173.5 | -173.5 | 178.8 | 178.7 | 178.7 | -179.0 |
| | $\chi_2$ | -100.9 | -100.9 | 61.3 | -100.8 | -100.8 | -101.2 |
| | $\chi_3$ | 19.3 | 19.3 | -4.1 | 179.0 | 179.0 | -179.9 |
| $Gly_2$ | $\phi$ | 78.5 | 78.5 | 73.8 | 177.8 | 177.8 | -173.9 |
| | $\psi$ | -86.5 | -86.5 | 47.6 | -179.9 | -180.0 | 177.1 |
| | $\omega$ | -177.3 | -177.3 | -179.2 | 180.0 | 180.0 | -179.8 |
| $Gly_3$ | $\phi$ | 162.4 | 162.4 | 167.6 | -180.0 | -180.0 | 179.6 |
| | $\psi$ | 92.2 | 92.2 | -145.2 | 179.9 | 179.9 | -179.3 |
| | $\omega$ | 172.6 | 172.6 | 175.2 | 179.7 | 179.7 | 179.6 |
| $Phe_4$ | $\phi$ | -150.3 | -150.3 | -149.3 | -155.3 | -155.4 | -155.4 |
| | $\psi$ | 159.8 | 159.8 | 135.8 | 147.2 | 149.5 | 149.3 |
| | $\omega$ | -178.1 | -178.1 | -176.6 | -176.8 | -178.3 | -178.3 |
| | $\chi_1$ | 65.8 | 65.8 | 177.3 | -179.5 | -179.5 | -179.7 |
| | $\chi_2$ | -87.4 | -87.4 | -108.1 | -111.7 | -105.6 | 74.4 |
| $Met_5$ | $\phi$ | -75.0 | -75.0 | -85.5 | -78.7 | -78.7 | -78.9 |
| | $\psi$ | 113.9 | 113.9 | -41.1 | -51.1 | 113.4 | 113.5 |
| | $\omega$ | -178.4 | -178.4 | 179.9 | 179.7 | -179.1 | -179.1 |
| | $\chi_1$ | -172.3 | -172.3 | -65.6 | -67.2 | -67.4 | -67.4 |
| | $\chi_2$ | 176.1 | 176.1 | -179.6 | -178.8 | -178.8 | -178.8 |
| | $\chi_3$ | -180.0 | -180.0 | -179.4 | -179.9 | -179.9 | -179.9 |
| | $\chi_4$ | 60.0 | 60.0 | 179.5 | -180.0 | 60.0 | -60.0 |
| G | | -50.060 | -41.896 | -34.566 | -28.604 | -22.828 | -17.166 |
| E | | -50.060 | -50.060 | -48.676 | -46.030 | -45.780 | -44.797 |

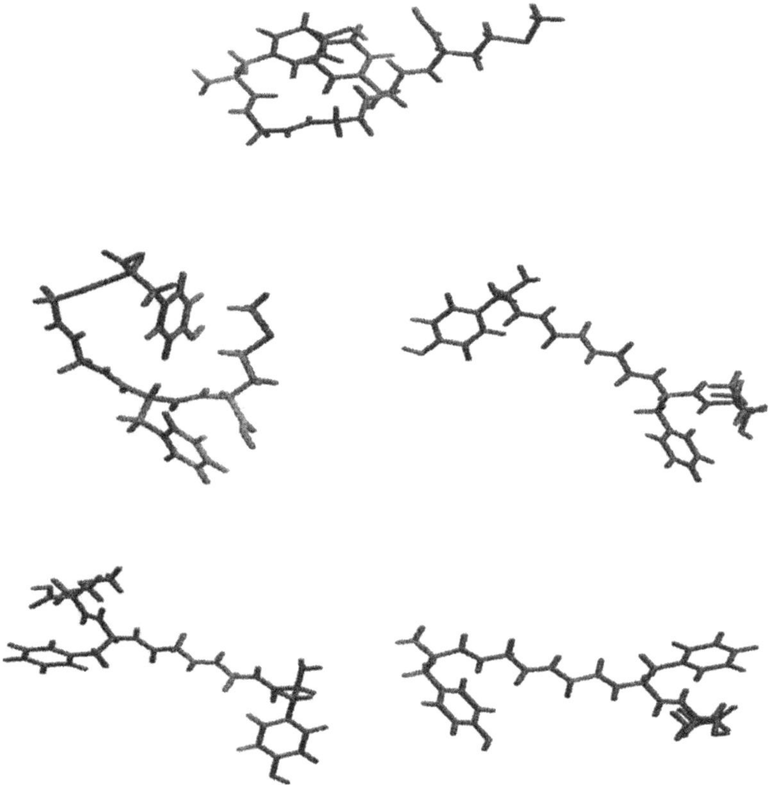

*Figure 3.4* FEGM structures for solvated met-enkephalin. The top figure is the PEGM and the FEGM for 100 K. The structures at other temperatures (200,300,400,500) are shown left to right, top to bottom.

# 5 STRUCTURE REFINEMENT WITH SPARSE RESTRAINTS

To effectively determine protein function it is important to predict the three dimensional structure of the macromolecule. Over the last several decades a number of experimental and theoretical approaches have been developed and refined in order to achieve this goal, such as the computational approaches outlined above. Experimentally, there now exist two basic techniques used to perform protein structure refinement. The first, X-ray crystallography, relies on the ability to crystallize the protein so that diffraction patterns can be used for sufficient resolution. These requirements have limited the applicability of this technique. A more powerful method, NMR (nuclear magnetic resonance) spectroscopy, is based on solution measurements of the system. Several key developments, including multidimensional NMR experiments, have resulted in the ability to determine solution structures for proteins consisting of over 200 residues.

This section focuses on the development of a novel approach for protein structure prediction via experimental NMR restraints. Traditionally, the protein folding global optimization problem involves a progression of unconstrained minimizations. However, the introduction of experimentally derived or artificial restraints can be used to recast the fundamental protein folding problem as a constrained global optimization problem. The constraints, through reduction of the feasible search space, serve two important purposes : 1) attempt to correct any deficiencies of the energy model, and 2) focus the efforts of the global optimization algorithm.

This constrained approach is applied to the NMR structure prediction problem, although a variety of restraint information could be used. The proposed constrained formulation differs from traditional NMR approaches in several fundamental ways. First, the energy model is represented by a detailed full atom force field, rather than simplified nonbonded potential terms. This should make the approach especially effective when the number of NMR restraints per residue decreases; that is, the accuracy of the energy model becomes more significant. In addition, traditional solution approaches apply target function distance geometry or simulated annealing to unconstrained problem formulations in which restraints are represented by penalty function approximations. The solution of the constrained formulation requires the use of constrained local optimization solvers and an overall global optimization framework for nonlinearly constrained problems. This is accomplished through the application of the ideas of the $\alpha$BB deterministic global optimization approach [3, 1, 4, 2, 7]. $\alpha$BB based global optimization techniques have also been applied to NMR type structure prediction problems [40, 61].

Because the location of the global minimum relies on effectively solving constrained local optimization problems, convergence to the global minimum can be enhanced by consistently identifying low energy solutions. These observations illustrate the need for reliably locating low energy feasible points for initializing the constrained local optimization routine. Along these lines, a combined torsion angle dynamics (TAD) and simulated annealing scheme has been implemented within the context of the global optimization framework. Torsion angle dynamics (TAD) has recently been shown to be more effective than Cartesian coordinate dynamics [26, 57]. In this case, the degrees of freedom are rotations around single bonds, which reduces the number of variables by approximately tenfold because bond lengths, bond angles, chirality and planarities are kept fixed at optimal values during the calculation.

## 5.1 ENERGY MODELING

Basic data obtained from NMR studies consist of distance and torsion angle restraints. Once resonances have been assigned, nuclear Overhauser effect (NOE) contacts are selected and their intensities are used to calculate interproton distances. Information on torsion angles are based on the measurement of coupling constants and analysis of proton chemical shifts. Together, this information is used to formulate a nonlinear optimization problem, the solution of which should provide the correct protein structure. Typically, a hybrid energy function of the following form is employed:

$$E = E_{\text{forcefield}} + W_{\text{nmr}} E_{\text{nmr}}. \tag{3.28}$$

The energy, $E$, specified by this target function includes a chemical description of the protein conformation through the use of a force field, $E_{\text{forcefield}}$. The force field potentials are generally much simpler representations of all atom force fields. The distance and dihedral angle restraints appear as weighted penalty, $E_{\text{nmr}}$, terms that should be driven to zero.

The second term of Equation (3.28) can be rewritten as :

$$E_{\text{nmr}} = E_{\text{distance}} + E_{\text{dihedral}}. \tag{3.29}$$

Here $E_{\text{distance}}$ and $E_{\text{dihedral}}$ represent the violation energies based on the distance and dihedral angle restraints, respectively. These functions can take several forms, although a simple square well potential is commonly used. The expressions also include a summation over both upper and lower distance violations; for example, $E_{\text{distance}} = E^{\text{upper}}_{\text{distance}} + E^{\text{lower}}_{\text{distance}}$. When considering upper distance restraints this becomes:

$$E^{\text{upper}}_{\text{distance}} = \sum_j \begin{cases} A_j(d_j - d_j^{\text{upper}})^2 & \text{if } d_j > d_j^{\text{upper}}, \\ 0 & \text{otherwise.} \end{cases} \tag{3.30}$$

The squared violation energy is considered only when the calculated distance $d_j$ exceeds the upper reference distance $d_j^{\text{upper}}$. This squared violation can then multiplied by a weighting factor $A_j$. A similar contribution is calculated for those distances that violate a lower reference distance, $d_j^{\text{lower}}$:

$$E^{\text{lower}}_{\text{distance}} = \sum_j \begin{cases} A_j(d_j - d_j^{\text{lower}})^2 & \text{if } d_j < d_j^{\text{lower}}, \\ 0 & \text{otherwise.} \end{cases} \tag{3.31}$$

For dihedral angle restraints the functional form is similar to that of Equations (3.30) and (3.31). As before, the total violation, $E_{\text{dihedral}}$, is a sum over upper and lower violations (i.e., $E_{\text{dihedral}} = E^{\text{upper}}_{\text{dihedral}} + E^{\text{lower}}_{\text{dihedral}}$). A dihedral angle $\omega_j$ can be restrained by employing a quadratic square well potential using upper ($\omega_j^{\text{upper}}$) and lower ($\omega_j^{\text{lower}}$) bounds on the variable values. However, due to the periodic nature of these variables, a scaling parameter must be incorporated to capture the symmetry of the system. Furthermore, by centering the full periodic region on the region defined by the allowable bounds, all transformed values will lie in the domain defined by $[\omega_j^{\text{lower}} - \Delta HW_{\omega_j}, \omega_j^{\text{upper}} + \Delta HW_{\omega_j}]$, where $\Delta HW_{\omega_j}$ is equal to half the excluded range of dihedral angle values (i.e., $\Delta HW_{\omega_j} = \pi - (\omega_j^{\text{upper}} - \omega_j^{\text{lower}})/2$).

The force field energy term, $E_{\text{forcefield}}$ of Equation (3.28), models the nonbonded interactions of the protein, which can consist of simple or more detailed energy functions. In practice, when considering NMR restraints, force field terms are often simplified to include only geometric energy terms such as quartic Van der Waals repulsions. Such functions neglect rigorous modeling of energetic terms in order to ensure that experimental distance violations are minimized. In fact, a basic representation for this target function would be :

$$E_{\text{S}} = E_{\text{distance}} + E_{\text{dihedral}}. \tag{3.32}$$

Here the $E_{\text{distance}}$ function includes additional distance restraints to avoid van der Waals contacts. Notice that when all restraints are satisfied, the objective function is driven to zero.

A detailed modeling approach is proposed by using the ECEPP/3 force field [52]. When considering an unconstrained minimization, this approach provides the following objective function :

$$E_{\mathrm{D}} = E_{\mathrm{distance}} + E_{\mathrm{dihedral}} + E_{\mathrm{ECEPP/3}}. \tag{3.33}$$

In contrast to Equation (3.32), the detailed energy modeling greatly increases the complexity of the objective function. It should also be noted that the transformation from Cartesian to internal coordinate space results in highly nonlinear functions. That is there is not a one-to-one correspondence between distances and internal coordinates. The advantage for working in dihedral angle space is that the variable set decreases, with the disadvantage being the increased nonconvexity of the energy hypersurface.

## 5.2 GLOBAL OPTIMIZATION

The determination of a three dimensional protein structure defines an optimization problem in which the objective function may correspond to one of the target functions outlined in the previous section. For the simple case, the formulation becomes :

$$\min_{\phi} \qquad E_{\mathrm{S}}(\phi) = E_{\mathrm{distance}} + E_{\mathrm{dihedral}}. \tag{3.34}$$

A standard procedure for addressing this global optimization problem consists of a combination of simulated annealing and molecular or torsional angle dynamics [12]. Generally, multiple initial conformers are generated and optimized to provide a set of acceptable structures. Typically, a set containing on the order of 100 acceptable conformers may be identified, from which a subset of similar structures (approximately 20) are used to characterize the system. The simulated annealing protocol is incorporated in an attempt to reduce trapping in local minimum energy wells.

However, the minimization of complex target functions necessitates the use of rigorous global optimization techniques. For the detailed target function, given by Equation (3.33), the unconstrained formulation is similar to formulation (3.34). Through the use of the constrained optimization approach, the dihedral angle bounds are implicitly included as box constraints. Furthermore, distance restraints are treated explicitly. This reformulation removes both $E_{\mathrm{dihedral}}$ and $E_{\mathrm{distance}}$ from the target function, leaving only $E_{\mathrm{forcefield}}$ :

$$\min_{\phi} \qquad E_{\mathrm{ECEPP/3}}, \tag{3.35}$$

$$\begin{aligned} \text{subject to} \qquad & E_l^{\mathrm{distance}}(\phi) \;\le\; E_l^{\mathrm{ref}} \quad l = 1, \ldots, N_{\mathrm{CON}}, \\ & \phi_i^L \;\le \phi_i \le\; \phi_i^U, \;\; i = 1, \ldots, N_{\phi}. \end{aligned}$$

Here $i = 1, \ldots, N_\phi$ corresponds to the set of dihedral angles, $\phi_i$, with $\phi_i^L$ and $\phi_i^U$ representing lower and upper bounds on these dihedral angles. In general, the lower and upper bounds for these variables are set to $-\pi$ and $\pi$, although appropriate bounds derived from NMR data are also suitable.

## 5.3 TORSION ANGLE DYNAMICS

Standard unconstrained molecular dynamics simulations have been used extensively to model the folding and unfolding of protein systems [18, 16, 13]. In addition, several methods for NMR structure calculation have been based on molecular dynamics in Cartesian space [12]. Torsion angle dynamics differs from traditional molecular dynamics in that bond lengths and bond angles are fixed at equilibrium values, thereby allowing for a transformation from the Cartesian to the internal coordinate system. The constraints on the systems also dampen high frequency motions, which permits the use of longer time steps during the numerical integration of the equations of motion. The use of TAD in place of conventional MD has been found to improve the efficiency of NMR structure prediction [26, 57].

A major disadvantage for employing TAD in place of Cartesian MD is that the equations of motion become much more complex for the constrained system. For unconstrained Cartesian MD the accelerations of the atoms can be calculated independently due to the decoupled nature of the equations of motion. The addition of constraints to the Cartesian system transforms the equations from a system of ODEs to a system of differential algebraic equations (DAEs). The alternative to solving this system of DAEs is to transform the equations of motion to the internal coordinate reference frame. In this case, the solution of a linear matrix equation in each time step is required, which, due to the highly coupled structure of the equations, scales as a cubic function of the number of degrees of freedom (torsion angles). To avoid the potentially prohibitive computational cost required for the solution of the equations of motion, a fast recursive algorithm, which scales linearly with the number of torsion angles, was implemented. The algorithm is based on spatial operator algebra which has been used to simulate the dynamics of astronautical and robotic equipment [30].

## 5.4 COMPUTATIONAL STUDY

The global optimization algorithm was tested on Compstatin, a synthetic 13-residue (ICVVQDWGHHRCT) cyclic peptide (disulfide bridge between the $Cys^2$ and $Cys^{12}$ residues) that binds to C3 (third component of complement) and inhibits complement activation [58]. Two-dimensional NMR techniques [51] yield a total of 30 backbone sequential (including $H^{\beta}$ - backbone), 23 medium and long range (including disulfide) and 82 intra-residue NOE restraints. In addition, 7 $\phi$ angle and 2 $\chi_1$ angle dihedral restraints are available. In previous work [51], traditional distance geometry–simulated annealing protocol was utilized to minimize the associated target function in the Cartesian coordinate space using the program X-PLOR [12]. NOE distance and dihedral angle restraints were modeled using a quadratic square well potential, while van der Waals overlaps were prevented through the use of a simple quartic potential function.

The NMR refinement protocols resulted in a family of 21 structures with similar geometries for the $Gln^5$–$Gly^8$ region. A representative structure was obtained by averaging the coordinates of the individually refined structures and then subjecting this structure to further refinement to release geometric strain produced by the averaging process. The formation of a type I $\beta$-turn was identified as a common characteristic for these structures.

The constrained global optimization approach was first applied to Compstatin structure prediction without the use of TAD. A subset of 26 (all $\phi$ and $\psi$) torsion angle, from a total of 73, were treated globally, while the remaining were allowed to vary locally. As was the case for local minimization, the same set of restraints were used to formulate the nonlinear constraint, with a constant 50 kcal/mol/$\AA$ weighting factor and a constraint parameter equal to 200 kcal/mol. The lowest energy structure satisfying the constraint functions provided an ECEPP/3 energy of -85.71 kcal/mol, an energy value more than 15 kcal/mol lower than those values provided by local minimization. The global minimization required approximately 40 CPU hours on a HP C160. The total distance violation equaled 6.690 $\AA$ which is near the average distance violation for the local minimum structures. Plots for superpositioning (backbone atoms) of the average local minimum energy structure $\overline{Compstatin}^{Local}$ and the global minimum energy structure are given in Figure 3.5.

## 5.5 GLOBAL OPTIMIZATION AND TORSION ANGLE DYNAMICS

A modified constrained global optimization was also applied to the Compstatin structure prediction problem using the same constraint function and parameters [39]. The goal of introducing TAD as a component of the upper bound solution approach is to increase the number of feasible points available for initialization of the constrained local minimiza-

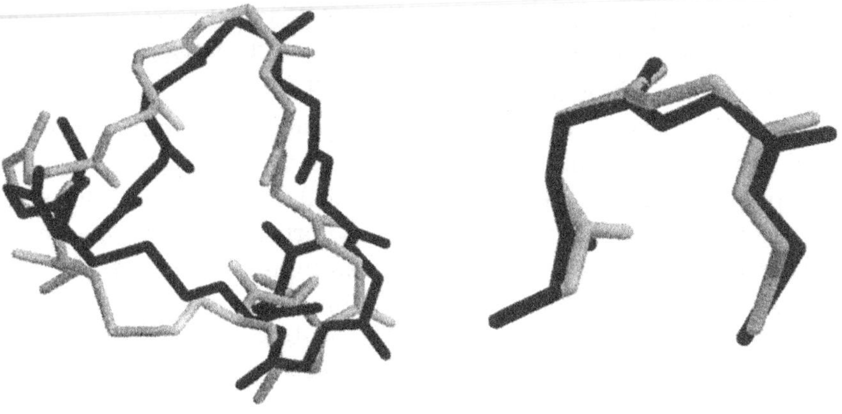

*Figure 3.5* Superposition of global minimum (in black) and $\overline{Compstatin}^{Local}$ (in light grey) structures. The left panel shows the full (backbone atom) structure, while the right panel compares only the $\beta$-turn region.

tion. Initially TAD is used in combination with simple van der Waals overlap restraints to drive the distance violations to zero. Taken independently, this methodology is comparable to the typical implementation of TAD for NMR structure prediction [26].

To gauge the performance of the combined $\alpha$BB and TAD constrained approach, a comparison was made to an independent TAD method (DYANA [26]) for solving distance restraint problems. The same dihedral angle and 53 medium and long range distance restraints were considered, along with additional distance restraints to prevent van der Waals overlaps. The coupled simulated annealing / TAD protocol was applied to a starting sample of 1000 randomly generated structures, from which a subset consisting of 20 conformers exhibiting the best target values were used to initialize a second set of runs. The five conformations with the best target function values were selected for further analysis, including initialization for constrained local minimizations with $E_{\mathrm{ECEPP/3}}$. The DYANA conformers satisfy the corresponding constraint, although their energy values are more than 70 kcal/mol higher than that of the global minimum energy structure. An analysis of the structural characteristics also indicates that the type I $\beta$-turn does not form along the $\mathrm{Gln}^5$-$\mathrm{Gly}^8$ b backbone for these structures. These results reflect the potential deficiencies of the independent TAD algorithm; that is, the simplified force field term is insufficient for sparse sets of distance restraints.

The use of TAD in the context of the global optimization approach surmounts this difficulty by using an iterative TAD scheme with two forms of the target function. The first set of TAD runs focuses on the reduction of the distance violations, while employing a simplified forcefield in the form of additional distance restraints to avoid atomic overlaps. This approach mimics the effects of a typical TAD approach for structure prediction. To ensure that these conformers provide low energy, this step is then followed by unconstrained minimization with a hybrid distance and ECEPP/3 energy objective function. If the ECEPP/3 energy is acceptably low, the algorithm proceeds to the constrained local minimization step, otherwise an iterative set of TAD runs are performed with readjustment of the relative weight of the distance and ECEPP/3 terms.

The results of the combined constrained global optimization and TAD algorithm can be assessed by examining the sequence of ECEPP/3 energies obtained from the solution of the upper bounding problems. When compared to the original algorithm, the TAD implementation augments the number of feasible starting points by more than a factor of two. This enhancement leads to earlier identification of low energy conformers. In particular, conformers with energies less than -70 kcal/mol, and thus lower in energy than the locally minimized PDB structures, are identified within the first 10 iterations of the global optimization approach. This property has important algorithmic implications, including the ability to fathom regions based on the current estimate of the global minimum. In general, the TAD enhanced search provides more consistent and denser population of low energy conformers.

Both experimental and theoretical methods exist for the prediction of protein structures. In both cases, additional restraints on the molecular system can be derived and used to formulate a nonconvex optimization problem. Here, the traditional unconstrained problem was recast as a constrained global optimization problem, and applied to protein structure prediction using NMR data. Both the formulation and solution approach of this method differ from traditional techniques, which generally rely on the optimization of penalty-type target function using SA/MD protocols.

As a first step, the penalty type restraint functions were replaced by nonlinear constraints, which can be individually enumerated for all or subsets of the distance restraints. In addition, the objective function was transformed to a full atom force field potential, a modification that should be particularly useful for systems possessing sparse set of restraints. To solve this reformulated molecular structure prediction problem the concepts of a deterministic global optimization approach, $\alpha$BB, were applied. This methodology can be used to develop theoretical guarantees for convergence to the global minimum of nonconvex constrained problems. The algorithm was further enhanced by modifying the upper bounding solution approach to include an iterative scheme involving TAD.

The approach was applied to the Compstatin structure prediction problem using both the original TAD approach and the coupled $\alpha$BB-TAD approach. When considering basic structural features, such as the formation of a type I $\beta$-turn, the predicted structure was found to agree with results based on X-PLOR [12]. However, constrained global optimization was able to identify conformers with significantly lower energies than those obtained from either local minimization or independent TAD algorithms. In particular, the coupled $\alpha$BB-TAD implementation consistently produced dense populations of low energy conformers.

# 6 CONCLUSIONS

The importance of the protein folding is evidenced by the large amount of experimental and theoretical research conducted in these areas. Although experimental studies of protein systems are necessary and insightful, the ability to computationally predict and understand the folding of proteins would greatly aid the advancement of the biological and chemical sciences. We have shown that both molecular modeling and global optimization are the dominant factors in the overall equation that will eventually provide a solution to these problems.

In particular, this chapter has focused on the use of ab-initio models, which give rise to a series of complex mathematical problems. The essential component has been the application of deterministic global optimization, namely the $\alpha$BB algorithm, for solving the resulting problems. Many issues related to the modeling of protein folding have been analyzed and discussed. These observations have highlighted the extreme difficulty of these problems and the crucial interdependence of ab initio modeling and deterministic global optimization approaches.

## Acknowledgments

The authors gratefully acknowledge financial support from the National Science Foundation and the National Institutes of Health (R01 GM52032, 1 F32 GM20007).

## References

[1] Adjiman, C. S., Androulakis, I. P., and Floudas, C. A. (1997). Global optimization of minlp problems in process synthesis and design. *Comput. Chem. Eng.*, 21:S445–S450.

[2] Adjiman, C. S., Androulakis, I. P., and Floudas, C. A. (1998a). A global optimization method for general twice-differentiable nlps - ii. implementation and computational results. *Comput. Chem. Eng.*, 22:1159–1179.

[3] Adjiman, C. S., Androulakis, I. P., Maranas, C. D., and Floudas, C. A. (1996). A global optimization method, $\alpha$BB, for process design. *Comput. Chem. Eng.*, 20:S419–S424.

[4] Adjiman, C. S., Dallwig, S., Floudas, C. A., and Neumaier, A. (1998b). A global optimization method for general twice-differentiable nlps - i. theoretical advances. *Comput. Chem. Eng.*, 22:1137–1158.

[5] Adjiman, C. S. and Floudas, C. A. (1996). Rigorous convex underestimators for general twice–differentiable problems. *J. Glob. Opt.*, 9:23–40.

[6] Al-Khayyal, F. A. and Falk, J. E. (1983). Jointly constrained biconvex programming. *Maths Ops Res.*, 8:273–286.

[7] Androulakis, I. P., Maranas, C. D., and Floudas, C. A. (1995). $\alpha$bb : A global optimization method for general constrained nonconvex problems. *J. Glob. Opt.*, 7:337–363.

[8] Androulakis, I. P., Maranas, C. D., and Floudas, C. A. (1997). Global minimum potential energy conformation of oligopeptides. *J. Glob. Opt.*, 11(1):1–34.

[9] Anfinsen, C. B., Haber, E., Sela, M., and Jr., F. H. W. (1961). The kinetics of formation of native ribonuclease during oxidation of the reduced polypeptide chain. *Proc. Natl. Acad. Sci. USA*, 47:1309–1314.

[10] Augspurger, J. D. and Scheraga, H. A. (1996). An efficient, differentiable hydration potential for peptides and proteins. *J. Comp. Chem*, 17:1549–1558.

[11] Becker, O. M. and Karplus, M. (1997). The topology of multidimensional potential energy surfaces: Theory and application to peptide structure and kinetics. *J. Chem. Phys.*, 106(4):1495–1517.

[12] Brünger, A. (1992). X-PLOR, version 3.1 a system for x-ray crystallography and nmr. Yale University Press, New Haven, USA.

[13] Caves, L. S. D., Evanseck, J. D., and Karplus, M. (1998). Locally accessible conformations of proteins: Multiple molecular dynamics simulations of cramb in. *Protein Sci.*, 7:649–666.

[14] Church, B. W., Orešič, M., and Shalloway, D. (1996). Tracking metastable states to free-energy global minima. In *DIMACS Series in Discrete Mathematics and Theoretical Computer Science*, volume 23, pages 41–64. American Mathematical Society.

[15] Czerminski, R. and Elber, R. (1990). Reaction path study of conformational transitions in flexible systems: Applications to peptides. *J. Chem. Phys.*, 92(9):5580–5601.

[16] Daggett, V., Li, A. J., and Fersht, A. R. (1998). Combined molecular dynamics and phi-value analysis of structure-reactivity relationships in the transition state and unfolding pathway of barnase: Structural basis of hammond and anti-hammon d effects. *J. Am. Chem. Soc.*, 120:12740–12754.

[17] Dejaegere, A. and Karplus, M. (1996). Analysis of coupling schemes in free energy simulations: A unified description of nonbonded contributions to solvation free energies. *J. Phys. Chem.*, 100:11148–11164.

[18] Duan, Y. and Kollman, P. A. (1998). Pathways to a protein folding intermediate observed in a 1-microsecond simulation in aqueous solution. *Science*, 282:740–744.

[19] Flory, P. J. (1974). Foundations of rotational isomeric state theory and general methods for generating configurational averages. *Macromolecules*, 7(3):381–392.

[20] Floudas, C. A. (2000). *Deterministic Global Optimization: Theory, Methods and Applications.* Nonconvex Optimization and its Applications. Kluwer Academic Publishers.

[21] Gerschgorin, S. (1931). Über die abgrenzung der eigenwerte einer matrix. *Izv. Akad. Nauk SSSR, Ser. fiz.-mat.*, 6:749–754.

[22] Gill, P. E., Murray, W., Saunders, M. A., and Wright, M. H. (1986). *NPSOL 4.0 User's Guide.* Systems Optimization Laboratory, Dept. of Operations Research, Stanford University, CA.

[23] Go, N. and Scheraga, H. A. (1969). Analysis of the contribution of internal vibrations to the statistical weights of equilibrium conformations of macromolecules. *J. Chem. Phys.*, 51(11):4751–4767.

[24] Go, N. and Scheraga, H. A. (1976). On the use of classical statistical mechanics in the treatment of polymer chain conformations. *Macromolecules*, 9(4):535–542.

[25] Goldberg, D. (1989). *Genetic algorithms in search, optimization and machine learning.* Addison-Wesley.

[26] Güntert, P., Mumenthaler, C., and Wüthrich, K. (1997). Torsion angle dynamics for nmr structure calculation with the new program dyana. *J. Mol. Biol.*, 273:283–298.

[27] Honig, B., Sharp, K., and Yang, A. (1993). Macroscopic models of aqueous solutions: Biological and chemical applications. *J. Phys. Chem.*, 97:1101–1109.

[28] Horst, R. and Pardalos, P. M., editors (1995). *Handbook of Global Optimization.* Kluwer Academic Publishers.

[29] Horst, R. and Tuy, H. (1993). *Global optimization: deterministic approaches.* Springer-Verlag, Berlin. 2nd. rev. edition.

[30] Jain, A., Vaidehi, N., and Rodriguez, G. (1993). A fast recursive algorithm for molecular dynamics simulation. *J. Comp. Phys.*, 106:258–268.

[31] Kang, Y. K., Gibson, K. D., Némethy, G., and Scheraga, H. A. (1988). Free energies of hydration of solute molecules 4. revised treatment of the hydration shell model. *J. Phys. Chem.*, 92(4739).

[32] Kang, Y. K., Némethy, G., and Scheraga, H. A. (1987a). Free energies of hydration of solute molecules 2. application of the hydration shell model to nonionic organic molecules. *J. Phys. Chem.*, 91:4109.

[33] Kang, Y. K., Némethy, G., and Scheraga, H. A. (1987b). Free energies of hydration of solute molecules 3. application of the hydration shell model to charged organic molecules. *J. Phys. Chem.*, 91:4118.

[34] Kim, P. S. and Baldwin, R. L. (1990). Intermediates in the folding reactions of small proteins. *Annu. Rev. Biochem.*, 59:631–660.

[35] Kirkpatrick, S., Jr., C. D. G., and Vecchi, M. P. (1983). Optimization by simulated annealing. *Science*, 220:671–680.

[36] Kitao, A., Hirata, F., and Go, N. (1993). Effects of solvent on the conformation and the collective motions of a protein. 2. structure of hydration in melittin. *J. Phys. Chem.*, 97:10223–10230.

[37] Klepeis, J. L., Androulakis, I. P., Ierapetritou, M. G., and Floudas, C. A. (1998). Predicting solvated peptide conformations via global minimization of energetic atom-to-atom interactions. *Comput. Chem. Eng.*, 22:765–788.

[38] Klepeis, J. L. and Floudas, C. A. (1999). Comparative study of global minimum energy conformations of hydated peptides. *J. Computational Chemistry*, 20:636.

[39] Klepeis, J. L. and Floudas, C. A. (2000). Deterministic global optimization and torsion angle dynamics for molecular structure prediction. *Comp. Chem. Eng.*, 24:1761–1766.

[40] Klepeis, J. L., Floudas, C. A., Morikis, D., and Lambris, J. D. (1999). Predicting peptide structures using nmr data and deterministic global optimization. *J Comp Chem*, 20:1354–1370.

[41] Kollman, P. (1993). Free energy calculations: Applications to chemical and biochemical phenomena. *Chem. Rev.*, 93:2395–2417.

[42] Leopold, P., Montal, M., and Onuchic, J. (1992). Protein folding funnels : A kinetic approach to the sequence-structure relationship. *Proc. Natl. Acad. Sci. USA*, 89:8721–8725.

[43] Levinthal, C. (1968). Are there pathways to protein folding ? *J. Chem. Phys.*, 65:44–45.

[44] Li, Z. and Scheraga, H. A. (1988). Structure and free energy of complex thermodynamic systems. *J. Mol. Struct. (Theochem.)*, 179:333–352.

[45] Maranas, C. D. and Floudas, C. A. (1994a). A deterministic global optimization approach for molecular structure determination. *J. Chem. Phys.*, 100(2):1247–1261.

[46] Maranas, C. D. and Floudas, C. A. (1994b). Global minimum potential energy conformations of small molecules. *J. Glob. Opt.*, 4:135–170.

[47] Maranas, C. D. and Floudas, C. A. (1995). Finding all solutions of nonlinearly constrained systems of equations. *Journal of Global Optimization*, 7(2):143–182.

[48] McCormick, G. P. (1976). Computability of global solutions to factorable nonconvex programs : Part i – convex underestimating problems. *Math. Programming*, 10:147–175.

[49] Meirovitch, H. and Meirovitch, E. (1997). Efficiency of monte carlo minimization procedures and their us e in analysis of nmr data obtained from flexible peptides. *J. Comput. Chem.*, 18:240–253.

[50] Meirovitch, H. and Vásquez, M. (1997). Efficiency of simulated annealing and the monte carlo minimization method for generating a set of low energy structures of peptides. *J. Mol. Struct. (Theochem.)*, 398-399:517–522.

[51] Morikis, D., Assa-Munt, N., Sahu, A., and Lambris, J. D. (1998). Solution structure of compstatin, a potent complement inhibitor. *Protein Sci.*, 7:619–627.

[52] Némethy, G., Gibson, K. D., Palmer, K. A., Yoon, C. N., Paterlini, G., Zagari, A., Rumsey, S., and Scheraga, H. A. (1992). Energy parameters in polypeptides. 10. *J. Phys. Chem.*, 96:6472–6484.

[53] Neumaier, A. (1990). *Interval Methods for Systems of Equations.* Encyclopedia of Mathematics and its Applications. Cambridge University Press.

[54] Ooi, T., Oobatake, M., Némethy, G., and Scheraga, H. A. (1987). Accessible surface areas as a measure of the thermodynamic parameters of hydration of peptides. *Proc. Natl. Acad. Sci. USA*, 84:3086.

[55] Perrot, G., Cheng, B., Gibson, K. D., J. Vila, K. A. P., Nayeem, A., Maigret, B., and Scheraga, H. A. (1992). Mseed: A program for the rapid analytical determination of accessible surface areas and their derivatives. *J. Comp. Chem*, 13:1–11.

[56] Ratschek, H. and Rokne, J. (1988). *Computer Methods for the Range of Functions.* Ellis Horwood Series in Mathematics and its Applications. Halsted Press.

[57] Rice, L. M. and Brünger, A. T. (1994). Torsion angle dynamics: Reduced variable conformational sampling enhances crystallographic structure refinement. *Proteins*, 19:277–290.

[58] Sahu, A., Kay, B., and Lambris, J. (1996). Inhibition of human complement by a c3-binding peptide isolated from a phage-displayed random peptide library. *J. Immunol.*, 157:884–891.

[59] Scheraga, H. (1996). *PACK: Programs for Packing Polypeptide Chains.* online documentation.

[60] Schiffer, C. A., Caldwell, J. W., Kollman, P. A., and Stroud, R. M. (1993). Protein structure prediction with a combined solvation free energy-molecular mechanics force field. *Mol. Sim.*, 10:121.

[61] Standley, D. M., Eyrich, V. A., Felts, A. K., Friesner, R. A., and McDermott, A. E. (1999). A branch and bound algorithm for protein structure refinement from sparse nmr data sets. *J Mol Bio*, 285:1691–1710.

[62] Stillinger, F. H. and Weber, T. A. (1984). *J. Chem. Phys.*, 80:4434.

[63] Straatsma, T. P. and McCammon, J. A. (1992). Computational alchemy. *Annu. Rev. Phys. Chem.*, 43:407–435.

[64] Vásquez, M., Némethy, G., and Scheraga, H. A. (1994). Conformational energy calculations on polypeptides and proteins. *Chemical Reviews*, 94:2183–2239.

[65] Šali, A., Shaknovich, E., and Karplus, M. (1996). Thermodynamics and kinetics of protein folding. In *DIMACS Series in Discrete*

*Mathematics and Theoretical Computer Science*, volume 23, pages 199–213. American Mathematical Society.

[66] Wesson, L. and Eisenberg, D. (1992). Atomic solvation parameters applied to molecular dynamics of proteins in solution. *Protein Science*, 1:227.

[67] Williams, R. L., Vila, J., Perrot, G., and Scheraga, H. A. (1992). Empirical solvation models in the context of conformational energy searches: Application to bovine pancreatic trypsin inhibitor. *Proteins*, 14:110–119.

Chapter 4

# SOME REMARKS ON MINIMUM PRINCIPLES

Franco Giannessi
*Dept. of Mathematics Univ. of Pisa*
*Via F. Buonarroti,2 - 56127 Pisa, Italy*
giannessi@dm.unipi.it

**Abstract** Image Space Analysis is reviewed and some properties of the image problem, corresponding to a given constrained extremum problem, are proved. Then, some remarks are made about optimality conditions and minimum principles.

**Keywords:** Image Space Analysis, Optimality Conditions, Maximum or Minimum Principles, Ekeland Variational Principle, Pontryagin Maximum Principle.

## 1 INTRODUCTION

Assume we are given the integers $m$ and $p$ with $m \geq 0$ and $0 \leq p \leq m$, the subset $X$ of a Banach space $B$, and the functions $f : X \to \mathbb{R}$, $g_i : X \to \mathcal{B}$, $i \in \mathcal{I} := \{1, \dots, m\}$, where $\mathcal{B}$ is another normed linear space. Consider the minimization problem:

(1.1)

$$\underline{f} : \min f(x) \quad , \quad \text{s.t.} \ \ g_i(x) = 0, \ i \in \mathcal{I}^0, \ g_i(x) \in \mathcal{B}_+, \ i \in \mathcal{I}^+, \ x \in X \ ,$$

where $\mathcal{I}^0 := \{1, \dots, p\}$, $\mathcal{I}^+ := \{p+1, \dots, m\}$, and $p = 0 \Rightarrow \mathcal{I}^0 = \emptyset$, $p = m \Rightarrow \mathcal{I}^+ = \emptyset$; $m = 0 \Rightarrow \mathcal{I} := \mathcal{I}^0 \cup \mathcal{I}^+ = \emptyset$; $\mathcal{B}_+$ is a convex cone of $\mathcal{B}$ with apex at the origin. Here and in the sequel, all the considered maxima and minima are supposed to exist. Let us set $g(x) := (g_1(x), \dots, g_m(x))$, $O_p := (0, \dots, 0) \in \mathcal{B}^p$, $D := O_p \times \mathcal{B}_+^{m-p}$, and $R := \{x \in X : g(x) \in D\}$, with the stipulation that $D = \mathcal{B}_+^m$ when $p = 0$, $D = O_m := (0, \dots, 0) \in \mathcal{B}^m$ when $p = m$; $m = 0$ does not require to define $D$.



*N. Hadjisavvas and P.M. Pardalos (eds.), Advances in Convex Analysis and Global Optimization,* 75–103.

When $\mathcal{B} = \mathbb{R}$ and $\mathcal{B}_+ = \mathbb{R}_+$, so that $O_p \in \mathbb{R}^p$ and $D = O_p \times \mathbb{R}_+^{m-p}$, then (1.1) embraces extremum problems in $\mathbb{R}^n$ as well as isoperimetric–type problems. For such problems – which share the characteristic of having a finite dimensional image – the Image Space Analysis (for short, ISA) has shown to be useful; we will deal with them in Sects. 2–9; see [6] for an overview. A first proposal on how to extend such an analysis to the problems having an infinite dimensional image is d escribed in [3,7]. In Sect. 10, such a proposal is outlined and some remarks are made on how ISA can produce, for the general problem (1.1), an unifying approach for deriving, among several theories, that of minimum principles.

## 2 FINITE DIMENSIONAL IMAGE. PRELIMINARY CONCEPTS.

Let us consider now the case $\mathcal{B} = \mathbb{R}$. The ISA arises naturally in as much as an optimality condition for (1.1) is achieved through the impossibility of a system. More precisely, by paraphrasing the very definition of global minimum we can say that $\bar{x} \in R$ is a global minimum point (or $f(\bar{x})$, with $\bar{x} \in X$, is a minimum) for (1.1) iff the system (in the unknown $x$):

$$f_{\bar{x}}(x) := f(\bar{x}) - f(x) > 0 \ , \ g(x) \in D \ , \ x \in X \tag{2.1}$$

is impossible, or

$$\mathcal{H} \cap \mathcal{K}_{\bar{x}} = \emptyset \ , \tag{2.2}$$

where $\mathcal{H} := \{(u,v) \in \mathbb{R} \times \mathbb{R}^m : u > 0, \ v \in D\} = (\mathbb{R}_+ \backslash \{0\}) \times D$ and $\mathcal{K}_{\bar{x}} := \{(u,v) \in \mathbb{R}^m : u = f_{\bar{x}}(x), \ v = g(x), \ x \in X\} = A(X)$, with $A := (f_{\bar{x}}, g)$. The space where $(u,v)$ runs, i.e. $\mathbb{R} \times \mathcal{B}^m$ – which, in the present case, is $\mathbb{R} \times \mathbb{R}^m$ – is the *Image Space* (for short, IS); the set $\mathcal{K}_{\bar{x}}$ is the *image* of (1.1). It is easy to see that (2.2) holds iff

$$\mathcal{H} \cap (\mathcal{K}_{\bar{x}} - c\ell\, \mathcal{H}) = \emptyset \ , \tag{2.3}$$

where the difference is in vector sense and $c\ell$ denotes topological closure. $\mathcal{E}_{\bar{x}} := \mathcal{K}_{\bar{x}} - c\ell\, \mathcal{H}$ is called the *conic extension* of $\mathcal{K}_{\bar{x}}$. The replacement of the image with its conic extension – which corresponds to modify $f$ and $g$ – does not affect the optimality conditions and has several advantages. For instance, if $X$, $f$ and $-g$ are convex (or, more generally, $(f,-g)$ is convexlike [2,13]), then $\mathcal{E}_{\bar{x}}$ is convex even if $calK_{\bar{x}}$ is not. Other kinds of extensions of $\mathcal{K}_{\bar{x}}$ are easily conceivable in the IS. Note that a change of $\bar{x}$

produces merely a translation of $\mathcal{K}_{\bar{x}}$ with respect to the $u$–axis. Hence, the properties of the image can be studied independently of the choice of $\bar{x}$.

The ISA must be viewed as a preliminary and auxiliary step for the analysis, in $B$, of (1.1), and not in competition with it. If this aspect is understood, then ISA may be higly fruitful. In fact, in the IS we may have a sort of "regularization": the image of (1.1), or its conic extension, may be convex or (semi)continuous or smooth when (1.1) does not enjoy the same property, so that convex or (semi)continuous or smooth analysis can be developed in the IS but not in $B$. Several results have alredy bee n achieved in this direction (see [6] and the references therein), and many others can be carried out; among these, by means of an example, in Sect.4, we mention a further direction related to Ekeland Variational Principle [9].

Let us note also that, in the present case – i.e., $\mathcal{B} = \mathbf{R}$ – the image is finite dimensional both when $B$ is finite and when it is infinite dimensional; hence, in the space $B$, these two types of (1.1) require substantially different mathematical tools, while, in the IS, the latter can be treated in the same way as the former, namely with finite dimensional arguments.

Starting from $\mathcal{K}_{\bar{x}}$, most of the theory of constrained extrema can be derived, like optimality conditions, duality, penalty [2,5,10]. In this paper we will address our attention to necessary optimality conditions and thus to minimum principles.

## 3 IMAGE PROBLEM AND SEPARATION

Continuing the analysis of the case $\mathcal{B} = \mathbf{R}$, it is easy to show that (1.1) is equivalent to the following problem[1]:

$$(3.1) \qquad \hat{u}_{\bar{x}} := \max(u) \quad , \quad \text{s.t. } (u, v) \in \mathcal{K}_{\bar{x}}, \ v \in D \ ,$$

which is called *image problem* associated to (1.1).

**Proposition** 3.1 Whatever $\bar{x} \in X$ may be, $\hat{x}$ is a global minimum point of (1.1) if and only if $(\hat{u}, \hat{v}) := (f(\bar{x}) - f(\hat{x}),\ g(\hat{x}))$ is a global maximum point of (3.1).

[1]For the sake of simplicity, when there is not any fear of confusion (like in Proposition 3.1), we understand the dependence of $\hat{u}$ on $\bar{x}$.

**Proof.** $\hat{x} \in R$ is a global minimum point of (1.1) iff the system

$$f(\hat{x}) - f(x) > 0 \ , \ g(x) \in D \ , \ x \in X$$

is impossible, or iff

$$f(\bar{x}) - f(x) - [f(\bar{x}) - f(\hat{x})] > 0 \ , \ g(x) \in D, \ x \in X$$

is impossible, or iff

$$f(\bar{x}) - f(\hat{x}) \geq f(\bar{x}) - f(x) \ , \quad \forall x \in R \ ,$$

or iff (recall from (2.2) that $u = f(\bar{x}) - f(x)$ and $v = g(x)$)

$$\hat{u} \geq u \ , \quad \forall (u, v) \in \mathcal{K}_{\bar{x}} \cap (\mathbf{R} \times D) \ ,$$

which means that $(\hat{u}, \hat{v})$ is a global maximum point of (3.1), where $\hat{v} = g(\hat{x})$. □

Hence, a maximum point $(\hat{u}, \hat{v})$ of (3.1) is the image – through the pair $(f_{\bar{x}}, g)$ – of a minimum point $\hat{x}$ of (1.1), and we have $f(\bar{x}) - \hat{u} = f(\hat{x})$ whatever $\bar{x} \in X$ may be; vice versa, $\hat{x} \in f_{\bar{x}}^{-1}(\hat{u}) := \{x \in R : f_{\bar{x}}(x) = \hat{u}\} \subseteq R$. Now consider the problem:

$$\text{(3.2)} \qquad \max(u) \ , \ \text{s.t.} \ (u, v) \in \mathcal{K}_{\bar{x}}, \ v_i = \xi_i, \ i \in \mathcal{I}^0, \ v_i \geq \xi_i, \ i \in \mathcal{I}^+ \ ,$$

where $\xi := (\xi_1, \ldots, \xi_m) \in \mathbf{R}^m$. Denote by $\hat{u}(\xi)$ the maximum in (3.2), and by $(1.1)_\xi$ the problem (1.1) in which $R$ is replaced by $R_\xi := \{x \in X : g_i(x) = \xi_i, \ i \in \mathcal{I}^0, \ g_i(x) \geq \xi_i, \ i \in \mathcal{I}^+\}$, so that $(1.1)_0$=(1.1) and $R_0 = R$; and denote by $\hat{x}(\xi)$ a global minimum point of $(1.1)_\xi$.

**Proposition** 3.2 Whatever $\bar{x} \in X$ may be, $\hat{x}(\xi)$ is a global minimum point of $(1.1)_\xi$, if and only if $(\hat{u}(\xi), \hat{v}(\xi)) := (f(\bar{x}) - f(\hat{x}(\xi)), \ g(\hat{x}(\xi)) - \xi)$ is a global maximum point of (3.2).

**Proof.** By means of the translation $v' = v - \xi$, $(1.1)_\xi$ and (3.2) are reduced to (1.1) and (3.1), respectively. Hence, Proposition 3.1 can be applied. Since the translation does not affect $u$, we have $f(\bar{x}) - \hat{u}(u) = f(\hat{x}(\xi))$. Besides $\hat{x}(\xi) \in R_\xi \Leftrightarrow \hat{x}(\xi) - \xi \in R; \hat{v}(\xi)$ is feasible for (3.2) iff $\hat{v}(\xi) - \xi$ is feasible for (3.1). □

The function $f(\hat{x}(\xi))$ is called *perturbation function* (or optimal value funtion) of (1.1), and – up to the obvious transformation $\hat{u}(\xi) = f(\bar{x}) - f(\hat{x}(\xi))$ – is given by $\hat{u}(\xi)$. The use of $\hat{u}(\xi)$ leads to deepen the analysis.

To prove directly whether or not (2.2) – or (2.3), and hence (2.1) – holds is generally impracticable. An indirect way of showing it consists in proving that $\mathcal{H}$ and $\mathcal{K}_{\bar{x}}$ lie, respectively, in two disjoint level sets of a same functional. This separation approach is exactly equivalent to finding a theorem of the alternative [2] related to (2.1); only the mathematical languages are different. The separation scheme enjoys mainly a geometrical appeal, while in studying alternative the algebraic characters are dominant. In general terms, a separation scheme begins by introducing a family, say $\mathcal{W}(\Omega)$, of functionals $w : \mathbb{R} \times \mathbb{R}^m \times \Omega \to \mathbb{R}$, such that:

$$(3.3_a) \qquad \text{lev}_{\geq 0}\ w(\cdot,\cdot;\omega) \supseteq cl\,\mathcal{H}\ ,$$

$$(3.3_b) \qquad \bigcap_{\omega\in\Omega} \text{lev}_{>0}\ w(\cdot,\cdot;\omega) \subseteq \mathcal{H}\ ,$$

where $\Omega$ denotes a set of parameters which describe $\mathcal{W}$; $\text{lev}_{\geq 0}$ and $\text{lev}_{>0}$ denote non–negative and positive level sets of $w(\cdot,\cdot;\omega)$, respectively. Conditions (3.3) do not imply

$$(3.4) \qquad \bigcap_{\omega\in\Omega} \text{lev}_{>0}\ w(\cdot,\cdot;\omega) = \mathcal{H}\ .$$

When also (3.4) holds, then $\mathcal{W}(\Omega)$ is called *regular* and denoted by $\mathcal{W}_R(\Omega)$. If a $w \in \mathcal{W}_R(\Omega)$ is found, whose nonpositive level set contains $\mathcal{K}_{\bar{x}}$, then (2.1) is impossible and $\bar{x}$ is a global minimum point of (1.1); this is discussed in Sect.4.

Let us introduce the column vector $v = (v_1, \dots, v_m)^T$ and the row vectors $\lambda = (\lambda_1, \dots, \lambda_m)$, $\mu = (\mu_1, \dots, \mu_m)$. A particular (but wide) family of type (3.3) is offered by the class, say $\mathcal{W}_{p\ell}(\Omega)$, of *parabolic–exponential functions*, defined (at $\Omega = \mathbb{R}_+ \times D^* \times \mathbb{R}^m_+$; $\omega = (\theta, \lambda, \mu)$) by:

$$(3.5) \qquad \begin{cases} w = w(u, v; \theta, \lambda, \mu) := \theta u + \underline{w}(v; \lambda, \mu), \\ \theta \in \mathbb{R}_+, \lambda \in D^* := \mathbb{R}^p \times \mathbb{R}^{m-p}_+,\ \mu \in \mathbb{R}^m_+,\ (\theta, \lambda) \neq 0\ , \end{cases}$$

where $*$ marks positive polar, and

$$\underline{w}(v; \lambda, \mu) := -\sum_{i\in\mathcal{I}^0} (\mu_i v_i^2 - \lambda_i v_i) + \sum_{i\in\mathcal{I}^+} \lambda_i v_i \exp(-\mu_i v_i)\ .$$

At $\mu = 0$, $\mathcal{W}_{\ell p}(\mathbb{R}_+ \times D^* \times \mathbb{R}^m_+)$ collapses to the family of linear functions:

$$(3.6) \qquad \ell = \ell(u, v; \theta, \lambda) := \theta u + \langle \lambda, v \rangle\ ,\ \theta \geq 0,\ \lambda \in D^*,\ (\theta, \lambda) \neq \emptyset\ ,$$

where $\langle \cdot, \cdot \rangle$ denotes the scalar product in $\mathbb{R}^m$. It is easy to check that $\mathcal{W}_{\ell p}(\mathbb{R}_+ \times D^* \times \mathbb{R}^m_+)$ – and hence the subfamily (3.6) – fulfils (3.3), but not (3.4); this is satisfied if $\theta > 0$; constraint qualification and regularity conditions [8,10] are assumptions on (1.1) which guarantee that the exclusion, from (3.5), of the functions having $\theta = 0$ does not decrease the possibilities of proving optimality.

$\mathcal{W}(\Omega)$ may be required to be at least a topological vector space; in such a case, it is called the *dual space.*

# 4 EKELAND VARIATIONAL PRINCIPLE

From the image of (1.1) or its conic extension we can derive, as already said, most of the topics which concern with the theory of constrained extrema; in particular, the optimality conditions. This can be done in several ways: by approximating sets and functions; by applying a separation scheme of type $\mathcal{W}(\Omega)$ or $\mathcal{W}_R(\Omega)$ to (1.1) or to (3.1) or to their approximations. We will discuss a few of them just as an instance of ISA.

Let us start with the well known Ekeland Variational Principle. It will be shown, in the case $\mathcal{B} = \mathbb{R}$, how to derive it in the IS; the extension to the case $\mathcal{B} \neq \mathbb{R}$ would be of interest.

Consider the function:

$$\Phi_{\bar{x}}(v) := \max_{(u,v) \in \mathcal{K}_{\bar{x}}} (u) \,.$$

**Proposition** 4.1 Let $\Phi$ be upper semicontinuous and bounded form above. $\forall \varepsilon > 0$, $\exists v_\varepsilon \in dom\, \Phi_{\bar{x}}(\cdot; \bar{x})$, such that:

$$(4.1_a) \qquad \Phi_{\bar{x}}(v_\varepsilon) \geq \max_{v \in D} \Phi_{\bar{x}}(v) - \varepsilon \,,$$

$$(4.1_b) \qquad \Phi_{\bar{x}}(v_\varepsilon) > \Phi_{\bar{x}}(v) - \varepsilon \|v - v_\varepsilon\|_2 \,, \quad \forall v \in D, \quad v \neq v_\varepsilon \,.$$

**Proof.** It is enough to apply Ekeland Theorem [9] to $\Phi_{\bar{x}}$. □

The above proposition is nothing more that Weak Ekeland Variational Principle applied to the image of (1.1). The following example shows that Proposition 4.1 is more general than the corresponding theorem in

$B$, even if in the "difference" there are problems which, perhaps, have only theoretical interest. However, the purpose is here to show that ISA can lead to generalize theorems which seem impossible to be extended. Of course, it would be interesting to extend the above approach to the strong Ekeland Theorem, as well as to the several derivations, like [9], and to vector extremum problems. Connections between the above principle and the perturbation function would be of interest.

**Example** 4.1 Let us set $B = \mathcal{B} = X = \mathbb{R}$, $p = 0, m = 1$, and

$$f(x) = \begin{cases} 4, & \text{if } -\infty < x \leq -2 \\ 4x+8, & \text{if } -2 < x \leq -1 \\ x^2+3, & \text{if } -1 < x < 1,\ x \neq 0 \\ 1, & \text{if } x = 0 \\ 0, & \text{if } x = 1 \\ x^2-2x, & \text{if } 1 < x < 2 \\ x-3, & \text{if } 2 \leq x < 3 \\ 0, & \text{if } 3 \leq x < +\infty \end{cases}$$

$$g(x) = \begin{cases} -x+1, & \text{if } -\infty < x \leq -2 \\ x+2, & \text{if } -2 < x \leq -1 \\ -x+2, & \text{if } -1 < x < 1, \\ 0, & \text{if } x = 1 \\ -x-1, & \text{if } 1 < x < 2 \\ 2x-6, & \text{if } 2 \leq x < 3 \\ -x, & \text{if } 3 \leq x < +\infty\ . \end{cases}$$

Set $\bar{x} = 0$. We find:

$$\mathcal{K}_0 = \left\{ (u,v) \in \mathbb{R}^2 : u = \left\{ \begin{array}{ll} 0 & , \text{if } -\infty < v \leq -3 \\ -v^2-4v-3 & , \text{if } -3 < v < -2 \\ -\frac{1}{2}v & , \text{if } -2 \leq v < 0 \\ 0 & , \text{if } v = 0 \\ -4v & , \text{if } 0 < v \leq 1 \\ -v^2+4v-7 & , \text{if } 1 < v < 3, \\ & \quad v \neq 2 \\ -1 & , \text{if } v = 2 \\ -4 & , \text{if } 3 \leq v < +\infty \end{array} \right\} = \Phi_0(v) \right\}.$$

Note that $\Phi_0$ is upper semicontinuous, while $f$ is not lower semicontinuous even in any neighbourhood of $x = 1$, which is the global minimum point of (1.1) in the present example. Hence, Proposition 4.1 can be applied, while Ekeland Theorem cannot. The restriction of $f$ to $R$ is lower semicontinuous, but to take this as assumption would be mathematically meaningless.

# 5 FROM A SUFFICIENT CONDITION TO NECESSARY ONES

Before going on in analysing minimum principles, let us state a sufficient condition. This is useful by itself and for the sequel. We continue to assume $\mathcal{B} = \mathbb{R}$.

Is is easy to show that, at[2] $\theta > 0$, (3.5) fulfils (3.4); i.e., at $\theta > 0$ $\mathcal{W}_{p\ell}(\mathbb{R}_+ \times D^* \times \mathbb{R}^m_+)$ is a subfamily of $\mathcal{W}_R(\mathbb{R}_+ \times D^* \times \mathbb{R}^m_+)$. Hence, if $\mathcal{K}_{\bar{x}}$ is included into the nonpositive level set of $w$, then (2.2) is satisfied. This proves the following sufficient optimality condition.

**Proposition** 5.1 If there exist $\theta = 1$, $\lambda \in D^*$, $\mu \in \mathbb{R}^m_+$, such that

$$(5.1) \qquad \theta u + \underline{w}(v; \lambda, \mu) \leq 0 \ , \quad \forall (u, v) \in \mathcal{K}_{\bar{x}} \ ,$$

then $\bar{x}$ is a global minimum point of (3.1) and hence of (1.1).

Note that (5.1) is equivalent to:

$$(5.1)' \qquad \theta f(\bar{x}) \leq \theta f(x) - \underline{w}(g(x); \lambda, \mu) := \mathcal{L}(x; \theta, \lambda, \mu), \quad \forall x \in X \ ,$$

which, up to formal changes, is equivalent to a saddle–point type condition. The study of (5.1)′ leads to consider the following problem[3]:

$$(5.2) \qquad f^* := \max_{\theta \geq 0, \lambda \in D^*, \mu \geq 0} \ \min_{x \in X} \mathcal{L}(x; \theta, \lambda, \mu) \ .$$

When $\theta > 0$, condition (5.1)′ is equivalent to find (5.2). In the case (3.6), (5.2) becomes:

$$(5.3) \qquad \max_{\theta \geq 0, \lambda \in D^*} \ \min_{x \in X} L(x; \theta, \lambda) \ ,$$

where $L(x; \theta, \lambda) := \mathcal{L}(x; \theta, \lambda, 0)$ is the *Lagrange Function.*

In the sequel we will restrict the analysis to the family (3.6), even if the wider family (3.5) or the abstract one (3.3) would be extremely interesting.

A way of obtaining a necessary condition is suggested by the separation condition (5.1). It expresses "disjunctive separation": among the separation functions of the given class, find one for which (2.2) holds. This search can be weakened by looking for a separation function (within

[2] If $\theta > 1$, then we can assume $\theta = 1$ by defining again $\underline{w}$.

[3] Here and in the sequel, we understand that max and min be replaced, respectively, by sup and inf, if they do not exist.

the given class) which makes the left–hand side of (5.1) "stationary"; this means to make "stationary", with respect to $x$, $\mathcal{L}(x;\theta,\Lambda,\mu)$, or in particular $L(x;\theta,\lambda)$. Such a condition is a generalized *Lagrange Minimum Principle.* At $\theta=1$, (5.2) – and in particular (5.3) – is called generalized *Lagrange Dual Problem*; obviously, $f^* \leq f; f-f^*$ is the *duality gap.*

Let us consider another way. To this end, consider again (5.1), where, now, we replace $\mathcal{K}_{\bar{x}}$ with $\mathcal{K}_{\bar{x}} \cap (\mathbf{R} \times D)$; namely,

$$\theta u + \underline{w}(v;\lambda,\mu) \leq 0\ , \quad \forall (u,v) \in \mathcal{K}_{\bar{x}} \cap (\mathbf{R} \times D)\ . \tag{5.4}$$

Such an alternative way corresponds to consider the image, through $A$, of $R$ istead of $X$. Of course, such a *restricted (feasible) image* (associated with (1.1)) has only analytical sense (to perform subsequent deductions more than to use it by itself), since any assumption on the restriction of $f$ or $g$ on $R$ has, in general, no sense from mathematical and numerical points of views.

While, in the general case, (5.1) is only sufficient, (5.4) is necessary and sufficient. The sufficiency is as for (5.1). With regard to the necessity, if $\bar{x}$ is a (global) minimum point of (1.1), so that (2.2) holds, then (5.4) is satisfied at least at $\theta=1$, $\lambda=\mu=0$, since now $\mathcal{K}_{\bar{x}} \cap (\mathbf{R} \times D) = \mathcal{K}_{\bar{x}} \cap (\mathbf{R}_- \times D)$. Hence, when we adopt (5.4), in the necessary part we need not to distinguish between $\theta>0$ and $\theta=0$, since a vector $(\theta,\lambda,\mu)$ with $\theta>0$ always exist.

From (5.1) we have been led to (5.2); in the same way, from (5.4) we are led to the problem:

$$\tilde{w}_{\bar{x}} := \min_{\theta \geq 0, \lambda \in D^*, \mu \geq 0} \ \max_{\substack{(u,v) \in \mathcal{K}_{\bar{x}} \\ v \in D}} [\theta u + \underline{w}(v;\lambda,\mu)]\ . \tag{5.5}$$

This problem enjoys the following equivalence:

**Proposition** 5.2 If $\theta=1$, then the maximum in (3.1) and the minimum in (5.5) are equal, if they exists.

**Proof.** Let $\theta=1$. Since $\underline{w}(v;0,0)=0$, at $\lambda=\mu=0$ the maxima in (3.2) and in (5.5) are equal; hence $\tilde{w}_{\bar{x}} \leq \hat{u}_{\bar{x}}$. If $v \in D \Rightarrow \underline{w}(v;\lambda,\mu) \geq 0$ and thus $u \leq u + \underline{w}(v;\lambda,\mu)\ \forall \lambda \in D^*$, $\forall \mu \in \mathbf{R}^m_+$, so that the maximum in (5.5) is not less than $\hat{u}_{\bar{x}}$, $\forall \lambda \in D^*$, $\forall \mu \in \mathbf{R}^m_+$; hence $\tilde{w}_{\bar{x}} \geq \hat{u}_{\bar{x}}$. It follows $\tilde{w}_{\bar{x}} = \hat{u}_{\bar{x}}$. □

**Example** 5.1 Let $B=\mathcal{B}=X=\mathbf{R}$, $p=0$, $m=1$, $f(x)=x$, $g(x)=-\sqrt{x^2}$, $D=\mathbf{R}_+$. Set $\bar{x}=1$. We find $\mathcal{K}_1 = \{(u,v) \in \mathbf{R} \times \mathbf{R} : v =$

$-\sqrt{(1-u)^2}\}$. Thus, $\mathcal{K}_1 \cap (\mathbb{R} \times \mathbb{R}_+) = \{(1,0)\}$. Hence $\hat{u}_1 = 1$, while (5.5) becomes:

$$\tilde{w}_1 = \min_{\theta,\lambda \geq 0} \max_{u=1,v=0} [\theta u + \lambda v \exp(-\mu v)] = \min_{\theta \geq 0}(\theta) = 0 .$$

The above example shows that (3.1) and (5.5) are not equivalent, when in a solution of (5.5) $\theta = 0$ necessarily.

Problem (5.5) can be equivalently written as:

$$\max_{\theta \geq 0, \lambda \in D^*, \mu \geq 0} \min_{x \in R} \mathcal{L}(x; \theta, \lambda, \mu) , \tag{5.6}$$

which, in the case (3.6), becomes:

$$\max_{\theta \geq 0, \lambda \in D^*} \min_{x \in R} L(x; \theta, \lambda) . \tag{5.7}$$

Proposition 5.2 shows the equivalence, at $\theta > 0$, between (5.5) or (5.6) and (1.1); it is easy to see that, if $(\tilde{u}, \tilde{v}; \tilde{\lambda}, \tilde{\mu})$ is a solution of (5.5), then $(\tilde{x} \in f^{-1}(f(\bar{x}) - \tilde{u}); \tilde{\lambda}, \tilde{\mu})$ is a solution of (5.6); vice versa, if $(\tilde{x}; \tilde{\lambda}, \tilde{\mu})$ is a solution of (5.6), then $(\tilde{u} := f(\bar{x}) - f(\tilde{x}), \tilde{v} := g(\tilde{x}); \tilde{\lambda}, \tilde{\mu})$ is a solution of (5.5). The maximum in (5.5) equals $\underline{f}$ .

The lack of equivalence shown by Example 5.1 at $\theta = 0$ is of no importance, when (5.5), (5.6), (5.7) are exploited under the assumption that $\bar{x}$ be a global minimum point, namely when we search for a necessary condition.

# 6 SEMISTATIONARITY

A necessary optimality condition for (1.1) can be achieved in several ways. Indeed, it consists in relaxing a necessary and sufficient condition. Starting from (5.1)–(5.2), two general ways consist in replacing $\mathcal{K}_{\bar{x}}$ with a suitable conic approximation or with the image of a positively homogeneous approximation of $f$ and $g$. In general, these way are not permutable, as simple examples show. Now, the latter will be carried on.

**Definition** 6.1 Let $G$ be a subset of the class of functions defined on $X \times (X - \bar{x})$ and positively homogeneous of 1st degree with respect to the 2nd argument. $f : X \to \mathbb{R}$ is called *G–differentiable*, iff $\exists \mathcal{D}_G f \in G$, such that:

$$\lim_{d \to 0} \frac{1}{\|d\|} [\varepsilon(\bar{x}; d) := f(x) - f(\bar{x}) - \mathcal{D}_G f(\bar{x}; d)] = 0 . \tag{6.1}$$

$\mathcal{D}_G f$ is the *G–derivative* of $f$ at $\bar{x}$ in the direction $d$.

Two important types of $G$ are that of linear ones and that of convex (and hence sublinear) ones; the latter will be denoted by $\mathcal{C}$. The class of $\mathcal{C}$–differentiable functions has been introduced in [2] (under the term generalized differentiable functions, and as a subclass of that of semidifferentiable ones) and developed in [4]. Independently of this, in the same period, the above class (under the term of *B–differentiable* functions) has been introduced in [11]. Assuming that $f$ and $g$ be $\mathcal{C}$–differentiable, it is possible to introduce the following approximation, at $\bar{x}$, of the image set $\mathcal{K}_{\bar{x}}$:

$$\mathcal{K}_{\bar{x}}^{(h)} := \{(u,v) : \mathbf{R} \times \mathbf{R}^m : u = -\mathcal{D}_{\mathcal{C}} f(\bar{x};d); v_i = g_i(\bar{x}) + \mathcal{D}_{\mathcal{C}} g_i(\bar{x};d),$$
$$i \in \mathcal{I},\ d \in X - \bar{x}\},$$

where $v = (v_1, \ldots, v_m)$. It is easy to see that $\mathcal{H}$ and $\mathcal{K}_{\bar{x}}^{(h)}$ are (linearly) separable (see [8]), iff $\exists \theta \in \mathbf{R}_+$ and $\lambda \in D^* = \mathbf{R}^p \times \mathbf{R}_+^{m-p}$, with $(\theta, \lambda) \neq 0$ , such that:

$$\theta u + \langle \lambda, v \rangle \leq 0 \ , \quad \forall (u,v) \in \mathcal{K}_{\bar{x}}^{(h)} \ . \tag{6.2}$$

**Definition** 6.2 Let $Y$ be a subset of $B$. $\bar{x} \in Y$ is called a *lower semistationary point* of $f$ on $Y$, if and only if

$$\liminf_{\substack{x \to \bar{x} \\ x \in Y \setminus \{\bar{x}\}}} \frac{f(x) - f(\bar{x})}{\|x - \bar{x}\|} \geq 0 \ , \tag{6.3}$$

where $\|\cdot\|$ is of course a norm in $B$.

**Proposition** 6.1 **(i)** If $\bar{x}$ is a local minimum point of (1.1), then (6.3) with $Y = R$ holds. **(ii)** If $f$ is differentiable, then (6.3) becomes:

$$\langle f'(\bar{x}), x - \bar{x} \rangle \geq 0 \ , \quad \forall x \in R \ , \tag{6.4}$$

which, if $x \in int\, R$, collapses to:

$$f'(\bar{x}) = 0 \ . \tag{6.5}$$

**(iii)** If $R$ and $f$ are convex, then (6.3) becomes:

$$f'(\bar{x}; x - \bar{x}) \geq 0 \ , \quad x \in R \ , \tag{6.6}$$

$f'$ denoting directional derivative in the direction $x - \bar{x}$, and every lower semistationary point of $f$ on $R$ is a global minimum point.

**Proof.** **(i)** It is not restrictive to assume that $R$ be not a singleton. By assumption, there exists a neighbourhood of $\bar{x}$, say $N(\bar{x})$, such that $f(x) \geq f(\bar{x})$ $\forall x \in R \cap N(\bar{x})$, so that $[f(x) - f(\bar{x})]/\|x - \bar{x}\| \geq 0$ $\forall x \in R \cap N(\bar{x}) \setminus \{\bar{x}\}$, and therefore (6.3) at $Y = R$ holds. **(ii)** We have $f(x) = f(\bar{x}) + \langle f'(\bar{x}), x - \bar{x}\rangle + \varepsilon(\bar{x}; x - \bar{x})$, where $\varepsilon$ is an infinitesimal as $x \to \bar{x}$. S et $S := \{d \in B : \|d\| = 1,\ \exists \tau \in \mathbb{R}_+ \setminus \{0\}$ s.t. $\bar{x} + td \in R\ \forall t \in [0, \tau]\}$. (6.3) with $Y = R$ becomes:

$$\sup_{\tau > 0} \inf_{\substack{d \in S \\ t \in ]0,\tau]}} [\langle f'(\bar{x}), d\rangle + \frac{1}{t}\varepsilon(\bar{x}; td)] \geq 0 \,. \tag{6.7}$$

(6.4) follows, if we show that (6.7)$\Rightarrow \ell := \inf_{d \in S} \langle f'(\bar{x}), d\rangle \geq 0$. Ab absurdo, suppose that $\ell < 0$. Then, $\forall \eta > 0\ \exists d_\eta \in S$, such that $\langle f'(\bar{x}), d_\eta\rangle < \ell + \eta$. Without any loss of generality, assume that $\ell + \eta < 0$. Select $\delta \in ]0, -(\ell + \eta)[$. If $\tau$ is positive and small enough, because of the differentiability, we have $\frac{1}{t}\varepsilon(\bar{x}; td) < \delta$, $\forall d \in S$, $\forall t \in ]0, \tau]$, and in particular at $d = d_\eta$. Therefore,

$$\langle f'(\bar{x}), d_\eta\rangle + \frac{1}{t}\varepsilon(\bar{x}; td_\eta) < \ell + \eta + \delta < 0 \ , \quad \forall t \in ]0, \tau] \,,$$

which shows that the infimum of the expression within the square brackets of (6.7) is less than a negative constant. This contradicts (6.7). If $\bar{x} \in \operatorname{int} R$, $\exists \varepsilon > 0$ small enough, such that $\bar{x} \pm \varepsilon f'(\bar{x}) \in R$, so that, at $x = \bar{x} \pm \varepsilon f'(\bar{x})$, (6.4) becomes $0 \leq \langle f'(\bar{x}), \pm f'(\bar{x})\rangle = \pm\|f'(\bar{x})\|^2$ and implies (6.5). **(iii)** To achieve (6.6), it is enough to repeat the proof of (ii) with the following changes: the expansion of $f$ is now $f(x) = f(\bar{x}) + f'(\bar{x}; x - \bar{x}) + \varepsilon(\bar{x}; x - \bar{x})$; the scalar product which appears in (6.7) and consequently in the sequel must be replaced by $f'(\bar{x}; d)$ and later by $f'(\bar{x}; d_\eta)$. Now, to show that $\bar{x}$ is a global minimum point, suppose, ab absurdo, that $\exists \hat{x} \in R$, such that $f(\hat{x}) < f(\bar{x})$. This fact and the convexity of $f$ imply:

$$\frac{1}{\|\alpha(\hat{x} - \bar{x})\|}[f(\bar{x} + \alpha(\hat{x} - \bar{x})) - f(\bar{x})] \leq \frac{1}{\|\hat{x} - \bar{x}\|}[f(\hat{x}) - f(\bar{x})] < 0 \,, \ \forall \alpha \in ]0, 1] \,.$$

From this we easily draw that the left–hand side of (6.3) is $\leq \frac{f(\hat{x}) - f(\bar{x})}{\|\hat{x} - \bar{x}\|} < 0$, which contradicts (6.3) at $Y = R$. □

When the norm in $B$ is not generated by a scalar product, then, of course, $\langle \cdot, \cdot \rangle$ must be understood as the application of a continuous linear functional to the variation.

Note that (6.4), which obviously implies the non–negativity of the directional derivative like (6.6), is a special case of a Variational Inequality. Now, let $\bar{z} := (\bar{u}, \bar{v}) := (f_{\bar{x}}(\bar{x}), g(\bar{x})) = (0, g(\bar{x}))$; $\bar{z}$ is of course the image of $\bar{x}$ through the map $A$. The positive polar of $\mathcal{K}_{\bar{x}}^{(h)} - \bar{z}$ (which is obviously a cone with apex at the origin $0_{1+m}$ of $\mathbb{R}^{1+m}$) is given by:

$$(\mathcal{K}_{\bar{x}}^{(h)} - \bar{z})^* := \{z^* \in \mathbb{R}^{1+m} : \langle z^*, z - \bar{z}\rangle \geq 0, \quad \forall z \in \mathcal{K}_{\bar{x}}^{(h)}\} =$$

$$\{(u^*, v^*) \in \mathbb{R} \times \mathbb{R}^m : u^*u + \langle v^*, v - \bar{v}\rangle \geq 0, \quad \forall (u,v) \in \mathcal{K}_{\bar{x}}^{(h)}\} .$$

When $\mathcal{K}_{\bar{x}}^{(h)} - \bar{z}$ is a linear manifold, then its polar collapses to the orthogonal complement and it is denoted by $(\mathcal{K}_{\bar{x}}^{(h)} - \bar{z})^{\perp}$. Also in the sequel, a $*$ as apex of a set will denote its positive polar. Recall that $L(x; \theta, \lambda) = \theta f(x) - \langle \lambda, g(x)\rangle$.

**Proposition** 6.2 (semistationarity). Let $f$ and $-g$ be $C$–differentiable. **(i)** If and only if

$$-(\theta, \lambda) \in (\mathcal{K}_{\bar{x}}^{(h)} - \bar{z})^* , \tag{6.8}$$

then

$$\liminf_{\substack{x \to \bar{x} \\ x \in X \setminus \{\bar{x}\}}} \frac{L(x; \theta, \lambda) - L(\bar{x}; \theta, \lambda)}{\|x - \bar{x}\|} \geq 0 . \tag{6.9}$$

If, in addition, the $C$–derivatives of $f$ and $-g_i$ $i \in \mathcal{I}$ are bounded from above by a constant independent of a neighbourhood of $\bar{x}$, or if $B = \mathbb{R}$, then liminf collapses to lim. **(ii)** If $\bar{x} \in int\, X$ and $f, g$ $i \in \mathcal{I}$ are differentiable (so that $C$ can be reduced to the set of linear functionals), then (i) becomes: if and only if

$$-(\theta, \lambda) \in (\mathcal{K}_{\bar{x}}^{(h)} - \bar{z})^{\perp} , \tag{6.8$'$}$$

then

$$L'_x(\bar{x}; \theta, \lambda) = 0 . \tag{6.9$'$}$$

**Proof.** **(i)** (6.8) is equivalent to:

$$-(\theta, \lambda) \in \{(u^*, v^*) \in \mathbb{R} \times \mathbb{R}^m : u^*u + \langle v^*, v - v^*\rangle \geq 0 \, , \ \forall (u,v) \in \mathcal{K}_{\bar{x}}^{(h)}\} ,$$

or

$$\theta \mathcal{D}_C f(\bar{x}; d) - \sum_{i \in \mathcal{I}} \lambda_i \mathcal{D}_C g_i(\bar{x}; d) \geq 0, \quad \forall d \in X - \bar{x} .$$

Since $\mathcal{C}$ is closed under addition, the left–hand side of the last inequality is the $\mathcal{C}$–derivative of $L$; namely $\mathcal{D}_{\mathcal{C}}L(\bar{x};d;\theta,\lambda) = \theta\mathcal{D}_{\mathcal{C}}f(\bar{x};d) - \sum_{i\in\mathcal{I}} \lambda_i\mathcal{D}_{\mathcal{C}}g_i(\bar{x};d)$. It follows that (6.8) is equivalent to:

$$\frac{1}{\|d\|}\mathcal{D}_{\mathcal{C}}L(\bar{x};d;\theta,\lambda) \geq 0\ , \quad \forall d \in X - \bar{x}\ .$$

This inequality, by adding to both sides the quantity:

$$\frac{1}{\|d\|}\bar{\varepsilon}(\bar{x};d;\theta,\lambda) := \frac{1}{\|d\|}[\theta\varepsilon_f(\bar{x};d) - \sum_{i\in\mathcal{I}}\lambda_i\varepsilon_{g_i}(\bar{x};d)]\ ,$$

$\varepsilon_f$ and $\varepsilon_{g_i}$ being the infinitesimals of the expansions, respectively, of $f$ and $g_i$, is equivalent to (recall that $d = x - \bar{x}$):

$$\frac{1}{\|d\|}[L(x;\theta,\lambda) - L(\bar{x};\theta,\lambda)] \geq \frac{1}{\|d\|}\bar{\varepsilon}(\bar{x};d;\theta,\lambda), \quad \forall x \in X\backslash\{\bar{x}\}\ .$$

Since $\bar{\varepsilon}$ is an infinitesimal as $d \to 0$, the above inequality implies (6.9); the vice versa follows, since we can prove that (6.9) implies the non–negativity of $\mathcal{D}_{\mathcal{C}}L(\bar{x};d;\theta,\lambda)$ by reasoning in a quite similar way as for (ii) of Proposition 6.1. The remaining part is obvious. **(ii)** Since now $\mathcal{K}_{\bar{x}}^{(h)}$ is an affine manifold, the polar of $\mathcal{K}_{\bar{x}}^{(h)} - \bar{z}$ becomes its orthogonal complement, and hence liminf collapses to lim, an d this is zero, as both $\geq$ and $\leq$ must hold. □

# 7 SEPARABILITY

As shown in Sect.2, optimality is equivalent to (2.2) or (2.3); an effective way of proving them is offered by (linear or nonlinear) disjunctive separation. If the image set $\mathcal{K}_{\bar{x}}$ is replaced by a suitable approximation, like $\mathcal{K}_{\bar{x}}^{(h)}$, and we merely ask for separation, then we achieve necessary conditions. Now, we will consider the linear separability between $\mathcal{H}$ and a generic set $K \subset \mathbb{R}^{1+m}$; later the special case $K = \mathcal{K}_{\bar{x}}^{(h)}$ will lead us to a necessary condition. Of course, nonlinear separability might allow one to sharpen the result.

For each $z = (u,\ v_i\ \ i \in \mathcal{I}^0,\ v_i\ \ i \in \mathcal{I}^+) \in \mathbb{R}^{1+m}$, denote by $proj\ z$ its projection on the coordinate subspace of the $v_i\ \in \mathcal{I}^0$; namely, $proj\ z := (v_i\ \ i \in \mathcal{I}^0) \in \mathbb{R}^p$.

**Proposition** 7.1 Let $K \subset \mathbb{R}^{1+m}$ and $k := \dim K$. If and only if for every set $\{z^1, \ldots, z^{k+1}\}$ of affinely independent vectors of $K$, such that[4]:

$$(7.1) \qquad O_p \in \mathit{int\,conv}\, \{\mathit{proj}\, z^1, \ldots, \mathit{proj}\, z^{k+1}\} \,,$$

we have:

$$(7.2) \qquad (\mathit{ri}\, \mathcal{H}) \cap \mathit{int\,conv}\{z^1, \ldots, z^{k+1}\} = \emptyset \,,$$

then $\mathcal{H}$ and $K$ are (linearly) separable.

**Proof.** **If.** Suppose that $k \leq p - 1$ (so that $p \geq 1$ and there cannot exist $p + 1$ affinely independent vectors of $K$), and let $B_{\mathcal{H}}$ and $B_K$ be bases for $\mathcal{H}$ and $K$, respectively; $|B_{\mathcal{H}}| = 1 + m - p$, $|B_K| = k$, $|B_{\mathcal{H}} \cup B_K| \leq 1 + m - p + k \leq m$. This shows that there exists an hyperplane of $\mathbb{R}^{1+m}$, which contains $\mathcal{H}$ and is parallel to $K$, so that separation holds. Now suppose $k \geq p \geq 1$, so that there exists at least one set of $k + 1 \geq p + 1$ affinely independent vectors of $K$; denote by $\{z^1, \ldots, z^{k+1}\}$ a generic one of such sets. Denote by $\mathit{proj}\, K$ the projection of $K$ into the coordinate subspace $\mathbb{R}^p$, whose origin is $O_p$. If for every affinely independent set $\{z^1, \ldots, z^{k+1}\} \subseteq K$ (7.1) does not hold, then $O_p \notin \mathit{int\,conv\,proj} K$ (otherwise, because of a well known theorem of Carathéodory, $O_p$ could be expressed as convex combination, with positive coefficients, of $p + 1$ affinely independent vectors of $\mathbb{R}^p$ which should be projections of $p + 1$ vectors of $K$; since $\mathit{dim}\, K = k$, these vectors could be augmented to form a set $\{z^1, \ldots, z^{k+1}\}$ of affinely independent vectors of $K$ which would satisfy (7.1)). The last relation implies:

$$(7.3) \qquad (\mathit{ri}\, \mathcal{H}) \cap \mathit{ri\,conv}\, K = \emptyset \,.$$

In fact, if ab absurdo we deny (7.3), so that $\exists \hat{z} := (\hat{u} > 0, \hat{v}_i = 0 \;\; i \in \mathcal{I}^0, \hat{v}_i > 0 \;\; i \in \mathcal{I}^+) \in \mathit{ri\,conv}\, K$, then $\hat{z}$ can be expressed as convex combination, with positive coefficients, of $k + 1$ affinely independent vectors of $\mathit{conv}\, K$, so that $O_p = \mathit{proj}\, \hat{z} \in \mathit{int\,conv\,proj}\, K$, contradicting the above relation. Because of a well known theorem, (7.3) implies separation (even proper) between $\mathcal{H}$ and $K$. Now suppose that (7.1)–(7.2) hold. (7.2) $\Rightarrow$ (7.3). In fact, if ab absurdo we deny (7.3) so that

[4]We stipulate that (7.1) be satisfied at $p = 0$; in such a case, both sides are "empty". When affinely independent $z^1, \ldots, z^{k+1} \in K$, such that (7.1) is fulfilled, do not exist, then of course the condition (7.1)–(7.2) is meant to be satisfied. *int*, *ri* and conv denote interior, relative interior and convex hull, respectively; of course, *int* denotes interior with respect to $\mathbb{R}^p$ in (7.1) and with respect to $\mathbb{R}^{1+m}$ in (7.2).

$\exists \hat{z} := (\hat{u} > 0, \hat{v}_i = 0 \;\; i \in \mathcal{I}^0, \hat{v}_i > 0 \;\; i \in \mathcal{I}^+) \in ri\, conv\, K$, then, because of a well known theorem of Carathéodory, $\hat{z}$ can be expressed as convex combination, with positive coefficients, of $k+1$ affinely independent vectors of $K$, and (7.2) is contradicted. Because of a well known theorem, (7.3) implies (even proper) separation between $\mathcal{H}$ and $K$. In the case $p = 0$, the proof is exactly the same as that of the case where $k \geq p$ and (7.1) holds. **Only if.** By assumption, there exist $\theta \in \mathbb{R}_+$ and $\lambda \in D^*$, with $(\theta, \lambda) \neq 0$, such that:

$$\theta u + \langle \lambda, v \rangle \leq 0 \quad , \quad \forall (u, v) \in K \; .$$

Denote by $\pi^-$ the halfspace (of $\mathbb{R}^{1+m}$) identified by the above inequality. Ab absurdo, suppose that there exist affinely independent vectors $z^1, \ldots, z^{k+1} \in K$, such that (7.1) holds [5] – so that $k \geq p$ –, while (7.2) is violated, so that there exist scalars $\alpha_j > 0 \;,\; j = 1, \ldots, k+1$, with $\sum_{j=1}^{k+1} \alpha_j = 1$, such that:

$$(7.3)_1 \qquad \begin{cases} proj\, z^j \neq O_p \;,\; j = 1, \ldots, k+1 \; ; \\ proj \left( \sum_{j=1}^{k+1} \alpha_j z^j \right) = \sum_{j=1}^{k+1} \alpha_j \, proj\, z^j = O_p \; ; \end{cases}$$

$$(7.3)_2 \qquad \sum_{j=1}^{k+1} \alpha_j z^j \in ri\, \mathcal{H} = (\mathbb{R}_+ \setminus \{0\}) \times O_p \times int\, \mathbb{R}_+^{m-p} \; .$$

Set $z^j = (u^j, v_i^j \;\; i \in \mathcal{I})$, $\lambda = (\lambda_i \;\; i \in \mathcal{I})$. Since $z^j \in K \subseteq \pi^-$, we have:

$$\theta u^j + \sum_{i \in \mathcal{I}^0} \lambda_i v_i^j + \sum_{i \in \mathcal{I}^+} \lambda_i v_i^j \leq 0 \;,\; j = 1, \ldots, k+1 \; .$$

Multiplying by $\alpha_1, \ldots, \alpha_{k+1}$ both sides of the above inequalities and summing up them side by side, we obtain the inequality:

$$\theta \sum_{j=1}^{k+1} \alpha_j u^j + \sum_{i \in \mathcal{I}^0} \lambda_i \sum_{j=1}^{k+1} \alpha_j v_i^j + \sum_{i \in \mathcal{I}^+} \lambda_i \sum_{j=1}^{k+1} \alpha_j v_i^j \leq 0 \; ,$$

which, because of $(7.3)_2$, implies $\theta = \lambda_i = 0, i \in \mathcal{I}^+$, and $(\lambda_i \;\; i \in \mathcal{I}^0) \neq 0$. Therefore, the inequality which identifies $\pi^-$ becomes:

$$(7.3)_3 \qquad \sum_{i \in \mathcal{I}^0} \lambda_i v_i \leq 0 \; .$$

[5] If (7.1) does not hold, then the thesis is trivial, and of course there is no need to proceed ab absurdo.

Because of $(7.3)_3$, the projections of $z^1, \ldots, z^{k+1}$ into the subspace $\mathbb{R}^p$, considered as points of $\mathbb{R}^{1+m}$, namely the points

$$y^j := (u = 0,\ proj\, z^j,\ v_i = 0\ \ i \in \mathcal{I}^+) = (u = 0,\ v_i = v_i^j\ \ i \in \mathcal{I}^0,$$

$$v_i = 0,\ i \in \mathcal{I}^+),\ j = 1, \ldots, k+1\ ,$$

belong to $\pi^-$ and, obviously, $conv\{y^1, \ldots, y^{k+1}\} \subseteq \pi^-$. From $(7.3)_1$ we deduce that $O_{1+m} \in ri\, conv\{y^1, \ldots, y^{k+1}\}$. Therefore, if $\beta > 0$ is small enough (for instance, less than $\min\{\|(v_1^j, \ldots, v_p^j)\|,\ j = 1, \ldots, k+1\}$), then $(u = 0,\ v_i = \beta\lambda_i\ \ i \in \mathcal{I}^0,\ v_i = 0\ \ i \in \mathcal{I}^+) \in ri\, conv\{y^1, \ldots, y^{k+1}\} \subseteq \pi^-$, so that we achieve the contradiction $\beta \sum_{i \in \mathcal{I}^0} \lambda_i^2 \leq 0$. □

**Hypothesis H**. Set $k := \dim \mathcal{K}_{\bar{x}}^{(h)}$. For every set $\{z^1, \ldots, z^{k+1}\}$ of affinely independent vectors of $\mathcal{K}_{\bar{x}}^{(h)}$, such that[6]:

$$O_p \in int\, conv\{proj\, z^1, \ldots, proj\, z^{k+1}\}\ , \tag{7.4}$$

we have

$$(ri\mathcal{H}) \cap ri\, conv\{z^1, \ldots, z^{k+1}\} = \emptyset\ . \tag{7.5}$$

Simple examples show that Hypothesis H may be satisfied without $\bar{x}$ being a minimum point of (1.1) or, vice versa, $\bar{x}$ may be a minimum point of (1.1) without Hypothesis H being fulfilled. The above hypothesis may be the sourse for deriving more practical assumptions; for instance, the proof of Proposition 7.1 suggests immediately the following ones:

**Hypothesis H'**. $p \geq 1$ and $\dim \mathcal{K}_{\bar{x}}^{(h)} \leq p - 1$.

**Hypothesis H"**. $k :=\dim \mathcal{K}_{\bar{x}}^{(h)}$. For every set $\{z^1, \ldots, z^{k+1}\}$ of affinely independent vectors of $\mathcal{K}_{\bar{x}}^{(h)}$, we have

$$O_p \notin int\, conv\{proj\, z^1, \ldots, proj\, z^{k+1}\}\ .$$

**Hypothesis H"'**. Let $p = m \geq 1$ and $\mathcal{K}_{\bar{x}}^{(h)}$ be a linear manifold (this happens if $X$ is open and $f, g$ differentiable).

[6]We stipulate that (7.4) be satisfied at $p = 0$; in such a case, both sides are "empty".

Because of Proposition 7.1 and its proof, we easily see that each of the above four hypotheses guarantees separation between $\mathcal{H}$ and $\mathcal{K}_{\bar{x}}^{(h)}$. Such hypotheses have been based on a particular approximation of $\mathcal{K}_{\bar{x}}$, namely $\mathcal{K}_{\bar{x}}^{(h)}$ (which is based on the $C$–differentiability of $f$ and $-g$); Proposition 7.1 allows one to work with any other kind of approximation; for instance, one might consider the Bouligand tangent cone or the hypertangent c one.

At $p = 0$, when (2.2) holds, then Hypothesis $H$ is verified, since the conic extension of $\mathcal{K}_{\bar{x}}^{(h)}$ is convex. This means that, by subsuming such an hypothesis, we can expect to achieve a necessary condition which embodies the Homogeneization Lemma for problems with unilateral constraints [5].

At $1 \leq p = m$, we have $g(\bar{x}) = 0$, so that $0 \in \mathcal{K}_{\bar{x}}^{(h)}$. Therefore, Hypothesis H is fulfilled if and only if

$$U := \{(u, v) \in \mathbb{R} \times \mathbb{R}^m : u > 0, v = 0\} \not\subset int\, \mathcal{K}_{\bar{x}}^{(h)}$$

If, in addition, $g$ is differentiable, then $\mathcal{K}_{\bar{x}}^{(h)}$ is a linear manifold, so that Hypothesis H is verified; this justifies Hypothesis H"'. If, furthermore, beside being differentiable, $\nabla g$ is surjective – this happens, for instance, in the classic Lagrange Multipliers Theorem, where the assumptions of Dini Implicit Functions Theorem are adopted – we achieve even more: separation with $\theta > 0$. A necessary and sufficient condition for having $\theta > 0$ is offered by Theorem 2. 1 of [2].

# 8 SIGN OF MULTIPLIERS AND COMPLEMENTARITY

Now we will establish two propositions, which are instrumental for achieving a necessary conditions.

**Proposition** 8.1 (sign of multipliers). Let $X$ be convex, $f$ and $-g$ be $C$–differentiable at $\bar{x} \in X$, and let Hypothesis H hold at least at $\bar{x}$. If $\bar{x}$ is a minimum point of (1.1), then

$$(\mathcal{K}_{\bar{x}}^{(h)} - \bar{z})^* \cap [\mathbb{R}_- \times \mathbb{R}^p \times \mathbb{R}_-^{m-p}] \backslash \{0\} \neq \emptyset \,. \tag{8.1}$$

If $f$ and $g$ are differentiable, then (8.1) becomes:

$$(\mathcal{K}_{\bar{x}}^{(h)} - \bar{z})^{\perp} \cap [\mathbb{R}_- \times \mathbb{R}^p \times \mathbb{R}_-^{m-p}] \backslash \{0\} \neq \emptyset \,. \tag{8.2}$$

**Proof.** Because of Hypothesis H, $\exists\omega \in \mathbb{R}^{1+m}\backslash\{0\}$, such that[7]:

$$\mathcal{H} \subset \{z \in \mathbb{R}^{1+m} : \langle \omega, z\rangle \leq 0\}\ ,\ \mathcal{E}(\mathcal{K}_{\bar{x}}^{(h)} - \bar{z}) \subseteq \{z \in \mathbb{R}^{1+m} : \langle \omega, z\rangle \geq 0\} := \pi^{+}\ .$$

The former of the above inclusions implies $\omega \in -\mathcal{H}^* = \mathbb{R}_- \times \mathbb{R}^p \times \mathbb{R}_+^{m-p}$; the latter implies:

$$\mathcal{E}^*(\mathcal{K}_{\bar{x}}^{(h)} - \bar{z})\ ,$$

which, account taken of the following well known polarity relationship

$$\mathcal{K}_{\bar{x}}^{(h)} - \bar{z}) \subseteq \mathcal{E}(\mathcal{K}_{\bar{x}}^{(h)} - \bar{z}) \ \Rightarrow\ \mathcal{E}^*(\mathcal{K}_{\bar{x}}^{(h)} - \bar{z}) \subseteq (\mathcal{K}_{\bar{x}}^{(h)} - z)^*\ ,$$

lets us achieve $\omega \in (\mathcal{K}_{\bar{x}}^{(h)} - \bar{z})^*$ and hence (8.1). (8.2) is obvious. □

**Proposition** 8.2 (complementarity). Let $X$ be convex, $f$ and $-g$ be $\mathcal{C}$–differentiable at $\bar{x}$, and let Hypothesis H hold at least at $\bar{x} \in X$. If $\bar{x}$ is a minimum point of (1.1), then

$$\exists(-\bar{\theta}, -\bar{\lambda}) \in (\mathcal{K}_{\bar{x}}^{(h)} - \bar{z})^* \cap [\mathbb{R}_- \times \mathbb{R}^p \times \mathbb{R}_-^{m-p}]\backslash\{0\},$$

such that:

$$\bar{\lambda}_i g_i(\bar{x}) = 0\ , \quad i \in \mathcal{I}^+\ . \tag{8.3}$$

**Proof.** Ab absurdo, suppose that (8.3) be false. Then,

$$\forall\omega := (-\theta, -\lambda) \in (\mathcal{K}_{\bar{x}}^{(h)} - \bar{z})^* \cap [\mathbb{R}_- \times \mathbb{R}^p \times \mathbb{R}_-^{m-p}]\backslash\{0\} \neq \emptyset$$

(where the inequality is consequence of Proposition 8.1, which can be applied because of the $\mathcal{C}$–differentiability and Hypothesis H) we have $\langle \lambda, g(\bar{x})\rangle > 0$ and hence $\langle \omega, -\bar{z}\rangle > 0$, since $\bar{z} = (0, g(\bar{x}))$, $g_i(\bar{x}) = 0\ i \in \mathcal{I}^0$, and $\sum_{i\in\mathcal{I}^+} \lambda_i g_i(\bar{x}) > 0$. This fact, the convexity of $\mathcal{E}(\mathcal{K}_{\bar{x}}^{(h)} - \bar{z})$ (implied by the $\mathcal{C}$–differentiability of $f$ and $-g$ at $\bar{x}$; see [5], Lemma 3(iii)), the inclusion $\mathcal{E}^*(\mathcal{K}_{\bar{x}}^{(h)} - \bar{z}) \subseteq (\mathcal{K}_{\bar{x}}^{(h)} - \bar{z})^*$, and $\omega \in \mathcal{E}^*(\mathcal{K}_{\bar{x}}^{(h)} - \bar{z})$ (see the proof of Proposition 8.1) imply:

$$-\bar{z} \in int\{\mathcal{E}^*(\mathcal{K}_{\bar{x}}^{(h)} - \bar{z}) \cap [\mathbb{R}_- \times \mathbb{R}^p \times \mathbb{R}_-^{m-p}]\backslash\{0\}\}^* =$$

[7] $\mathcal{E}(\mathcal{K}_{\bar{x}}^{(h)} - \bar{z})$ is the conic extension (Sect. 2) of the cone $\mathcal{K}_{\bar{x}}^{(h)} - \bar{z}$, whose apex is the origin of $\mathbb{R}^{1+m}$.

$$= int\{\mathcal{E}^*(\mathcal{K}_{\bar{x}}^{(h)} - \bar{z}) \cap [\mathbb{R}_- \times \mathbb{R}^p \times \mathbb{R}_-^{m-p}]\}^* =$$
$$= int\, cl\, conv[\mathcal{E}(\mathcal{K}_{\bar{x}}^{(h)} - \bar{z}) \cup (-\mathcal{H})] = int\, \mathcal{E}(\mathcal{K}_{\bar{x}}^{(h)} - \bar{z}) \ .$$

Denote by $S(z)$ an open sphere with centre at $z$ and suitably small radius. From the last relations we deduce that $S(-\bar{z}) \subset \mathcal{E}(\mathcal{K}_{\bar{x}}^{(h)} - \bar{z})$, and thus $S(O_{1+m}) \subset \mathcal{E}(\mathcal{K}_{\bar{x}}^{(h)})$; this inclusion implies $\mathcal{E}(\mathcal{K}_{\bar{x}}^{(h)}) \cap ri\, \mathcal{H} \neq \emptyset$, which contradicts the minimality of $\bar{x}$, because of the equivalence between (2.2) and (2.3). □

# 9 A NECESSARY CONDITION

**Proposition** 9.1 (sign of directional derivative). **(i)** Let $X$ be convex, $f$ and $-g$ be $\mathcal{C}$–differentiable at $\bar{x} \in X$. If $\bar{x}$ is a lower semistationary point of $L(\bar{x}; \theta, \lambda)$ with $\theta \geq 0$, $\lambda \in D^*$, $(\theta, \lambda) \neq 0$, then

$$\text{(9.1)} \qquad \inf_{d \in S} \mathcal{D}_{\mathcal{C}} L(\bar{x}; d, \theta, \lambda) \geq 0 \ ,$$

where $S := \{d \in X - \bar{x} : \|d\| = 1\}$, and $\mathcal{D}_{\mathcal{C}} L = \theta \mathcal{D}_{\mathcal{C}} f(\bar{x}; d) - \sum_{i \in \mathcal{I}} \lambda_i \mathcal{D}_{\mathcal{C}} g_i(\bar{x}; d)$ is the $\mathcal{C}$–derivative of the Lagrangian function $L$ at $\bar{x}$. **(ii)** If $\bar{x} \in int\, X$, $f$ and $g$ are differentiable, then (9.1) becomes:

$$\text{(9.2)} \qquad V_L(\bar{x}; d; \theta, \lambda) = 0 \quad , \quad \forall d \in S \ ,$$

where $V_L$ is a (continuous) linear functional (1st variation of $L$). If, in addition, $X = B = \mathbb{R}^n$, then, instead of (9.2), we find:

$$\text{(9.3)} \qquad \langle L'(\bar{x}; \theta, \lambda), d \rangle = 0 \quad , \quad \forall d \in S \ ,$$

where $L'_x$ denotes the gradient of $L$ at $\bar{x}$ with respect to $x$.

**Proof.** **(i)** Set $x = \bar{x} + \alpha d \in X$ with $\alpha \in \mathbb{R}_+$. (6.3) with $L$ replacing $f$ holds[8] iff

$$\text{(9.4)} \qquad \sup_{\tau > 0} \inf_{\substack{d \in S \\ \alpha \in ]0, \tau]}} [\mathcal{D}_{\mathcal{C}} L(\bar{x}; d; \theta, \lambda) + \frac{1}{\|\alpha d\|} \varepsilon_L(\bar{x}; \alpha d; \theta, \lambda)] \geq 0 \ ,$$

where $\varepsilon_L := \theta \varepsilon_f - \sum_{i \in \mathcal{I}} \varepsilon_{g_i}$ is the remainder of the expansion of $L$. Ab absurdo, suppose that

$$\ell := \inf_{d \in S} \mathcal{D}_{\mathcal{C}} L(\bar{x}; d; \theta \lambda) < 0 \ .$$

[8] If $S \not\subseteq X - \bar{x}$, then we can define $S$ as a sphere of suitably small and positive radius, and intersect it with $X - \bar{x}$.

Since $\varepsilon_L/\|\alpha d\| \to 0$ as $\alpha \downarrow 0$, we find that the infimum in (9.4) is less than a negative constant, so that its supremum is still negative; this fact, which holds a fortiori if $\ell = -\infty$, contradicts (9.4) and proves (9.1). **(ii)** To achieve (9.2) and (9.3) it is enough to note that $d \in X - \bar{x} \Rightarrow -d \in X - \bar{x}$, and that the left–hand sides of (9.2) and (9.3) are linear. □

When $X = B = \mathbb{R}^n$, then the infimum in (9.1) is achieved, since $\mathcal{D}_C L$ is a linear combination of continuous functions ($\mathcal{D}_C f$ and $-\mathcal{D}_C g_i$ are convex) and $S$ is compact; the same happens, when $X = B \neq \mathbb{R}^n$, if in a neighbourhood of $\bar{x}$ $\mathcal{D}_C f$ and $-\mathcal{D}_C g_i$ are bounded from above by a constant.

**Definition** 9.1 Let $X$ be convex and $f : X \to \mathbb{R}$. The *C–subdifferential* of a $C$–differentiable function at $\bar{x}$ is given by:

(9.5) $$\partial_C f(\bar{x}) := \{t \in B^* : \mathcal{D}_C f(\bar{x}; d) \geq \langle t, d\rangle \ , \quad \forall d \in X - \bar{x}\} \ ,$$

where $B^*$ is the continuous dual of $B$; $t$ is called the *C–subgradient* of $f$ at $\bar{x}$. If $f$ is convex, then (9.5) collapses to the classic subdifferential of Convex Analysis:

(9.5)′ $$\partial f(\bar{x}) := \{t \in B^* : f'(\bar{x}; d) \geq \langle t, d\rangle \ , \quad \forall d \in X - \bar{x}\} \ .$$

**Theorem** 9.1 (necessary conditions). Let $X$ be convex, $f$ and $-g_i$ $i \in \mathcal{I}$ be $C$–differentiable at $\bar{x}$, and such that:

(9.6) $$ri\, dom\mathcal{D}_C f \cap ri\, dom\mathcal{D}_C(-g_1) \cap \ldots \cap ri\, dom\mathcal{D}_C(-g_m) \neq \emptyset \ .$$

Let Hypothesis H be fulfilled. If $\bar{x}$ is a minimum point of (1.1), then there exist multipliers $\bar{\theta} \in \mathbb{R}$, $\bar{\lambda} \in \mathbb{R}^m$, such that:

(9.7a) $$\inf_{d \in S} \mathcal{D}_C L(\bar{x}; d; \bar{\theta}, \bar{\lambda}) \geq 0 \ ,$$

(9.7b) $$g(\bar{x}) \in D \ , \quad \bar{\theta} \in \mathbb{R}_+ \ , \quad \bar{\lambda} \in D^* \ , \quad (\bar{\theta}, \bar{\lambda}) \neq 0 \ ,$$

(9.7c) $$\bar{\lambda}_i g_i(\bar{x}) = 0 \ , \quad i \in \mathcal{I}^+$$

where $\mathcal{D}_C L = \mathcal{D}_C f(\bar{x}; d) - \sum_{i \in \mathcal{I}} \mathcal{D}_C g_i(\bar{x}; d)$ is the $C$–derivative of the Lagrangian function at $\bar{x}$, and $S := \{d \in X - \bar{x} : \|d\| = 1\}$. (9.7a) is equivalent to:

(9.7a)′ $$O \in \bar{\theta}\partial_C f(\bar{x}) + \sum_{i \in \mathcal{I}} \bar{\lambda}\partial_C(-g_i(\bar{x})) \ ,$$

which becomes:

$$(9.7a)'' \qquad O \in \bar{\theta}\partial f(\bar{x}) + \sum_{i \in \mathcal{I}} \bar{\lambda}\partial(-g_i(\bar{x})) \ ,$$

if, in particular, $f$ and $g_i$ $i \in \mathcal{I}$ are convex. When $f$ and $g$ are differentiable at $\bar{x} \in int\, X$, then (9.7a) collapses to $V_L = 0$ along $x = \bar{x}$, where $V_L$ is the first variation of $L$, and becomes:

$$(9.7a)''' \qquad \Psi'_x(t, \bar{x}, \bar{x}'; \bar{\theta}, \bar{\lambda}) - \frac{d}{dt}\Psi'_{x'}(t, \bar{x}, \bar{x}'; \bar{\theta}, \bar{\lambda}) = 0 \ ,$$

if $f(x) = \int_T \Psi_0(t, x, x')dt, g_i = \int_T \Psi_i(t, x, x')dt, T := [t_0, t_1]$ with $-\infty < t_0 < t_1 < +\infty$, and $\Psi$ is the integrand of the Lagrangian function. If $X = B = \mathbb{R}^n$, then (9.7a) collapses to:

$$(9.7a)'''' \qquad L'_x(\bar{x}; \bar{\theta}, \bar{\lambda}) = 0 \ ,$$

where $L'_x$ is the gradient of $L$ with respect to $x$.

**Proof.** Because of Hypothesis H, Proposition 8.2 can be applied and implies the existence of $(-\bar{\theta}, -\bar{\lambda}) \in (\mathcal{K}^{(h)}_{\bar{x}} - \bar{z})^* \cap [\mathbb{R}_- \times \mathbb{R}^p \times \mathbb{R}^{m-p}] \backslash \{0\}$ (so that, by Proposition 6.2, $\bar{x}$ is a lower semistationary point of $L$ at $\bar{x}$ and, by Proposition 9.1, (9.7a) holds), such that $\bar{\theta} \geq 0$, $\bar{\lambda} \in D^*$, $(\bar{\theta}, \bar{\lambda}) \neq 0$, $\bar{\lambda}_i g_i(\bar{x}) = 0$ $i \in \mathcal{I}^+$, and thus (9.7 b,c) hold too. $\mathcal{D}_C$ is sublinear, so that $\partial_C L \neq \emptyset$. (9.7a) $\Rightarrow O \in \partial_C L$. Because of (9.6), this relation becomes (9.7a)$'$. (9.7a)$''$ is obvious, since in that case $\partial_C L = \partial L$. Let $f$ and $g$ be differentiable at $\bar{x} \in int\, X$; since both $d$ and $-d$ are in $S$, $\mathcal{D}_C L$ collapses to $V_L$ or to $L'_x$ and (9.7a) to $V_L = 0$ or to (9.7a)", respectively when $B$ is infinite or finite dimens! ional. In the former case it is well known that – the functionals being differentiable – the variation can be equivalently replaced by Euler Variation, so that classic deductions lead to (9.7a)$''''$. □

In the above condition, the multiplier associated with the objective function $f$ is merely non–negative. Several conditions are known [8], under which $\bar{\theta}$ of Theorem 9.1 exists positive and hence $\bar{\theta} = 1$. A necessary and sufficient condition is given by Theorem 2.1 of [2].

## 10 FURTHER DEVELOPMENTS

Now, we will make some comments on how to extend the above ISA to the problems having infinite dimensional image. To this end, we will consider a class of such problems and outline a possible way of extending Sects. 2–9. Within this class, we will consider a particular classic control problem, and will give some hints on how ISA might be exploited to achieve the classic Pontryagin Maximum Principle and more general results, hoping that this will stimulate research in this field. The present section does n ot contain any claim or statement, but proposals of statements only.

Let us consider a special case of (1.1): $B$ is the set of all continuous functions $x(t) := (x_1(t), \dots, x_n(t))$, $t \in T := [t_0, t_1] \subset \mathbb{R}, -\infty < t_0 < t_1 < +\infty, n \in \mathbb{N}$, equipped with the norm $\|x\| := \max_{z \in T} \|x(t)\|_2$, $x \in B$, where $\|\cdot\|_2$ denotes the Euclidean norm in $\mathbb{R}^n$. $X$ is the subset of $B$ having continuous derivatives $x'(t) := (x'_1(t), \dots, x'_n(t))$, $t \in T$, except at most for a finite number of points $\bar{t}$ at which there exist and are finite $\lim_{t \downarrow \bar{t}} x'(t)$ and $\lim_{t \uparrow \bar{t}} x'(t)$; fixed end–point conditions are included in the definition of $X$. Consider the functions $\psi_i : \mathbb{R} \times \mathbb{R}^n \times \mathbb{R}^n \to \mathbb{R}$ $i \in \{0\} \cup \mathcal{I}$, and the special case of (1.1):

$$\text{(10.1a)} \qquad \min f(x) := \int_T \psi_0(t, x(t), x'(t))dt \ ,$$

subject to

$$\text{(10.1b)} \qquad \psi_i(t, x(t), x'(t)) = 0 \quad \forall t \in T \ , \quad i \in \mathcal{I}^0 \ ,$$

$$\text{(10.1c)} \qquad \psi_i(t, x(t), x'(t)) \geq 0 \quad \forall t \in T \ , \quad i \in \mathcal{I}^+ \ ,$$

$$\text{(10.1d)} \qquad x \in X \ .$$

In the preceding sections, where $\mathcal{B} = \mathbb{R}$, each constraining function $g_i$ sends an element of $X$ into a real number, so that the IS is finite dimensional. In the present case $\mathcal{B} \neq \mathbb{R}$, $g_i$ is the function which sends an element of $X$, i.e. a function $x(t)$, $t \in T$, into another function, namely $\tilde{\psi}_i(t) := \psi_i(t, x(t), x'(t))$ defined on $T$. Hence, an obvious extension of the approach of Sects. 2–9 would consist in introducing the IS as $\mathbb{R} \times \mathcal{B}^m$ (indeed, this is the general definition of IS given in Sect. 2); the analysis should be carried on in an infinite dimensional space ($\mathcal{H}$ and $\mathcal{K}_{\bar{x}}$ would be infinite dimensional), and could not be done by elementary mathematical tools like in Sects, 2–9. Such an infinite dimensionality is an obvious drawback; it cannot be eliminated, but can be postponed (indeed, the great advantage of Lagrange idea about multipliers consists exactly in postponning the handling of constraining equations to the writing of a necessary conditions; the enormous effect of such a postposition is that we are no longer obliged to solve "analytically" – i.e., to find *all* the solutions – the system of constraining equations, but we merely have to find, in general, one or a few solutions of it, by means of an enlarged system). This can be done by introducing a multifunction approach (proposed in [3]), which will allow us to circumvent the infinite dimensionality and to be reduced, in order to study the image of (10.1), to handle finite dimensional sets, and thus to exploit the scheme described in Sects.2–9 for the case $\mathcal{B} = \mathbb{R}$. Such an approach will be now briefly outlined.

As said above, the image of function $x$, through $g_i$, is again a function defined on $T$, i.e. $\tilde{\psi}_i$; the image of $\tilde{\psi}_i$ is a subset of $\mathbb{R}$. Hence, we can introduce the multifunction, which sends $x$ into the subset of $\mathbb{R}^{1+m}$ which is the image of $(f_{\bar{x}}, \tilde{\psi})$ with $\tilde{\psi} := (\tilde{\psi}_i \in \mathcal{I})$. Let $\bar{x} \in X$ be given, and consider the multifunction $A \to 2^{\mathbb{R}^{1+m}}$, defined by:

$$A(x) := \{(u,v) \in \mathbb{R} \times \mathbb{R}^m : u = f(\bar{x}) - f(x) \text{ and } \exists t \in T \text{ s.t.}$$
$$v = \psi(t, x(t), x'(t))\} ,$$

where $\psi := (\psi_i \ i \in \mathcal{I})$. $A(X)$ will be called the *image* of (10.1) and corresponds to $\mathcal{K}_{\bar{x}}$ of Sect.2. By means of such a definition we reduce ourseleves to work in a finite dimensional space, namely $\mathbb{R}^{1+m}$, as happens in Sects.2–9: we replace, as IS, $\mathbb{R} \times \mathcal{B}^m$ with $\mathbb{R}^{1+m}$. However, in Sects.2–9 $A$ is merely a function, so that $A(x)$ is an element of the IS, while now $A$ is a multifunction and $A(x)$ is a set of elements of $\mathbb{R}^{1+m}$. Therefore, we must expect a more complex development than that of Sects.2–9. However, the present approach has the advantage of postponing the infinite dimensionality to the introduction of the IS.

Let us consider any $\bar{x} \in X$. Obviously, an $\bar{x}$ which satisfies (10.1 b,c) is a global minimum point for (10.1) iff the system (in the unknown $x$):

$$(10.2) \qquad \begin{cases} f_{\bar{x}}(x) := f(\bar{x}) - f(x) = \\ \qquad \int_T [\psi_0(t, \bar{x}(t), \bar{x}'(t)) - \psi_0(t, x(t), x'(t))]dt > 0 , \\ \\ \psi(t, x(t), x'(t)) \in D \ , \ \ \forall t \in T \ ; \ \ x \in X \end{cases}$$

is impossible. The impossibility of (10.2) – which corresponds to (2.1) – is equivalent to that of the following condition:

$$(10.3) \qquad \exists x \in X \ , \ \text{ s.t. } \ A(x) \subseteq \mathcal{H} ,$$

which replaces (2.2) and, unlike (2.2), cannot be proved by means of separation arguments. In other words, we have overcome the infinite dimensionality of the IS, but we have lost the possibility of expressing optimality as disjunction of two sets. This new drawback can be overcome by means of the selection theory for multifunctions.

We want to introduce a vector function from $X$ to $\mathbb{R}^{1+m}$ with the propose to select an element of $A(x)$. This can be done, obviously, in many ways; we propose the following one. Consider the functions $\sigma_i : T \to \mathbb{R}$ , $i \in \mathcal{I}$; denote by $\Sigma$ the set of vectors $\sigma := (\sigma_1, \dots, \sigma_m)$, whose elements are not all zero on $T$ and such that $\sigma_i \geq 0$ , $i \in \mathcal{I}^+$; here $\Sigma$ represents a class of functional parameters which satisfy a suitable condition (depending on the assumptions made about $\psi_i$) to be rendered more precise, under which the integral in next (10.4) makes sense. The selection in this case is obta ined by means of the parameter $\sigma$, so that the above mentioned vector function is of type[9] $\Phi : X \times \Sigma \to \mathbb{R}^{1+m}$, defined $\forall x \in X$, by:

$$(10.4)\quad \Phi(x;\sigma) := \int_{A(x)} \sigma dt := (f_{\bar{x}}(x), \int_T \sigma_i(t) \cdot \psi(t, x(t), x'(t)) dt \; i \in \mathcal{I}) ,$$

where the 1st integral is a short writing to mean selection of an element of $A(x)$ by means of a weighted integration.

According to Sect. 2 of [3], $\Phi$ is called *generalized selection function* (for short, GSF) iff $\forall x$ we have:

$$(10.5)\qquad A(x) \subseteq \mathcal{H} \;\Leftrightarrow\; \Phi(x;\sigma) \in \mathcal{H} \; , \quad \forall \sigma \in \Sigma \; ;$$

$\sigma$ is called *selection multiplier* (for short, SM). The inclusion in (10.5) is obviously equivalent to the impossibility of (10.2).

It has been shown [3, Th. 5.1] that, under assumptions which make valid the so–called Fundamental Lemma of Calculus of Variations, a function of type (10.4) is a GSF for (10.1) without (10.1b) and with (10.1c) independent of $x'$. The extension of such a proposition to the present case (10.1 b,c) has been obtained in [7, Sect. 3].

Note that, instead of (10.5), we can consider the equivalent relationship:

$$(10.6)\qquad A(x) \not\subseteq \mathcal{H} \;\Leftrightarrow\; \exists \sigma \in \Sigma \text{ s.t. } \Phi(x;\sigma) \notin \mathcal{H} \, .$$

This shows the dependence of the SM on $x$, while in (10.5) such a dependence his hidden: in (10.5) we have to show "something" $\forall \sigma \in \Sigma$, while in (10.6) we have to prove the existence of a $\sigma$, which of course varies with $x$. Such a class of multipliers, which might be called *quasi-multipliers* – by borrowing the terminology from Variational Inequality Theory –, does not appear at least explicitly in the Calculus of Variations literature. Quasimultipliers might turn out to be useful to improve sufficient optimality conditions or to decrease a positive duality gap.

[9]Here the symbol $\Phi$ is independent of that of Sect. 4.

By following the above outlined approach, it should be possible to revisit and, perhaps, extend several topics of the Calculus of Variations. One of them, the deduction of the celebrated Pontryagin Maximum Principle, is now shortly touched, just to show an instance[10]. Consider the following special case of (10.1)[11]:

$$\min f(x) := \int_T f^0(x(t), \xi(t))dt \;, \tag{10.7a}$$

subject to:

$$\frac{dx_i}{dt} = f^i(x;\xi) \;, \quad i \in I := \{1, \dots, n\} \;, \tag{10.7b}$$

$$x(t_0) = x^0 \;, \; x(t_1) = x^1 \;, \; f^i, \frac{dx_i}{dt} \;, \; \frac{\partial f^i}{\partial x_j} \in C^0(T) \;, \tag{10.7c}$$

$$\xi \in \Xi \;, \tag{10.7d}$$

where $x = (x_1, \dots, x_n)$ and, $\Xi \subseteq \mathbb{R}^q$ is a given set of "controls"; $f^i : \mathbb{R}^{n+q} \to \mathbb{R}$, $x^0, x^1 \in \mathbb{R}^n$ are given. At $p = m = n$, $\mathcal{I}^0 = \mathcal{I} = I$, with $x$ replaced by $(x, \xi)$, and with further obvious positions, (10.1) collapses to (10.7). We have now:

$$\mathcal{H} := \{(u, v) \in \mathbb{R} \times \mathbb{R}^n : u > 0 \;, \; v = 0\} \;,$$

$$\mathcal{K}_{\bar{x}}(x, \xi; \sigma) := \{(u, v) \in \mathbb{R} \times \mathbb{R}^n : u = \textstyle\int_T [f^0(\bar{x}; \xi) - f^0(x; \xi)]dt,$$
$$v_i = \textstyle\int_T \sigma_i(t)[\frac{dx_i}{dt} - f^i(x; \xi)]dt, \; i \in I, \; x \in C^1(T)^n\} \;, \; \xi \in \Xi \;, \; \sigma \in \Sigma.$$

Note that $\mathcal{K}_{\bar{x}}(x, \xi; \sigma)$ is a special case of $\Phi$, and that $\mathcal{K}_{\bar{x}}(X, \Xi; \Sigma)$, which we call *selected image*, is finite dimensional and plays the same role of $\mathcal{K}_{\bar{x}}$ of Sects. 2–9. Therefore, what has been done for $\mathcal{K}_{\bar{x}}$ can be immediately extended to $\mathcal{K}_{\bar{x}}(X, \Xi; \Sigma)$; this set, for the sake of simplicity, will be denoted, in the sequel, by $\mathcal{K}_{\bar{x}}$. Of course, this depends on $\sigma$; indeed, the infinite d imensionality of the IS has not disappeared: it has been transformed into the choice of $\sigma$, and hence postponed to the writing of a necessary condition.

In place of (3.1), we find now the following *image problem:*

$$\inf_{\sigma \in \Sigma} \; \sup_{\substack{(u,v) \in \mathcal{K}_{\bar{x}} \\ v=0}} (u) \;, \tag{10.8}$$

[10] In the classic version of Pontryagin Maximum Principle, the multipliers can be considered independent of the unknown $(x, \xi)$.

[11] The present $\xi$ is independent of that of Sect. 3.

which, due to Proposition 5.2, at $\theta = 1$, is equivalent to:

$$\inf_{\substack{\sigma\in\Sigma \\ \lambda\in\mathbf{R}^n}} \sup_{(u,v)\in\mathcal{K}_{\bar{x}}} [\theta u - \sum_{i\in I} \lambda_i \sigma_i(t) v_i)] \ . \tag{10.9}$$

As noted in Sect. 5, the lack of equivalence at $\theta = 0$ necessarily, between (10.8) and (10.9) is of no importance when we are looking for a necessary condition; this happens if we want to achieve a minimum (or maximum) principle.

If we replace $u$ and $v$ with their expressions in terms of the given data of (10.1), and if we neglect the constant term, then (10.9) becomes:

$$\sup_{\substack{\sigma\in\Sigma \\ \lambda\in\mathbf{R}^n}} \inf_{(x,\xi)\in R} \int_T \{\theta f^0(x;\xi) - \sum_{i\in I} \lambda_i \sigma_i(t)[\frac{dx_i}{dt} - f^i(x;\xi)]\}dt \ , \tag{10.10}$$

where $R$ denotes now the set of $(x, \xi)$ which fulfil (8.1 b–d). By setting:

$$\omega_0 := -\theta \ , \ \omega_i(t) := -\lambda_i \cdot \sigma_i(t) \ i \in I \ , \ H(x;\xi;\omega) := \sum_{i=0}^{n} \omega_i(t) f^i(x;\xi) \ ,$$

(10.10) becomes:

$$- \inf_{\omega\in\Omega} \sup_{(x,\xi)\in R} \int_T [H(x;\xi;\omega) - \sum_{i\in I} \omega_i \frac{dx_i}{dt}]dt \ , \tag{10.11}$$

where $\omega := (\omega_0, \dots, \omega_n)$ , $\Omega := \{\omega \in \mathbf{R} \times C^0(T)^n : \omega_0 = -\theta, \omega_i = -\lambda_i \sigma_i(t) \ , \ \lambda_i \in \mathbf{R} \ , \ i \in I\}$. $H$ is called *Hamiltonian function,* while that in square brackets of (10.11) is called *Lagrangian auxiliary function.*

Since $R$ is a Cartesian product between the set of $x$ which satisfy (10,7 b,c) and $\Xi$, the supremum in (10.11) is not achieved if, for each $x$ and $\omega$, $H$ does not achieve – in (10.11) – its supremum with respect to $\xi$, i.e.

$$\mathcal{M}(x;\omega) := \sup_{\xi\in\Xi} H(x;\xi;\omega) \ . \tag{10.12}$$

Furthermore, the gradient, with respect to $\omega$, of the integrand in (10.11) must vanish, or

$$\frac{\partial H}{\partial \omega_i} - \frac{dx_i}{dt} = 0 \ , \ i \in I \ . \tag{10.13}$$

By integrating by parts (10.11), we obtain:

$$- \inf_{\omega\in\Omega} \sup_{(x,\xi)\in R} \int_T [H(x;\xi,\omega) + \sum_{i\in I} x_i \frac{d\omega_i}{dt}]dt - \sum_{i\in I} [\omega_i x_i]_{t_0}^{t_1} \ . \tag{10.14}$$

The gradient of the integrand of (10.14), with respect to $x$, must vanish:

$$\frac{\partial H}{\partial x_i} + \frac{\partial \omega_i}{dt} = 0 \; , \; i \in I \; . \tag{10.15}$$

(10.12), (10.13) and (10.15) are the celebrated Pontryagin Maximum Principle. It would be interesting to revisit this principle as well as other ones in terms of ISA. Note that the above approach introduces a wider class of multipliers; namely, those coming from the nonlinear separation functions. Therefore, other topics – like sufficient conditions, duality, penalization – might be treated, and not only necessary conditions. Moreover, note that in the above approach, unlike the existing licterature on t he Calculus of Variations, the "multiplier function" turns out to be split into a number, i.e. $\lambda_i$, and a function, i.e. $\sigma_i(t)$; they play different roles: gradient of a separation hyperplane and selector, respectively. This might improve both the theory – for instance, in studying the perturbation function –, and the applications – for instance, in drawing information from the dual variables, shadow prices, potentials –. Furthermore, in the above short c! omments, the possibility has been outlined of considering selection multipliers depending on the unknown, namely of type $\sigma(t;x)$; these might be useful in the nonsmooth cases.

## References


[1] V. Dem'yanov and A. Rubinov, "Quasidifferentiability and related topics". Kluwer, 2000

[2] F. Giannessi, "Theorems of the alternative and optimality conditions". Jou. Optimization Theory Appls., Vol. 42, N. 3, 1984, pp. 331–365.

[3] F. Giannessi, 'Theorems of the alternativefor multifunctions with applications to Optimization: General results". Jou. Optimization Theory Appls., Vol. 55, N. 2, 1987, pp. 233–256.

[4] F. Giannessi, "Semidifferentiable functions and necessary optimality conditions". Jou. Optimization Theory Appls., Vol. 60, N. 2, 1989, pp. 191–241.

[5] F. Giannessi, "General optimality conditions via a separation scheme". In "Algorithms for continuous optimization", E. Spedicato Ed., Kluwer Ac. Publ., 1994, pp. 1–23.

[6] F. Giannessi, "Image space approach to Optimization", In "Encyclopedia of Optimization", C.A. Floudas and P.M. Pardalos Eds., Kluwer, 2001.

[7] F. Giannessi and A. Uderzo, "A multifunction approach to extremum problems having infinite dimensional image. I: Composition and selection". Atti Seminario Matematico e Fisico, Università di Modena, Suppl. Vol. XLVI, 1998, pp. 771–785.

[8] O.L. Mangasarian, "Nonlinear Programming". SIAM, Philadelphia, 1994.

[9] W. Oettli and M. Théra, "Equivalents of Ekeland's principle". Bull. Austral. Math. Sc., Vol. 48, 1993, pp. 385–392.

[10] T. Rapcsák, "Smooth Nonlinear Optimization in $\mathbb{R}^n$". Kluwer, Acad. Publ., 1997.

[11] S.M. Robinson, "Local structure of feasible sets in nonlinear programming, part III: stability and sensitivity". Mathematical Programming Study, N. 30, 1987, pp. 45–66.

[12] R.T. Rockafellar and R.J.-B. Wets, "Variational Analysis". Springer–Verlag, 1998.

[13] F. Tardella, "On the image of a constrained extremum problem and some applications to the existence of a minimum". Jou. Optimization Theory Appls., Vol. 60, N. 1, 1989, pp. 93–104.

# Chapter 5

# TRANSVERSAL HYPERGRAPHS AND FAMILIES OF POLYHEDRAL CONES

Leonid Khachiyan*

*Department of Computer Science*

*Rutgers University, USA*

leonid@athos.rutgers.edu

**Abstract** We discuss the complexity of generating certain geometric configurations related to the classical theorems of Carathéodory and Helly. Given a set $\mathcal{K}$ of rational cones in $n$ dimensions and a rational $n$-vector $b$, we consider the families $MIN(\mathcal{K}, b)$ and $MAX(\mathcal{K}, b)$ of all minimal and all maximal subsets of $\mathcal{K}$ that generate or avoid $b$, respectively. We show that the problems of generating $MIN(\mathcal{K}, b)$ and $MAX(\mathcal{K}, b)$ are both $NP$-hard already for systems of dihedral cones, while the case where all cones in $\mathcal{K}$ are rays remains open and includes the well-known problem of enumerating the vertices of a given polytope. On the other hand, for any set of rational cones, the generation of $MIN(\mathcal{K}, b) \cup MAX(\mathcal{K}, b)$ is polynomially equivalent to the hypergraph transversal problem, and hence can be carried out in a quasi-polynomial time. We provide other examples of mutually transversal $NP$-hard families of cones and polytopes.

**Keywords:** Incremental vertex generation, Transversal hypergraph, Incremental polynomial time

*Research supported in part by NSF grant CCR-9618796.



*N. Hadjisavvas and P.M. Pardalos (eds.), Advances in Convex Analysis and Global Optimization,* 105–118.

# 1 MINIMAL/MAXIMAL CONIC FAMILIES CONTAINING/AVOIDING A PRESCRIBED VECTOR

Let $\mathcal{K}$ be a finite set of polyhedral cones in $\Re^n$ and let $b \in \Re^n$ be a given vector such that

$$b \in \sum_{K \in \mathcal{K}} K.$$

Denote by $MIN(\mathcal{K}, b)$ the family of all minimal sets $\mathcal{S} \subseteq \mathcal{K}$ such that

$$b \in \sum_{K \in \mathcal{S}} K.$$

By Carathéodory's theorem, the cardinality of each set $\mathcal{S} \in MIN(\mathcal{K}, b)$ does not exceed $n$. We assume that $b$ and all the cones $K \in \mathcal{K}$ are rational and discuss the complexity of generating the set family $MIN(\mathcal{K}, b)$. We start with the special case where each cone $K \in \mathcal{K}$ is a ray.

Let $\mathcal{K} = \{K_1, \ldots, K_m\}$, where $K_i = cone\{a_i\}$ for given rational vectors $a_i \in \Re^n$. Suppose that the polyhedron $P = \{x \in \Re^m \mid \sum_{i=1}^m a_i x_i = b,\ \ x = (x_1, \ldots, x_m) \geq 0\}$ is bounded. Then there is a natural one-to-one correspondence between $MIN(\mathcal{K}, b)$ and $VERT(P)$, the vertex set of $P$:

> $\mathcal{S} \in MIN(\mathcal{K}, b)$ *if and only if* $\mathcal{S}$ *is the set of positive components of a vertex* $v \in VERT(P)$.

In particular, let $P$ be the flow circulation polytope for a directed graph $G = (V, E)$, i.e., the set of all vectors $x = \{x_e\} \in \Re^E$ such that

$$\sum_{e \text{ incident to } v} x_e = \sum_{e \text{ incident from } v} x_e \quad \text{for any vertex } v \in V,$$

$$\sum_{e \in E} x_e = 1, \qquad x_e \geq 0 \quad \text{for any arc } e \in E.$$

Then for any simple circuit $C$ in $G$, the expressions

$$x_e = \begin{cases} 1/|C| & \text{if } e \in C \\ 0 & \text{otherwise} \end{cases}$$

define the components of a vertex $x = \{x_e\} \in VERT(P)$. Conversely, any vertex of $P$ is the characteristic vector of some simple circuit $C$, normalized by the length of $C$. Letting $n = |V|$ and $m = |E|$, we can write $P = \{x \in \Re^m \mid Ax = b,\ x \geq 0\}$, where $b = (0, \ldots, 0, 1)^t \in \Re^n$ and where $A$ is the $m \times n$ matrix obtained by replacing the last row of the incidence matrix of $G$ by the vector of all ones. We thus conclude that if $P = \{x \in \Re^m \mid Ax = b,\ x \geq 0\}$ is the flow circulation polytope for a digraph $G$, then the set family $MIN(\mathcal{K}, b)$ can be identified with the set of all simple circuits in $G$. It is well known [14] that uniform sampling from the set of simple circuits of a digraph is a hard computational problem. More precisely, a randomized algorithm $\mathcal{G}$ is said to be an *$\alpha$-uniform generator* for a finite set $\Omega$ if

- $\mathcal{G}$ outputs only elements $\omega \in \Omega$, unless it stops with no output;
- $\sum\{p(\omega) | \omega \in \Omega\} \geq 1/2$, where $p(\omega)$ is the probability that $\mathcal{G}$ outputs $\omega \in \Omega$;
- $\max\{\ p(\omega)/p(\omega') \mid \omega, \omega'\} \leq \alpha$.

It is trivial to construct a polynomial-time $2^m$-uniform generator for the vertex set of any rational polytope $P = \{x \in \Re^m | Ax = b,\ x \geq 0\}$ in $\Re^m$. For instance, this can be done as follows:

1) Deterministically compute a vertex $v$ of $P$;

2) Pick uniformly at random one of the $2^m$ subsets $I \subseteq \{1, \ldots, m\}$;

3) If the linear system

$$Ax = b, \quad x_i = 0,\ i \in I$$

defines a vertex $x$ of $P$, output $x$. Otherwise, output $v$.

On the other hand, the amplification technique of [14] leads to the following result:

**Theorem 1.** *Let $\alpha = 2^{m^{1-\varepsilon}}$, where $\varepsilon$ is a positive constant. Suppose there exists a randomized polynomial-time algorithm which, given a rational polytope $P = \{x \in \Re^m | Ax = b,\ x \geq 0\}$, acts as an $\alpha$-uniform generator on $VERT(P)$. Then any $NP$-complete problem can be solved in polynomial time by a randomized algorithm with arbitrarily small one-sided failure probability.*

**Proof.** Let $G = (V, E)$ be a given digraph with $n = |V|$ vertices and $m = |E|$ arcs, and let $l$ be a positive integer. Following [14], consider the directed multigraph $G_l$ obtained by replacing each arc $u \to v$ of $G$ by the chain $u \rightrightarrows \bullet \rightrightarrows \bullet \ldots \bullet \rightrightarrows v$ of $l$ pairs of parallel arcs (we can turn $G_l$ into a digraph by placing an auxiliary vertex on each of its $2ml$ arcs). Since each arc of the original digraph can be traversed in $2^l$ different ways in $G_l$, a circuit $C$ in $G$ generates $2^{|C|l}$ circuits in $G_l$. In particular, if $G$ is Hamiltonian, then $G_l$ contains at least $2^{nl}$ circuits of length $nl$. On the other hand, since $G$ has $m$ arcs, the total number of circuits in $G$ cannot exceed $2^m$, and hence at most $2^{m+(n-1)l}$ circuits in $G_l$ are not Hamiltonian. Assume that $G$ is Hamiltonian and let $l$ be the minimum integer such that $2^{nl} \geq \alpha 2^{m+(n-1)l}$. Then, with probability at least 1/2, each application of an $\alpha$-uniform generator to the circuit set of $G_l$ yields a Hamiltonian circuit. However, for $\alpha = 2^{(2ml)^{1-\varepsilon}}$, the condition $2^{nl} \geq \alpha 2^{m+(n-1)l}$ is equivalent to the inequality $l \geq (2ml)^{1-\varepsilon} + m$. The latter inequality can be satisfied with $l = (3m)^{1/\varepsilon}$. This implies that for any fixed $\varepsilon$, the transformation $G \to G_l$ can be carried out in polynomial time, which completes the proof of the theorem.

Theorem 1 shows that efficient algorithms for generating random elements of $MIN(\mathcal{K}, b)$ are unlikely to exists. In view of this fact, it is natural to ask whether all elements of $MIN(\mathcal{K}, b)$ can be enumerated by an *output-polynomial* algorithm, i.e., in time polynomial in $|MIN(\mathcal{K}, b)|$ and the size of the input representation of $\mathcal{K}, b$. (Note that $|MIN(\mathcal{K}, b)|$ may be exponential in $n$ and $|\mathcal{K}|$.) For the special case where $\mathcal{K}$ is a system of rays, we arrive at the well-known question:

> *Can the vertex set of a given polytope* $P = \{x \in \Re^m \mid Ax = b,\ x \geq 0\}$ *be constructed in time bounded by a polynomial in* $|VERT(P)|$ *and the input size of* $A, b$?

A polynomially-equivalent decision version of the same problem is as follows:

> *Given two rational polytopes* $P = \{x \in \Re^m \mid Ax = b,\ x \geq 0\}$ *and* $P' = con.hull\{v_1, \ldots, v_k\}$, *can we check in polynomial time whether the two input representations are equivalent:* $P = P'$?

Even thought the above questions can be answered in the affirmative for some classes of polytopes, including simple and 0,1-polytopes (see e.g. [1] and [8]), the existence of an output-polynomial algorithm for enumerating all vertices of an arbitrary rational polytope remains an important open problem in linear programming. However, for more general systems of cones we can show the following negative result.

**Theorem 2.** *Suppose that each cone in* $\mathcal{K} = \{K_1, \ldots, K_m\}$ *is given by two generators, i.e.,* $K_i = cone(a_i, a'_i)$, *where* $a_i$ *and* $a'_i$ *are given rational* $n$*-vectors,* $i = 1, \ldots, m$. *Then, given a rational vector* $b$ *and a collection of sets* $\mathcal{M} \subseteq MIN(\mathcal{K}, b)$, *it is coNP-complete to decide whether the given collection is exhaustive:* $\mathcal{M} = MIN(\mathcal{K}, b)$.

**Proof.** We use a polynomial-time transformation from the following well-known coNP-complete problem:

($\mathcal{D}$): *Given a disjunctive normal form* $f(X_1, \ldots, X_N) = C_1(X_1, \ldots, X)\vee \ldots \vee C_M(X_1, \ldots, X_N)$, *test whether the given form is tautological, i.e.,* $f(X) \equiv 1$ *for all* $X = (X_1, \ldots, X_N) \in \{0,1\}^N$.

In the sequel we shall assume without loss of generality that $M \geq 2$.

The first step of our transformation is a modification of the transformation used in [5] to reduce ($\mathcal{D}$) to the enumeration of so-called relay paths in a relay circuit with two terminals. We start by replacing each pair of literals $X_i$ and $\neg X_i$ in each conjunction $C_j$ by new binary variables $Y_{ij}$ and $Z_{ij}$, respectively. This gives a monotone disjunctive normal form $g(Y,Z) = C_1(Y,Z) \vee \ldots \vee C_m(Y,Z)$ such that

$$g(Y,Z) \equiv f(X) \quad \text{for } Y_{ij} = \neg Z_{ij} = X_i, \ i = 1, \ldots, N, \ j = 1, \ldots, M,$$

where $Y = (Y_{ij}) \in \{0,1\}^{NM}$ and $Z = (Z_{ij}) \in \{0,1\}^{NM}$. Then we construct the following *relay circuit* $\mathcal{G}$ with two terminals $s$ and $t$:

1) Connect $s$ to $t$ by $M$ disjoint directed paths $L_1, \ldots, L_m$, where the length of $L_j$ is the number of literals in $C_j(Y,Z)$, $j = 1, \ldots, m$. Label the $k$-th arc in $L_j$ by the $k$-th literal in $C_j(Y,Z)$, $k = 1, \ldots, |L_i|$.

2) In addition to the $M$ paths above, connect $s$ to $t$ by the chain of the form

$$s < \begin{matrix} Y_{11} \to Y_{12} \to \cdots \to Y_{1M} \\ Z_{11} \to Z_{12} \to \cdots \to Z_{1M} \end{matrix} > \bullet < \begin{matrix} Y_{21} \to Y_{22} \to \cdots \to Y_{2M} \\ Z_{21} \to Z_{22} \to \cdots \to Z_{2M} \end{matrix} > \bullet$$

$$< \cdots > \bullet < \begin{matrix} Y_{N1} \to Y_{N2} \to \cdots \to Y_{NM} \\ Z_{N1} \to Z_{N2} \to \cdots \to Z_{NM} \end{matrix} > t$$

Note that each of the literals $Y_{11}, \ldots, Y_{NM}, Z_{11}, \ldots, Z_{NM}$ is assigned to at most two of the $2NM + \sum_{j=1}^{M} |C_j|$ arcs of $\mathcal{G}$. By definition, a *relay s-t path* in $\mathcal{G}$ is a minimal collection of literals $\mathcal{L} \subseteq \{Y_{11}, \ldots, Y_{NM}, Z_{11}, \ldots, Z_{NM}\}$ such that $\mathcal{L}$ connects $s$ and $t$. Clearly, the set $\mathcal{L}_j$ of literals in any conjunction $C_j(Y,Z)$, $j = 1, \ldots, M$, forms a relay path in $\mathcal{G}$. Furthermore, it is not hard to see that if $\mathcal{L}$ is any other relay path in $G$, then

(i) For each $i = 1, \ldots, N$, the path contains either all $M$ literals $Y_{i1}, Y_{i2}, \ldots, Y_{iM}$, or all $M$ literals $Z_{i1}, Z_{i2}, \ldots, Z_{iM}$, but not both, and

(ii) Letting $X^* = (X_1^*, \ldots, X_N^*)$, where

$$X_i^* = \begin{cases} 1 & \text{if } Y_{i1}, Y_{i2}, \ldots, Y_{iM} \in \mathcal{L} \\ 0 & \text{if } Z_{i1}, Z_{i2}, \ldots, Z_{iM} \in \mathcal{L}, \end{cases} \tag{5.1}$$

we obtain $f(X^*) = 0$.

Conversely, for any vector $X^* \in \{0,1\}^N$ such that $f(X^*) = 0$, the set of $NM$ literals

$$\mathcal{L} = \cup\{ Y_{ij} \mid X_i^* = 1,\ j = 1, \ldots, M\} \bigcup \cup\{ Z_{ij} \mid X_i^* = 0,\ j = 1, \ldots, M\}$$

is a relay $s$-$t$ path in $\mathcal{G}$ and $\mathcal{L} \neq \mathcal{L}_j$, $j = 1, \ldots, m$. We thus conclude that

*It is coNP-complete to determine whether the set $\{\mathcal{L}_1, \ldots, \mathcal{L}_M\}$ contains all possible relay $s$-$t$ paths in $\mathcal{G}$.*

Let us now erase all the literals assigned to the arcs of $\mathcal{G}$ and consider the resulting digraph $G = (V, E)$ with $|E| = 2N + \sum_{i=1}^M |C_i|$ arcs. Let $P$ be the polytope of unit $s$-$t$ flows on $G$:

$$\sum_{e \text{ incident to } v} x_e = \sum_{e \text{ incident from } v} x_e \quad \text{for all } v \in V \setminus \{s, t\},$$

$$\sum_{e \text{ incident from s}} x_e = 1, \qquad x_e \geq 0 \quad \text{for all } e \in E.$$

We can write $P = \{x \in \Re^E \mid \sum_{e \in E} a_e x_e = b,\ x \geq 0\}$, where $x_e \geq 0$ is the flow on arc $e$ and $a_e$ is the incidence vector of $e$. Let $\mathcal{R} = \{Y_{11}, \ldots, Y_{NM}, Z_{11}, \ldots, Z_{NM}\}$ be the set of all literals in $\mathcal{G}$. Recalling that each arc $e \in E$ is assigned a literal $R = R(e) \in \mathcal{R}$, consider the family $\mathcal{K}$ of $|\mathcal{R}| = 2NM$ cones $K_R = cone(\ a_e \mid R(e) = R)$, where $R \in \mathcal{R}$. Since each literal $R$ is assigned to a maximum of two arcs in $G$, each cone $K_R \in \mathcal{K}$ is generated by at most two $\{0, \pm 1\}$-vectors $a_e$. Moreover, it is easy to see that

*A set $\mathcal{L} \subseteq \mathcal{R}$ is a relay $s$-$t$ path if and only if $\mathcal{L} \in MIN(\mathcal{K}, b)$, i.e., $\sum_{R \in \mathcal{L}} K_R$ is a minimal cone containing $b$.*

We thus conclude that generating $MIN(\mathcal{K}, b)$ is as hard as enumerating relay paths in $\mathcal{G}$. This completes the proof of the theorem.

We mention, in passing, that the conic system $\{K_R,\ R \in \mathcal{R}\}$ constructed in the proof of Theorem 2 consists of pairwise disjoint cones whose $\{0, \pm 1\}$ generators $a_e$ all belong to a common open half-space. In addition, dropping the nonnegativity conditions $x_e \geq 0$ in the description of the polytope of unit $s$-$t$ flows, one can easily modify the above proof to show that Theorem 2 also holds for 2-dimensional linear subspaces:

> *Given a set $\mathcal{L}$ of two-dimensional linear spaces $L_i = lin.hull(a_i, a_i') \subseteq \Re^n$, $i = 1, \ldots, m$, and a target vector $b \in \Re^n$, it is coNP-complete to decide whether $\mathcal{M} = MIN(\mathcal{L}, b)$, where $MIN(\mathcal{L}, b)$ is the family of all minimal subsets of $\mathcal{L}$ whose direct sum contains $b$ and $\mathcal{M}$ is an explicitly given subset of $MIN(\mathcal{L}, b)$.*

Theorem 2 shows that unless $P = NP$, the construction of $MIN(\mathcal{K}, b)$ cannot be realized in output-polynomial time already for systems of dihedral cones. Out next goal is to show that this negative result also holds for the family $MAX(\mathcal{K}, b)$ of all maximal sets $\mathcal{S} \subseteq \mathcal{K}$ such that $b \notin \sum_{K \in \mathcal{S}} K$.

**Theorem 3.** *Given a system of rational dihedral cones $K_i = cone(a_i, a_i')$, $i = 1, \ldots, m$, a rational vector $b$, and a collection $\mathcal{M} \subseteq MAX(\mathcal{K}, b)$, it is coNP-complete to determine whether $\mathcal{M} = MIN(\mathcal{K}, b)$.*

**Proof.** We use the transformation dual to that used in the proof of Theorem 2. Consider the coNP-complete version of the classical satisfiability problem: *Test whether a given conjunctive normal form $f(X_1, \ldots, X_N) = D_1(X_1, \ldots, X_) \wedge \ldots \wedge D_M(X_1, \ldots, X_N)$ is not satisfiable, i.e., $f(X) \equiv 0$ for all $X = (X_1, \ldots, X_N) \in \{0, 1\}^N$.* We reduce this decision problem to the enumeration of *relay cuts* in $\mathcal{H} = \mathcal{G}^d$, the circuit dual to the circuit $\mathcal{G}$ constructed in the proof of Theorem 2. To this end, we first replace each occurrence of literals $X_i$ or $\neg X_i$ in each disjunction $D_j$ by new variables $Y_{ij}$ and $Z_{ij}$. This gives a monotone conjunctive normal form $g(Y, Z) = D_1(Y, Z) \wedge \ldots \vee D_m(Y, Z)$ such that $g(Y, Z) \equiv f(y)$ for $Y_{ij} = \neg Z_{ij} = X_i$. Then we realize $g(Y, Z)$ by the series-parallel relay circuit $\mathcal{H}$ of the form $s \to \mathcal{D}_1 \to \ldots \to \mathcal{D}_M \to \mathcal{C} \to t$, where

> $\mathcal{D}_j$ consists of $|D_j(Y, Z)|$ parallel arcs labeled by the variables occuring in disjunction $D_j(Y, Z)$, and
>
> $\mathcal{C}$ is built of $N$ parallel blocks of the form $Y_i \to Z_i$, where $Y_i$ and $Z_i$ are the collections of $N$ parallel arcs $Y_{i1}, \ldots, Y_{iM}$ and $Z_{i1}, \ldots, Z_{iM}$, respectively.

Note that each literal $R \in \{Y_{11}, \ldots, Y_{NM}, Z_{11}, \ldots, Z_{NM}\}$ still occurs in $\mathcal{H}$ at most twice. A *relay s-t cut* in $\mathcal{H}$ is a minimal collection of literals

$\mathcal{L} \subseteq \{Y_{11}, \ldots, Y_{NM}, Z_{11}, \ldots, Z_{NM}\}$ whose removal disconnects $s$ and $t$. Clearly, the set $\mathcal{L}_j$ of literals in any disjunction $D_j(Y, Z)$, $j = 1, \ldots, M$, is a relay $s$-$t$ cut in $\mathcal{H}$. Let $\mathcal{L}$ be any other relay $s$-$t$ cut in $G$, then for each $i = 1, \ldots, N$, the cut contains either all $M$ literals $Y_{i1}, Y_{i2}, \ldots, Y_{iM}$, or all $M$ literals $Z_{i1}, Z_{i2}, \ldots, Z_{iM}$, but not both, and (5.1) gives a binary vector $X^*$ such that $f(X^*) = 1$. The converse is also true: any binary vector $X^*$ satisfying $f(X^*) = 1$ yields a relay $s$-$t$ cut $\mathcal{L} \neq \mathcal{L}_j$, $i = j, \ldots, M$. This implies that *determining whether* $\{\mathcal{L}_1, \ldots, \mathcal{L}_M\}$ *are all possible relay s-t cuts in* $\mathcal{H}$ *is coNP-complete.*

To finish the proof of the theorem, consider the polytope $P = \{x \in \Re^E \mid \sum_{e \in E} a_e x_e = b,\ x \geq 0\}$ of unit $s$-$t$ flows on $H$, the underlying digraph for $\mathcal{H}$. As in the proof of Theorem 2, define $K_R = cone(\ a_e \mid e$ labeled by $R)$, where $R$ is a literal in $\mathcal{R} \doteq \{Y_{11}, \ldots, Y_{NM}, Z_{11}, \ldots, Z_{NM}\}$. Then each cone $K_R$ is generated by at most two vectors $a_e$, and it is easy to see that *a set* $\mathcal{L} \subseteq \mathcal{R}$ *is a relay s-t cut in* $\mathcal{H}$ *if and only if* $\mathcal{R} \setminus \mathcal{L} \in MAX(\mathcal{K}, b)$. The proof of Theorem 3 is completed.

Note that a simple modification of the above prove shows that Theorem 3 is valid for 2-dimensional linear spaces:

> *Given a family* $\mathcal{L}$ *of two-dimensional linear spaces* $L_i = lin.hull(a_i, a_i')$ $\subseteq \Re^n$, $i = 1, \ldots, m$, *and a target vector* $b \in \Re^n$, *it is coNP-complete to tell whether* $\mathcal{M} = MAX(\mathcal{L}, b)$ *for a given set* $\mathcal{M} \subseteq MAX(\mathcal{L}, b)$.

## 2 TRANSVERSAL HYPERGRAPHS AND JOINT GENERATION OF $MIN(K, B)$ AND $MAX(K, B)$

Even though constructing $MIN(\mathcal{K}, b)$ and $MAX(\mathcal{K}, b)$ is NP-hard, enumerating all elements of $MIN(\mathcal{K}, b) \cup MAX(\mathcal{K}, b)$ is equivalent to a simpler combinatorial problem known as *TRANSVERSAL HYPERGRAPH* or *HYPERGRAPH DUALIZATION* (See e.g. [10, 11, 3, 6, 13, 12].)

Let $V$ be a finite set, and let $\mathcal{E} \subseteq 2^V$ be a set family (i.e., hypergraph) on $V$. A set $X \subseteq V$ is called a *transversal* to $\mathcal{E}$ if $X$ has a nonempty intersection with each set $E \in \mathcal{E}$. The hypergraph $\mathcal{E}^d$ consisting of all the minimal transversals of $\mathcal{E}$ is called the *transversal* or *dual* hypergraph of $\mathcal{E}$. It is easy to see that $\mathcal{E}^d$ is a Sperner hypergraph , i.e., $E \not\subseteq E'$ for any two distinct sets $E, E' \in \mathcal{E}^d$. If $\mathcal{E}$ is also Sperner then $(\mathcal{E}^d)^d = \mathcal{E}$. For instance, $MIN(\mathcal{K}, b)$ and the complementarity family $\{\mathcal{S} \subseteq \mathcal{K} \mid \mathcal{K} \setminus \mathcal{S} \in MAX(\mathcal{K}, b)\}$ are mutually transversal Sperner hypergraphs on $\mathcal{K}$. The transversal hypergraph problem can be stated as follows:

TRANSVERSAL HYPERGRAPH: *Given a complete list of all sets in $\mathcal{E}$ and a subfamily of minimal transversals $\mathcal{M} \subseteq \mathcal{E}^d$, either prove that $\mathcal{M} = \mathcal{E}^d$, or find a new transversal $X \in \mathcal{E}^d \setminus \mathcal{M}$.*

Note that all the minimal transversals to $\mathcal{E}$ can be enumerated by initializing $\mathcal{M} = \emptyset$ and iteratively solving the transversal hypergraph problem $|\mathcal{E}^d| + 1$ times. In particular, we say that the transversal hypergraph $\mathcal{E}^d$ can be constructed in *incremental polynomial time* if the above problem can be solved in time bounded by a polynomial in $|V|, |\mathcal{E}|$ and $\mathcal{M}|$. Clearly, any algorithm that constructs $\mathcal{E}^d$ in incremental polynomial time is also output-polynomial.

It is not known whether the transversal hypergraph problem can be solved in polynomial time for arbitrary input hypergraphs. This problem, however, is very unlikely to be NP-hard because it can be solved in *quasi-polynomial time* $O(|V|m) + m^{o(\log m)}$, where $m = |\mathcal{E}| + |\mathcal{F}|$ (See [11].) In particular, $\mathcal{E}^d$ can always be constructed in output-quasi-polynomial time $(|\mathcal{E}| + |\mathcal{E}^d|)^{o((|\mathcal{E}|+|\mathcal{E}^d|))}$. In addition, many classes of hypergraphs can be dualized in incremental polynomial time. These include hypergraphs of bounded dimension ($\max\{|E| \mid E \in \mathcal{E}\} \leq \text{const}$), 2-monotonic, threshold, matroid, read-bounded, acyclic and some other hypergraphs, see e.g. [2, 5, 9, 10, 16]. For hypergraphs of bounded dimension, the transversal hypergraph problem can also be solved efficiently in parallel [4].

The transversal hypergraph problem can be easily embedded in the problem of constructing $MIN(\mathcal{K}, b) \cup MAX(\mathcal{K}, b)$ for an appropriate system of polyhedral cones $\mathcal{K}$. This can be done as follows. Given a Sperner hypergraph $\mathcal{E} \in 2^V$, consider the family $\mathcal{K}$ of $|V|$ cones

$$K(v) = cone\,(e_E \mid E \in \mathcal{E},\ E \text{ contains } v), \quad v \in V,$$

where $e_E = (0, \ldots, 1, \ldots, 0)$ is the $E$-th coordinate vector in $\Re^{\mathcal{E}}$. Let $b = (1, \ldots, 1) \in \Re^{\mathcal{E}}$ be the vector of all ones. Then constructing $MAX(\mathcal{K}, b)$ is trivial, since each set $S \in MAX(\mathcal{K}, b)$ has the form $S = \{K(v) \mid v \in V \setminus E\}$. On the other hand, constructing $MIN(\mathcal{K}, b)$ is easily seen to be identical to the enumeration of all the minimal transversals to $\mathcal{E}$.

Theorem 4 below asserts that under very mild conditions (always satisfied for rational polyhedral cones), constructing $MIN(\mathcal{K}, b) \cup MAX(\mathcal{K}, b)$ can in turn be reduced to the transversal hypergraph problem.

**Theorem 4.** *Consider the following problem:*

($\mathcal{J}$) *Given a conic system $\mathcal{K}$, a target vector $b$, and two explicitly listed set families $\mathcal{M} \subseteq MIN(\mathcal{K}, b)$ and $\mathcal{N} \subseteq MAX(\mathcal{K}, b)$, either find a new set $\mathcal{S} \in [MIN(\mathcal{K}, b) \setminus \mathcal{M}] \cup [MAX(\mathcal{K}, b) \setminus \mathcal{N}]$, or show that no such set exists.*

*Suppose that*

(*) *For any input set $\mathcal{S} \subseteq \mathcal{K}$, we can determine in polynomial time whether or not $b \in \sum_{K \in \mathcal{S}} K$.*

*Then problem $\mathcal{J}$ can be reduced in polynomial time to TRANSVERSAL HYPERGRAPH. .*

Note that condition (*) is clearly satisfied for rational $b$ and families of rational polyhedral cones defined by any number of extremal directions and/or linear inequalities. In particular, we conclude that for such conic families, all elements of $MIN(\mathcal{K}, b) \cup MAX(\mathcal{K}, b)$ can be enumerated in incremental quasi-polynomial time. Specifically, for each $k \leq |MIN(\mathcal{K}, b) \cup MAX(\mathcal{K}, b)|$, we can find $k$ elements of $MIN(\mathcal{K}, b) \cup MAX(\mathcal{K}, b)$ in $poly(size(\mathcal{K}, b)) + k^{o(\log k)}$ time. Theorem 4 also implies that in fixed dimension, all elements of $MAX(\mathcal{K}, b)$ can be constructed in incremental polynomial time.

Theorem 4 is a special case of the following general reduction [3, 12, 13]:

> *Let $f : \{0, 1\}^m \to \{0, 1\}$ be a monotone Boolean function whose value at any point $x \in \{0, 1\}^m$ can be evaluated in polynomial time. Then constructing the set of all minimal true and maximal false vectors of $f$ reduces in polynomial time to the transversal hypergraph problem.*

We obtain Theorem 4 by letting

$$f(x) = \begin{cases} 1, & \text{if } b \in \sum_{i=1}^{m} x_i K_i \\ 0, & \text{otherwise} \end{cases}$$

## 3 OTHER NP-HARD TRANSVERSAL FAMILIES

Suppose that each cone $K \in \mathcal{K}$ is pointed, i.e., contains no line, and let $MAX(\mathcal{K}, L)$ be the family of all maximal subsets $\mathcal{S} \subseteq \mathcal{K}$ for which $\sum_{K \in \mathcal{S}} K$ is also a pointed cone. Let us further denote by $MIN(\mathcal{K}, L)$ all the minimal subsets $\mathcal{S} \subseteq \mathcal{K}$ such that $\sum_{K \in \mathcal{S}} K$ is not pointed. Theorem

5 below can be regarded as a homogeneous version of Theorems 3 and 4, and its proof easily follows from these theorems.

**Theorem 5.** *Given a system $\mathcal{K}$ of rational dihedral cones, it is coNP-complete to tell whether a given subfamily of $MIN(\mathcal{K}, L)$ (or $MAX(\mathcal{K}, L)$) is exhaustive.*

Note that, given a system of rational polyhedral cones $\mathcal{S} \subseteq \mathcal{K}$, we can decide in polynomial time whether the cone $\sum_{K \in \mathcal{S}} K$ is pointed. Hence, as in Section 3, we conclude that constructing $MIN(\mathcal{K}, L) \cup MAX(\mathcal{K}, L)$ is polynomially equivalent to the transversal hypergraph problem.

One more example of transversal NP-hard conic families which, however, are jointly reducible to TRANSVERSAL HYPERGRAPH is as follows. Assume that $\sum_{K \in \mathcal{K}} K = \Re^n$, and let $MIN(\mathcal{K}, \Re^n)$ be all the minimal subsystems $\mathcal{S} \subseteq \mathcal{K}$ that generate the entire space: $\sum_{K \in \mathcal{S}} K = \Re^n$. Let $MAX(\mathcal{K}, \Re^n)$ denote all the maximal subsystems $\mathcal{S} \subseteq \mathcal{K}$ for which $\sum_{K \in \mathcal{S}} K \neq \Re^n$ (equivalently, for which all cones $K \in \mathcal{S}$ are contained in a closed half-space).

**Theorem 6.** *Assume that each cone $K \in \mathcal{K}$ is represented by a list of rational generators. Then the problems of determining the completeness of a given subset in $MIN(\mathcal{K}, \Re^n)$ or in $MAX(\mathcal{K}, \Re^n)$ are both coNP-complete.*

Again, generating all elements of $MIN(\mathcal{K}, \Re^n) \cup MAX(\mathcal{K}, \Re^n)$ is poly-time reducible to the transversal hypergraph problem. This is because given a set $\mathcal{S} \subseteq K$, we can check in polynomial time whether $\sum_{K \in \mathcal{S}} K = \Re^n$.

Our final example of NP-hard transversal geometric families deals with a system $\mathcal{P}$ of $m$ polytopes $P_i = \{x \in \Re^n \mid A_i x \leq b_i\}$ such that $\cap_{i=1}^m P_i = \emptyset$. Let $MIN(\mathcal{P})$ be the Helly sybsystems of $\mathcal{P}$, i.e., $\mathcal{S} \in MIN(\mathcal{P})$ if $S$ is a minimal infeasible subsystem of $\mathcal{P}$. Next, let $MAX(\mathcal{P})$ denote the collection of all the maximal feasible subsystems of $\mathcal{P}$.

**Theorem 7** [13]. *It is coNP-complete to determine whether a given list of Helly's systems is exhaustive. Determining whether $\mathcal{M} = MAX(\mathcal{P})$ is also coNP-complete.*

Needless to say that as before, the generation of $MIN(\mathcal{P}) \cup MAX(\mathcal{P})$ reduces to the transversal hypergraph problem and hence can be carried out in incremental quasi-polynomial time. Note also that Theorem 6 can be obtained from Theorem 7 by considering the polar cones for $K_i = \{(x, x_{n+1}) \in \Re^{n+1} \mid A_i x \leq x_{n+1} b_i,\ x_{n+1} \geq 0\},\quad i = 1, \ldots, m.$

We close with an example of mutually transversal hypergraphs only one of which is hard to generate. Let $A$ and $b$ be non-negative $n \times m$ matrix and $n$-vector, respectively. Denote by $MIN(A, b)$ the family of all minimal solutions to the Boolean system of linear inequalities

$$Ax \geq b, \quad x \in \{0,1\}^m. \tag{5.2}$$

Let $MAX(A, b)$ be the maximal infeasible solutions to (5.2). The Boolean family $MIN(A, b)$ and the family of all complements to $MAX(A, b)$ can be regarded as a pair of transversal hypergraphs. Since the monotone Boolean function

$$f(x) = \begin{cases} 1, & \text{if } Ax \geq b \\ 0, & \text{otherwise} \end{cases}$$

is computable in polynomial time, we conclude that the generation of $MIN(A, b) \cup MAX(A, b)$ can be carried out in incremental quasi-polynomial time. As shown in [15], generating $MAX(A, b)$ is hard even for 0,1-matrices $A$: Given a subfamily $\mathcal{M} \subseteq MAX(A, b)$, it is coNP-complete to determine if $\mathcal{M} = MAX(A, b)$. However, it turns out [6] that the generation of $MIN(A, b)$ reduces in polynomial time to the transversal hypergraph problem and hence is very unlikely to be NP-hard. The latter reduction in based on the fact that the hypergraph $MIN(A, b)$ is dual-bounded: *for any Boolean system* (5.2) *we have the inequality*

$$|MAX(A, b)| \leq nm|MIN(A, b)|,$$

see [6] for more detail.

In particular, for each $k \leq |MIN(A, b)|$, we can find $k$ minimal Boolean solutions to (5.2) in $poly(size(A, b)) + (knm)^{o(\log knm)}$ time. Other examples of dual-bounded hypergraphs can be found in [7].

**Acknowledgment** The author is thankful to David Avis, Vladimir Gurvich, and Raimund Seidel for helpful discussions.

## References

[1] D. Avis, B. Bremner and R. Seidel, How good are convex hull algorithms. *Computational Geometry: Theory and Applications*, 7 (1997), 265-302.

[2] P. Bertolazzi and A. Sassano, An $O(mn)$ time algorithm for regular set-covering problems. *Theoretical Computer Science* 54 (1987) 237-247.

[3] J. C. Bioch and T. Ibaraki, Complexity of identification and dualization of positive Boolean functions. *Information and Computation* 123 (1995) 50-63.

[4] E. Boros, K. Elbassioni, V. Gurvich and L. Khachiyan, An incremental RNC algorithm for generating all maximal independent sets in hypergraphs of bounded dimension. DIMACS Technical Report 2000-21, August 2000, *http://dimacs.rutgers.edu/TechnicalReports/2000.html*, to appear in *Parallel Processing Letters.*

[5] E. Boros, P. L. Hammer, T. Ibaraki and K. Kawakami, Polynomial time recognition of 2-monotonic positive Boolean functions given by an oracle. *SIAM Journal on Computing* 26 (1997) 93-109.

[6] E. Boros, V. Gurvich, L. Khachiyan, and K. Makino, Generating partial and multiple transversals in a hypergraph. In: *Proceedings of the 27th International Colloquium on Automata, Languages and Programming (ICALP)*, (U. Montarani, J.D.P. Rolim and E. Welzl, eds.) Lecture Notes in Computer Science, 1853, 588-599, Springer Verlag, Berlin, Heidelberg, New York, 2000.

[7] E. Boros, V. Gurvich, L. Khachiyan, and K. Makino, Generating weighted transversals in a hypergraph. DIMACS Technical Report 2000-17, June 2000, *http://dimacs.rutgers.edu/TechnicalReports/2000.html.*

[8] M. R. Bussieck and M. E. Lübbecke, The vertex set of a 0/1 polytope is strongly $\mathcal{P}$-enumerable. *Computational Theory and Applications* 11(2) (1998), 103-109.

[9] Y. Crama, Dualization of regular Boolean functions. *Discrete Applied Mathematics*, 16 (1987) 79-85.

[10] T. Eiter and G. Gottlob, Identifying the minimal transversals of a hypergraph and related problems. *SIAM Journal on Computing*, 24 (1995) 1278-1304.

[11] M. L. Fredman and L. Khachiyan, On the complexity of dualization of monotone disjunctive normal forms. *Journal of Algorithms*, 21 (1996) 618-628.

[12] D. Gunopulos, R. Khardon, H. Manilla and H. Toivonen, Data mining, hypergraph transversals and machine learning. In: *Proceedings of the 16th ACM-SIGACT-SIGMOD-SIGART Symposium on Principles of Database Systems* 1997, 12-15.

[13] V. Gurvich and L. Khachiyan, On generating the irredundant conjunctive and disjunctive normal forms of monotone Boolean func-

tions. *Discrete Applied Mathematics*, 1996-97, issue 1-3, (1999) 363-373.

[14] M. R. Jerrum, L. G. Valiant and V. V. Vazirani, Random generation of combinatorial structures from a uniform distribution. *Theoretical Computer Science* 43 (1986) 169-188.

[15] K. Makino and T. Ibaraki, Interior and exterior functions of Boolean functions. *Discrete Applied Mathematics*, 69 (1996) 209-231.

[16] U. N. Peled and B. Simeone, An $O(nm)$-time algorithm for computing the dual of a regular Boolean function. *Discrete Applied Mathematics* 49 (1994) 309-323.

Chapter 6

# SDP RELAXATIONS IN COMBINATORIAL OPTIMIZATION FROM A LAGRANGIAN VIEWPOINT

Claude Lemaréchal, François Oustry
*INRIA, 655 avenue de l'Europe, Montbonnot, 38330 Saint Ismier, France*
{claude.lemarechal, francois.oustry}@inrialpes.fr

**Abstract** A successful technique for some problems in combinatorial optimization is the so-called SDP relaxation, essentially due to L. Lovász, and much developed by M. Goemans and D.P. Williamson. As observed by S. Poljak, F. Rendl and H. Wolkowicz, this technique can be interpreted from the point of view of Lagrangian duality. A central tool for this is dualization of quadratic constraints, an operation pioneered by N.Z. Shor. We synthesize these various operations, in a language close to that of nonlinear programming. Then we show how the approach can be applied to general combinatorial problems.

**Keywords:** Combinatorial optimization, stable set, max cut, Lagrangian relaxation, SDP relaxation, SDP duality, duality, duality gap, quadratic constraints.

## 1 INTRODUCTION

Let us symbolize a combinatorial problem by

$$\min f(x), \quad x \in S;$$

usually $f$ is linear, sometimes quadratic; while $S$ is a (finite but) "nasty" set, which can be fruitfully viewed as a subset of $R^n$. One of the fundamental tools to solve such a problem is *relaxation*. It consists in replacing $S$ by a *larger* set, say $S'$, more amenable (closed convex). As a result, the optimal value $v'$ of $\min_{x \in S'} f(x)$ underestimates the optimal value of the original problem.

**Remark 1** *This bounding operation is to be understood within a branch and bound process. It goes with the companion operation that extracts a*



*N. Hadjisavvas and P.M. Pardalos (eds.), Advances in Convex Analysis and Global Optimization,* 119–134.

*smaller portion from the feasible set (branching). With some luck, the lower bound $v'$ will be high enough to guarantee that no solution can belong to the extracted portion – a precious information.* •

The most common relaxation is when $f$ is linear and $S$ is defined with the help of affine constraints as follows:

$$\min b^\top x\,, \quad Ax \leq a\,, \quad x \in D\,. \tag{6.1}$$

Here $D$ is (finite but) amenable, in the sense that minimizing a linear function over $D$ is easy. Lagrangian relaxation then suggests itself: by assumption, solving

$$\min_{x \in D} b^\top x + u^\top (Ax - a)$$

for fixed $u \geq 0$ is easy. This provides an underestimate of the optimal value. The best such underestimate, given by duality, amounts to replacing $D$ by its convex hull (see [3, 7], or also [10, Cor. XII.5.1.4]):

$$S := \{x \in D : Ax \leq a\}\,, \quad S' = \{x \in \operatorname{co} D : Ax \leq a\}\,.$$

It can be seen that $S' \supset \operatorname{co} S \supset S$, so the present operation is a relaxation indeed, which introduces the possible duality gap $v - v' \geq 0$.

**Scope of the paper.** A more recent technique, called SDP relaxation, has been introduced for some particular combinatorial problems: max-stable and max-cut; this is the subject of §§ 2.1, 2.2. In §2.3, we suggest a general "recipe" (terminology of [23]) putting SDP relaxation in the perspective of general optimization problems with quadratic objective and constraints. SDP relaxation turns out to be a standard application of Lagrangian duality, and this is shown in §3 Finally, we come back in §4 to general optimization problems, showing how new SDP schemes can be proposed, as valid alternatives to standard Lagrangian relaxation.

**Notation.** Here is a glossary of some symbols used in this paper.

- $R^n$ is equipped with the standard dot-product $x^\top y = \sum_i x_i y_i$;
- $e = (1, \ldots, 1)^\top$ is the vector of all 1's (say in $R^n$);
- $\mathcal{S}^n$ is the set of symmetric $n \times n$ matrices; it is equipped with the standard scalar product

$$\langle M, N \rangle := \sum_{i,j=1}^{n} M_{ij} N_{ij}\,;$$

- $M \succ 0$ [resp. $M \succeq 0$] denotes a positive [resp. semi-]definite matrix $M \in \mathcal{S}^n$;

- we also write $M \in \mathcal{S}^n_{++}$ [resp. $\mathcal{S}^n_+$];
- $R^n_{++}$, $R^n_+$ denote the same operations with $n$-vectors;
- $D : R^n \to \mathcal{S}^n$ is the operator which makes a diagonal matrix with the coordinates of a vector;
- $d : \mathcal{S}^n \to R^n$ is its transpose, making a vector with the diagonal of a matrix
- $K^\circ := \{u : u^\top z \leq 0 \text{ for all } z \in K\}$ is the polar of the set $K$; it is always a closed convex cone.

# 2 THE SDP RELAXATION

First we introduce the technique of SDP relaxation for two known combinatorial problems. Then we give in §2.3 a "recipe" allowing an easy and mechanical development of SDP relaxation for general quadratic optimization problems.

## 2.1 MAX-STABLE

An important problem in combinatorial optimization is the so-called max-stable set. We formulate it with notation suited for our purpose:

$$\min b^\top x\,, \quad x_i x_j = 0\,,\ (i,j) \in E\,, \quad x_i^2 - x_i = 0\,,\ i = 1,\dots,n\,. \tag{6.2}$$

Here $b \in R^n$ is given, as well as $E \subset \{1,\dots,n\} \times \{1,\dots,n\}$. A relevant observation is that this problem has only quadratic constraints. It is not clear how to apply the standard Lagrangian relaxation alluded to in §1

**Remark 2** *Of course, the constraints $x_i^2 - x_i = 0$ just express that $x_i \in \{0,1\}$. Then observe that $x_i x_j = 0$ can equivalently be written $x_i + x_j \leq 0$. Thus, our problem could be formulated as*

$$\min b^\top x\,, \quad x_i + x_j \leq 0\,,\ (i,j) \in E\,, \quad x \text{ binary in } R^n\,,$$

*a problem in the general form (6.1), to which standard Lagrangian relaxation can be applied. This technique has never had good reputation, though: it is known to produce a rather wide duality gap.* •

This is why L. Lovász proposed in [19] a rather novel mechanism opening the way to *SDP relaxations*. The space $R^n$ of control variables is *lifted* to the space $\mathcal{S}^n$ of symmetric matrices, each scalar variable $x_i$ being transformed to an $n$-vector $X_i \in R^n$. The resulting $n$ vectors altogether form an $n \times n$ matrix – call it $X$, with generic term $X_{ij}$ –, imposed to be symmetric. The relaxed problem is then formulated in terms of the control variable $X$, via a constraint of positive definiteness:

(6.2) is transformed to (recall the diagonal operator $d$ defined in §1)

$$\min b^\top d(X)\,, \quad X_{ij}=0\,,\ (i,j)\in E\,, \quad X\succeq d(X)d(X)^\top\,. \qquad (6.3)$$

**Remark 3** *Although more convenient for our purpose, (6.3) is not the original proposal of [19]; it was rather given in [20]. See also [13] for various other equivalent formulations of the above Lovász' relaxation.* •

For each $x$ feasible in (6.2), set $X_{ij}(x) := x_i x_j$, producing $X(x) := xx^\top \in \mathcal{S}^n$. This $X(x)$ (a rank-one matrix) is obviously feasible in (6.3); observe in particular that $X(x) = d(X(x))d(X(x))^\top$; (6.3) is really a relaxation of (6.2). It can even be said that, if we imposed $X$ in (6.3) to have rank one, then we would just obtain an equivalent formulation of (6.2).

No matter what the motivation of Lovász was, §2.3 below will start from the above observation, allowing fruitful generalizations in §4

## 2.2 MAX-CUT

Another well-known combinatorial problem is max-cut, which we write

$$\min \sum_{i,j=1}^{n} Q_{ij}x_i x_j = x^\top Qx\,, \quad x_i^2=1,\ i=1,\dots,n\,. \qquad (6.4)$$

Here again, the quadratic functions make it difficult to apply standard Lagrangian relaxation. A slight simplification occurs with respect to (6.2), though: there are no linear terms in (6.4).

**Remark 4** *A technique does exist to transform a quadratic combinatorial problem into a linear one: setting $y_{ij} := x_i x_j$, the 0-1 values of $y$ can be simulated by imposing constraints such as $y_{ij}\le x_i$, $y_{ij}\le x_j$, $x_i+x_j\le 2y_{ij}$. However, techniques of this type have no good reputation either.* •

Now a Lovász-type relaxation of this problem was proposed in [8]. Lifting again the control variables from $\mathbb{R}^n$ to $\mathcal{S}^n$, the relaxed problem is now

$$\min\langle Q,X\rangle\,, \quad X\succeq 0\,, \quad d(X)=e\,. \qquad (6.5)$$

As before take $x$ feasible in (6.4) and set $X_{ij}(x) := x_i x_j$. The resulting matrix $X(x) := xx^\top$ is again feasible in (6.5). Besides observe the general relation $x^\top Qx = \left\langle Q, xx^\top\right\rangle$; hence $\langle Q, X(x)\rangle = x^\top Qx$, the two objective functions coincide. Conclusion again: (6.5) is a relaxation of (6.4), which would rather become an equivalent formulation if $X$ were imposed to have rank one.

Thus we have two examples of combinatorial problems for which relaxations have been defined, based on

- replacing $n$-vectors by symmetric $n \times n$ matrices,
- and introducing some positive definiteness.

These relaxations result in problems of semidefinite programming (SDP). The approach has been used for some other combinatorial problems; let us cite [12] for graph-coloring, [30] for graph partitioning, [31] for quadratic assignment, [9] for quadratic knapsack; more generally, see [22] and the review [21]. As testified by [6], the technique was even known fairly long ago in quantum chemistry.

In the next subsection we follow [23] to give a "recipe" allowing a systematic construction of such relaxations.

## 2.3 THE TENSORIAL LIFT FOR QUADRATIC OPTIMIZATION

It is not by chance that SDP relaxation was motivated by problems involving quadratic functions. In fact, this technique is intimately related to quadratic forms. Things turn out to be clearer if one thinks in general terms, so we consider

$$\left\{ \begin{array}{l} \inf x^\top Q_0 x + 2b_0^\top x + c_0 \,, \\ x^\top Q_k x + 2b_k^\top x + c_k = 0 \,, \quad k = 1, \ldots, m \,. \end{array} \right. \tag{6.6}$$

This problem could also have inequality constraints, the following "recipe" would work just the same:

- As already observed, a quadratic form $x^\top Qx$ can equivalently be written as $\left\langle Q, xx^\top \right\rangle$.
- Setting $X := xx^\top$, (6.6) can therefore equivalently be written

$$\left\{ \begin{array}{ll} \inf \langle Q_0, X\rangle + 2\langle b_0, x\rangle + c_0 \,, & (x, X) \in R^n \times \mathcal{S}^n \,, \\ \langle Q_k, X\rangle + 2\langle b_k, x\rangle + c_k = 0 \,, & k = 1, \ldots, m \,, \\ X = xx^\top \,. & \end{array} \right.$$

- In this last problem, everything is linear except the last constraint, which is nonconvex.
- *Relax* this last constraint $X = xx^\top$ to $X \succeq xx^\top$, which is now convex with respect to $(x, X)$ (in fact, it can be seen as the infinite system $c_v(x, X) \geq 0$ for each $v$, where the functions $c_v(x, X) := v^\top(X -$

$xx^\top)v = v^\top Xv - \|v^\top x\|^2$ are concave with respect to $(x, X)$). Needless to say, this widens the feasible domain.

In a word, the present recipe relaxes (6.6) to

$$\left\{\begin{array}{ll} \inf \langle Q_0, X\rangle + 2\langle b_0, x\rangle + c_0\,, & (x,X) \in R^n \times S^n\,, \\ \langle Q_k, X\rangle + 2\langle b_k, x\rangle + c_k = 0\,, & k = 1,\ldots,m\,, \\ X \succeq xx^\top\,. & \end{array}\right. \tag{6.7}$$

We mention here that the above relaxation essentially amounts to taking the convex hull of rank-one matrices in $S^n_+$: in $R^n \times S^n$,

$$\mathrm{co}\,\{(x,X) : X = xx^\top\} = \{(x,X) : X \succeq xx^\top\}\,;$$

see [16] for details. Such a lifting procedure can be qualified "tensorial": its key operation is to lift an element $x$ of the original space $H = R^n$ to the tensor from $H$ to $H$ defined by $v \mapsto (v^\top x)x$.

Let us illustrate the mechanism on the max-stable set problem (6.2):

- The objective function is linear and remains unchanged, namely to minimize the linear form $b^\top x$.
- For $k = (i,j) \in E$, the quadratic form $x_i x_j$ becomes the linear form $X_{ij}$; this was done directly in §2.1 but we can introduce the matrix $Q_k$ with all entries null except $(Q_k)_{ij} = (Q_k)_{ji} = 1$; then $x^\top Q_k x = 2x_i x_j$ and $\langle Q_k, X\rangle = 2X_{ij}$.
- For $k = i$, the quadratic functions $x_i^2 - x_i$ become likewise $X_{ii} - x_i$.
- In summary, (6.2) is transformed to (6.3), with the notation $x = d(X)$.

The illustration is just as easy for max-cut and is left to the reader. One comment, though: there is no linear term in (6.4), so $x$ in the lifted problem appears only in the constraint $X \succeq xx^\top$: we can set $x = 0$ without harm, which yields (6.5).

Now comes an important observation: via the trick $x \in \{0,1\} \Leftrightarrow x^2 = x$, any combinatorial problem can be formulated as a quadratically constrained optimization problem. Conclusion: any combinatorial problem is amenable to SDP relaxation (and even any polynomially constrained problem: adding extra variables and constraints, polynomials can be represented with the help of quadratic functions in a sufficiently larger space; see [25, 27] for example). This will be the content of §4 However, we start by giving to the above recipe the status of a real *method*, resulting from the application of a general methodology: Lagrangian duality.

**Remark 5** *Relating the tensorial lift with duality allows a deeper insight into this technique, hardly possible with the sole help of* (6.2) *or* (6.4). *In fact, the message of* §4 *will be that affine constraints need a special treatment, even though they are just particular cases of quadratic constraints.*

*This could be guessed neither from max-stable nor from max-cut, which have no affine constraints.* •

# 3 THE LAGRANGIAN POINT OF VIEW

The recipe of §2.3 allows a very easy derivation of the form (6.7) relaxing (6.6). It can be summarized as: introduce the new variable $X = xx^\top$ and bring convexity where necessary. The aim of the present section is to show that something really profound is involved.

## 3.1 ABSTRACT LAGRANGIAN RELAXATION

First we recall some basic principles. Consider an abstract optimization problem

$$\inf f(x), \quad g(x) = 0 \in R^m, \quad x \in D, \tag{6.8}$$

with associate Lagrangian and dual function

$$\begin{array}{lcl} D \times R^m \ni (x,u) & \mapsto & L(x,u) := f(x) + u^\top g(x), \\ R^m \ni u & \mapsto & \theta(u) := \inf_{x \in D} L(x,u). \end{array} \tag{6.9}$$

The dual function provides a relaxation of (6.8), in the sense that $\theta(u) \leq f(x)$ for all feasible $x$ in (6.8). Computing the best such relaxation is the dual problem $\sup_u \theta(u)$. The dual optimal value thus obtained is usually smaller than the optimal value of (6.8) (this is the duality gap): it corresponds to a certain convexification of (6.8), easy to characterize when $f$ and $g$ are linear – see §1 For general problem-functions, the characterization is much more intricated, we refer to [2, § III.1], [1, 4, 18].

**Remark 6** *At this point, it is important to understand that*

- (6.8) *may have several "primal-equivalent" formulations, which give the same optimal value (and possibly the same solution-set);*
- *but these formulations may result in different Lagrangians, and thereby not be "dual-equivalent".*

*Take for example the problem* $\min x^2$ *subject to* $2x = 0$. *The dual function* $\min_x x^2 + 2ux = -u^2$ *has maximal value 0; there is no duality gap. Nevertheless, an equivalent formulation is* $\min x^3$ *subject to* $2x = 0$, *whose associated dual function is identically* $-\infty$*; the duality gap is infinite.*

*The quality of a Lagrangian relaxation can be measured by the tightness of the convexification, more precisely by the duality gap: a relaxation is better when* $\sup\theta$ *is higher. When designing a duality scheme for a given problem, two criteria must therefore be taken into account: how easy it is to minimize the Lagrangian, and how high* $\sup\theta$ *will be.* •

The aim of this section is to apply Lagrangian relaxation to (6.6). As is obvious from first sight, the resulting dual problem will be SDP ($Q_0 + \sum_k u_k Q_k$ must be positive semidefinite! see §3.3 below). Then we will show that the SDP-dual of the latter – a sort of bidual – is just the recipe of §2.3.

## 3.2 AN EXAMPLE: MAX-CUT

Duality computations involving quadratic functions are rather fussy. They notably simplify when there is no linear term. Since this is the case of max-cut, we use this problem as an appetizer.

The Lagrangian associated with (6.4), defined for $(x,u) \in R^{n\times n}$, is

$$L(x,u) = x^\top Qx + \sum_{i=1}^{n} u_i(x_i^2 - 1) = x^\top(Q + D(u))x - e^\top u\,.$$

Its minimization with respect to $x$ (unconstrained!) is trivial. It produces $-\infty$ if $Q + D(u) \not\succeq 0$; otherwise, any $x$ in $\operatorname{Ker}(Q + D(u))$ will do, producing $x^\top(Q + D(u))x = 0$ anyway. In a word,

$$\theta(u) = \begin{cases} -e^\top u & \text{if } x^\top(Q + D(u))x \succeq 0\,,\\ -\infty & \text{otherwise}\,,\end{cases}$$

and the dual problem is

$$\sup -e^\top u\,, \quad Q + D(u) \succeq 0\,. \tag{6.10}$$

Now let us apply Lagrangian relaxation to this last problem, formulated as

$$\inf e^\top u\,, \quad U = Q + D(u)\,, \quad U \succeq 0\,.$$

In the terminology of (6.8), the "primal" variables are $(u, U) \in R^n \times S^n_+$ and the constraints are $U - Q - D(u) = 0$. Call $X \in S^n$ the ("bi")dual variable; the ("bi")Lagrangian is then

$$L'(u,U,X) = e^\top u + \langle U - Q - D(u), X\rangle\,,$$

which is the sum of three terms:

- the term $-\langle Q, X\rangle$, independent of $(u, U)$;
- the term $(e - d(X))^\top u$, whose minimization with respect to $u \in R^n$ imposes the constraint $e - d(X) = 0$ (otherwise we get $-\infty$);

- the term $\langle U, X \rangle$, whose minimization with respect to $U \succeq 0$ requires a result from matrix theory: the polar cone of $\mathcal{S}_+^n$ is $-\mathcal{S}_+^n$ ([11, Cor. 7.5.4] for example); thus,

$$\inf_{U \succeq 0} \langle U, X \rangle = \begin{cases} 0 & \text{if } X \succeq 0, \\ -\infty & \text{otherwise}. \end{cases}$$

In summary, our ("bi")dual function is

$$\theta'(X) = \begin{cases} -\langle Q, X \rangle & \text{if } X \succeq 0 \text{ and } d(X) = e, \\ -\infty & \text{otherwise}. \end{cases}$$

Flipping signs, we recognize the objective function of (6.5).

Conclusion: to obtain the SDP relaxation of (6.4), one can apply Lagrangian duality twice:

- first dualize the binary constraints in (6.4); this produces the SDP problem (6.10), as well as a duality gap;
- second apply standard SDP duality to (6.10); this produces (6.5) and, since standard convex duality is involved, there is no duality gap: (6.5) and (6.10) have the same optimal value (just check that the Slater condition holds).

**Remark 7** *We could go further and relate argmins of lagrangians with subdifferentials of dual functions. This requires some more technicalities from convex analysis and is skipped here.* •

The same exercise could be done with the max-stable set problem (6.2). Calculations are more tedious but do produce (6.3) as bidual. Instead, we consider here directly the general case (6.6).

## 3.3 DUALIZING AND BIDUALIZING QUADRATIC CONSTRAINTS

Now Lagrangian duality can be applied to the general form (6.6). The work has already been done in various places, see for example [15, 28, 25, 26, 14], and the more recent [29, 5].

Set

$$Q(u) := Q_0 + \sum_{k=1}^m u_k Q_k , \quad b(u) := b_0 + \sum_{k=1}^m u_k b_k , \quad c(u) := c_0 + \sum_{k=1}^m u_k c_k ,$$

to obtain the Lagrangian

$$R^n \times R^m (x, u) \mapsto L(x, u) := x^\top Q(u) x + 2b(u)^\top x + c(u) . \qquad (6.11)$$

Its minimization with respect to $x$ is a matter of algebraic calculations. As a quadratic function, $L(\cdot, u)$ is known to have a lower bound if and only if it has a minimum ([24, Thm. 1.5] for example). This explains that the dual domain is the set of $u$ such that $Q(u) \succeq 0$ and $b(u) \in \operatorname{Im} Q(u)$. Invoking in particular Shur's lemma ([11, Thm. 7.7.6] for example), we can characterize the hypograph of the dual function: $u_0 \leq \theta(u)$ if and only if the matrix $\begin{bmatrix} c(u) - u_0 & b(u)^\top \\ b(u) & Q(u) \end{bmatrix}$ is positive semidefinite. As a result, the dual problem of (6.6) is

$$\left\{ \begin{array}{l} \sup u_0 , \quad (u_0, u) \in R^{1+m} \\ \begin{bmatrix} c(u) - u_0 & b(u)^\top \\ b(u) & Q(u) \end{bmatrix} \succeq 0 . \end{array} \right. \tag{6.12}$$

Bidualization is then a mechanical application of SDP duality, which can follow the canvas of §3.2. The SDP constraint of (6.12) gives raise to a ("bi")dual matrix in $\mathcal{S}^{1+m}$ and we form the ("bi")Lagrangian

$$L(u_0, u, x_0, x, X) = u_0 + \left\langle \begin{bmatrix} c(u) - u_0 & b(u)^\top \\ b(u) & Q(u) \end{bmatrix} , \begin{bmatrix} x_0 & x^\top \\ x & X \end{bmatrix} \right\rangle .$$

Its maximization with respect to $(u_0, u)$ (unconstrained in $R^{1+m}$) produces the bidual problem of (6.6):

$$\left\{ \begin{array}{ll} \inf \langle Q_0, X \rangle + 2\langle b_0, x \rangle + c_0 , & (x, X) \in R^n \times \mathcal{S}^n , \\ \langle Q_k, X \rangle + 2\langle b_k, x \rangle + c_k = 0 , & k = 1, \ldots, m , \\ \begin{bmatrix} 1 & x^\top \\ x & X \end{bmatrix} \succeq 0 . & \end{array} \right. \tag{6.13}$$

One more use of Shur's lemma shows that the last SDP constraint can be written $X - xx^\top \succeq 0$, so the above bidual is just (6.7).

Let us summarize this section: the relaxation obtained by the recipe of §2.3 can also be obtained by applying Lagrangian duality twice. Besides, the optimal values of (6.7) and of (6.12) are equal, provided that there is no duality gap between the convex mutually dual problems (6.12) and (6.13). This occurs for example when a Slater condition holds, say

$$\exists u \in R^m \quad \text{such that} \quad Q_0 + \sum_{k=1}^{m} u_k Q_k \succ 0$$

(which means $Q(u) \succ 0$; then take $u_0 < c(u)$ in (6.12) and use Shur's lemma to obtain the required Slater condition).

# 4 SDP RELAXATION FOR GENERAL COMBINATORIAL PROBLEMS

Let us consider a special form of (6.6):

$$\min b^\top x\,,\quad Ax = a \in R^m\,,\quad x_i^2 - x_i = 0,\; i = 1,\ldots,n\,, \tag{6.14}$$

which includes most combinatorial optimization problems. The case of affine inequality constraints is more delicate and is postponed to §4.3. Our aim is to apply the previous ideas, thereby exploring the way to new relaxation schemes in combinatorial optimization. What follows is indeed a summary of the results in [17], where in particular proofs are given for the various statements below.

## 4.1 STRAIGHTFORWARD RELAXATIONS

Section 1 alluded to the "standard" duality scheme for (6.14): dualize the affine constraints. In order to connote the linear character of this scheme, we will use the subscript L for it:

$$\theta_{\mathrm{L}}(u) := \inf_{x\in\{0,1\}^n} L_{\mathrm{L}}(x,u) := (b + A^\top u)^\top x - a^\top u\,. \tag{6.15}$$

As already said, the resulting duality gap amounts to convexifying the unit cube: $\{0,1\}^n$ is replaced by $[0,1]^n$ (note that this does not change $\theta_{\mathrm{L}}$).

Now let us apply mechanically to (6.14) the duality scheme (6.11) of §3.3. Call $u \in R^m$ [resp. $v \in R^n$] the multipliers of the affine [resp. binary] constraints and denote by C (for complete) the resulting dualization:

$$\theta_{\mathrm{C}}(u,v) := \inf_{x\in R^n} L_{\mathrm{C}}(x,u,v) := x^\top D(v)x + (b + A^\top u - v)^\top x - a^\top u\,. \tag{6.16}$$

Here come a series of crucial observations:

- If some $v_i$ is negative then $\theta_{\mathrm{C}}(u,v) = -\infty$.
- Hence, the maximization of $\theta_{\mathrm{C}}$ contains the "hidden" constraint $v \geq 0$.
- This means that the constraints $x_i^2 - x_i = 0$ come in the dual as if they were inequalities $x_i^2 - x_i \leq 0$,
- ... which means $x_i \in [0,1]$.
- As a result, the convexification effect is the same as in the standard duality scheme:

**Claim 8** *The* L- *and* C-*duality gaps are equal:* $\theta_{\mathrm{L}} = \sup_v \theta_{\mathrm{C}}(\cdot, v)$. •

This negative result seems to limit SDP relaxation to combinatorial problems containing explicitly quadratic functions; i.e. problems not amenable to standard relaxation. This is not true, however: we proceed to demonstrate that SDP relaxation is relevant even for purely linear problems.

## 4.2 THE BOOLEAN RELAXATION

Section 3.3 was based on the observation that minimizing a quadratic function is an easy operation (despite its algebraic technicalities). Now suppose we dualize only the Boolean constraints in (6.14); the resulting Lagrangian will be a quadratic function, again easy to minimize over the affine subspace $Ax = a$. Thus we consider the B-duality scheme

$$\theta_{\mathrm{B}}(v) := \inf_{Ax=a} L_{\mathrm{B}}(x,v) := x^{\top} D(v) x + (b-v)^{\top} x\,. \tag{6.17}$$

Here, less constraints are dualized than in the C-scheme. As is generally known, this can only reduce the duality gap (see [18] for example):

**Claim 9** *The dual optimal values satisfy* $\sup \theta_{\mathrm{B}} \geq \sup \theta_{\mathrm{C}} = \sup \theta_{\mathrm{L}}$. •

It is interesting to reproduce here the reasoning of §4.1 explaining why dualizations L and C are the same:

- Maximize $\theta_{\mathrm{C}}(u,v)$ with respect to $u$ alone. In other words, solve the problem $\theta(v) := \sup_u \inf_x L_{\mathrm{C}}(x,u,v)$. Needless to say, $\sup \theta_{\mathrm{C}} = \sup \theta$.
- Observe that $\theta$ is the dual optimal value associated with the problem (6.17) defining $\theta_{\mathrm{B}}(v)$. From weak duality, $\theta_{\mathrm{B}} \geq \theta$, which "explains" that B-duality is better.
- If $v \geq 0$, (6.17) is a nicely convex linear-quadratic problem, with no duality gap: $\theta(v) = \theta_{\mathrm{B}}(v)$.
- On the other hand, take $v \not\geq 0$; then $\inf_x L_{\mathrm{C}}(x,\cdot,v) \equiv -\infty$, so $\theta(v) = -\infty$.
- By contrast, $\theta_{\mathrm{B}}(v)$ may well be defined outside $R^n_+$. Indeed, $\theta_{\mathrm{B}}(v) > -\infty$ for

$$v \in V := \{v \in R^n : D(v) \succ 0 \text{ on } \operatorname{Ker} A\}\,.$$

In words: the C-scheme somehow *restricts* $\theta_{\mathrm{B}}$: it chops off the domain of $\theta_{\mathrm{B}}$ every point with a negative component. As a result, when there are more equality constraints, $V$ becomes larger and the above restriction becomes more severe; the efficiency of B-duality increases.

Let us now turn to actual B-computations. The constraints imposed to the minimization of $L_{\mathrm{B}}(\cdot,v)$ complicate substantially the algebra. In particular the simplicity of the tensorial lift is spoiled, and the bidual is no longer so easy to derive. However a notable simplification was observed in [23]: the constraints $Ax = a$ can be "quadraticized" to

$\|Ax - a\|^2 = 0$ – and then dualized (or tensor-lifted). In other words, formulate (6.14) as

$$\min b^\top x\,, \quad x^\top A^\top Ax - 2a^\top Ax + \|a\|^2 = 0\,, \quad x_i^2 - x_i = 0\,, \qquad (6.18)$$

which has only quadratic constraints and enters the framework of §3.3. Calling $\pi$ the multiplier corresponding to $\|Ax - a\|^2 = 0$, we obtain the "penalized B-dualization scheme ":

$$L_{\mathrm{B}}^{\pi}(x, v) := x^\top (D(v) + \pi A^\top A)x + (b - 2\pi a^\top A - v)^\top x + \pi\|a\|^2\,,$$

and its associated dual function $\theta_{\mathrm{B}}^{\pi}(v) := \inf_{x \in R^n} L_{\mathrm{B}}^{\pi}(x, v)$.

**Claim 10** *This technique leaves the* B*-duality gap unchanged. In other words,* $\sup_{\pi,v} \theta_{\mathrm{B}}^{\pi}(v) = \sup_v \theta_{\mathrm{B}}(v)$. •

Now applying to (6.18) the tensorial lift of §2.3 is straightforward and gives the "B-bidual relaxation of (6.14) in aggregate form":

$$\min b^\top d(X)\,, \quad \left\langle A^\top A, X\right\rangle - 2a^\top Ad(X) + \|a\|^2 = 0\,, \quad X \succeq d(X)d(X)^\top .$$

## 4.3 THE CASE OF INEQUALITIES

Finally consider a combinatorial problem with inequality constraints:

$$\min b^\top x\,, \quad Ax \le a \in R^m\,, \quad x_i^2 - x_i = 0,\ i = 1, \ldots, n\,. \qquad (6.19)$$

The L- and C-duality schemes, which both dualize the affine constraints, are little affected; simply, the corresponding $u$ becomes a nonnegative dual variable.

On the other hand, the B-scheme breaks down: now we have to minimize $L_{\mathrm{B}}(\cdot, v)$ under inequality constraints, which is NP-hard (unless $v \ge 0$, but this brings us back to the standard technique; remember §4.1). However one is not completely disarmed against inequality constraints, providing that artificial manipulations are called for. Just as affine constraints $Ax = a$ were "quadraticized" in §4.2 with no harm, affine inequality constraints can be here fruitfully quadraticized.

One such artifact can be found in [20]: because $x_i$ is a nonnegative variable, an affine constraint such as $A_j x - a_j \le 0$ is equivalent to the quadratic constraint $x_i(A_j x - a_j) \le 0$, which can be B-dualized (or tensor-lifted). The idea is rich enough to cancel out the duality gap if repeated many times enough.

Another artifact is possible when a lower bound is known on the linear form; say $a_j' \le A_j x$. Then the affine constraint $A_j x - a_j \le 0$ is equivalent to the quadratic constraint $(A_j x - a'_j)(A_j x - a_j) \leq 0$; again this last constraint can be dualized (or tensor-lifted).

**Claim 11** *Compared to standard dualization of the affine constraint $A_j x \leq a_j$, this last technique can only improve the duality gap.* •

To conclude this paper, we can say that SDP is indeed a general relaxation scheme for combinatorial optimization.

- It is especially suitable for quadratic functions (objective or constraints).
- Its chances to improve standard Lagrangian relaxation increase when there are more equality constraints.
- By contrast, it is somewhat clumsy with affine inequalities, for which some other tools are required.

## References

[1] J.P. Aubin and I. Ekeland. Estimates of the duality gap in nonconvex optimization. *Mathematics of Operations Research*, 1(3):225–245, 1976.

[2] I. Ekeland and R. Temam. *Convex Analysis and Variational Problems.* Classics in Applied Mathematids 28. SIAM Publications, 1999.

[3] J.E. Falk. Lagrange multipliers and nonconvex programs. *SIAM Journal on Control*, 7(4):534–545, 1969.

[4] S. Feltenmark and K. C. Kiwiel. Dual applications of proximal bundle methods, including lagrangian relaxation of nonconvex problems. *SIAM Journal on Optimization*, 10(3):697–721, 2000.

[5] T. Fujie and M. Kojima. Semidefinite protramming relaxation for nonconvex quadratic programs. *Journal of Global Optimization*, 10:367–380, 1997.

[6] C. Garrod and J.K. Percus. Reduction of the N-particle variational problem. *Journal of Mathematical Physics*, 5(12), 1964.

[7] A.M. Geoffrion. Lagrangean relaxation for integer programming. *Mathematical Programming Study*, 2:82–114, 1974.

[8] M. X. Goemans and D. P. Williamson. Improved approximation algorithms for maximum cut and satisfiability problems using semidefinite programming. *Journal of the ACM*, 6:1115–1145, 1995.

[9] C. Helmberg, F. Rendl, and R. Weismantel. Quadratic knapsack relaxations using cutting planes and semidefinite programming.

In W. H. Cunningham, S. T. McCormick, and M. Queyranne, editors, *Integer Programming and Combinatorial Optimization V*, volume 1084 of *Lecture Notes in Computer Science*, pages 175–189. Springer Verlag, Berlin, 1996.

[10] J.-B. Hiriart-Urruty and C. Lemaréchal. *Convex Analysis and Minimization Algorithms.* Springer-Verlag, 1993.

[11] R.A. Horne and Ch.R. Johnson. *Matrix analysis.* Cambridge University Press, 1989. (New edition, 1999).

[12] D. Karger, R. Motwani, and M. Sudan. Approximate graph coloring by semidefinite programming. *Journal of the ACM*, 45(2):246–265, 1998.

[13] D.E. Knuth. The sandwich theorem. *Electronic Journal of Combinatorics*, 1(A1), 1994.

[14] F. Körner. A tight bound for the boolean quadratic optimization problem and its use in a branch and bound algorithm. *Optimization*, 19(5):711–721, 1988.

[15] F. Körner and C. Richter. Zur effektiven lösung von booleschen, quadratischen optimierungsproblemen. *Numerische Mathematik*, 40:99–109, 1982.

[16] C. Lemaréchal, Yu. Nesterov, and F. Oustry. Duality gap analysis for problems with quadratic constraints, 2001. In preparation.

[17] C. Lemaréchal and F. Oustry. Semi-definite relaxations and lagrangian duality with application to combinatorial optimization. Rapport de Recherche 3710, INRIA, 1999. Submitted to Mathematical Programming.

[18] C. Lemaréchal and A. Renaud. A geometric study of duality gaps, with applications. *Mathematical Programming*, 2001. to appear.

[19] L. Lovász. On the Shannon capacity of a graph. *IEEE Transactions on Information Theory*, IT 25:1–7, 1979.

[20] L. Lovász and A. Schrijver. Cones of matrices and set-functions and 0-1 optimization. *SIAM Journal on Optimization*, 1(2):166–190, 1991.

[21] Yu.E. Nesterov, H. Wolkowicz, and Y.Y Ye. Semidefinite programming relaxations of nonconvex quadratic optimization. In Lieven Vandenberghe R. Saigal and H. Wolkovicz, editors, *Hanbook on Semidefinite Programming. Theory, Algorithms and Applications.* Kluwer, 2000.

[22] P. Pardalos and H. Wolkowicz. *Topics in Semidefinite and Interior-Point Methods.* Fields Institute Communications Series 18. American Mathematical Society, 1998.

[23] S. Poljak, F. Rendl, and H. Wolkowicz. A recipe for semidefinite relaxation for (0,1)-quadratic programming. *Journal of Global Optimization*, 7:51–73, 1995.

[24] B.N. Pshenichnyi. *The Linearization Method for Constrained Optimization.* Number 22 in Computational Mathematics. Springer-Verlag, 1994.

[25] N.Z. Shor. Class of global minimum bounds of polynomial functions. *Cybernetics*, 23(6):731–734, 1987.

[26] N.Z. Shor. Quadratic optimization problems. *Soviet Journal of Computer and Systems Sciences*, 25:1–11, 1987.

[27] N.Z. Shor. Dual estimates in multiextremal problems. *Journal of Global Optimization*, 2:411–418, 1992.

[28] N.Z. Shor and A.S. Davydov. Method of opbtaining estimates in quadratic extremal probems with boolean variables. *Cybernetics*, (2):207–211, 1985.

[29] L. Vandenberghe and S. Boyd. Semidefinite programming. *SIAM Review*, 38(1):49–95, 1996.

[30] H. Wolkowicz and Z. Zhao. Semidefinite relaxations for the graph partitioning problem. *Discrete Applied Mathematics*, 96-97:461–479, 1999.

[31] Q. Zhao, S.E. Karisch, F. Rendl, and H. Wolkowicz. Semidefinite programming relaxations for the quadratic assignment problem. *Journal of Combinatorial Optimization*, 2(1):71–109, 1998.

Chapter 7

# CONVEX ANALYSIS IN THE CALCULUS OF VARIATIONS

R. T. Rockafellar
*Department of Mathematics*
*University of Washington*
*Seattle, WA 98195-4350*
rtr@math.washington.edu

**Abstract** Convexity properties are vital in the classical calculus of variations, and many notions of convex analysis, such as the Legendre-Fenchel transform, have their origin there. Conceptual developments in contemporary convex analysis have in turn enriched that venerable subject by making it possible to treat a vastly larger class of problems effectively in a "neoclassical" framework of extended-real-valued functions and their subgradients.

**Keywords:** convex analysis, calculus of variations, generalized Euler-Lagrange equations, Legendre-Fenchel transform, generalized Hamiltonian equations, neoclassical problems of Bolza, optimal control

## 1 CLASSICAL LAGRANGE PROBLEMS

The calculus of variations is the oldest branch of optimization, dating from over three hundred years ago. Most books on that subject are steeped in traditional thinking and show little influence of convex analysis as we now know it elsewhere in optimization. Nevertheless convexity properties have an essential role in the calculus of variations. They can be seen as deeply affecting the statement of necessary and sufficient conditions for optimality and, through investigations of the existence of solutions, the very way that problems ought to be posed. Concepts emerging from convex analysis have furthermore led to the development of a mathematical framework so sturdy and broad that traditional statements in the calculus of variations can be stretched in their interpreta-



*N. Hadjisavvas and P.M. Pardalos (eds.), Advances in Convex Analysis and Global Optimization,* 135–151.

tion so as to cover problems of optimal control and much more than ever might have been imagined. This article aims at explaining in basic terms why—and how—that has come about.

The fundamental *Lagrange problem* in the calculus of variations, with respect to a fixed interval $[\tau_0, \tau_1]$ and fixed endpoints $\xi_0$ and $\xi_1$, has the form:

$$(\mathcal{P}_0) \qquad \begin{array}{c} \text{minimize } J_0[x] := \int_{\tau_0}^{\tau_1} L(t, x(t), \dot{x}(t))dt \\ \text{subject to } x(\tau_0) = \xi_0, \ x(\tau_1) = \xi_1, \end{array}$$

for a function $L : [\tau_0, \tau_1] \times \mathbb{R}^n \times \mathbb{R}^n \to \mathbb{R}$ which is assumed to be continuously differentiable. The minimization takes place in a space of functions $x : [\tau_0, \tau_1] \to \mathbb{R}^n$, called *arcs*, with $\dot{x}(t)$ denoting the derivative of $x(t)$. But what space should that be? The simplest and seemingly most natural choice is

$$\mathcal{C}_n^1[\tau_0, \tau_1] \;=\; \text{space of continuously differentiable arcs.}$$

The integral functional $J_0$ is well defined on this space and even continuously Fréchet differentiable under a standard choice of norm. In terms of the gradients of $L(t, x, v)$ with respect to $x$ and $v$, the directional derivatives are given by

$$J_0'[x; w] = \int_{\tau_0}^{\tau_1} \Big[ \langle \nabla_x L(t, x(t), \dot{x}(t)), w(t) \rangle + \langle \nabla_v L(t, x(t), \dot{x}(t)), \dot{w}(t) \rangle \Big] dt, \tag{1}$$

where $\langle \cdot, \cdot \rangle$ denotes the canonical inner product in $\mathbb{R}^n$. A classical argument yields the fact that if an arc $x$ is optimal in $(\mathcal{P}_0)$ (where for purposes of this article we always interpret optimality as *global* optimality), the *Euler-Lagrange condition* must hold:

$$\frac{d}{dt} \nabla_v L(t, x(t), \dot{x}(t)) = \nabla_x L(t, x(t), \dot{x}(t)), \tag{2}$$

along with the *Weierstrass condition*:

$$L(t, x(t), v) \geq L(t, x(t), \dot{x}(t)) + \langle \nabla_v L(t, x(t), \dot{x}(t)), \, v - \dot{x}(t) \rangle \ \text{ for all } v, \tag{3}$$

which in turn entails the *Legendre condition*:

$$\nabla_{vv}^2 L(t, x(t), \dot{x}(t)) \ \text{ is positive semi-definite} \tag{4}$$

when $L(t, x(t), v)$ is twice differentiable with respect to $v$, so that the Hessian matrix $\nabla_{vv}^2 L(t, x(t), \dot{x}(t))$ exists. Already here, convexity enters implicitly in important ways.

First and most obvious is the connection between the Legendre condition and possible convexity of the function $f(v) = L(t, x(t), v)$. When $f$ is twice differentiable, it is a convex function if and only if its Hessian $\nabla^2 f(v)$ is positive semi-definite for all $v$. Of course, the Legendre condition (4) asserts positive semi-definiteness only for $v = \dot{x}(t)$, but still it has suggestive potential. The strengthened form of the Legendre condition in which positive semi-definiteness is replaced by positive definiteness, would, in the presence of continuous twice differentiability, imply that $f$ is strongly convex on a neighborhood of $\dot{x}(t)$. In this sense, convexity is close by. In fact the Weierstrass condition, which supersedes the Legendre condition, brings it even closer.

To see this aspect of the Weierstrass condition, let $y(t) = \nabla_v L(t, x(t), \dot{x}(t))$ and again consider $f(v) = L(t, x(t), v)$ for fixed $t$, so that $y(t) = \nabla f(\dot{x}(t))$ and (3) can be written as

$$f(v) \geq f(\dot{x}(t)) + \langle y(t), v - \dot{x}(t) \rangle \text{ for all } v. \tag{5}$$

The right side of the inequality is an affine function $a(v)$; the inequality says that $f(v) \geq a(v)$ everywhere, and that these functions agree at $v = \dot{x}(t)$. Unlike the Legendre condition, this is not just a one-point property but a global property. Convexity is brought in because (5) would automatically hold if $f$ is a convex function, whereas it seems puzzlingly strong if $f$ is not convex. Indeed, let

$$\bar{f} = \text{convex hull of } f,$$

this being the greatest convex function majorized by $f$. From our knowledge that (5) holds for $y(t) = \nabla f(\dot{x}(t))$, it can be verified that $\bar{f}$ is a finite convex function which is differentiable at $\dot{x}(t)$ and has the same gradient there as $f$, i.e., the vector $y(t)$. Thus, the real content of the Weierstrass condition (3) is in the relations

$$f(\dot{x}(t)) = \bar{f}(\dot{x}(t)), \qquad \nabla f(\dot{x}(t)) = \nabla \bar{f}(\dot{x}(t)). \tag{6}$$

In other words, it says something about how the optimality of an arc $x$ implies that $L$ acts along $x$ as if it were replaced by its convex hull in the $v$ argument.

The issue of convexity or convexification of $L(t, x, v)$ with respect to $v$ has significance also from a different perspective. When can we be sure that an optimal arc even exists in problem $(\mathcal{P}_0)$? It may be anticipated that the existence of an optimal arc depends on some kind of compactness property of the sublevel sets $\{x \mid J_0[x] \leq \alpha\}$, $\alpha \in \mathbb{R}$, but such compactness must be tuned to the space over which the minimization takes place, and $C^1_n[\tau_0, \tau_1]$ is for that reason a poor choice. Often in textbooks, $\mathcal{C}_n^1[\tau_0, \tau_1]$ is replaced by the space $\bar{\mathcal{C}}_n^1[\tau_0, \tau_1]$ consisting of the arcs $x : [\tau_0, \tau_1] \to I\!R^n$ that are *piecewise* continuously differentiable; elementary examples abound in which solutions can be found in $\bar{\mathcal{C}}_n^1[\tau_0, \tau_1]$ but not in $\mathcal{C}_n^1[\tau_0, \tau_1]$, and this has long been recognized. But $\bar{\mathcal{C}}_n^1[\tau_0, \tau_1]$ fares no better than $\mathcal{C}_n^1[\tau_0, \tau_1]$ in attempts to establish the existence of solutions in general. We know now that to get existence we have to pass to a far more general space and introduce both coercivity and *convexity* assumptions on the behavior of $L(t, x, v)$ with respect to $v$.

Let $|v|$ stand for the Euclidean norm of $v \in I\!R^n$. The function $L$ is said to satisfy the *Tonelli growth condition* if

$$L(t, x, v) \geq \theta(|v|) \text{ for all } (t, x, v) \in [\tau_0, \tau_1] \times I\!R^n \times I\!R^n \tag{7}$$

for a nondecreasing function $\theta : [0, \infty) \to (-\infty, \infty)$ that is coercive, i.e., has $\theta(s)/s \to \infty$ as $s \to \infty$. For $p \in [1, \infty]$ let

$$\mathcal{A}_n^p[\tau_0, \tau_1] = \text{space of absolutely continuous arcs } x \text{ with } \dot{x} \in \mathcal{L}_n^p[\tau_0, \tau_1].$$

Thus, $\mathcal{A}_n^1[\tau_0, \tau_1]$ consists simply of all absolutely continuous $x : [\tau_0, \tau_1] \to I\!R^n$, whereas $\mathcal{A}_n^\infty[\tau_0, \tau_1]$ consists of all Lipschitz continuous $x : [\tau_0, \tau_1] \to I\!R^n$. Of course an arc $x$ in any of these spaces is only differentiable almost everywhere, so that the integrand $L(t, x(t), \dot{x}(t))$ itself is only defined almost everywhere—but that is enough to make sense of the functional $J_0$, provided that its value is interpreted as $\infty$ in the eventuality that $L(t, x(t), \dot{x}(t))$ is not bounded from above by an integrable function of $t$.

**Theorem 1.** *Let problem $(\mathcal{P}_0)$ be placed in the arc space $\mathcal{A}_n^1[\tau_0, \tau_1]$. If the function $L$ satisfies the Tonelli growth condition, and $L(t, x, v)$ is convex with respect to $v$, then the minimum in $(\mathcal{P}_0)$ is attained.*

This fact suggests that $\mathcal{A}_n^1[\tau_0, \tau_1]$ should be adopted as the arc space without further ado, but unfortunately there are impediments to that. On this larger space the functional $J_0$ lacks adequate differentiability, except in very special circumstances, and the traditional derivation of optimality conditions collapses. A fallback instead of $\mathcal{A}_n^1[\tau_0, \tau_1]$ is $\mathcal{A}_n^\infty[\tau_0, \tau_1]$, on which the directional derivative formula (1) persists under our working assumptions on $L$.

**Theorem 2.** *Let problem $(\mathcal{P}_0)$ be placed in the arc space $\mathcal{A}_n^\infty[\tau_0, \tau_1]$. Then, for an arc $x$ that is optimal in $(\mathcal{P}_0)$, the Euler-Lagrange equation (2) and the Weierstrass condition (3) must hold in the following interpretation: there is an arc $y$, likewise in $\mathcal{A}_n^\infty[\tau_0, \tau_1]$, such that for almost every $t \in [\tau_0, \tau_1]$ one has*

$$y(t) = \nabla_v L(t, x(t), \dot{x}(t)), \qquad \dot{y}(t) = \nabla_x L(t, x(t), \dot{x}(t)), \tag{8}$$

$$L(t, x(t), v) \geq L(t, x(t), \dot{x}(t)) + \langle y(t), v - \dot{x}(t) \rangle \text{ for all } v \tag{9}$$

Note that the formulation of the Euler-Lagrange condition in (8) is preferable to the version in (2) even when $(\mathcal{P}_0)$ is placed in $C_n^1[\tau_0, \tau_1]$, in which case $y$ too belongs to $C_n^1[\tau_0, \tau_1]$. It makes explicit the differentiability property of $\nabla_v L(t, x(t), \dot{x}(t))$ in $t$ that is tacit in the original statement (2). The arc $y$, said to be *adjoint* to $x$, deserves the added attention anyway.

Note also that the Weierstrass condition (9), which still implies the Legendre condition (4), once more holds automatically when $L(t, x, v)$ is convex in $v$, for the reasons already noted. Without such convexity, (9) can again be seen as a condition relating the function $f(v) = L(t, x(t), v)$ to its convex hull. But the discussion of this issue is incomplete without bringing the so-called canonical from of the Euler-Lagrange condition into the picture, which we do next.

In further analysis of the first equation in (8), let us step back from the consideration of arcs x and y and look at the equation $y = \nabla_v L(t, x, v)$ as a relation between vector variables $y$ and $v$ in $\mathbb{R}^n$ for given $(t, x) \in [\tau_0, \tau_1] \times \mathbb{R}^n$. Suppose we are able to invoke the implicit function theorem in order to solve this equation uniquely for $v$ as a function of $(t, x, y)$; the technical assumption needed for this locally is that the mapping $v \mapsto \nabla_v L(t, x, v)$ is continuously differentiable and has nonsingular Jacobian, or in other words, that $L(t, x, v)$ is twice differentiable in $v$, with the Hessian $\nabla^2_{vv} L(t, x, v)$ being nonsingular and depending continuously on $(t, x, v)$. Let the solution mapping be denoted by $v = h(t, x, y)$, and let

$$H(t, x, y) = \langle y, v \rangle - L(t, x, v) \text{ with } h(t, x, y) \text{ substituted for } v. \tag{10}$$

Formulas for the derivatives of $H$ can be obtained from the implicit function theorem; one has

$$\begin{cases} \nabla_y H(t, x, y) = h(t, x, y), \\ \nabla_x H(t, x, y) = -\nabla_x L(t, x, h(t, x, y)). \end{cases}$$

Then, on returning to the context of arcs, we can express the Euler-Lagrange condition (8) in the form

$$\dot{x}(t) = \nabla_y H(t, x(t), y(t)), \qquad \dot{y}(t) = -\nabla_x H(t, x(t), y(t)), \tag{11}$$

and recognize that this constitutes an ordinary differential equation in the form of $\dot{z}(t) = F(t, z(t))$ for $z(t) = (x(t), y(t))$. The conversion of (8) to an ODE is obviously both illuminating and advantageous.

The function $H$ is known as the *Hamiltonian* associated with $L$, which in contrast is itself often called the *Lagrangian*; the process of obtaining

$H(t,x,y)$ from $L(t,x,v)$ by way of the implicit function theorem in the manner described is the *Legendre transform* (with respect to the $v$ variable with $(t,x)$ fixed). The relations (11) are the *Hamiltonian equations*.

Unfortunately, in this traditional development there is a serious flaw which textbooks seldom address. The nonsingularity assumption on $\nabla^2_{vv}L(t,x,v)$ on which the Legendre transform depends is local. Indeed, it is often envisioned merely along the $x$ trajectory being tested for optimality, where it comes down to the nonsingularity of $\nabla^2_{vv}L(t,x(t),\dot{x}(t))$ and therefore the positive definiteness of this Hessian, since we already know from the Legendre condition (4) that $\nabla^2_{vv}L(t,x(t),\dot{x}(t))$ must be positive semi-definite. Thus, if this approach through the Legendre transform is adopted, the Hamiltonian $H$ threatens to be defined only in a local sense in relation to a particular $x$ trajectory, and only to the extent that $L(t,x(t),v)$ is strongly convex in $v$ around $v=\dot{x}(t)$. Furthermore, it can be seen from such local convexity that the prescription for $H$ in (10) then implies

$$H(t,x(t),y(t)) = \text{local max of } \langle y(t),v\rangle - L(t,x(t),v) \text{ in } v \text{ around } \dot{x}(t),$$

with the local max being attained when $v=\dot{x}(t)$. Another angle, however, is that the Weierstrass condition says

$$\text{global max of } \langle y(t),v\rangle - L(t,x(t),v) \text{ in } v \text{ is attained at } \dot{x}(t).$$

Obviously this situation is perplexing and unsatisfactory. An answer in part could be to insist that $\nabla^2_{vv}L(t,x,v)$ be positive definite for all $(t,x,v)\in[\tau_0,\tau_1]\times I\!R^n\times I\!R^n$, so as to avoid the awkward localization, but that would be unpleasantly restrictive.

Convex analysis will come to the rescue with a superior definition of the Hamiltonian function $H$. That definition will rely instead on the Legendre-Fenchel transform, which reduces to the Legendre transform in the presence of continuous second-order differentiability with positive definite Hessians, but actually makes sense even without any differentiability. Before getting to that, however, it is good to complement the discussion of necessary conditions by taking a brief look at how convexity supports sufficient conditions for optimality.

**Theorem 3.** *Let problem* $(\mathcal{P}_0)$ *be placed in the arc space* $\mathcal{A}^p_n[\tau_0,\tau_1]$ *for any* $p\in[1,\infty]$, *and suppose that* $L(t,x,v)$ *is convex with respect to* $(x,v)$. *If the Euler-Lagrange condition in form (8) holds for arcs* $x$ *and* $y$ *with* $x\in\mathcal{A}^p_n[\tau_0,\tau_1]$ *and* $y\in\mathcal{A}^1_n[\tau_0,\tau_1]$, *then* $x$ *is optimal in* $(\mathcal{P}_0)$.

The proof of this result is elementary but also illuminating because it opens doors to generalization. The convexity of $L(t,x,v)$ not just in $v$ but with respect to $(x,v)$ yields the gradient inequality for the function

$L(t,\cdot,\cdot)$ that, when used to compare the pair $(x(t),\dot{x}(t))$ from the arc $x$ with a pair $(x'(t),\dot{x}'(t))$ coming from an alternative arc $x'$, says in the context of (8) that

$$L(t,x'(t),\dot{x}'(t)) \geq L(t,x(t),\dot{x}(t))+\langle\dot{y}(t),x'(t)-x(t)\rangle+\langle y(t),\dot{x}'(t)-\dot{x}(t)\rangle.$$

On integrating both sides of this inequality we get

$$J_0[x'] \geq J_0[x] + \langle y(\tau_1),\, x'(\tau_1)-x(\tau_1)\rangle - \langle y(\tau_0),\, x'(\tau_0)-x(\tau_0)\rangle.$$

When $x'$, like $x$, satisfies the endpoint constraints $x'(\tau_0)=\xi_0$, $x'(\tau_1)=\xi_1$, we obtain $J_0[x'] \geq J_0[x]$ and thus the global optimality of $x$.

Observe that under the convexity hypothesis on $L$ in Theorem 3, $J_0$ is a convex functional on the arc space in question.

Theorem 3 does not give the most general form of sufficient condition in the calculus of variations. In the classical theory that would be a hard-to-verify criterion known as the Jacobi condition, involving second derivatives of the functional $J_0$ for which there is an integral expression based on second derivatives of $L$. By contrast, the condition in Theorem 3, although not as comprehensive, is simple to use and still covers many applications. But books usually neglect to mention it.

Variants of the optimality conditions in Theorems 2 and 3 for problems beyond $(\mathcal{P}_0)$, having boundary conditions other than $x(\tau_0)=\xi_0$ and $x(\tau_1)=\xi_1$, are well known and easily obtained. If the first endpoint of $x$ is fixed but the second endpoint is free, the condition $y(\tau_1)=0$ has to be added, whereas if the free endpoint is the first instead of the second, the requirement is $y(\tau_0)=0$. If both endpoints are free but a cost term $l(x(\tau_0),x(\tau_1))$ is added to the integral in the functional $J_0$ being minimized, one has to have

$$(y(\tau_0),-y(\tau_1)) = \nabla l(x(\tau_0),x(\tau_1)).$$

Mixtures of such possibilities can also be handled, involving an endpoint cost term in $J_0$ but also a system of constraints on $x(\tau_0)$ and $x(\tau_1)$, not only equations but perhaps also inequalities. In every such case, optimality entails a certain relation between $(x(\tau_0),x(\tau_1))$ and $(y(\tau_0),y(\tau_1))$, known historically as a *transversality condition.* Classical theory does not provide a universal way of expressing transversality conditions, but a simple, unifying scheme does come out of the modern developments described below.

Besides these variants on how to treat $x(\tau_0)$ and $x(\tau_1)$ in $(\mathcal{P}_0)$ there can be constraints of equality or inequality type, or both, imposed on $x(t)$ and $\dot{x}(t)$, separately or jointly. When all such complications are allowed simultaneously in the problem statement, one gets a formulation that marked the high point in the calculus of variations before the advent of optimal control, called a *Bolza problem.*

It is also possible to consider problems in which the interval $[\tau_0, \tau_1]$ is not fixed but subject to optimization as well, but we will skip over that in the interests of simplicity.

## 2 NEOCLASSICAL BOLZA PROBLEMS

The classical framework for the calculus of variations has led to many successes, most notably in applications to physics. Its shortcomings, such as have been pointed out above, were not really apparent until "optimization" arose in the 1950's as a practical subject with a numerical orientation. Inequality constraints came to be viewed not just as an occasional necessity in problem formulation, but rather as a dominant feature affecting the analysis of solutions and the design of algorithms for finding them. Penalty expressions with discontinuities in derivatives entered the stage that way too. Methods of differential calculus were no longer enough, whether in connection with such expressions or in coping with the geometry of sets specified by systems of inequality constraints. On the other hand, convexity properties of sets and functions took on a new significance, and fascinating phenomena of duality came to light.

For the calculus of variations, the most revolutionary development was the emergence of optimal control theory. A typical problem of optimal control is

$(\mathcal{P}_1)$

$$\text{minimize } J_1[x,u] := \int_{\tau_0}^{\tau_1} f_0(t,x(t),u(t))dt + h(x(\tau_1)) \text{ subject to}$$
$$x(\tau_0) = \xi_0, \;\; \dot{x}(t) = f(t,x(t),u(t)), \;\; u(t) \in U(t,x(t)), \;\; x(\tau_1) \in E,$$

where the minimization takes place over pairs $(x, u)$ comprised of arcs $x : [\tau_0, \tau_1] \to I\!R^n$ and *control* functions $u : [\tau_0, \tau_1] \to I\!R^m$ that satisfy the specified constraints. For instance, one could require $x \in \mathcal{A}_n^\infty[\tau_0, \tau_1]$ and $u \in \mathcal{L}_m^\infty[\tau_0, \tau_1]$, interpreting the constraints $\dot{x}(t) = f(t,x(t),u(t))$ and $u(t) \in U(t,x(t))$ as holding for almost every $t$. The sets $U(t,x(t))$ and $E$ could be defined by equations and inequalities, but we need not go into that. There could likewise be a condition $x(t) \in X(t)$, called a state constraint, but we may regard that as implicit in the description of the region where $U(t,x(t)) \neq \emptyset$.

The challenge of working with such formulations involving controls led to new results like the Pontriagin "maximum principal," which expresses necessary conditions for optimality akin to the Euler-Lagrange equation and Weierstrass condition, but is limited to situations where $U(t,x(t))$ does not actually depend on $x(t)$.

The attitude in the early days of control theory was that problems in optimal control could be regarded as generalizations of problems in the calculus of variations in which the differential equation $\dot{x}(t) = f(t, x(t), u(t))$ provides additional interest and capabilities. In the elementary case where $f(t, x, u) \equiv u$, the differential equation reduces to $\dot{x}(t) = u(t)$, problem $(\mathcal{P}_1)$ comes down to a certain classical problem of Bolza.

The utterly different point of view that we explain next, about how optimal control and the calculus of variations can be seen as fitting together, emerged instead from discoveries made in applying convex analysis to optimization more generally. It focuses on the seemingly much simpler problem model

$$(\mathcal{P}) \qquad \text{minimize } J[x] := \int_{\tau_0}^{\tau_1} L(t, x(t), \dot{x}(t))dt + l(x(\tau_0), x(\tau_1)),$$

where the minimization is over a space of arcs, e.g. $x \in \mathcal{A}_n^p[\tau_0, \tau_1]$ for some $p \in [1, \infty]$. This is called a *generalized* Bolza problem, but at first that may sound paradoxical. Shouldn't $(\mathcal{P})$ rather be called a simplified Bolza problem, since no contraints are apparent in it?

The key source of generality in $(\mathcal{P})$ lies in the great breadth of the class of functions $L$ and $l$ that are now admitted. No longer is it tacitly assumed that these functions are differentiable to whatever order might be deemed useful. To the contrary, they need not be continuous or even finite everywhere. They are allowed to take $+\infty$ as a value, in particular as a means of representing constraints by imposing infinite penalties when they are violated.

For that reason, $(\mathcal{P})$ covers not only classical Bolza problems but also problems in optimal control like $(\mathcal{P}_1)$. To capture $(\mathcal{P}_1)$, one would take

$$l(x_0, x_1) = \begin{cases} g(x_1) & \text{if } x_0 = \xi_0 \text{ and } x_1 \in E, \\ \infty & \text{if not,} \end{cases} \tag{12}$$

and on the other hand

$$L(t, x, v) = \inf\{f_0(t, x, u) \,|\, u \in U(t, x) \text{ satisfying } f(t, x, u) = v\}, \tag{13}$$

where the convention is used that the infimum of an empty set of numbers is $+\infty$. The suppression of control variables in passing in this way from $(\mathcal{P}_1)$ to $(\mathcal{P})$ loses nothing because, once an arc $x$ has been identified as solving $(\mathcal{P})$, a corresponding control function $u$ can be obtained by selecting, for each $t$ at which $\dot{x}(t)$ exists, a vector $u(t)$ in the set

$$\operatorname{argmin}\{f_0(t, x(t), u) \,|\, u \in U(t, x(t)) \text{ satisfying } f(t, x(t), u) = \dot{x}(t)\}. \tag{14}$$

Of course, in validating such an approach a number of technical issues have to be resolved. The functional $J$ in $(\mathcal{P})$ has to be well defined when $l$ and $L$ come out of formulas like those in (12) and (13), and there has to be assurance that, when $u(t)$ is selected from the set in (14) this can be done so as to make the resulting function $u$ belong to the right function space. But such issues have by now all been worked out to satisfaction.

The radical departure from classical theory in investigating generalized Bolza problems $(\mathcal{P})$ without the customary restrictions on $L$ and $l$ started in 1970 in [10] under the alternative assumption that $L(t,\cdot,\cdot)$ and $l$ are convex, which makes the Bolza functional $J$ itself be convex. By that time, researchers in convex analysis were well accustomed to working with infinite penalty representations of constraints (see e.g. [8]) and had realized that many conditions usually stated in terms of gradients could, for convex functions without differentiability, be articulated instead with "subgradients." It was a natural step to try this out in the calculus of variations by studying the fully convex case of $(\mathcal{P})$ and looking for subgradient versions of the Euler-Lagrange condition and the Hamiltonian equations.

In the intervening years, a huge effort has gone into expanding the territory of this type of "nonsmooth" analysis beyond convex functions, so that subgradient conditions in Bolza problems could be established more generally. The doctoral thesis of Clarke [1] gave the first big advance, which eventually led to his book [2]. Many others have also gotten involved, and much research is still ongoing; see [14] and its commentaries.

To explain this major development further, let us review some basic ideas that underlie that literature of convex and nonsmooth analysis, in which notation and terminology are oriented asymmetrically toward minimization. A function $\varphi : I\!R^n \to \overline{I\!R}$ (where $\overline{I\!R} = [-\infty, \infty]$) has *effective domain*

$$\operatorname{dom}\varphi = \{v \mid \varphi(v) < \infty\}$$

and is called *proper* if it is finite on this set, and this set is nonempty (or in other words, if $\varphi(v) < \infty$ for some $v$ and $\varphi(v) > \infty$ for every $v$). The proper functions are thus the extended-real-valued functions on $I\!R^n$ obtained by taking a finite function on a nonempty subset $C$ of $I\!R^n$ and extending it by giving it the value $\infty$ everywhere outside of $C$.

The *epigraph* of $\varphi$ is the set $\operatorname{epi}\varphi = \{(v,\alpha) \in I\!R^n \times I\!R \mid \varphi(v) \geq \alpha\}$. It is a closed set if and only if $\varphi$ is lower semicontinuous (lsc) and, more importantly for indicating the motivation behind epigraphs, it is a convex set if and only if $\varphi$ is a convex function. For extended-real-valued functions, the epigraph is a better carrier of geometric information than the graph, because it belongs to $\mathbb{R}^n \times \mathbb{R}$, a vector space, whereas the graph merely belongs to $\mathbb{R}^n \times \overline{\mathbb{R}}$.

Minimizing $\varphi$ over $\mathbb{R}^n$ is equivalent to minimizing $\varphi$ over $\operatorname{dom}\varphi$; this is the *principle of infinite penalization.* The minimum is attained if $\varphi$ is proper and its sublevel sets $\{v \mid \varphi(x) \le \alpha\}$ are compact. When $\varphi$ is lsc, these sets are closed, so for them to be compact only boundedness is required, and that can be enforced by a growth condition. The same notions can be invoked for functions in infinite-dimensional spaces rather than just $\mathbb{R}^n$, but then a more sophisticated assessment of compactness is needed.

In $(\mathcal{P})$ it is natural to take $L(t,\cdot,\cdot)$ and $l$ to be proper functions on $\mathbb{R}^n \times \mathbb{R}^n$ that are lsc, but how should $L(t,\cdot,\cdot)$ depend on $t$? This question, more subtle than might be anticipated, was the first big challenge in putting problem $(\mathcal{P})$ on a firm technical footing. For one thing, integrand $L(t,x(t),\dot{x}(t))$ has to be Lebesgue measurable as a function of $t$ for any $x \in \mathcal{A}^1_n[\tau_0,\tau_1]$, but that can fail under the assumption merely that $L(t,x,v)$ is Lebesgue measurable in $t$ for each $(x,v) \in \mathbb{R}^n \times \mathbb{R}^n$, or that $L$ is Lebesgue measurable with respect to $(t,x,v)$. The answer, confirmed from many different angles, has turned out to be that, in combination with $L(t,x,v)$ being lsc with respect to $(x,v)$, $L$ should be measurable with respect to the hybrid $\sigma$-field generated by the Lebesgue sets in the $t$ argument and the Borel sets in the $(x,v)$ argument. With these properties, $L$ is said to be a *normal integrand.*

The concept of a normal integrand was developed originally in the case of convex functions dependent on an additional parameter belonging to a measure space. Thus, it is an innovation attributable to the rise of convex analysis. A full discussion of normal integrands, their history and properties (including various equivalent definitions), is given in Chap. 14 of [14].

With $L$ taken to be a normal integrand, the functional $J$ in $(\mathcal{P})$ can be given the following rigorous definition when recalling that, for any Lebesgue measurable function $\beta$ on $[\tau_0,\tau_1]$ that is majorized by a function $\gamma \in \mathcal{L}^1_1[\tau_0,\tau_0]$ majorizing $\beta$, the integral $\int_{\tau_0}^{\tau_1}\beta(t)dt$ has a well defined value in $[-\infty,\infty)$. This mirrors the fact that the sum of two numbers in $[-\infty,\infty)$ is a well defined number in $[-\infty,\infty)$. Accordingly, for any $x \in \mathcal{A}^1_n[\tau_0,\tau_1]$, the value of $J[x]$ in $(\mathcal{P})$ is a well defined number in $[-\infty,\infty)$ as long as $L(t,x(t),\dot{x}(t))$ is majorized by an integrable function and $l(x(\tau_0),x(\tau_1)) < \infty$. When these conditions are not met, $J[x]$ is defined as equal to $\infty$. It is immediate then that

$$J[x] < \infty \quad \Longrightarrow \quad \begin{cases} (x(t),\dot{x}(t)) \in \operatorname{dom} L(t,\cdot,\cdot) \text{ for a.e. } t, \\ (x(\tau_0),x(\tau_1)) \in \operatorname{dom} l. \end{cases}$$

The principle of infinite penalization reveals therefore that the constraints on the right are implicit in the minimization in problem $(\mathcal{P})$.

Once $(\mathcal{P})$ has been interpreted in this manner, it is possible to move on to the question of whether a solution arc exists. In view of the classical result in Theorem 1, it may be expected that the right space for this is $\mathcal{A}^1_n[\tau_0, \tau_1]$ and that the convexity of $L(t, x, v)$ in $v$ will be required. Something additional will be needed to reflect the relaxation of the fixed endpoint constraints in $(\mathcal{P}_0)$ by the endpoint term in $(\mathcal{P})$. On the other hand, while the Tonelli growth condition (7) might still be serviceable, it is really too severe for many of the targeted applications. For example, Lagrangian functions $L$ coming from control problems, via (13), have a hard time satisfying it.

**Theorem 4.** *Let problem $(\mathcal{P})$ be placed in the arc space $\mathcal{A}^1_n[\tau_0, \tau_1]$. Suppose the function $l$ is lsc and the function $L$ is a normal integrand such that $L(t, x, v)$ is convex with respect to $v$ and the following growth condition is fulfilled:*

$$\begin{cases} L(t,x,v) \geq \theta(\max\{0, |v| - \alpha(t)|x| - \beta(t)\}) - \gamma(t)|x| - \delta(t), \\ l(x_0, x_1) \geq \theta_0(x_0) - \theta_1(x_1), \end{cases} \tag{15}$$

*for integrable functions $\alpha, \beta, \gamma, \delta : [\tau_0, \tau_1] \to [0, \infty]$ and nondecreasing functions $\theta, \theta_0, \theta_1 : [0, \infty) \to [0, \infty]$ such that*

$$\lim_{s \to \infty} \frac{\theta(s)}{s} = \infty, \qquad \lim_{s \to \infty} \frac{\theta_0(s)}{s} = \infty, \qquad \lim_{s \to \infty} \frac{\theta_1(s)}{s} < \infty. \tag{16}$$

*If at least one arc $x \in \mathcal{A}^1_n[\tau_0, \tau_1]$ exists with $J[x] < \infty$, then the minimum in $(\mathcal{P})$ is finite and attained.*

This result comes out of [9], where in fact the growth condition on $L$ is stated in a slightly broader dual form which allows $\theta$ to depend on $t$. Note that this condition imposes a coercive penalty on $\dot{x}(t)$ to the extent that a bound of the form $|\dot{x}(t)| \leq \alpha(t)|x(t)| + \beta(t)$ is transgressed by the arc $x$. It covers the Tonelli condition (7) as a very special case, namely the one where the functions $\alpha$, $\beta$, $\gamma$ and $\delta$ are $\equiv 0$. In the growth condition on $l$, the coercivity in the first argument but counter-coercivity in the second argument could be reversed, i.e., one could assume instead that $l(x_0, x_1) \geq \theta_0(x_1) - \theta_1(x_0)$ without undermining the conclusions.

The cited paper [9] furthermore provides criteria under which the generalized Bolza problem $(\mathcal{P})$ can rightly be considered to reflect, through formulas like (13), an optimal control problem with respect to arcs $x$ and control functions $u$. The means of recovering an optimal $u$ from an optimal $x$ are furnished there as well.

For a "Lagrangian" function $L$ in the vast class envisioned for $(\mathcal{P})$, is there an associated "Hamiltonian" function $H$? A powerful and convincing answer to this question is available through convex analysis. To understand it in clearest terms, let us begin by considering again an arbitrary function $\varphi : I\!R^n \to \overline{I\!R}$. The *Legendre-Fenchel transform* of convex analysis assigns to $\varphi$, as its *conjugate* the function $\varphi^* : I\!R^n \to \overline{I\!R}$ defined by

$$\varphi^*(y) = \sup_v\{\langle v, y\rangle - \varphi(v)\},$$

and as its *biconjugate* the function conjugate to $\varphi^*$, which is

$$\varphi^{**}(v) = \sup_y\{\langle v, y\rangle - \varphi^*(y)\}.$$

Regardless of any assumptions on $\varphi$, both $\varphi^*$ and $\varphi^{**}$ are convex and lsc. If $\varphi \equiv \infty$, then obviously $\varphi^* \equiv -\infty$ and $\varphi^{**} \equiv \infty$, whereas if $\varphi \not\equiv \infty$ but $\varphi$ fails to majorize any affine function, one has $\varphi^* \equiv \infty$ and $\varphi^{**} \equiv -\infty$. In the remaining and most important case, where $\varphi$ is proper and majorizes an affine function (the latter being known to follow from the former when $\varphi$ is convex), both $\varphi^*$ and $\varphi^{**}$ are proper and moreover $\varphi^{**}$ is the greatest proper, lsc, convex function majorized by $\varphi$. Thus in particular, if $\varphi$ is lsc, proper and convex, the same holds for $\varphi^*$, and then $\varphi^{**} = \varphi$. In this manner the Legendre-Fenchel transform induces a one-to-one correspondence within the collection of all lsc, proper, convex functions on $I\!R^n$. A further observation is that if $\varphi$ is twice continuously differentiable with Hessian matrices that are positive definite, the Legendre-Fenchel transform reduces essentially to the older Legendre transform. The details behind these facts can be found in [8] (or [14]).

It is natural from this perspective to define, for any $L : [\tau_0, \tau_1] \times I\!R^n \times I\!R^n \to \overline{I\!R}$, the associated Hamiltonian $H : [\tau_0, \tau_1] \times I\!R^n \times I\!R^n \to \overline{I\!R}$ by

$$H(t, x, y) = \sup_v\{\langle v, y\rangle - L(t, x, v)\}. \tag{17}$$

Then $H(t, x, y)$ is convex in $y$ and, for any $(t, x)$ such that $L(t, x, v)$ is lsc, proper and convex with respect to $v$, it obeys the reciprocal formula

$$L(t, x, v) = \sup_y\{\langle v, y\rangle - H(t, x, y)\}. \tag{18}$$

On the other hand, when $L(t, x, \cdot)$ lacks convexity, but at least is proper and majorizes an affine function, the expression in $v$ given by the right side of (18) is the "lsc convex hull" of $L(t, x, \cdot)$ and is proper. In that case we encounter, therefore, convexification of the kind already viewed in the classical context in connection with the Weierstrass condition.

This approach to what the Hamiltonian for problem $(\mathcal{P})$ should be was first proposed in [11], in the days when only convex analysis offered a means by which optimality conditions could be stated without resorting to gradients, which of course for functions like $L$ and $l$ might well not exist. It was essential then to assume $L(t, x, v)$ to be convex not only with respect to $v$ but with respect to $(x, v)$. In Hamiltonian terms, that corresponds to having $H(t, x, y)$ be concave in $x$ in addition to being, as always, convex in $y$.

It was for convex functions that robust substitutes for gradients were initially developed, so we focus now on that case here. A vector $y \in \mathbb{R}^n$ is called a *subgradient* of a convex function $\varphi : \mathbb{R}^n \to \overline{\mathbb{R}}$ at a point $v$ if

$$\varphi(v') \geq \varphi(v) + \langle y, v' - v \rangle \text{ for all } v' \in \mathbb{R}^n. \tag{19}$$

The set of such subgradients $y$ is denoted by $\partial\varphi(v)$ and is always closed and convex, but perhaps empty. Interestingly $\partial\varphi(v)$ consists of a single $y$ if and only if $\varphi$ is finite and differentiable at $v$, in which case $y = \nabla\varphi(v)$. Furthermore, when $\varphi$ is also lsc and proper, the set-valued subgradient mapping $v \mapsto \partial\varphi(v)$ and the corresponding mapping $y \mapsto \partial\varphi^*(y)$ for the lsc, proper, convex function $\varphi^*$ conjugate to $\varphi$, are the inverses of each other:

$$y \in \partial\varphi(v) \iff v \in \partial\varphi^*(y).$$

An extensive calculus is available for determining subgradients of convex functions in taking advantage of the formulas that may be used to construct such functions. Subgradients can similarly be defined for a function $\varphi$ that is concave instead of convex by reversing the inequality in (19); the notation $\tilde{\partial}\varphi$ then supplants $\partial\varphi$.

To save words, let us now speak formally of the *fully convex* case of problem $(\mathcal{P})$ as the one in which $L(t, \cdot, \cdot)$ and $l$ are lsc, proper and convex on $\mathbb{R}^n \times \mathbb{R}^n$, with $L$ being a normal integrand that in addition satisfies the following minor technical condition: $L(t, x, v) \geq a(t, x, v)$ for an expression $a(t, x, v)$ that is affine in $(x, v)$ and summable in $t$. (The latter is assured for instance under the growth condition on $L$ in (15).) Then in particular, $J$ is a well defined *convex* functional from $\mathcal{A}^1_n[\tau_0, \tau_1]$ to $\overline{\mathbb{R}}$ that nowhere has the value $-\infty$.

Consider further in this setting the subsets $C_l$ and $C_L$ of $\mathbb{R}^n \times \mathbb{R}^n$ that express the two kinds of endpoint constraints implicit in $(\mathcal{P})$; specifically, let $C_l$ consist of all $(x_0, x_1)$ such that $l(x_0, x_0) < \infty$ (i.e., $C_l = \operatorname{dom} l$) and let $C_L$ consist of all $(x_0, x_1)$ such that

$$\exists\, x \in \mathcal{A}^1_n[\tau_0, \tau_1] \text{ with } x(\tau_0) = x_0,\ x(\tau_1) = x_1,\ \int_{\tau_0}^{\tau_1} L(t, x(t), \dot{x}(t))dt < \infty.$$

Note that these sets $C_l$ and $C_L$ are convex, and that

$$J[x] < \infty \implies (x(\tau_0), x(\tau_1)) \in C_l \cap C_L \neq \emptyset. \tag{20}$$

We will want to look also at the *relative interiors* of $C_l$ and $C_L$, which are their interiors relative to their affine hulls (cf. [8] or [14]). In a constraint qualification introduced below, the issue will be whether not only $C_l$ and $C_L$ themselves have nonempty intersection, as in (20), but also their relative interiors have nonempty intersection.

**Theorem 5.** *In the fully convex case of problem* $(\mathcal{P})$, *placed in the arc space* $\mathcal{A}_n^1[\tau_0, \tau_1]$, *let* $x \in \mathcal{A}_n^1[\tau_0, \tau_1]$ *be an arc that is feasible in the sense of having* $J[x] < \infty$. *If there is an arc* $y \in \mathcal{A}_n^1[\tau_0, \tau_1]$ *satisfying the generalized Euler-Lagrange condition*

$$(\dot{y}(t), y(t)) \in \partial_{x,v} L(t, x(t), \dot{x}(t)) \tag{21}$$

*for almost every* $t \in [\tau_0, \tau_1]$ *and also the generalized transversality condition*

$$(y(\tau_0), -y(\tau_1)) \in \partial l(x(\tau_0), x(\tau_1)), \tag{22}$$

*then* $x$ *is optimal. Conversely, the existence of such an arc* $y$ *is necessary for the optimality of* $x$ *if the relative interiors of* $C_l$ *and* $C_L$ *have nonempty intersection.*

*Here the Euler-Lagrange condition (21) can be written equivalently as the generalized Hamiltonian condition*

$$\dot{x}(t) \in \partial_y H(t, x(t), y(t)), \qquad -\dot{y}(t) \in \tilde{\partial}_x H(t, x(t), y(t)). \tag{23}$$

The sufficiency of the generalized Euler-Lagrange and transversality conditions in this theorem was brought out in [10], and the necessity (much harder to prove) in [12]. The equivalence between the Euler-Lagrange condition and the Hamiltonian condition (23) was demonstrated in [11]. These works further reveal that the adjoint arc $y$ solves a certain *dual* problem, which developed in a certain way out of the convex functions conjugate to $L(t, \cdot, \cdot)$ and $l$. This dual problem again fits the generalized Bolza format. (One of the many virtues of the notion of normal integrand, incidentally, is that in passing to the conjugate of $L(t, \cdot, \cdot)$ for each $t$ one gets another normal integrand.)

The optimality conditions in Theorem 5 can be elaborated in fine detail when particular structures are given for $L$ and $l$. A remarkable feature is the way that a very wide range of endpoint formulations and corresponding transversality conditions can all be combined in the single subgradient relation (22). But Theorem 5 only covers problem $(\mathcal{P})$ in the fully convex case.

Can subgradients be defined in for a much broader class of functions in such a manner that the subgradients of convex analysis, as just described, are recovered when the functions are convex, but ordinary gradients are obtained when the functions are smooth? And is it possible that way to derive necessary conditions for optimality resembling those in Theorem 5 even for Bolza problems $(\mathcal{P})$ that are not fully convex? Yes.

A breakthrough in that direction was made in Clarke's thesis [1], as already mentioned. Subsequently the topic were extensively developed further by him and many other researchers, with modifications here and there in the concepts, the most important being the recognition of the need to relinquish certain convexifications that seemed altogether natural at the start of the theory, but later got in the way. In that respect key contributions were made by Mordukhovich; cf. for instance [7].

This larger subject and its history are too much to explain here, but some references can be given. The theory of subgradients in current form is presented comprehensively in the book [14]. Recent achievements in characterizing optimality in terms of subgradient conditions that extend the ones in Theorem 5 can be found for example in [13], [5], [4], [6], [3].

Even today, though, there continue to be new developments in the setting of full convexity, too. In this vein can be mentioned the recent results in the Hamilton-Jacobi theory associated with generalized Bolza problems $(\mathcal{P})$, cf. [15], [16], where the duality theory in [12] has strongly been utilized.

## References

[1] F.H. Clarke (1973), *Necessary Conditions for Nonsmooth Problems in Optimal Control and the Calculus of Variations*, thesis, Univ. Washington.

[2] F.H. Clarke (1983), *Optimization and Nonsmooth Analysis*, Wiley.

[3] A.D. Ioffe (1997), "Euler-Lagrange and Hamiltonian formalisms in dynamic optimization," *Trans. Amer. Math. Soc.* 349, 2871–2900.

[4] A.D. Ioffe and R.T. Rockafellar (1996), "The Euler and Weierstrass conditions for nonsmooth variational problems," *Calculus of Variations and PDE* 4, 59–87.

[5] P.D. Loewen and R.T. Rockafellar (1996) "New necessary conditions for the generalized problem of Bolza," *SIAM J. Control Opt.* 34, 1496–1511.

[6] P.D. Loewen and R.T. Rockafellar (1997) "Bolza problems with general time constraints," *SIAM J. Control Opt.* 35, 2050–2069.

[7] B.S. Mordukhovich (1988), *Approximation Methods in Problems of Optimization and Control*, Nauka, Moscow (in Russian).

[8] R.T. Rockafellar (1970), *Convex Analysis*, Princeton University Press, Princeton, New Jersey (available from 1997 also in paperback in the series Princeton Landmarks in Mathematics).

[9] R.T. Rockafellar (1975), "Existence theorems for general control problems of Bolza and Lagrange," *Advances in Math.* 15, 312–333.

[10] R.T. Rockafellar (1970), "Conjugate convex functions in optimal control and the calculus of variations," *J. Math. Analysis Appl.* 32, 174–222.

[11] R.T. Rockafellar (1970), "Generalized Hamiltonian equations for convex problems of Lagrange," *Pacific J. Math.* 33, 411–428.

[12] R.T. Rockafellar (1971), "Existence and duality theorems for convex problems of Bolza," *Trans. Amer. Math. Soc.* 159, 1–40.

[13] R.T. Rockafellar (1996), "Equivalent subgradient versions of Hamiltonian and Euler-Lagrange equations in variational analysis," *SIAM J. Control Opt.* 34, 1300–1315.

[14] R.T. Rockafellar and R.J-B Wets (1998), *Variational Analysis*, Springer Verlag, Berlin.

[15] R.T. Rockafellar and P.R. Wolenski (2000), "Convexity in Hamilton-Jacobi theory I: dynamics and duality," *SIAM J. Control Opt.*

[16] R.T. Rockafellar and P.R. Wolenski (2001), "Convexity in Hamilton-Jacobi theory II: envelope representations," *SIAM J. Control Opt.*

Chapter 8

# GLOBAL MINIMIZATION AND PARAMETER ESTIMATION IN COMPUTATIONAL BIOLOGY

J.B. Rosen
*Computer Science and Engineering Department*
*University of California, San Diego*
*San Diego, CA 92023*
jbrosen@cs.ucsd.edu

A.T. Phillips
*Department of Computer Science*
*University of Wisconsin - Eau Claire*
*Eau Claire, WI 54702*
phillipa@uwec.edu

S.Y. Oh
*Department of Mathematics*
*Chungnam National University*
*Taejon, Korea 305-764*
soh@math.chungnam.ac.kr

K.A. Dill
*Department of Pharmaceutical Chemistry*
*University of California, San Francisco*
*San Francisco, CA 94143*
dill@maxwell.ucsf.edu

**Abstract** Two important problems in computational biology are the prediction of the structure of a protein, given its sequence of amino acids (the pro-



*N. Hadjisavvas and P.M. Pardalos (eds.), Advances in Convex Analysis and Global Optimization,* 153–180.

tein folding problem), and the docking of a ligand on a protein (drug design). Both of these problems require the computation of the global minimum (or near global minimum) of a differentiable function with many local minima. The Convex Global Underestimator (CGU) algorithm for solving this type of problem has been developed and implemented and successfully tested on a variety of structure prediction and docking problems. The function representing the potential energy of a protein molecule in terms of its dihedral (backbone) angles (a vector $\Phi$) has approximately $2n$ independent variables for a protein consisting of $n$ amino acids.

Also, the function contains parameters ($\alpha$) which must be adjusted so that the structure (determined by the dihedral angles) at the global minimum corresponds to known protein molecular structures. A new algorithm has been developed and implemented which has successfully solved this problem for a number of small test molecules. A structural error function $\rho(\Phi, \alpha)$ is defined which measures the error between the global minimum $\Phi$ and the known native state value of $\Phi$. This global minimum is a function of $\alpha$, and the algorithm finds a minimum in the parameter space of the structural error function $\rho$.

**Keywords:** Global optimization, Parameter estimation, Computational biology, Protein folding, Drug docking.

# 1 INTRODUCTION

Two important problems in computational biology are the prediction of the structure of a protein, given its sequence of amino acids (the protein folding problem), and the docking of a ligand on a protein (drug design). Both of these problems require the computation of the global minimum (or near global minimum) of a differentiable function with many local minima. The Convex Global Underestimator (CGU) algorithm for solving this type of problem has been developed and implemented and successfully tested on a variety of structure prediction and docking problems ([39, 35]). The function representing the potential energy of a protein molecule in terms of its dihedral (backbone) angles (a vector $\Phi$) has approximately $2n$ independent variables for a protein consisting of $n$ amino acids.

Also, the function contains parameters ($\alpha$) which must be adjusted so that the structure (determined by the dihedral angles) at the global minimum corresponds to known protein molecular structures. A new algorithm has been developed and implemented which has successfully solved this problem for a number of small test molecules. A structural error function $\rho(\Phi, \alpha)$ is defined which measures the error between the global minimum $\Phi$ and the known native state value of $\Phi$. This global minimum is a function of $\alpha$, and the algorithm finds a minimum in the parameter space of the structural error function $\rho$.

A major goal of computational biology is to develop a computer algorithm for ab initio protein folding to predict native structures from amino acid sequences without other information ([4]). Such a folding algorithm would be an important complement to experiments. For determining protein structures, crystallography and NMR are slow in some cases and impossible in others. The number of known amino acid sequences is nearly $10^6$ and growing exponentially, while the number of known native structures is only in the thousands. And some amino acid sequences are too unrecognizable to benefit from homology modeling. Despite a number of important advances in computational modeling ( [3, 5, 6, 18, 22, 28, 30, 36, 46, 47, 55, 56, 57]), no general algorithm yet exists.

Protein folding is a difficult interdisciplinary problem, requiring a combination of protein biochemistry, global optimization, and high performance computing. A fast global optimization search method, detailed protein-like models, and accurate potential energy functions are essential to the successful development of a computational protein folding algorithm. Attempting to develop all of these at once is very difficult; such approaches have not yet folded proteins successfully. Instead, our approach is to concentrate on the global optimization search algorithm, and then we and others can use it to improve models. With a fast global search, we can quickly test any energy function model change to determine if it is indeed an improvement. For the purpose of testing our search strategy we use the simple energy function of Sun ([52]). While this function is not an accurate model, it includes many features required in a more realistic (and more complicated) energy function.

## 2 OVERVIEW OF THE CGU SEARCH METHOD

Our global optimization algorithm, the CGU (Convex Global Underestimator) algorithm ([40]), is very different than Molecular Dynamics (MD), Monte Carlo (MC), Simulated Annealing (SA), or Genetic Algorithms (GA). The CGU method does not search the tops of energy landscapes, does not get caught in kinetic traps, and its speed does not depend on the shapes of energy landscapes (amino acid sequences), but only on the sizes of landscapes (for a chain having $n$ degrees of freedom, the computer time is on the order of $n^3$).

Using hundreds of different starting points, for chains of lengths up to 36 monomers, we know that the CGU finds the same unique structure in every run, indicating: (1) that it is probably finding global optima, and (2) that it is doing so with great reliability. Having tried this for many different sequences of different chain lengths, we know that finding global minima is not restricted to certain sequences or specific native structures. Our tests, summarized later, show that the performance of the CGU method is much better than Simulated Annealing ([21]) on these problems, and the advantage grows with chain length. SA seldom finds global optima, so it is not even known how the computer time for SA scales with chain length. Using 32 nodes of a Cray T3E, the CGU finds global minima for 50-mers in 9 hours.

In addition to the global minimum, the CGU method also computes a large number of low energy local minima. This is useful for learning the shapes of energy landscapes and the nature of energy gaps, which is helpful in computational studies of folding kinetics.

# 3 SUMMARY OF RECENT RESULTS

## 3.1 CURRENT SEARCH METHODS ARE MUCH TOO SLOW

Conformational searching is often done by Monte Carlo/Simulated Annealing ([31, 38, 41, 43]), Genetic Algorithms ([51, 54]), Molecular Dynamics ([2, 37]), and by transformation methods, for example based on the diffusion equation ([25, 26, 27, 49]). SA, GA, and MD are much too slow, arguably by 3-10 orders of magnitude. What is the basis for this estimate? First, search methods need to provide some assurance that they reach global, not local, minima. Biomolecules often achieve thermodynamic stability, implying that they usually do not get stuck in kinetic traps. Proteins fold in milliseconds to seconds, while MD all-atom simulations are limited to nanoseconds. Second, search methods must be fast enough to reach the large sizes of real molecules. Current methods often aim to fold crambin (46 amino acids), but most real proteins of interest have hundreds of monomers or more. Search methods must scale gracefully with chain length. Third, search methods must be fast enough to deal with the many degrees of freedom in realistic chain representations in continuum space; lattice models are not sufficient. Current methods are not close to satisfying these requirements, and factor-of-2 type fixes are insufficient.

## 3.2 CURRENT SEARCH METHODS GET STUCK IN KINETIC TRAPS

What is wrong with current search methods? They get stuck in kinetic traps. MC, MD, SA, and GA methods all search the tops of conformational energy landscapes, which have deep valleys separated by hills. When searches fall into local minima, it can take a long time to climb over barriers to proceed toward global minima. The symptom of kinetic trapping is that search times depend more strongly on the shape of the landscape (amino acid sequence) than on its size (number of amino acids). Because it is seldom known whether a search method reaches a true global minimum, or just local minima, remarkably little is known about the speed properties of standard search methods ([1]). The only scaling law that is known for conformational searching is the exact enumeration time, which scales exponentially with chain length ( [4, 44]).

Our CGU method is very different. It is an ab initio method which searches under the landscape, not on top of it. We have shown that it does not get caught in traps: search time depends very little on the amino acid sequence or the native structure of the fold. Search time depends only on the chain length, and the scaling is on the order of $n^3$ (on average) where $n$ is the number of degrees of freedom.

## 3.3 HOW DOES THE CGU METHOD WORK?

The CGU method searches for the globally optimal lowest energy conformation $\Phi_G$. Our approach involves the construction of a convex function which underestimates all known local minima and does so by the least possible amount (Figure 8.1).

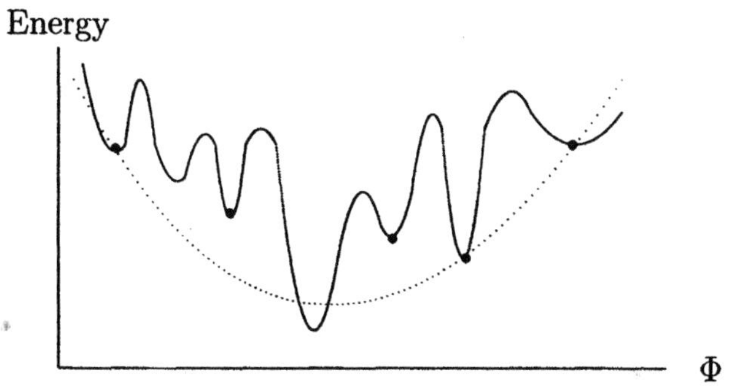


*Figure 8.1* The Convex Global Underestimator (CGU)

Based on the premise that protein folding energy landscapes are funnels with bumps ([8, 29, 48, 58]), there is information about where to find the native structure distributed everywhere throughout the landscape. To find the bottom of a funnel, even a bumpy one, you need to head generally downhill. In general, we know that most well-understood proteins must have such landscapes. The idea of a funnel is nothing more than the statement that proteins fold much faster than the "Levinthal time", the exhaustive search time, and always to the same unique state.

The CGU simply makes use of the overall funnel-like shape to guide and localize the search to regions that are estimated to be near the native structure. This is done iteratively. Specifically, we are given a primary sequence of amino acids with $n$ degrees of freedom $\Phi \in \Re^n$, and a potential energy function $F(\Phi)$. Typically the $n$ degrees of freedom will include at least the backbone dihedral angles $\phi/\psi$, and possibly others, such as the sidechain $\chi$ angles, so that $n \geq 2*$ (the number of residues). Our strategy for searching for the global energy minimum $F_G \equiv F(\Phi_G)$ of $F(\Phi)$ involves an iterative process of three phases during each iteration: (I) sampling the landscape, (II) forming the convex global underestimator surface, a parabolic surface under the lowest minima found so far, and (III) finding the minimum on this underestimator surface.

**Phase I**: $k \geq 2n+1$ local energy minimum conformations $\Phi^{(j)}$ are generated in the search region of interest (a minimum of $2n+1$ conformations are required for construction of the convex underestimator in $n$ dimensions as described below). These conformations may be generated in many ways, but presently they are sampled from a uniform random distribution (over the desired search region) and then relaxed to a local energy minimum state by a Quasi-Newton (QN) continuous minimization technique. This QN approach is well known as a robust general continuous minimization method for this type of problem, and it has the additional benefit of providing an approximation to the Hessian of $F(\Phi)$ at each local solution.

**Phase II**: The CGU function $U(\Phi)$ is then constructed as a more global surface to "fit" these $k$ local minima by underestimating all of them in the least possible amount (i.e. the L1 norm) by solving the optimization problem:

$$\min \sum_{j=1}^{k} \delta_j \tag{8.1}$$

where $\delta_j = F(\Phi^{(j)}) - U(\Phi^{(j)}) \geq 0$ is required for all conformations $j = 1, \ldots, k$. While many choices for the convex underestimating function

$U(\Phi)$ are possible, our approach is to use a separable quadratic function of the form

$$U(\Phi) = c_0 + \sum_{i=1}^{n} (c_i \Phi_i + \frac{1}{2} d_i \Phi_i^2). \tag{8.2}$$

This choice is not essential but has many important benefits. First, convexity of $U(\Phi)$ is easily guaranteed by simply requiring $d_i \geq 0$ in Eq 8.2. Second, since $c_i$ and $d_i$ appear only linearly in the constraints of Eq 8.2, the solution to Eq 8.1 can be computed by a simple linear programming technique, the complete details of which are given in ([40]). Third, the minimum energy conformation of $U(\Phi)$, denoted $\Phi_{PRED}$, is very easily computed by $(\Phi_{PRED})_i = -c_i/d_i$. This conformation then serves as a prediction for $\Phi_G$. In this way the CGU searches under the landscape of $F(\Phi)$ and provides a prediction $\Phi_{PRED}$ which can then be used in Phase III.

**Phase III**: Given the predicted structure $\Phi_{PRED}$ and the best known local minimum structure computed so far, denoted $\Phi_L$, the search region is now localized around $\Phi_{PRED}$ while also including $\Phi_L$ (the boxed region in Figure 8.2).

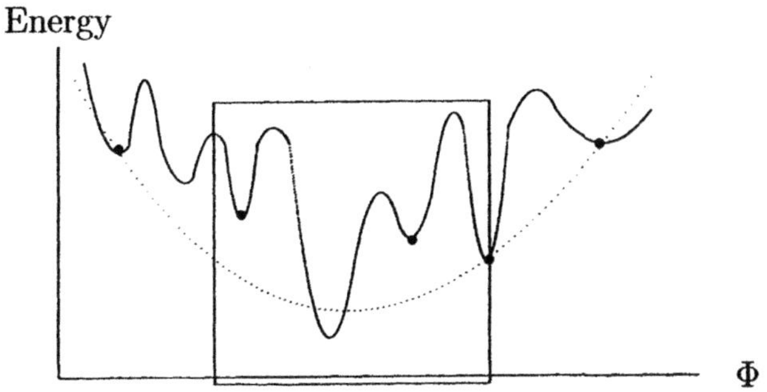


*Figure 8.2* The Reduced Search Region

Phases I-III are repeated over the continually reduced search regions until $\Phi_{PRED} = \Phi_L$. That is, when the CGU predicts $\Phi_{PRED} = \Phi_L$, then the method terminates, and $\Phi_L$ is declared the global minimum energy conformation, denoted $\Phi_{CGU}$.

This CGU method is clearly not *guaranteed* to find the global minimum $\Phi_G$ of the potential function $F(\Phi)$. We are aware of no practical method that makes such a claim for models having continuous degrees of freedom. In fact, we can construct examples for which the CGU method finds $\Phi_{CGU} \neq \Phi_G$. Nevertheless we believe the CGU method may quite robustly find global minima of reasonable models of proteins, based on growing evidence that protein energy landscapes are "folding funnels" with bumps ([8, 29, 48, 58]). Since the lateral area of an energy landscape at a given depth represents the number of conformations having the same internal free energy, the funnel idea is simply that as folding progresses toward lower energies, the chain's conformational options become increasingly narrowed, ultimately resulting in the one native structure. This is fundamentally a consequence of the fact that proteins are heteropolymers. Such landscapes are ideal for the CGU method since the convex quadratic underestimator closely approximates the funnel, and ignores the bumps, as the algorithm narrows its search region. In these cases the CGU method is very likely to succeed and find $\Phi_{CGU} = \Phi_G$.

## 3.4 EVIDENCE THAT THE CGU METHOD DOES WORK

We have evidence that the CGU method does in fact find the global minimum conformations, $\Phi_G$, in the model, for different sequences. Our current tests involve the simple Sun model, which has the form shown in Figure 8.3 where each sidechain $C_s$ is represented by a single united atom (the size for which is dictated by the specific amino acid) and is classified as either hydrophobic or polar. The only degrees of freedom are the pairs of $\phi/\psi$ backbone dihedral angles. The potential function used is of the form

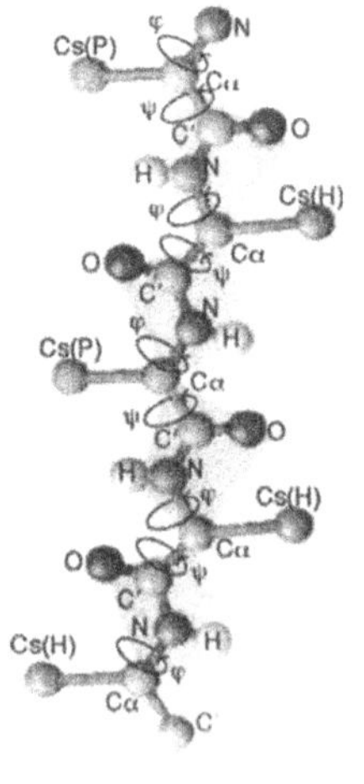


*Figure 8.3* The Simplified Molecular Model

$$F(\Phi) = \sum \alpha_i F_i(\Phi) \tag{8.3}$$

where $\sum \alpha_i = 1$, and $\alpha_i > 0$ are energy function parameters determining the relative weight of each energy term $F_i(\Phi)$. In our tests, we chose a modified version of the Sun energy function ([52]) which includes $F_i$ terms representing hydrogen bond formation, hydrophobic attraction, steric repulsion, and $\phi/\psi$ restrictions based on the Ramachandran maps of the twenty individual amino acids, parameterized from the Protein Databank. Since the CGU method requires a differentiable potential function (in order to apply our continuous minimization method), we approximate the discrete Ramachandran map data by a continuous function of $\phi$ and $\psi$ (see [9] for complete details). However, the CGU search method is still quite general; it depends neither on the level of detail of the molecular model nor on the functional forms of the energy terms $F_i(\Phi)$.

The following evidence suggests that the CGU method is a successful search strategy:

1. Having completed more than 100 trials for each sequence tested (including a 36 residue avian pancreatic polypeptide, a 30 residue zinc finger motif, a 23 residue beta-beta-alpha motif ([50]), 9 residue oxytocin, and 5 residue met-enkephalin), the CGU method consistently finds the same global solution $\Phi_{CGU}$ for each one, from any starting point. The CGU method is not deterministic; that is, the local minimum conformation generation phase (Phase I) is "randomized" so that every trial of the method will sample a different set of starting conformations. Although this is not proof, we take this as good evidence that the solution $\Phi_{CGU}$ is also the global one $\Phi_G$.

2. The native structures found by the CGU, $\Phi_{CGU}$ have lower potential energies than the best ones found by simulated annealing. As the chain length increases, the energy discrepancy, $(F_{SA} - F_{CGU})/|F_{CGU}|$, between the best SA structure, $\Phi_{SA}$, and the best CGU structure, $\Phi_{CGU}$, typically widens as shown in Figure 8.4 (see also [12]). Our tests were based on a widely available simulated annealing code, ASA ([20]). As a fair test, we provided the SA method with the same number of starting conformations as used in all of the conformation generation phases of the CGU method.

   Our model natives (in the Sun model) always have lower energies than the true natives (PDB). In the context of this work, this is good news because it shows that the flaws are in the energy

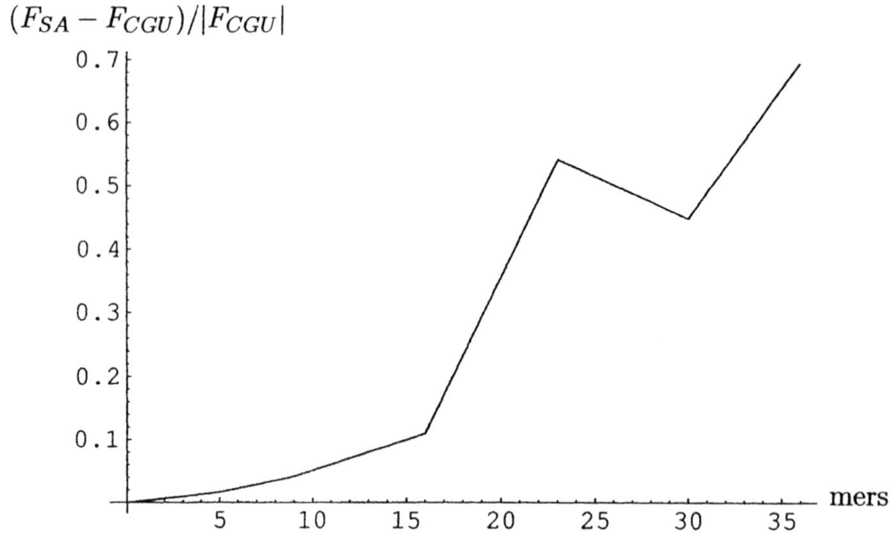


*Figure 8.4* Energy Gap between SA and CGU

function, not the search strategy. The model native structures are not correct, but they are protein-like ([7]): they are compact, with hydrophobic cores and hydrogen-bonded secondary structure (Figure 8.5). Hence, if we can push the search speeds even higher, we could then refine the energy functions, or include more degrees of freedom, to improve the model.

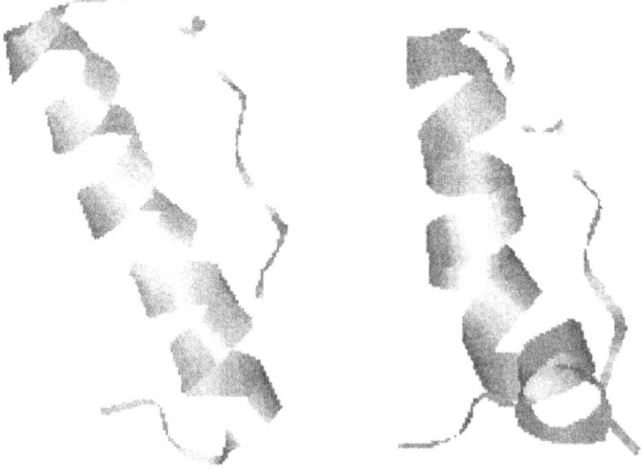

*Figure 8.5* True Native $\Phi_N$ (left) vs Model Native $\Phi_{CGU}$ (right) Structures for PPT

3. The computer time required to find the native state is independent of monomer sequence, monomer composition, and native chain fold. The CGU search time doesn't depend on the shape of the landscape, and the CGU method doesn't get caught in kinetic traps. As shown in ([10]), the running time of the CGU algorithm is insensitive to the monomer sequence. For example, permutations of a 30 residue sequence (in which the percent hydrophobicity remained fixed across all cases) resulted in various model native structures. These computations also showed that the computation time required by the CGU method is approximately independent of monomer sequence. Furthermore, for a given monomer composition, i.e. sequence of hydrophobic (H) and polar (P) type residues, the CGU algorithm is also time invariant (see [10]) with respect to the specific monomer sequence, even though different sequences fold to very different model native structures.

# 4 ENERGY FUNCTION IMPROVEMENT

There are many models in computational chemistry, biology, and materials science, in which parameterized energy functions are used to predict 3-dimensional structures. Folding, threading, docking, protein-protein recognition, and loop refinement methods are examples from computational biology. It is seldom clear whether failures of such models are attributable to the form of the model's mathematical functions, or to poor choices of the parameters used in them. Such models are usually so computationally expensive that it is impossible to be systematic about finding the "optimal" parameters, i.e., the parameters that predict the most accurate structures for the largest number of molecules. To find the optimal parameters, say for a folding algorithm, it would be necessary to: (1) compute the folded structures for many proteins, (2) determine the errors, (3) change the parameters, then iterate this whole process for many different sets of parameters to find the best ones. This is not computationally feasible at the present time. Instead, model parameters are usually chosen to varying degrees by physical estimates, guesswork, and arbitrary trial and error. Such efforts involve small searches through large parameter spaces. As with other types of search problems, parameter optimization can depend on the order in which the parameters are chosen, and it can get caught in traps from which it cannot escape.

It could be argued that the inability to find optimal parameters for folding algorithms is the reason for the emergence of the field of *protein threading*, or fold recognition. The key simplification in threading is that conformational searching takes place in a finite and discrete space that is known to contain the global optimum. Because the searching and comparison process is controlled and discrete, error criteria can be readily devised that can distinguish right from wrong conformations rapidly enough that optimal parameters can be found. Hence there are many methods that optimize parameters for threading ([17, 32, 14, 53, 34, 19, 24]) and for lattice models of folding ( [16, 14, 45, 15]), which also involve discrete and controlled conformational searches.

But as far as we know, no algorithm currently finds optimal parameters for models having continuous degrees of freedom. Of course, there are many approaches to systematic parameter improvement ([11, 33, 23, 16]). But the challenge is to know the global optimum of an energy landscape and how it changes throughout the parameter optimization process. In traditional conformational search methods, it is seldom known whether an energy minimum is local or global, so it is impossible to study how global minima shift with changes in parameters. The CGU method is able to find global minima of energy landscapes, at least for short enough chains ([10, 9]). This makes the CGU an enabling technology for energy function parameter improvement.

We describe here a new computational method that searches for globally optimal parameters for continuous models. It takes as input: (1) a given model and its associated adjustable parameters, (2) the incorrect best structure for each of $m$ different molecules that is predicted by the starting parameters, (3) the $m$ correct (true) structures that the model *should produce*, and (4) some measure of structural error between predicted structures and true, known, or correct best structures.

Our approach is general, and guarantees improved, and sometimes even optimal, parameters. It should be applicable to a wide range of models in computational biology and chemistry. It allows *any* differentiable measure of "structural error" to be minimized over the continuous space of both potential function parameters and molecular conformations, while placing no restrictions on the form of the potential function other than that it be differentiable. Our method simultaneously "tracks" the global minimum for each of the proteins in the model as the parameters are changed, while providing a *guaranteed reduction* in the structural error between model-native and true-native states. For this we do not require any additional global minimization (beyond that needed to get the initial global minima), so the method is computationally efficient. With an appropriate test set selection, it is then expected that the parameters determined for the selected test set of proteins would also apply to proteins not in the test set. Our hope is that by finding better parameters for any given model, this method will ultimately allow computational biologists to develop improved models, for folding, binding, docking, etc.

# 5 PARAMETER OPTIMIZATION

We consider some model for an energy landscape, $F(\Phi, \alpha)$. $F(\Phi, \alpha)$ is the conformational energy or free energy as a function of the $n$ degrees of freedom (coordinates or bond angles, etc.) that are given by the vector $\Phi \in \Re^n$ and as a function of the $k$ parameters of the model (relative strengths of each type of interaction, etc.) that are given by the parameter vector $\alpha \in \Re^k$. Although the method is general, we will make the discussion more concrete by focusing on protein folding. The predicted native state is given by the global minimum vector $\Phi_G$, which has a free energy $F(\Phi_G)$. If the model energy function were correct, it would predict the true native structure; that is, we would have $\Phi_G = \Phi_N$, where $\Phi_N$ is the correct dihedral angle vector for the native state of this protein.

We will consider the prediction accuracy in terms of the error in the degrees of freedom $\|\Phi_G - \Phi_N\|^2$ for each protein. Other measures, such as RMS, could be used with minor modification. To make explicit the dependence of $F$ on the parameters, we let $\alpha$ be a vector of $k$ parameters of the energy function that are common for all the proteins. These parameters might include the relative weightings of hydrophobic, hydrogen bonding, steric, and bond angle terms, or the power-law exponents of interaction potentials, for example (see the test problem in the last section). In the energy functions we are using, we typically have $k \leq 15$.

Suppose we wish to improve the native structure predictions for a set of $m$ proteins, with energy functions $F^{(j)}(\Phi^{(j)}, \alpha)$ and native states $\Phi_N^{(j)}$, $j = 1, \ldots, m$. Note that each potential function $F^{(j)}$ and set of $n_j$ independent variables $\Phi^{(j)}$ will be different, and will depend on the number and sequence of residues in the $j^{th}$ protein. But the vector $\alpha$ is independent of $j$ and will be the same for *all* proteins. For any fixed value of $\alpha$, we can use the CGU algorithm to find the corresponding global minimum $\Phi_G^{(j)}(\alpha)$ for each protein. Each such global minimum is characterized by the conditions

$$F^j(\Phi_G^{(j)}, \alpha) \leq F^{(j)}(\Phi^{(j)}, \alpha), \tag{8.4}$$

$$\nabla_\Phi F^{(j)}(\Phi_G^{(j)}, \alpha) = 0, \text{ and} \tag{8.5}$$

$$H^{(j)}(\Phi_G^{(j)}, \alpha) \text{ positive definite} \tag{8.6}$$

where $H^{(j)}$ is the $(n_j \times n_j)$ Hessian matrix with respect to $\Phi$ of $F^{(j)}(\Phi^{(j)}, \alpha)$. While unlikely, it is possible that for some initial choice $\alpha_I$, some Hessian $H^{(j)}$ may only be positive semidefinite at the corresponding global minimum. If this occurs, a different value of $\alpha_I$ should be chosen. The total conformational error is defined by

$$\rho(\alpha) = \sum_{j=1}^{m} w_j \rho_j(\alpha) \tag{8.7}$$

where $w_j \geq 0$ are any arbitrary weighting factors that the user wants to include, and $\rho_j(\alpha)$ is the $j^{th}$ molecule conformational error, given by

$$\rho_j(\alpha) = \|\Phi_G^{(j)}(\alpha) - \Phi_N^{(j)}\|^2. \tag{8.8}$$

The weights $w_j$ can be selected to give more/less importance to certain proteins, for example based on known accuracies or reliabilities of their structures, but otherwise might normally be set equal to one. The parameter adjustment method proposed here can now be stated as a $k$-dimensional minimization problem, with simple bounds on the parameters

$$\min_\alpha \ \rho(\alpha) \text{ subject to } \alpha_{min} \leq \alpha \leq \alpha_{max}. \tag{8.9}$$

The bounds $(\alpha_{min})_j$ and $(\alpha_{max})_j$ should be chosen to appropriately restrict the range on the parameter $\alpha_j$ to values meaningful to the problem.

The minimization problem given in Eq 8.9, with $k \leq 15$, is small in terms of the number of variables. The major difficulty is that computing the function $\rho(\alpha)$, and its gradient $\nabla_\alpha \rho$, are computationally expensive. Therefore we use an efficient minimization method which requires relatively few function and gradient calls, the Sequential Quadratic Programming (SQP) method ([13]). It is crucial for this minimization problem that we can accurately compute $\Phi_G^{(j)}(\alpha)$ as $\alpha$ is adjusted. We do this by requiring that the system of $n_j$ nonlinear equations in Eq 8.5 remains satisfied as $\alpha$ changes, starting with an initially chosen parameter vector $\alpha_I$. This requirement ensures that $\Phi_G^{(j)}(\alpha)$ will remain a *local* minimum of $F^{(j)}(\Phi^{(j)}, \alpha)$ as $\alpha$ is modified. Note that there is no guarantee that the updated value of $\Phi_G^{(j)}(\alpha)$ will also remain the *global* minimum of

$F^{(j)}(\Phi^{(j)}, \alpha)$, although this is easily checked at the conclusion of the method by another CGU global optimization step.

$\Phi_G^{(j)}(\alpha)$ can be updated without recomputing the global minimum, by using the implicit function theorem, provided that the Hessians $H^{(j)}$ are positive definite. This condition is satisfied at $\Phi_G^{(j)}$, by Eq 8.6. The implicit function theorem uses a linearized approximation to Eq 8.5 at its current values of $\Phi$ and $\alpha$, and therefore gives the required small change $\Delta\alpha$ to balance any small change $\Delta\Phi$. It is valid if the neglected terms in $\|\Delta\alpha\|^2$ and $\|\Delta\Phi\|^2$ are much smaller than the first order term in $\|\Delta\alpha\|$ and $\|\Delta\Phi\|$. For a small change $\Delta\Phi_G^{(j)}$, the corresponding change $\Delta\alpha$ required to keep Eq 8.5 satisfied is given approximately by

$$H^{(j)}(\Phi_G^{(j)}, \alpha)\Delta\Phi_G^{(j)} + J_\alpha^{(j)}(\Phi_G^{(j)}, \alpha)\Delta\alpha = 0 \tag{8.10}$$

where $J_\alpha^{(j)} = J_\alpha^{(j)}(\Phi_G^{(j)}, \alpha)$ are the $n_j \times k$ Jacobians of $\nabla_\Phi F^{(j)}(\Phi_G^{(j)}, \alpha)$. Since the Hessians $H^{(j)} = H^{(j)}(\Phi_G^{(j)}, \alpha)$ are non-singular, we can write this as

$$\Delta\Phi_G^{(j)} = -[H^{(j)}]^{-1} J_\alpha^{(j)} \Delta\alpha. \tag{8.11}$$

Since $\nabla_\alpha \rho_j(\alpha) = -[J_\alpha^{(j)}]^T [H^{(j)}]^{-1} \nabla_\Phi \rho_j(\alpha)$, it then follows that the desired gradient $\nabla_\alpha \rho(\alpha)$ of the total conformation error as a function of parameters is given by

$$\nabla_\alpha \rho(\alpha) = -2 \sum_{j=1}^{m} w_j [J_\alpha^{(j)}]^T [H^{(j)}]^{-1} (\Phi_G^{(j)}(\alpha) - \Phi_N^{(j)}). \tag{8.12}$$

Corresponding to each $\Phi_G^{(j)}$, good approximations to the Hessians $H^{(j)}$ are obtained from the Sequential Quadratic Programming (SQP) implementation of the local minimization, an important phase of the CGU method. Also, the Jacobians $J_\alpha^{(j)}$ are relatively easy to compute since $J_\alpha^{(j)}$ has only $k$ columns.

# 6 THE ENPOP ALGORITHM

We call our method ENPOP (ENergy function Parameter OPtimization). Here is the general algorithm:

1. Guess an initial $\alpha_I$.

2. Run the CGU method successively on the functions $F^{(j)}, j = 1, \ldots, m$ with $\alpha = \alpha_I$ fixed. Denote the result of each CGU run by $\Phi_G^{(j)}$.

3. Using each $\Phi_G^{(j)}$ from step (2), perform the minimization over $\alpha$ described in Eq 8.9. To perform this step, use $\alpha_I$ as a starting point and use Eq 8.12 for the gradient of $\rho$.

   To correctly compute the change $\Delta\Phi_G^{(j)}$ corresponding to the change (at each step of the local minimization) $\Delta\alpha$, we use the implicit function theorem which states that (see Eq 8.10) $\Delta\Phi_G^{(j)} = -[H^{(j)}]^{-1} J_\alpha^{(j)} \Delta\alpha$. This step will result in a new parameter set $\alpha_{new}$ with corresponding global minima $\Phi_G^{(j)}(\alpha_{new}), j = 1, \ldots, m$.

The ENPOP algorithm is quite general and can be applied to any set of molecular structures with a given model and potential function. All that is required is a suitable representation for the Hessian $H^{(j)}$ and Jacobian $J_\alpha^{(j)}$ specific to the model.

# 7 GUARANTEEING REDUCTION IN STRUCTURAL ERROR

The method described above optimizes parameters while enforcing that the initial global minimum must remain at least a local minimum of the model energy function. But what guarantees that the original global minimum will not shift to become just a local minimum? In this section, we describe criteria that insure that the original minimum remains global. The purpose of such criteria is computational efficiency: when the global minimum is no longer assured, the CGU can be run again to check whether or not the current minimum is global. The goal is to run the CGU only the minimum possible number of times, to save computational cost.

When attempting to improve the parameter values for the set of potential functions $F^{(j)}(\Phi_G^{(j)}(\alpha), \alpha)$, for $j = 1, \ldots, m$, it is useful to have prior information about the amount of reduction in the error $\rho(\alpha)$ that can be expected. We now show that a lower bound on the decrease in $\rho(\alpha)$ can be given, based on information available at the initial value $\alpha = \alpha_I$. That is, we show that we can always obtain a parameter vector $\alpha_1$ such that

$$\rho(\alpha_1) \leq \rho(\alpha_I) - \Delta\rho \tag{8.13}$$

where $\Delta\rho$ is given by Eq 8.17. Since $\Delta\rho$ is a guaranteed lower bound, the actual decrease in $\rho(\alpha)$ obtained by ENPOP may typically be much greater.

In order to determine $\Delta\rho$, we need to know several quantities that are available once the global minima $\Phi_G^{(j)}$ have been computed for each protein by the CGU method. For this purpose, we define the gradient in parameter space, $g = \nabla_\alpha \rho(\alpha_I)$, as given by Eq 8.12, and the initial energy gaps $\Delta F^{(j)}$ for $j = 1, \ldots, m$ and $\alpha = \alpha_I$, by

$$\Delta F^{(j)} = F_{LM}^{(j)} - F_G^{(j)} > 0. \tag{8.14}$$

In Eq 8.14, $F_G^{(j)} \equiv F^{(j)}(\Phi_G^{(j)}(\alpha_I), \alpha_I)$ represents the global minimum energy found by the CGU method using the parameter vector $\alpha_I$, and $F_{LM}^{(j)}$ is the corresponding next lowest local minimum energy, both for the $j^{th}$ protein. If the gradient $\|g\| = 0$, then $\alpha_I$ is already a stationary point of $\rho(\alpha)$, and a new value of $\alpha_I$ must be selected. Similarly, if $\Delta F^{(j)} = 0$, for any $j$, then any change in $\alpha$ may cause the coordinates of the global minimum for $F^{(j)}(\Phi(\alpha), \alpha)$ to move discontinuously from its current conformation $\Phi_G^{(j)}$ to the conformation corresponding to the alternate global minimum (since $F_{LM}^{(j)} = F_G^{(j)}$). In this case, the $j^{th}$ protein should be replaced (or removed) in the test set. Therefore, we can assume that Eq 8.14 holds for all $j = 1, \cdots, m$, and also that $\|g\| > 0$.

In addition, we will also require the quantities $\Delta F_{min} \equiv \min_j \Delta F^{(j)} > 0$, the Hessian $H_\rho(\alpha)$ of $\rho(\alpha)$, and the curvature of $\rho$ in the direction of $g$ as given by

$$\lambda = g^T H_\rho g / g^T g. \tag{8.15}$$

The value of $\lambda$ is bounded by $\lambda_{min} \leq \lambda \leq \lambda_{max}$, where $\lambda_{min}$ and $\lambda_{max}$ are the minimum and maximum eigenvalues of $H_\rho(\alpha_I)$. Since $H_\rho(\alpha)$ may be indefinite at $\alpha_I$, $\lambda$ may be negative.

Finally, we need a uniform bound on the gradient of $F(\Phi(\alpha), \alpha)$ with respect to $\alpha$:

$$\|\nabla_\alpha F(\Phi(\alpha), \alpha)\| \leq \beta. \tag{8.16}$$

In terms of these quantities, it can be shown that:

$$\Delta\rho = \begin{cases} \|g\|\Delta F_{min}/2\beta & \lambda \leq 0 \\ \|g\|\Delta F_{min}/4\beta & \lambda > 0 \text{ and } \lambda\Delta F_{min} \leq 2\beta\|g\| \\ \|g\|^2/2\lambda & \lambda > 0 \text{ and } \lambda\Delta F_{min} > 2\beta\|g\| \end{cases} \tag{8.17}$$

In the first two bounds in Eq 8.17, the change in $\alpha$ is limited by the possibility that the current global minimum is replaced by one of the other local minima (it is assumed that a new global minimum will arise only from existing local minima). In the third bound, the decrease in $\rho(\alpha)$ is at least that obtained by a single steepest descent step in the direction $-g$ in $\alpha$-space.

We now show the results for a computational test of ENPOP. The test involves three short protein molecules, in two-dimensional conformational space, using a simple energy function with four parameters. ENPOP reduces the total conformation error from its initial value $\rho(\alpha_I) = 6.98$, to its minimum $\rho(\alpha_N) = 0.045$.

## 8 TEST PROBLEM: 2D PROTEIN FOLDING MODEL

We consider a problem involving 3 short chain molecules, in two-dimensional conformation space, using a simple energy function. The energy function consists of 4 terms, a bond length penalty term, a Lennard-Jones (LJ) attraction/repulsion term, a hydrophobic attraction term, and a hydrogen bond term. The energy function contains 4 adjustable parameters. The test molecules have either 12 or 13 residues and differ from each other in the specification of which residues are hydrophobic, and which pairs are hydrogen bonded. The native state conformations are given. The 4 parameters are adjusted by ENPOP so that the global minimum of the energy function, for each molecule, gives a corresponding conformation as close as possible to its known native conformation.

In order to simplify both the calculation and the discussion, the degrees of freedom are the $(x, y)$ coordinates of each residue, $(x_i, y_i), i = 1, \cdots, n, n = 12$ or 13. For each molecule, residue 1 is fixed at the origin. We let $\mathbf{X}$ denote the vector with $2n$ elements $(x_i, y_i), i = 1, \cdots, n$. The conformation of a molecule is then specified by the vector $\mathbf{X}$ . Since the distance $r_{ij}$ between residues $i$ and $j$ is given by ${r_{ij}}^2 = (x_i - x_j)^2 + (y_i - y_j)^2$, the vector $\mathbf{X}$ completely specifies all pairwise distances between residues. We also let $\alpha \in \Re^4$, denote the parameter vector specifying the 4 parameter values $\alpha_i, i = 1, \cdots, 4$.

The energy function is:

$$
\begin{aligned}
F(\mathbf{X}, \alpha) = \; & 50 \sum_{i=1}^{n-1} [r_{i,i+1} - 1.0]^2 + \sum_{i=1}^{n-1} \sum_{j>i} L(r_{ij}, \alpha_1, \alpha_2) \\
& - \alpha_3 \sum_{i,j \in H} {r_{ij}}^{-6} - \alpha_4 \sum_{i,j \in HB} {r_{ij}}^{-6}
\end{aligned}
\tag{8.18}
$$

where, $\alpha_i > 0$, and

$$
L(r, \alpha_1, \alpha_2) = \alpha_1[-2r^{-6} + 0.1\alpha_2 r^{-12}]. \tag{8.19}
$$

$H$ denotes the set of hydrophobic residues, and $HB$ denotes the set of hydrogen bonded pairs. Note that the minimum in the LJ term occurs at $r_{min} = (0.1\alpha_2)^{\frac{1}{6}}$, where it has the value $\frac{-10\alpha_1}{\alpha_2}$.

For each test molecule the sets of hydrophobic and hydrogen bonded residues are chosen to be different, while the parameter values are the same for all molecules. We represent the energy function for the $j$th test molecule by $F_j(\mathbf{X}_j, \alpha), j = 1, 2, 3$. For each test molecule we have a known native state conformation $\mathbf{X}_{Nj}, j = 1, 2, 3$.

For Molecule 1 there are 13 residues, with residues 11 and 13 hydrophobic: $H = (11, 13)$; and hydrogen bonded pairs: $HB = (1, 4), (3, 6), (5, 8), (7, 10), (9, 12)$. For Molecule 2 there are 13 residues, with $H = (4, 13)$; $HB = (2, 5), (6, 10)$. For Molecule 3, there are 12 residues, with $H = (5, 12)$; $HB = (1, 8), (3, 6), (7, 10)$.

We start with an arbitrary initial choice $\alpha_I$ of the parameter vector. Corresponding to $\alpha_I$, are global minimum energy conformations $\mathbf{X}_{Ij}$, obtained by minimizing $F_j(\mathbf{X}_j, \alpha_I), j = 1, 2, 3$. These initial and native conformations are shown in Figures 8.6, 8.7, and 8.8. Let $\mathbf{X}_j(\alpha)$ denote the minimum energy conformation for any $\alpha$. We have $\mathbf{X}_j(\alpha_I) = \mathbf{X}_{Ij}$. We aim to find $\alpha_N$, so that the conformation error, $\rho(\alpha)$ is minimized, where

$$
\rho(\alpha) = \sum_{j=1}^{3} \|\mathbf{X}_j(\alpha) - \mathbf{X}_{Nj}\|_2. \tag{8.20}
$$

To illustrate the process of parameter optimization, we explore trajectories in parameter space, from the initial to final parameter vectors. We follow the steepest descent path in parameter space to find the native parameter vector $\alpha_N$, such that $\rho(\alpha_N)$ is a minimum. At each descent step $\Delta\alpha$ we reduce $\rho(\alpha)$ by a small amount. The corresponding changes in the function values $F_j$, and corresponding conformations $\mathbf{X}_j(\alpha+\Delta\alpha)$, are then obtained by local minimizations of $F_j(\mathbf{X}_j, \alpha+\Delta\alpha)$, starting with $\mathbf{X}_j(\alpha)$. The Sequential Quadratic Programming code NPSOL ([13]) is

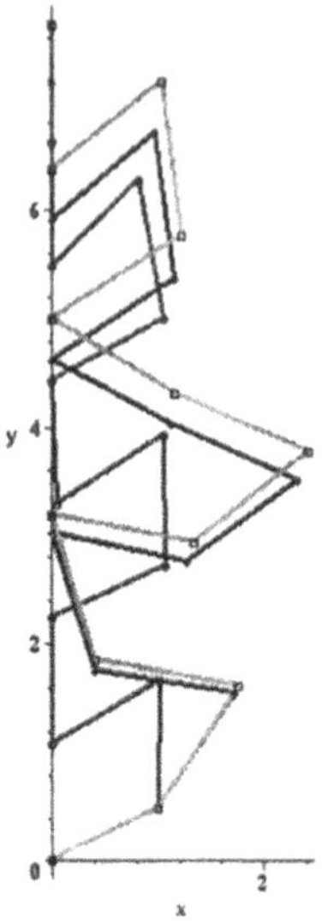


*Figure 8.6* Conformation change in test molecule 1. Green: Initial, Blue: Intermediate, Red: Final

used for this local minimization, and since the conformation change due to $\Delta\alpha$ is small, this computation is fast. A typical steepest descent cal-

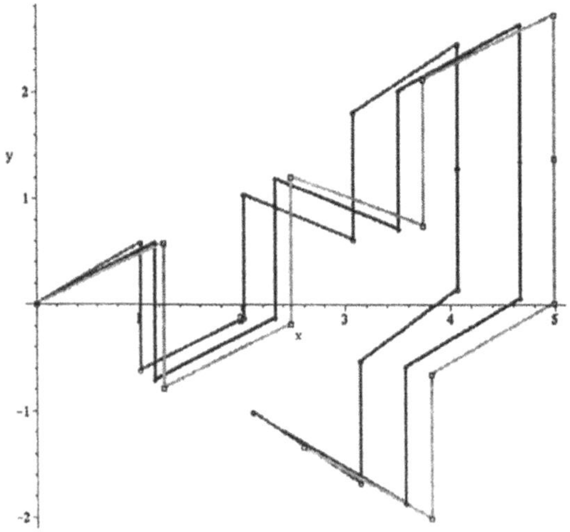


*Figure 8.7* Conformation change in test molecule 2. Green: Initial, Blue: Intermediate, Red: Final

culation in parameter space requires several hundred steps and takes 10 - 30 minutes on a current workstation. Note that the more efficient SQP minimization method could also have been used in parameter space, but it would have given much less information about the nature of the contours on the $\rho(\alpha)$ surface.

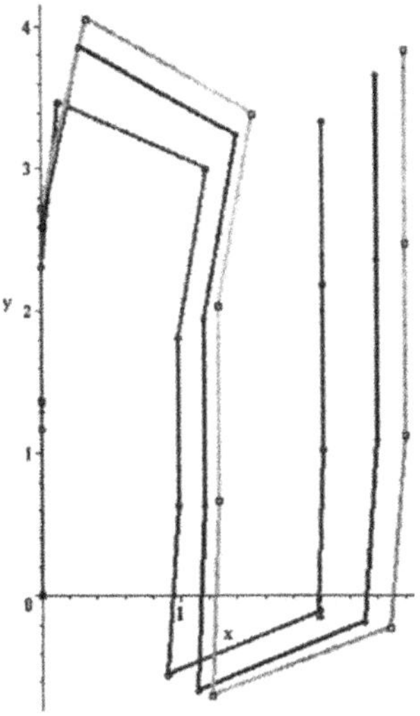


*Figure 8.8* Conformation change in test molecule 3. Green: Initial, Blue: Intermediate, Red: Final

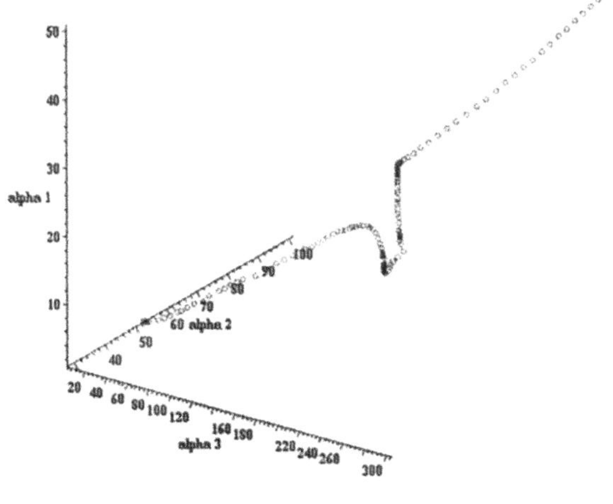


*Figure 8.9* Parametric space steepest descent path

For this example, the conformation changes for each of the 3 molecules as $\alpha$ changed from $\alpha_I$ to $\alpha_N$, are shown in Figures 8.6, 8.7, and 8.8. The initial conformations $\mathbf{X}_{Ij}$, $j = 1, 2, 3$, are shown in green, intermediate conformations are shown in blue, and the final conformations $\mathbf{X}_j(\alpha_N) \cong \mathbf{X}_{Nj}$, are shown in red. The initial values of the parameters were $\alpha_I = (50, 100, 300, 15)$. The parameter values that minimized $\rho(\alpha)$ were $\alpha_N = (1.89, 50.6, 9.80, 9.58)$. The initial value of the conformation error was $\rho(\alpha_I) = 6.98$, and its final value was $\rho(\alpha_N) = 0.045$. The difference between the desired native conformations $\mathbf{X}_{Nj}$ and the final computed conformations $\mathbf{X}_j(\alpha_N)$ are so small that they cannot be distinguished in the Figures 8.6, 8.7, and 8.8. The manner in which the 4 parameters changed along the steepest descent path is shown in Figures 8.9 and 8.10, which show the trajectory in parameter space in moving from $\alpha_I$ to $\alpha_N$, along the steepest descent path for $\rho(\alpha)$. Figure 8.9 shows the trajectory in $(\alpha_3, \alpha_2, \alpha_1)$ space, and Figure 8.10 shows the trajectory in $(\alpha_3, \alpha_2, \alpha_4)$ space. It was found that $\alpha_3$ decreased monotonically from 300 to 9.80 along the descent path, whereas $\alpha_1, \alpha_2, \alpha_4$, each had at least one decrease and increase along the steepest descent path. In fact, $\alpha_4$ initially increases rapidly from 15 to about 78, before decreasing more slowly to 9.58. This shows that while a descent path exists from $\alpha_I$ to $\alpha_N$, it is somewhat circuitous in the parameter space. Since $\alpha_3$ decreases monotonically along the descent path we can use it to represent distance along the path. Therefore, we show in Figure 8.11 the conformation error $\rho(\alpha)$ as a function of $\alpha_3$. It is seen that $\rho(\alpha)$ decreases monotonically (although very quickly at first) from $\rho(\alpha_I) = 6.98$ to $\rho(\alpha_N) = 0.045$, as $\alpha_3$ changes from 300 to 9.80.

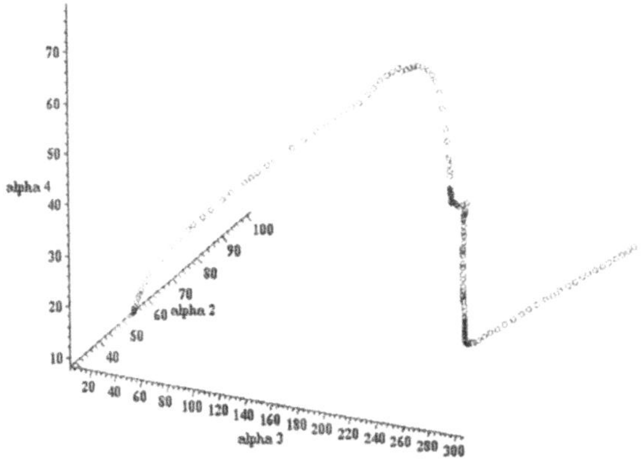


*Figure 8.10* Parametric space steepest descent path

# 9 SUMMARY

We have described the need for efficient methods for global optimization and parameter estimation in computational biology, and have presented two new algorithms. The CGU algorithm usually finds the global minimum of a continuous energy function for small protein-like molecular models. The ENPOP algorithm improves the set of parameter values in an energy function so that the global minimum conformation of a molecular model is as close as possible to its known native state. Computational results for both algorithms have been summarized. Additional computational results for the ENPOP algorithm are given in ( [42]). Based on the results presented here both algorithms should be applicable to larger and more realistic computational models in protein folding and docking.

# References


[1] Beutler, T. and Dill, K. (1996). A fast comformational search strategy for finding low energy structures of model proteins. *Protein Science*, 5:2037–2043.

[2] Bishop, T., Heller, H., and Schulten, K. (1997). Molecular dynamics on parallel computers: applications for theoretical biophysics.


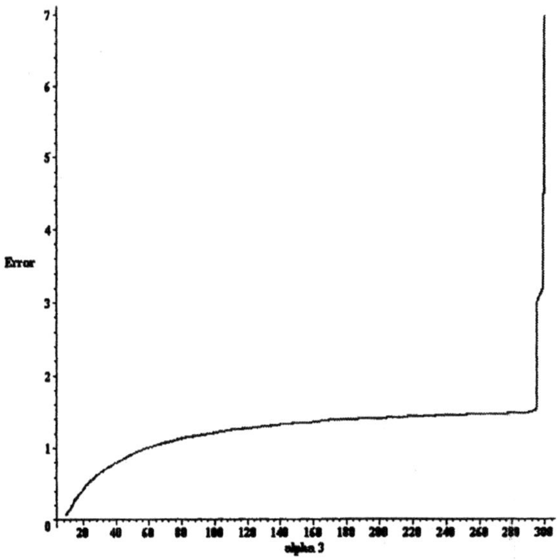


*Figure 8.11* Decrease in conformation error along steepest descent path

In Kalia, R. and Vashishta, P., editors, *Toward Teraflop Computing and New Grand Challenge Applications*, pages 129–138.

[3] Boczko, E. and Brooks, C. (1995). First-principles calculation of the folding free energy of a three-helix bundle protein. *Science*, 269:393–396.

[4] Chan, H. and Dill, K. (1993). The protein folding problem. *Physics Today*, February 1993:24–32.

[5] Covell, D. (1992). Folding protein alpha-carbon chains into compact forms by Monte Carlo methods. *Proteins: Struct Funct Genet*, 14:409–420.

[6] Covell, D. (1994). Lattice model simulations of polypeptide chain folding. *J Mol Biol*, 235:1032–1043.

[7] Dill, K. (1990). Dominant forces in protein folding. *Biochemistry*, 29(31):7133–7155.

[8] Dill, K. and Chan, H. (1997). From Levinthal to pathways to funnels. *Nature Structural Biology*, 4(1):10–19.

[9] Dill, K., Phillips, A., and Rosen, J. (1997a). Molecular structure prediction by global optimization. In et al, I. B., editor, *Developments in Global Optimization*, pages 217–234.

[10] Dill, K., Phillips, A., and Rosen, J. (1997b). Protein structure and energy landscape dependence on sequence using a continuous energy function. *Journal of Computational Biology*, 4(3):227–239.

[11] Esposito, W. and Floudas, C. (1998). Global optimization in parameter estimation of nonlinear algebraic models via the error-in-variable approach. *Ind. Eng. Chemistry & Research*, 37:1841–1858.

[12] Foreman, K., Phillips, A., Rosen, J., and Dill, K. (1999). Comparing search strategies for finding global optimia on energy landscapes. *Journal of Computational Chemistry*, 20:1527–1532.

[13] Gill, P., Murray, W., Saunders, M., and Wright, M. (1986). User's guide for NPSOL (Version 4.0): a Fortran package for nonlinear programming. Technical report, Department of Operation Research, Stanford University.

[14] Goldstein, R., Luthy-Schulten, Z., and Wolynes, P. (1992). Protein tertiary structure recognition using optimized Hamiltonians with local interactions. *Proc. Natl. Acad. Sci. USA*, 89:9029–9033.

[15] Govindarajan, S. and Goldstein, R. (1995). Searching for folding protein structures using optimixed energy functions. *Biopolymers*, 36:43–51.

[16] Hao, M. and Scheraga, H. (1996). How optimization of potential functions affects protein folding. *Proc. Natl. Acad. Sci. USA*, 93:4984–4989.

[17] Hendlich, M., Lackner, M., Weitckus, S., and Floeckner, H. (1990). Identification of native protein folds amongst a large number of incorrect models - the calculation of low energy conformations from potentials of mean force. *J. Mol. Biol.*, 216:167–180.

[18] Hinds, D. and Levitt, M. (1994). Exploring conformational space with a simple lattice model for protein structure. *J Mol Biol*, 243:668–682.

[19] Huber, T. and Torda, A. (1998). Protein fold recognition without Boltzmann statistics or explicit physical basis. *Prot. Sci.*, 7:142–149.

[20] Ingber, L. (1989). Very fast simulated re-annealing. *J. Mathl. Comput. Modeling*, 12:967–973.

[21] Kirkpatrick, S., Gelatt, C., and Vecchi, M. (1983). Optimization by simulated annealing. *Science*, 220(4598):671–680.

[22] Kolinski, A. and Skolnick, J. (1994). Monte Carlo simulations of protein folding. I. Lattice model and interaction scheme. *Proteins: Struct Funct Genet*, 18:338–352.

[23] Koretke, K., Luthey-Schulten, Z., and Wolynes, P. (1998). Self-consistently optimized energy function for protein structure prediction by molecular dynamics. *Proc. Natl. Acad. Sci. USA*, 95:2932–2937.

[24] Koretke, K., Luthy-Schulten, Z., and Wolynes, P. (1996). Self-consistency optimized statistical mechanical energy functions for sequence structure alignment. *Prot. Sci.*, 5:1043–1059.

[25] Kostrowicki, J. and Piela, L. (1991). Diffusion equation method of global minimization: performance for standard test functions. *JOTA*, 69:269–284.

[26] Kostrowicki, J., Piela, L., Cherayil, B., and Scheraga, H. (1991). Performance of the diffusion equation method in searches for optimum structures of clusters of Lennard-Jones atoms. *J Phys Chem*, 95:4113–4119.

[27] Kostrowicki, J. and Scheraga, H. (1992). Application of the diffusion equation method for global optimization to oligopeptides. *J Phys Chem*, 96:7442–7449.

[28] Kuntz, I., Crippen, G., Kollman, P., and Kimelman, D. (1976). Calculation of protein tertiary structure. *J Mol Biol*, 106:983–994.

[29] Leopold, P., Montal, M., and Onuchic, J. (1992). Protein folding funnels: A kinetic approach to the sequence structure relationship. *Proc Natl Acad Sci USA*, 89:8721–8725.

[30] Levitt, M. and Warshel, A. (1975). Computer simulation of protein folding. *Nature*, 253:694–698.

[31] Li, Z. and Scheraga, H. (1987). Monte Carlo minimization approach to the multiple minima problem in protein folding. *Proc Natl Acad Sci USA*, 84:6611–6615.

[32] Maiorov, V. and Crippen, G. (1992). Contact potential that recognizes the correct folding of globular proteins. *J. Mol. Biol.*, 227:876–888.

[33] Maiorov, V. and Crippen, G. (1994). Learning about protein folding via potential functions. *Proteins Struct. Funct. Genet.*, 20:167–173.

[34] Mirny, L. and Shakhnovich, E. (1996). How to derive a protein folding potential? A new approach to an old problem. *J. Mol. Biol.*, 264:1164–1179.

[35] Mitchell, J., Phillips, A., Rosen, J., and Eyck, L. T. (2000). A coupled scanning optimization scheme for analyzing molecular interactions. In Floudas, C. and Pardalos, P., editors, *Optimization in Computational Chemistry and Molecular Biology*, pages 191–207.

[36] Monge, A., Friesner, R., and Honig, B. (1994). An algorithm to generate low-resolution protein tertiary structures from knowledge of secondary structure. *Proc Natl Acad Sci USA*, 91:5027–5029.

[37] Nelson, M., Humphrey, W., Gursoy, A., Dalke, A., Kalé, L., Skeel, R., and Schulten, K. (1997). NAMD – A parallel, object oriented molecular dynamics program. *International Journal of Supercomputing Applications and High Performance Computing*, 10:251–268.

[38] O'Toole, E. and Panagiotopoulos, A. (1992). Monte Carlo simulation of folding transitions of simple model proteins using a chain growth algorithm. *J Chem Phys*, 97:8644–8652.

[39] Phillips, A., Rosen, J., and Dill, K. (2000). Energy landscape projections of molecular potential functions. In Floudas, C. and Pardalos, P., editors, *Optimization in Computational Chemistry and Molecular Biology*, pages 47–55.

[40] Phillips, A., Rosen, J., and Walke, V. (1995). Molecular structure determination by global optimization. *Dimacs Series in Discrete Mathematics and Theoretical Computer Science*, 23:181–198.

[41] Ripoll, D. and Thomas, S. (1990). A parallel Monte Carlo search algorithm for the conformational analysis of proteins. *Proc IEEE/ACM Supercomputing '90*, pages 94–102.

[42] Rosen, J., Phillips, A., Oh, S., and Dill, K. (2000). A method for parameter optimization in computational biology. *Biophysics Journal*, 79:2818–2824.

[43] Shakhnovich, E., Farztdinov, G., Gutin, A., and Karplus, M. (1991). Protein folding bottlenecks: a lattice Monte Carlo simulation. *Phys Rev Lett*, 67:1665–1668.

[44] Shakhnovich, E. and Gutin, A. (1990). Enumeration of all compact conformations of copolymers with random sequence of links. *J Chem Phys*, 93:5967–5971.

[45] Shrivastava, I., Vishveshwara, S., Cieplak, M., et al. (1995). Lattice models for rapidly folding protein-like heteropolymers. *Proc. Natl. Acad. Sci. USA*, 92:9206–9209.

[46] Sippl, M., Hendlich, M., and Lackner, P. (1992). Assembly of polypeptide and protein backbone conformations from low energy ensembles of short fragments: Development of strategies and construction of models for myoglobin, lysozyme, and thymosin beta 4. *Protein Science*, 1:625–640.

[47] Skolnick, J. and Kolinski, A. (1990). Simulations of the Folding of a Globular Protein. *Science*, 250:1121–1125.

[48] Socci, N. and Onuchic, J. (1994). Folding kinetics of protein-like hetropolymers. *J Chem Phys*, 100:1519–1528.

[49] Stillinger, F. (1985). Role of potential-energy scaling in the low-temperature relaxation behavior of amorphous materials. *Phys. Rev. B*, 32:3134–3141.

[50] Struthers, M., Cheng, R., and Imperiali, B. (1996). Design of a monomeric 23-residue polypeptide with defined tertiary structure. *Science*, 271:342–345.

[51] Sun, S. (1993). Reduced representation model of protein structure prediction: statistical potential and genetic algorithms. *Protein Science*, 2:762–785.

[52] Sun, S., Thomas, P., and Dill, K. (1995). A simple protein folding algorithm using binary code and secondary structure constraints. *Protein Engineering*, 8(8):769–778.

[53] Thomas, P. and Dill, K. (1996). An iterative method for extracting energy-like quantities from protein structure. *Proc. Natl. Acad. Sci. USA*, 93:11628–11633.

[54] Unger, R. and Moult, J. (1993). Genetic algorithms for protein folding simulations. *J Mol Biol*, 231:75–81.

[55] Vajda, S., Jafri, M., Sezerman, O., and DeLisi, C. (1993). Necessary conditions for avoiding incorrect polypeptide folds in conformational search by energy minimization. *Biopolymers*, 33:173–192.

[56] Wallqvist, A., Ullner, M., and Covell, D. (1994). A simplified amino acid potential for use in structure predictions of proteins. *Proteins: Struct Funct Genet*, 18:267–280.

[57] Wilson, C. and Doniach, S. (1989). A computer model to dynamically simulate protein folding – Studies with crambin. *Proteins: Struct Funct Genet*, 6:193–209.

[58] Wolynes, P., Onuchic, J., and Thirumalai, D. (1995). Navigating the folding routes. *Science*, 267:1619–1620.

Chapter 9

# LAGRANGIAN QUADRATIC BOUNDS IN POLYNOMIAL NONCONVEX AND BOOLEAN MODELS WITH SUPERFLUOUS CONSTRAINTS

N.Z. Shor
*Institute of Cybernetics, NAS of Ukraine*
*Kiev, Ukraine*
nzshor@d120.icyb.kiev.ua

**Abstract** The article is devoted to the Lagrangian methods of obtaining bounds for optimal value of nonlinear optimization problems, especially for that of quadratic type problems. Calculation of such bounds is reduced in many cases to the semidefinite programming problems which can be solved by modern methods of nondifferentiable optimization. For obtaining more exact bounds the author proposed to use the so-called superfluous constraints which do not change the optimal value of original problem but lead to modification of Lagrange function. The presented technique is illustrated on examples of finding global minimum of polynomial functions and two combinatorial optimization problems: finding an optimal weighted stable set in a graph and the maximal cut problem.

**Keywords:** Lagrangian bounds, nondifferentiable optimization, quadratic-type problems, polynomial problems, superfluous constraints, extremal graph problems, semidefinite programming.

## 1 LAGRANGIAN BOUNDS AND NONDIFFERENTIABLE OPTIMIZATION

Lagrangian bounds are the source of many interesting problems of nondifferentiable optimization. Consider the problem of nonlinear programming in general form: find

$$f^* = \inf f_0(x), \quad x \in X \subseteq E^n,$$



*N. Hadjisavvas and P.M. Pardalos (eds.), Advances in Convex Analysis and Global Optimization,* 181–204.

$$\text{subject to } f_i(x) = 0, \quad i = 1, \ldots, m,$$

where $E^n$ is $n$-dimensional Euclidean space, $X$ is a closed set in this space, $f_0, f_1, \ldots, f_m$ are continuous functions, determined on $E^n$. (If the problem (9.0) has no feasible solution we set $f^* = \infty$.) Let us form usual Lagrange function:

$$L(x, u) = f_0(x) + \sum_{i=1}^{m} u_i f_i(x).$$

where $u = (u_1, \ldots, u_m)$ is a vector of Lagrange multipliers.

For each $\overline{u} \in R^m$ we obtain the local problem: to find

$$\psi(\overline{u}) = \inf_{x \in X} L(x, \overline{u}). \tag{9.1}$$

It is easy to verify that $\psi(\overline{u}) \leq f^*$ for arbitrary $\overline{u}$.

We try to find the best lower bound for $f^*$ in this class of Lagrangian estimates and obtain the coordinating problem: to find

$$\psi^* = \sup_u \psi(u).$$

It is clear, that $\psi^* \leq f^*$. $\psi(u)$ is a concave function, but not always proper. Let $\psi(u)$ be a proper function and *dom*$\psi$ has dimension $m$. If $\overline{u} \in$ *intdom*$\psi$ and $x(\overline{u})$ is a solution of local problem (9.1) when $u = \overline{u}$. Then the supergradient $g_\psi(\overline{u})$ of $\psi$ in the point $\overline{u}$ can be chosen as

$$\{f_i(x(\overline{u}))\}_{i=1}^{m}.$$

If $x(\overline{u})$ is a unique solution of the problem (9.1), then $g_\psi(\overline{u})$ is the gradient of $\psi$ in the point $\overline{u}$. As a rule, the points $\overline{u}$ where $x(\overline{u})$ is not single are the points of nondifferentiability of function $\psi$. Thus if we have an algorithm for solving local problems (9.1) we can at once calculate a supergradient of $\psi$ in the corresponding point. So there are good possibilities for using the subgradient-type methods for calculation $\psi^*$. One of the most effective methods for solving various complex problems of nondifferentiable optimizations is the subgradient-type method with space dilation in the direction of difference of two successive subgradients (so-called $r$-algorithm). Proposed by the author in 1970 (see [2]) family of algorithms becomes one of the main tools for solving various problems: the realization of some decomposition schemes for solving large-scale problems; finding of Lagrangian bounds for many problems of multiextremal and boolean type; for solving complex problems with constraints by using exact nonsmooth penalty functions; for obtaining the optimal parameters of inscribed ellipsoids and superscribed ellipsoids and so on (see [1]).

Now we shall give a short description of $r$-algorithm.

Let $f(x)$ be a convex function defined on $n$-dimensional Euclidean space $E^n$, and $f(x) \to +\infty$, when $\|x\| \to \infty$. By $g_f(x)$ we denote a subgradient of function $f$ in the point $x$.

Consider a general scheme of $r$-algorithm.

**0-step.** Initial point $x_0 \in E^n$ and nonsingular matrix $n \times n$ $B_0$ are given (for example, $B_0 = I_n$ (identity matrix)).

**1-st step.** Calculate:

$$g_0 = g_f(x_0);\ \tilde{g}_0 = B_0^T g_0;\ \xi_0 = \frac{\tilde{g}_0}{\|\tilde{g}_0\|};$$

$$x_1 = x_0 - h_0 B_0 \xi_0;\ (h_0 > 0 \text{ is the step multiplier }). \tag{9.2}$$

Remember $x_1$, $g_f(x_0)$, $B_0$. After $k$-steps we have:

$$x_k \in E^n,\ g_f(x_{k-1}) \in E^n,\ n \times n \text{ matrix } B_{k-1}.$$

**($k$+1)-st step.** Calculate:

$$(a)\quad g_f(x_k);\ d_k = g_f(x_k) - g_f(x_{k-1}); \tag{9.3}$$

$$(b)\quad r_k = B_{k-1}^T d_k;\ \xi_k = \frac{r_k}{\|r_k\|}; \tag{9.4}$$

$$(c)\quad B_k = B_{k-1} \cdot R_{\beta_k}(\xi_k),\ 0 < \beta_k < 1; \tag{9.5}$$

$$(d)\quad x_{k+1} = x_k - h_k B_k p_k, \tag{9.6}$$

where

$$p_k = \frac{\tilde{g}_k}{\|\tilde{g}_k\|};\ \tilde{g}_k = B_k^T g_f(x_k),$$

$h_k > 0$ is $k$-th step multiplier.

Remember $x_{k+1}$, $g_f(x_k)$, $B_k$. Stop if stopping criteria are satisfied; otherwise go to the next step.

There are many modifications of this general scheme, so we have the family of $r$-algorithms which differ in the rules of step-size regulation $h_k$ and dilation coefficient sequence $\{\alpha_k\}_{k=1}^{\infty}$ ($\beta_k = \alpha_k^{-1}$).

$R_\alpha(\xi)$, $\|\xi\| = 1$, is a linear operator of space dilation in direction $\xi$, which transforms a vector $x \in E^n$ by formula:

$$R_\alpha(\xi)x = x + (\alpha - 1)(x, \xi)\xi. \tag{9.7}$$

The specific property of this operator is that it needs only $O(n)$ arithmetic operations (not $O(n^2)$ as in general case). Detail description of $r$-algorithm one can find in [1], [2].

## 2 LAGRANGIAN BOUNDS FOR QUADRATIC-TYPE PROBLEMS

Polynomial and many combinatorial optimization problems can be easily transformed by introducing new variables and constraints to quadratic-type problems of the form: find

$$Q^* = \inf_{x \in E^n} Q_0(x) \text{ subject to } \quad Q_i(x) = 0, \quad i = 1, \ldots, m, \tag{9.8}$$

where $Q_\nu(x)$, $\nu = 0, \ldots, m$ are quadratic or linear functions determined on $n$-dimensional Euclidean space $E^n$. Using usual Lagrange function $L(x, u)$, $u = \{u_1, \ldots, u_m\}$, one can obtain Lagrangian lower bounds for such problems by finding $\psi^* = \sup \psi(u)$, where $\psi(u) = \inf_{x \in E^n} L(x, u)$. If $dom\psi$ is nonempty, then $\psi(u)$ is a concave function. Otherwise we obtain a trivial bound $\psi^* = -\infty$.

Consider the problem of finding dual (Lagrangian) estimates for quadratic-type problems of the form (9.8) in more detail. Let quadratic functions $Q_\nu(x)$, $\nu = 0, \ldots, m$ have the following description:

$$Q_\nu(x) = (K_\nu x, x) + (c_\nu, x) + d_\nu,$$

where $K_\nu$ be symmetric quadratic $n \times n$ matrices, $c_\nu$ be the $n$-dimensional vectors, $d_\nu$ be numbers. So, $(K_\nu x, x)$ is a quadratic part of $Q_\nu(x)$ and $(c_\nu, x)$ is a linear part of $Q_\nu(x)$, $\nu = 0, \ldots, m$.

Usual Lagrange function $L(x, u)$ can be represented as

$$L(x, u) = Q_0(x) + \sum_{i=1}^{m} u_i Q_i(x) = (K(u)x, x) + (c(u), x) + d(u),$$

where

$$K(u) = K_0 + \sum_{i=1}^{m} u_i K_i;$$

$$c(u) = c_0 + \sum_{i=1}^{m} u_i c_i;$$

$$d(u) = d_0 + \sum_{i=1}^{m} u_i d_i;$$

(here $u = \{u_1, \ldots, u_m\}$ is a $m$-dimensional vector of Lagrange multipliers).

Consider $\psi(u) = \inf_x L(x, u)$. If $K(u)$ is a positive definite matrix then $\psi(u) = L(x(u), u)$, where $x(u)$ is a solution of the linear system of equations

$$2K(u)x + c(u) = 0,$$

i.e.

$$x = -\frac{1}{2}\Big(K(u)\Big)^{-1} c(u).$$

If minimal eigenvalue of $K(u)$ $\lambda_n\Big(K(u)\Big) < 0$, then $\psi(u) = -\infty$.

In the case $\lambda_n\Big(K(u)\Big) = 0$ the matrix $K(u)$ is a positive semidefinite but singular. Let $s_1, \ldots, s_n$ be the orthonormal basis in $E^n$, corresponding to eigenvectors ordered in decreasing order of their eigenvalues (with taking in account their multiplicity). Then $c(u)$ can be represented in the form:

$$c(u) = \sum_{i=1}^{n-r} \alpha_i(u)s_i + \sum_{i=n-r}^{r} \alpha_i(u)s;$$

where $r$ is multiplicity of minimal eigenvalue being equal to zero. If all $\alpha_k(u) = 0$ for $n - r + 1 \le k \le n$, then $x(u)$ exists, i.e. $u \in \text{dom}\psi$. Otherwise $\psi(u) = -\infty$.

Let dom$\psi$ be non-empty. Then $\psi$ is a proper concave function as result of minimization on $x$ of functions

$$\psi_u(x) = Q_0(x) + \sum_{i=1}^{m} u_i Q_i(x),$$

which are linear in $u$ for all $x$.

For each $u \in \text{dom}\psi$ $\psi(u) \le Q^*$ ($Q^*$ is an optimal value of initial problem (9.8)).

Let $\overline{u} \in E^m$ and $K(\overline{u}) \succ 0$. Then $Q(\overline{u})$ is an interior point of dom$\psi$. Denote by $\Omega^+$ the set

$$\{u \in E^m \;/\; K(u) \succ 0\}.$$

Boundary $\overline{\Omega}$ of $\Omega^+$ consists of $u$ for which $K(u)$ has the minimal eigenvalue equal to zero. It is easy to prove that all $u \in \overline{\Omega}$ are limit points of the set $\Omega^+$. So if $\Omega^+$ is nonempty then dom$\psi$ is closed in $E^m$.

Let $\psi^* = \sup_{u \in E^m} \psi(u)$ and there exists a point $u^*$ such that $\psi^* = \psi(u^*)$. If $u^* \in \Omega^+$ then grad $\psi(u^*) = 0$, i.e. $Q_i(x^*) = 0$, $i = 1, \ldots, m$. In this case $\psi^* = Q^*$. Otherwise, when $x^* \in \overline{\Omega}$, it may be positive defect of duality:

$$\Delta = Q^* - \psi^* > 0.$$

For many interesting problems we may improve the dual bounds by using so-called functionally superfluous constraints in the form of quadratic equations (inequalities) which do not change the optimal value of initial polynomial problem but lead to modification of Lagrange function of corresponding quadratic problem. This modification may give significant increasing of new Lagrangian bound for a modified quadratic-type problem in comparison with $\psi^*$ for the old one.

Note that if we add to the initial problem (9.8) new quadratic constraints $Q_{m+1}(x) = 0, \ldots, Q_{m+r}(x) = 0$, $r \geq 1$ and form more long vector of Lagrange multipliers $U = \left\{\{u\}, u_{m+1}, \ldots, u_{m+r}\right\}$ then for $\inf_{x \in E^n} Q_0(x)$ subject to constraints

$$Q_i(x) = 0, \quad i = 1, \ldots, n, n+1, \ldots, n+r \tag{9.9}$$

corresponding Lagrange function

$$L_1(x, U) = Q_0(x) + \sum_{i=1}^{m+r} u_i Q_i(x) = L(x, u) + \sum_{i=m+1}^{m+r} u_i Q_i(x),$$

so

$$L(x, u) = L_1\Big(x, (\{u\}, 0, \ldots, 0)\Big),$$

$$\psi_1(U) = \inf_x L_1(x, U) \geq \inf_x L(x, u) = \psi(u),$$

and

$$\psi_1^* = \sup_u \psi_1(U) \geq \psi^*.$$

We shall demonstrate the possible improving of dual bounds by introducing superfluous constraints on a simple example in the next section.

# 3 SUPERFLUOUS QUADRATIC CONSTRAINTS FOR THE PROBLEM OF FINDING GLOBAL MINIMUM OF POLYNOMIAL FUNCTIONS

**Example.** Let $P_6(x_1)$ be a polynomial of sixth degree in one variable $x_1$:

$$p_6(x_1) = x_1^6 + a_5 x_1^5 + a_4 x_1^4 + a_3 x_1^3 + a_2 x_1^2 + a_1 x_1 + a_0. \tag{9.10}$$

and $\min_{x_1} p_6(x_1) = p^*$. One may transform this problem of obtaining global minimum $p_6(x_1)$ in a quadratic-type problem by introducing new variables $x_2 = x_1^2$; $x_3 = x_1 x_2$.

Consider a vector $x = \{x_1, x_2, x_3\}$ and obtain the problem:
to minimize

$$Q_0(x) = x_3^2 + a_5 x_2 x_3 + a_4 x_3 x_1 + a_3 x_1 x_2 + a_2 x_2 + a_1 x_1 + a_0 \tag{9.11}$$

subject to constraints:

$$Q_1(x) = x_1^2 - x_2 = 0 \tag{9.12}$$

$$Q_2(x) = x_1 x_2 - x_3 = 0 \tag{9.13}$$

The Lagrange function $L(x, u)$ of this problem has the following form ($u = \{u_1, u_2\}$ is the vector of Lagrange multipliers):

$$L(x, u) = Q_0(x) + u_1 Q_1(x) + u_2 Q_2(x).$$

Consider the matrix $K(u)$ which define the quadratic part of $L(x, u)$.

$$K(u) = \begin{pmatrix} u_1 & \frac{u_2+a_3}{2} & \frac{a_4}{2} \\ \frac{u_2+a_3}{2} & 0 & \frac{a_5}{2} \\ \frac{a_4}{2} & \frac{a_5}{2} & 1 \end{pmatrix}$$

One may see that if $a_5 \neq 0$, for all $u = \{u_1, u_2\}$, $K(u)$ cannot be positive semidefinite, so we obtain a trivial lower bound $\psi^* = -\infty$. So, it seems an attempt to use dual quadratic bounds for our example failed. But if we add to our model a superfluous constraint

$$Q_3(x) = x_2^2 - x_1 x_3 = 0$$

and correspondigly modify Lagrange function we radically change the situation.

New Lagrange function $L_1$ has 3 Lagrange multipliers

$$u^{(1)} = \{u_1, u_2, u_3\}$$

and

$$L_1(x, u^{(1)}) = L(x, u) + u_3(x_2^2 - x_1 x_3)$$

and $K_1(u^{(1)})$ has the following form:

$$K_1(u^{(1)}) = \begin{pmatrix} u_1 & \frac{u_2+a_3}{2} & \frac{a_4-u_3}{2} \\ \frac{u_2+a_3}{2} & u_3 & \frac{a_5}{2} \\ \frac{a_4-u_3}{2} & \frac{a_5}{2} & 1 \end{pmatrix}.$$

One may easily prove that if $u_2$ is fixed, the matrix $K_1(u^{(1)})$ becomes positive definite for sufficiently large $u_1$ and $u_3$, so modified quadratic-type problem with superfluous constraint has nontrivial dual bound $\psi_1^*$. Moreover, $\psi_1^* = p_6^*$, that we show later.

Further we show that $\psi_1(u^{(1)}) = \inf_x L_1(x, u^{(1)})$ has a nonempty domain and $\psi_1^* = p^*$ (see Corollary of the Main Theorem).

In general we can prove that for each polynomial of even degree $2k$ in one variable $x_1$ with positive elder coefficient, one may define new variables $x_i$ by quadratic formula:

$$x_{i+1} = x_i \times x_1, \quad i = 1, \dots, k-1$$

and generate some family of quadratic superfluous constraints with a dual lower bound equals to minimal value of a given polynomial.

For polynomials in several variables situation is more complex. Further we shall give a review of main results in the theory of dual quadratic bounds (with using of superfluous constraints) for polynomials of several variables.

Let $R^n$ be $n$-dimensional linear space of real vectors $x = \{x_1, \dots, x_n\}$, $P(x) = P(x_1, \dots, x_n)$ be a polynomial real function determined on $R^n$. Consider the problem of finding

$$f^* = \inf_{z \in R^n} P(x_1, \dots, x_n).$$

We will be interested in nontrivial case where $f^* > -\infty$, i.e. $P(x)$ is bounded from below. Such polynomials will be called $BB$-polynomials. It is clear that if $P(x)$ belongs to $BB$-class, then for any $i$, $1 \leq i \leq n$, the highest degrees $S_i$ of variables $x_i$ must be even. Note that the problem of the $BB$-property is in general similar by its computational complexity to the problem of finding $f^*$.

Let $S_i = 2l_i$, $i = 1, \ldots, n$, and $P(x)$ be recorded in a standard form as a sum of monomials with some real nonzero coefficients. For compact record of monomials we use a vector of degree $\alpha = (\alpha_1, \ldots, \alpha_n)$ with nonnegative integer entries and symbols $R[\alpha]$ of the corresponding monomials.

$$R[\alpha] = x_1^{\alpha_1}, \ldots, x_n^{\alpha_n}; \quad \alpha_i \le S_i, \quad i = 1, \ldots, n.$$

So

$$P(x) = \sum_{\alpha} c_\alpha R[\alpha] \text{ in new variables,}$$

$$0 \le \alpha_i \le 2l_i, i = 1, \ldots, n. \tag{9.14}$$

Let all monomials of polynomial $P(x) = P(x_1, \ldots, x_n)$ have maximal degree on variable $x_i$ equal $2l_i$ $(i = 1, \ldots, n)$. Consider "feasible" monomials $R[\alpha] = \prod_{i=1}^{n} x_i^{\alpha_i}$, where $\alpha_i \le l_i$, $i = 1, \ldots, n$.

For each monomial $R[\alpha]$ choose the "standard" representation of monomial $R[\alpha]$ as a product of two feasible monomials:

$$R[\alpha] = R[\alpha_1(\alpha)] \cdot R[\alpha - \alpha_1(\alpha)],$$

$$\alpha_1(\alpha), (\alpha - \alpha_1(\alpha)) \ge 0.$$

Moreover, integer vector

$$\alpha_1(\alpha) \ge \alpha - \alpha_1(\alpha) \ge 0,$$

or

$$\alpha_1(\alpha) \ge \frac{\alpha}{2}.$$

Consider for each monomial $R[\alpha], \alpha_i \le 2l_i, i = 1, \ldots, n$ the full set of quadratic-type equalities in feasible $R$-variables of the form:

$$R[\alpha_1(\alpha)] \cdot R[\alpha - \alpha_1(\alpha)] - R[\beta] \cdot R[\alpha - \beta] = 0, \tag{9.15}$$

where $\beta$ runs all possible values of integer feasible vectors, not equal $\alpha_1(\alpha)$, such that

$$\beta \ge \alpha - \beta \ge 0, \text{ so } \beta \ge \frac{\alpha}{2}.$$

We obtain the "full" set of constraints if we record the equations of the type (9.15) for all $\alpha = (\alpha_1, \ldots, \alpha_n), \alpha_i \le 2l_i, i = 1, \ldots, n$.

Using the full family of equalities of the type (9.15) one can receive all possible representations of polynomial $P(x_1, \ldots, x_n)$ as quadratic function in feasible $R$-variables:

$$\begin{aligned} &P(x_1, \ldots, x_n) = L(R, u) = \\ &= \textstyle\sum_\alpha c_\alpha R[\alpha_1(\alpha)] \cdot R[\alpha - \alpha_1(\alpha)] + \\ &+ \textstyle\sum_{\alpha,\beta} u_{\alpha\beta} (R[\alpha_1(\alpha)] \cdot R[\alpha - \alpha_1(\alpha)] - \\ &- R[\beta] \cdot R[\alpha - \beta]), \end{aligned} \tag{9.16}$$

where $u_{\alpha\beta}$ are arbitrary numbers which multiply on left part of equalities (9.15) with corresponding $\{\alpha, \beta\}, c_\alpha$ are coefficients in usual representation $P(x)$ in (9.14).

By the other hand, we can consider $L(R, u)$ as a Lagrange function of quadratic-type problem in feasible variables $R(\alpha)$:

to minimize

$$P(x_1, \ldots, x_n) = \sum_\alpha c_\alpha \prod_{i=1}^n x_i^{\alpha_i} = \sum_\alpha c_\alpha (R[\alpha_1(\alpha)] \cdot R[\alpha - \alpha_1(\alpha)])$$

subject to the full set of constraints of the form (9.15), $u = \{u_{\alpha\beta}\}$ are Lagrange multipliers.

Our aim is to find conditions when

$$\psi^* = \sup_u \left[ \inf_R L(R, u) \right] = P^* = \min_x P(x).$$

These conditions are formulated in

**Main Theorem.** *Let polynomial function in $n$ variables $P(x)$ has the global minimum in a point $x^*$ and $P(x^*) = P^*$.*

*Then a dual quadratic bound $\psi^* = P^*$ if and only if the nonnegative polynomial $\overline{P}(x) = P(x) - P^*$ can be represented as sum of squares of polynomials, which have only feasible monomials.*

**Proof of the Main Theorem.** We say that polynomial $P(x)$, $\min_{x \in E^n} P(x) = P(x^*) = P^*$, posesses $E$-property if Lagrangian quadratic bound is exact i.e. $\psi^* = P^*$. To prove the Main Theorem we use the next

**Lemma.** *If polynomial $P(x)$ posesses $E$-property then for arbitrary $a \in E^n$ polynomial $P_a(x) = P(x + a)$ also posesses this property (see [1]).*

Due to the previous Lemma one can suppose without loss of generality that optimal point is $x^* = 0$.

Let $BB$-polynomial $\overline{P}(x) - P^*$ is represented as a sum of squares of real polynomials $R_i(x)$, $i = 1, \ldots, k$, i.e.

$$\overline{P}(x) = \sum_{i=1}^k [R_i(x)]^2.$$

Replace each monomial $M_{\alpha^i}$ contained in polynomial $R_i(x)$ by corresponding feasible variable $R[\alpha^{(i)}]$. The output of $[R_i(x)]^2$ for each $i$ can be represented as a sum of monomials of the form $C_s^{(i)} C_t^{(t)} R[\alpha^{(it)}] R[\alpha^{(is)}]$ after summing of similar members, one obtains for each possible $R[\alpha]$

the corresponding coefficient

$$C_\alpha = \sum_i \sum_{(s,t)} C_s^{(i)} C_t^{(i)},$$

where pairs $(s,t)$ are such that $\alpha^{(i,s)} + \alpha^{(i,t)} = \alpha$.

Choose Lagrange multipliers in expression for $L(R,u)$ (9.16) equal to zero, then value of corresponding Lagrange function $L(x,u)$ coinsides with objective function. Since objective function minus $P^*$ is a sum of squares, the corresponding representation of a quadratic part in $R[\alpha]$ variables is positive semidefinite, so $\overline{u} = 0$ belongs to dom $\psi$. $x^*(0)$ is a solution of linear system of equations in $R[\alpha]$ variables

$$Q_i[R] = 0, \quad i = 1, \ldots, m.$$

The optimal value of the function $\overline{P}(x)$ equals to zero, i.e. $\overline{P}(0)$. So due to the Lemma, $P(x)$ can be represented as a sum of squares of real polynomials. $P(x)$ posesses $E$-property.

Continue the proof of the Main Theorem. Let polynomial $P(x)$ posesses $E$-property and $\min\limits_{x \in E^n} P(x) = P(x^*) = P^*$. This polynomial $\overline{P}(x) = P(x) - P^*$ takes its minimum in the same point $x^*$ and $\overline{P}(x^*) = 0$. Since $P(x)$ posesses $E$-property there exists $u^*$ such that for the corresponding quadratic-type problems in variables $R[\alpha]$ Lagrange function $L(R,u)$ when $u = u^*$ is positive-semidefinite in $R$-variables, so it can be represented as a sum of squares of linear functions in feasible $R$-variables

$$L(R, u^*) = \sum_{i=1}^{k} \{l_i^{u^*}[R]\}^2. \tag{9.17}$$

Instead of each member of such determined linear functions in $R[\alpha]$ variables one can substitute . corresponding possible feasible monomials in $x$ variables. So we obtain the representation $\overline{P}(x)$ in the form

$$\overline{P}(x) = \sum_{i=1}^{k} \left\{ [\sum_{i=1}^{n} c_{ij} M_{ij}(x)]^2 \right\},$$

where $\{c_{ij}\}_{i=1}^{k}$ are vectors of coefficients of linear functions $l_i^{u^*}$, $i = 1, \ldots, k$, $M_{ij}(x)$ are corresponding monomials. So $\overline{P}(x)$ can be represented as a sum of squares of real polynomials.

The Main Theorem is proved.

**Corollary.** *If $n = 1$, the polynomial $\overline{P}(x_1) = P(x_1) - P^*$ of even degree $2m$ with positive eldest coefficient $c$ is nonnegative and can be represented as a sum of squares of real polynomials.*

**Proof.** All complex roots of such polynomials, if they exist, can be joint in pairs of the form $a_\alpha + ib_\alpha$ and $a_\alpha - ib_\alpha$, here $a_\alpha$ and $b_\alpha$ are real numbers. The real roots must have the even multiplicity otherwise $P(x)$ cannot be positive. Let

$$f_1(x_1) = \prod_\alpha [x_1 - (a_\alpha + ib_\alpha)] = R(x_1) + iQ(x_1)$$

$$f_2(x_1) = \prod_\alpha [x_1 + (a_\alpha - ib_\alpha)] = R(x_1) - iQ(x_1).$$

Then

$$R(x_1) = f(x_1)f(x_2)\left[\Pi(x_1 - a_\alpha)^{2\delta(\alpha)}\right] =$$

$$\left[R^2(x_1) + Q^2(x_1)\right]\left[\Pi(x_1 - a_\alpha)^{\delta(\alpha)}\right]^2$$

So the nonnegative polynomial $P(x_1)$ can be represented as a sum of not more than two squares. Due to the Main Theorem each polynomial in one variable of even degree and elder positive coefficient posesses $E$-property. It is easy to calculate that the number of superfluous constraints for polynomials $P_{2m}(x_1)$ of even degree $2n$ is equal to

$$s = \frac{m(m-1)}{2} - 1.$$

Particularly for $m = 3$, $s = 1$; $m = 4$, $s = 5$ and so on.

A great mathematician D.Hilbert considered the problem of representation of nonnegative polynomial forms as a sum of squared polynomials more than hundred years ago. He proved that if dimension $n = 1$ or 2, arbirtrary nonnegative forms (homogenious polynomials) can be represented as a sum of squares. If $n = 3$ and the number of variables $\leq 3$, also the corresponding nonnegative forms can be represented as a sum of squares. But if $n = 3$ and degree $2m \geq 4$ there exist nonnegative polynomial forms which cannot be represented as a sum of squares. Our approach gives a possibility not only to determine whether a given nonnegative polynomial can be decomposed in a sum of squares, but to find such decomposition using the described above algorithm of finding Langrangian quadratic bounds with use of full set of superfluous constraints.

## 4 THE MAXIMUM WEIGHT INDEPENDENT SET IN A GRAPH

Let an undirecred graph $G(V, E)$ be given: $V = \{1, \ldots, n\}$ be the set of vertices, $E$ be the set of edges; $(i, j) \in E$ is the edge with end points $i$

and $j$ belonging to $V$, $(i,j)$ and $(j,i)$ are equivalent symbols. The subset $I \subseteq V$ is called independent (stable) if there is no pair $i$ and $j$ such that $(i,j) \in I$ and $(i,j) \in E$, $i \neq j$.

The subset $K \subseteq V$ is called a clique if for all pairs $(i,j) \in K$, $(i,j) \in E$, $i \neq j$.

Denote by $A_G = \{a_{ij}\}_{i,j=1}^n$ the adjacency matrix of $G(V,E)$:

$$a_{ij} = 1 \text{ if } (i,j) \in E; \quad a_{ij} = 0, \quad \text{if } (i,j) \notin E.$$

The **complement** graph to $G(V,E)$ is $\overline{G}(V,\overline{E})$ with the same vertex set and

$$\overline{E} = \{(i,j), \quad i \neq j/(i,j) \in \overline{E}, \quad \text{if} \quad (i,j) \notin E\}.$$

The graph $G(V,E)$ may be vertex weighted if for every $i \in V$ the weight $w_i \geq 0$ is given.

Let $w = \{w_i\}_{i \in V}$. For a subset $S \subseteq V$ we define $W(S)$ (weight of $S$) as

$$W(S) = \sum_{i \in S} w_i.$$

We call

$$G(S) = G(S, E \cap S * S)$$

the subgraph of $G(V,E)$ induced by $S$. The maximum (weight) clique problem is to find a clique of maximum weight.

The maximum (weight) independent (stable) set problem is to find an independent set of maximum cardinality (of maximum weight). The size of a maximum independent set is the stability number of $G$, denoted by $\alpha(G)$. The maximum weight independent set is denoted by $\alpha_w(G)$. It is easy to see that $S$ is the clique of $G$ if and only if $S$ is an independent set of complement graph $\overline{G}$. So, any result obtained for one of the mentined problems can be reformulated for another problem. Both of these problems are $NP$-complete for the class of arbirtrary graphs. But for some specific classes of graphs the polynomial-time algorithms were constructed. We shall consider two such classes of graphs: ***t*-perfect** and **perfect graphs**.

The definition of these classes is tightly linked with the properties of integer polytopes for the corresponding stable set problems.

The weighted maximum stable set problem in graphs can be formulated as the following 0–1 problem: to find

$$\alpha_w(G) = \quad \max(w,x), \; x = (x_1, \ldots, x_n), \tag{9.18}$$

$$x_i + x_j \leq 1 \quad \text{for all } (i,j) \in E \tag{9.19}$$

$$x_k \in \{0,1\} \quad \text{for all } k \in V \tag{9.20}$$

We introduce the stable set polytope

$$\mathrm{STAB(G)} := \mathrm{conv}\{\chi^S \in R^V | S \subseteq V \text{ is a stable set } \} \tag{9.21}$$

defined as the convex hull of the incidence vectors of all stable sets of vertices of $G$. $\alpha_w(G)$ is equal to the maximum value of linear function $(w, x)$ on convex polytope STAB(G). Of course, it is very useful to represent the STAB(G) by a system of linear inequalities. Unfortunately, in general case, it is a very hard problem. Therefore we consider some particular cases.

The linear relaxation of the problem (9.18)–(9.20) is to find:

$$\alpha_1(G, V) = \max(w, x) \tag{9.22}$$

$$x_i + x_j \leq 1, \forall (i, j) \in E \tag{9.23}$$

$$0 \leq x_k \leq 1, \ k = 1, \ldots, n \tag{9.24}$$

**Theorem** [3]. *The inequalities (9.23), (9.24) give full description of STAB(G) if and only if G is bipartite. Hence, for bipartite graphs the problem (9.18)–(9.20) can be solved in polynomial time.*

The minimal graphs for which inequalities (9.23), (9.24) are not sufficient to describe STAB(G) are the odd circuits. In fact, if $G(V, E)$ is an odd circuit then $|V| = 2m + 1$, and the point $x_i = \frac{1}{2}, i \in V$ satisfies the inequalities (9.23), (9.24) but does not belong to STAB(G). So we can propose a new class of linear inequalities valid for STAB(G), so-called **odd circuits constraints.**

$$\sum_{i \in V(C)} x_i \leq \frac{|V(C)| - 1}{2} \quad \text{for each odd circuit C} \tag{9.25}$$

($V(C)$ is the set of vertices that lie in the circuit $C$).

Let us call the graph $t$-perfect if (9.23), (9.24) and (9.25) are enough to describe STAB(G).

In general case we do not know whether the problem of checking the $t$-perfectness is in NP or in P. Despite this fact a maximum weight stable set in a $t$-perfect graph can be found in polynomial time by using a slight modification of the ellipsoid method (see [3], p.276). This problem is reduced to the LP-problem (9.22)–(9.24), (9.25) with possibly exponentially many odd circuits constraints (9.25). But for obtaining cutting plane in the ellipsoid method it is enough to have one constraint of type (9.25) that is not satisfied for a given $x$. The search of such constraint is equivalent to finding the shortest weight odd circuit. For the last problem there exist polynomial time algorithms ([3], p.236).

A partition of $V$ into stable sets (cliques, respectively) is called a **coloring (clique covering**, respectively) of $G$. The coloring number (respectively, clique covering number) is the smallest number of stable sets in a coloring (respectively, cliques in a clique covering) of $G$, and is denoted by $\chi(G)$ (respectively, $\overline{\chi}(G)$). It is clear that

$$\chi(G) = \overline{\chi}(\overline{G}).$$

Each stable set in $G$ has no more than one representative in each clique, hence, we have the inequality:

$$\alpha(G) \leq \overline{\chi}(G) \tag{9.26}$$

Similary we obtain

$$\omega(G) \leq \chi(G) \tag{9.27}$$

where $\omega(G)$ is a clique number of graph $G$. Berge called a graph $G$ **perfect** if the equality

$$\omega(G') = \chi(G') \tag{9.28}$$

holds for every induced subgraph $G'$ of $G$. The first Berge conjecture about perfect graphs was the following [4], [5]: *The complement of a perfect graph is also perfect.* This was proved by well known Hungarian mathematician Lovász in 1972 [6]. In 1962 Berge also proposed the **Strong Perfect Graph Conjecture.** *The graph is perfect if and only if it or its complement, does not contain an odd circuit of length at least five as an induced subgraph.* This conjecture is still unsolved.

In [10] L. Lóvasz proposed and proved some interesting upper bounds for the $\alpha(G)$, which are exact for the class of perfect graphs. The interest of L. Lóvasz to the problem of finding $\alpha(G)$ was stimulated by works of great founder of information theory C.Shannon. In 1956 Shannon published the article ([11]) about information capicity of the system of symbols some of which can be mixed up (the graph of possible mixing $G(S, E)$ be given). In this graph the vertex set are symbols, which connected by edges in the case, when corresponding symbols may be mixed up.

Shannon proposed a linear upper bound $L(G)$ for information capacity of a graph. It is interesting that for a class of perfect graphs such bounds are exact. For other graphs Shannon capacity is an irrational number which is in general case very hard to calculate. The symplest nonperfect graph is 5-cycle $C_5$ ("Pentagon"). The linear bound equals $\frac{3}{2}$ for this graph.

Shannon proposed a conjecture that the capacity of $C_5$ $\alpha_w(C_5) = \sqrt{5}$. Many years young Hungarian mathematician Lóvasz tried to prove this cojecture. In [10] he gave a positive solution of this problem but there were also much more important results in this article, namely, a series of upper bounds for independence and Shannon capacity numbers of graph. Some of proposed bounds can be considered as solutions of extremal matrix parametric problems.

In a paper [10] Lóvasz proposed an upper bound on stability number $\alpha(G)$ known as the theta function $\vartheta(G)$. This function can be expressed in many equivalent ways: as an eigenvalue bound for certain class of matrices, as a semidefinite program, or in terms of orthonormal representation of graphs. Theta functions can be generalised on class of the problems of finding an upper bound $\vartheta_w(G)$ for maximal weighted stable set, where $w = \{w_1, \ldots, w_n\}$ in the weight vector of graph vertices (see [3], §9.3).

For obtaining good bounds for $\alpha(G)$ Lóvasz used the technique of **orthonormal representation** of graphs [10], [3].

Let $G = (V, E)$ be a graph. An orthonormal representation of $G$ is a sequence of $\{u_i | i \in V\}$ vectors $u_i \in E^N$, where $N$ is some positive integer, such that $\|u_i\| = 1$ for all $i \in V$ and $(u_i, u_j) = 0$ for all pairs $i, j$ of non adjacent vertices.

Let $(u_i | i \in V), u_i \in E^N$ be any orthonormal representation of $G$ and let $c \in R^N$ with $\|c\| = 1$. Then for any stable set $S \subseteq V$ we have inequalities

$$\sum_{i \in S} (c, u_i)^2 \leq 1, \tag{9.29}$$

and

$$\sum_{i \in V} (c, u_i)^2 x_i \leq 1 \tag{9.30}$$

holds for the incidence vector $X(S)$ of any stable set $S$ of nodes of $G$.

We call (9.30) the orthonormal representation constraints, which are valid for polytope STAB(G) (the convex hull of all incidence vectors of stable sets of $G$). Using mentioned above technique Lóvasz proposed five different extremal problems for finding upper bounds for $\alpha_w(G)$ which have the same values. All of these problems can be easily reduced to semidefinite programming problems and solved by the methods of nondifferentiable optimization (particulary, by $r$-algorithm).

Below we shall formulate two more general extremal matrix problems studied while finding upper bounds for the weighted independent set problem (see [3], §9.3).

**I.** Let $G = \{V, E\}$ be a graph, $|V| = n$, $\{w_i\}_{i=1}^n$ be a vector of vertex weights. Consider the family $F_w(y)$ of matrices $n \times n$ with the following properties: $Y = \{y_{ij}\}_{i,j=1}^n$ are entries of such matrices.

**(1)** $y_{ij} = 0$, if $\{i, j\} \in E$

**(2)** the matrix $Y$ is positive semidefinite

**(3)** $\sum_{i=1}^{n} y_{ii} = 1$

The extremal matrix problem has the following form:

to find

$$\vartheta_w(G) = \max \sum_{i,j} \sqrt{w_i w_j} y_i y_j$$

subject to mentioned constraints (1)-(3).

**II.** The other of the upper bounds has the form:

$$\vartheta_w(G) = \min_A \lambda_{\max}[A + W] \tag{9.31}$$

where $A$ is the class of the matrices $n \times n$, $A = \{a_{ij}\}$, $a_{ii} = 0$ for all $i \in V$, $a_{ij} = 0$ for all $i, j$ nonadjacent in $G$, $W = \{\sqrt{w_i \cdot w_j}, i, j \in V\}$, $\lambda_{\max}(\cdot)$ denotes the maximal eigenvalue.

Since $\lambda_{\max}(X)$ is a nondifferentiable convex function of entries $X$, the problem (9.31) is a typical problem of nondifferentiable optimization.

We can obtain the sàme bounds using quadratic-type formulation of the problem of finding maximal weighted stable set in graph $G$ and calculating corresponding Lagrangian bounds.

Namely, let the graph $G = (V, E)$, $|V| = n$, $W = \{w_i\}_{i=1}^{n}$ (the vector of weights) be given. Consider the quadratic-type problem:

to find

$$\alpha_w(G) = \max \sum_{i=1}^{n} w_i x_i \tag{9.32}$$

subject to constraints:

$$x_k^2 - x_k = 0, \forall k = 1, \ldots, n \text{ (boolean property)} \tag{9.33}$$

$$x_i x_j = 0, \forall (i, j) \in E. \tag{9.34}$$

Let $U = (\{u_k\}_{k=1}^{n}, \{u_{ij}\}_{i,j} \in E)$ be the vector of Lagrange variables. Consider the optimal Lagrangian bound

$$\rho_w(G) = \inf_u \sup_x L(x, u).$$

In our case the function $\psi(u) = \sup L(x, u)$ has nonempty domain. As proved in [1], $\rho_w(G)$ has the same value, as Lóvasz's bounds.

These boùnds are exact for arbitrary weights if $G$ is a perfect graph. For nonperfect graphs one can find such weights, that corresponding bounds are not exact.

Thus, using natural quadratic-type formulation of the problem of finding maximal weight stable number in a graph and applying to this problem standard technique of finding Lagrangian bounds we get the same results as obtained by using "specific" approach of orthonormal representation of graphs.

In some cases we can even improve the Lagrangian bounds by adding by adding to model (9.32)–(9.34) trivial superfluous constraints

$$x_i x_j - 1 \leq 0 \text{ for all } i, j \notin E. \tag{9.35}$$

For example, let $G = (V, E)$ be a graph, where $V$ is the set of boolean vectors of the length 6, $\|V\| = 2^6$. Two vertices are linked by edge if Hamming distance between the corresponding vectors is greater than 3. For this graph $\alpha(G) = 4$ (example of optimal independent set : $\{0,0,0,0,0,0\}$ $\{1,1,1,1,0,0\}$, $\{1,1,0,0,1,1\}$, $\{0,0,1,1,1,1\}$). One can calculate that Lagrangian bound for model (9.32)–(9.34) equals $\frac{16}{3}$, for model (9.32)–(9.35) we have the exact bound 4 (see [14]).

Thus, the use of superfluous constraints substantially improves Lagrangian bound in this example.

## 5 LAGRANGIAN BOUNDS FOR THE MAXIMUM CUT PROBLEM

One of the most bright examples of using quadratic superfluous constraints for improving dual bounds in quadratic-type problems are connected to the max- cut problem.

Let $G(V, E)$ be an ordinary graph with the vertex set $V = \{1, \ldots, n\}$ and the edge set $E = \{(i,j) = (j,i)\}$, where $(i,j)$ is the edge, linking vertices $(i,j)$ or $(j,i)$. The weight function $W$ is given by symmetric $n \times n$ matrix

$$W = \begin{cases} 0 & \text{for } (i,j) \notin E \\ w_{ij} & \text{for } (i,j) \in E \end{cases}.$$

Let the vertex set be devided on two nonempty nonintersecting parts $V_1$ and $V_2$: $V = V_1 \bigcup V_2$. We say that edge $(i,j)$ belongs to the cut $R(V_1, V_2)$ if this edge have its ends in different subsets of subdivision $V = V_1 \bigcup V_2$.

Let $x_k \in \{-1, 1\}$ be a binary variable corresponding to the vertex $k$ $(k \in \{1, 2, \ldots, n\})$,

$$x_k = \begin{cases} -1 & \text{if } x_k \in V_1 \\ +1 & \text{if } x_k \in V_2 \end{cases}.$$

We must find such division $V = V_1 \bigcup V_2$ that the sum of weights of all edges, belonging to the corresponding cut, be maximal.

Without loss of generality one can suppose that the graph $G(V, E)$ is full. In this case the value of cut may be represented in terms of binary variables as a quadratic function:

$$f(x, W) = \frac{1}{8} \sum_{(i,j), i \neq j} w_{ij}(x_i - x_j)^2 = \frac{1}{4}(\sum_{(i,j)} w_{ij} - \sum_{(i,j)} w_{ij} x_i x_j) \qquad (9.36)$$

Since $\max f(x, W)$ subject to constraints $x_k^2 - 1 = 0, \quad k = 1, \ldots, n$ equal to $1/4 \sum_{(i,j)} w_{ij} - \min_x \sum_{(i,j)} w_{ij} x_i x_j$ subject to the same constraints, the max-cut problem can be reduced to the quadrtic-type problem: to find

$$S(W) = \min_x \sum_{i,j=1}^{n} w_{ij} x_i x_j \qquad (9.37)$$

$$\text{subject to constraints} \quad x_k^2 - 1 = 0, \quad k = 1, \ldots, n$$

Denote by $R^*(W)$ the optimal value of the max-cut. Due to (9.37)

$$R^*(W) = \frac{1}{4}\left(\sum_{(i,j)} w_{ij} - S(W)\right) \qquad (9.38)$$

Let $u = \{u_1, \ldots, u_n\}$ be the vector of Lagrange multipliers corresponding to equalities (9.37). Then Lagrange function $L(x, u)$ of reduced problem (9.37)-(9.38) has the form

$$L(x, u) = (W(u)x, x) - \sum_{k=1}^{n} u_k,$$

where

$$W(u) = W + \text{diag } u,$$

(diag $u$ is a diagonal matrix with the components $d_{ii} = u_i, i = 1, \ldots, n$). Let $\lambda_{\min}(A)$ denotes the minimal eigenvalue of matrix $A$.

$L(x, u)$ is a gomogeneous quadratic in $x$ function, so

$$\inf_x L(x, u) = \begin{cases} -\infty, & \lambda_{min}(W(u)) < 0 \\ -\sum_{k=1}^{n} u_k, & W(u) \succeq 0, \text{ i.e. } \lambda_{min}(W(u)) \geq 0. \end{cases} \qquad (9.39)$$

By using an exact nonsmooth penalty function, the problem (9.39) is reduced to an unconditional optimization of nondifferential function

$$f(u, s) = \sum_{k=1}^{n} u_k + s\left[\lambda_{min}^{-}(W(u))\right], \qquad (9.40)$$

where

$$\lambda_{min}^{-} = \begin{cases} 0, & \text{if } \lambda_{min} \geq 0 \\ \lambda_{min}, & \text{if } \lambda_{min} \leq 0 \end{cases},$$

$s$ is a penalty multiplier. One can prove, that if $s \geq n$, the problem of minimization of $f(u, s)$ is equivalent to the problem (9.40) (see [1], §1.4, §8.1).

One can solve the problem (9.40) by using $r$-algorithm. After obtaining an optimal value of (9.40) $S(W)$ we can obtain the upper bound for the optimal cut value, using (9.38).

The alternative approaches for finding upper bounds of the max-cut problems are represented in [15],[16], [17]. In articles [19]-[20] the bounds similar to our ones are obtained by using another technique. In [18] it is shown how to describe the max-cut problem in terms of semidefinite programming.

Let $G = (V, E)$ be a graph with even number of vertices: $|V| = 2k$. Consider the modification of the max-cut problem when subdivisions $V = V_1 \cup V_2$ are such that $|V_1| = |V_2| = k$ (so-called bisection problem). Let $x_i \in \{-1, +1\}$ are binary variables. Then for bisection problem

$$\sum_{i=1}^{2k} x_i = 0,$$

so one may exclude $x_{2k}$ and obtain equalities

$$x_{2k} = -\sum_{i=1}^{2k-1} x_i \qquad (9.41)$$

and $(\sum_{i=1}^{2k-1} x_i)^2 = 1$. Substituting (9.41) instead of $x_{2k}$ we obtain a quadratic-type problem in binary variables $\{x_i\}_{i=1}^{2k-1}$, corresponding Lagrange function and Lagrangian bound, which coincides with a bound, proposed in [15], [22].

The max-cut problem may be formulated also as the problem of linear programming in edge boolean variables $y = \{y_{ij}\}_{i,j=1,i\neq j}^{n}$, $y_{ij} \in \{0, 1\}(y_{ij} = 1$, if $(i, j)$ belongs to the cut $R(V_1, V_2)$, and 0 in the opposite case) (see [7], BaMa.86, GrPu.81).

Let $M(G)$ be the convex hull of all feasible solutions of the max-cut problem in edge variables $\{y_{ij}\}$. So, one may reformulate the max-cut problem as a linear programming problem: to find

$$l^*(G, W) = \max_{y \in M(G)} \sum_{(i,j)} w_{ij} y_{ij}.$$

Edge variables $y_{ij}$ and vertex binary variables are linked by simple formula:

$$y_{ij} = \frac{1 - x_i x_j}{2} \text{ for all } y_{ij}, i \neq j. \tag{9.42}$$

In a large graph the set $M(G)$ has tremendous number of faces. But for special graph all its faces can be described in compact form. For example, if graph is planar for each triangle $\{i, j, k\}$, one may write down the inequality

$$y_{ij} + y_{jk} + y_{ik} \leq 0. \tag{9.43}$$

Denote by $y(N)$ the sum $\sum_{(i,j) \in N} y_{ij}$ for the set $N$ of edges. Using this notation we can record more general family of inequalities of the following form:

$$y(F) - y(C \backslash F) \leq |F| - 1 \text{ for all circuits } C \text{ of } G \text{ and } F \subseteq C \text{ with } |F| \text{ odd.} \tag{9.44}$$

These inequalities are valid for $M(G)$.

Barahona and Mahjoub [7], [8] have shown that the solution set of (9.44) is equal $M(G)$ if and only if $G$ has no subgraphs contractible to $K_5$ ($K_5$ is 5-clique). There is a simple algorihm (see [3], §8.3, p.235-236) for verifying validation of system (9.44) for given $\overline{y}$ ($0 \leq \overline{y}_{ij} \leq 1$ for all $(i, j)$).

We construct a new graph $H = (V' \bigcup V'', E' \bigcup E'' \bigcup E''')$, consisting of disjoint copies $G'(V', E')$ and $G''(V'', E'')$ of $G$ and an additional edge set $E'''$ that contains the edges $u'v'', u''v'$ for each $uv \in E$ . The edges $u'v' \in E'$ and $u''v'' \in E''$ get the weight $y_{uv'}$ while the edges $u'v''$ and $u''v' \in E'''$ get the weight $1 - y_{uv}$. For each node $u \in V$ we calculate a shortest path (with respect to weights just defined). Such a path contains an odd number of edges of $E'''$ and corresponds to a closed walk in $G$, containing $u$. Clearly, if the shortest of these $[u', u'']$-paths has the length equal at least to one, then $y$ satisfies (9.44), otherwise there exists a cycle $C$ and a set $F \subseteq C$, $|F|$ odd, such that $\overline{y}$ violates the corresponding inequality. Such linear in $\overline{y}$ inequalities can be converted by substitution (9.42) in quadratic inequalities in binary variables $x$ which may be added to the set of obtained earlier inequalities in $x$ as superfluous constraints. So we described one of the possible ways to generate the superfluous inequalities for model (9.37). Further adding of superfluous constraints as a rule essentially improves dual quadratic upper bounds for the max-cut problem. In combination with heuristic methods of constructing feasible cuts such approach allows to obtain an exact solution for graphs with integer weight matrix which subgraphs are not contractable to $K_5$

([7], [3]). More detail description of corresponding algorithm one can find in [1], §8.1.

# 6 CONCLUSION

We showed that multiextremal and combinatorial optimization problems can be often reduced to quadratic-type models. Finding of Lagrangian bounds for such problems as generates as a rule the semidefinite programming problems (see [1], [16], [18]) which can be successively solved by modern methods of nondifferentiable optimization, particularly, by $r$-algorithm. In some cases adding of superfluous quadratic constraints gives more exact dual bounds and creates a possibility for effective solving of some subclasses of NP-hard problems.

As it is shown in some articles [19], [21], [13], [23], the Lagrangian bounds are the most exact bounds in a certain more spread class of estimates for combinatorial problems.

In the work [18] another approach for obtaining the dual bounds for some combinatorial problems on graphs is described which is based on the models of semidefinite programming and using of interior-point methods. It is very interesting to compare the numerical effectivness of finding the dual bounds by described technique using the modern algorithms of nondifferentiable optimization and by the interior-point methods on the same classes of problems.

## Acknowledgments

I am very grateful to the organisers of the Conference on "Advances in Convex Analysis and Global Optimization" honoring the memory of C. Caratheodory (1873-1950) for the invitation and excellent conditions at the Samos.

## References

[1] N.Z. SHOR Nondifferentiable Optimization and Polynomial Problems, Dordrecht, Kluwer, 1998.

[2] N.Z. SHOR, N.G. ZHURBENKO A minimization method using the operation of space dilation in the direction of the difference of two successive gradients, Kibernetika (Kiev), 3:51-59, 1971.

[3] M.GRÖTSCHEL, L. LÓVASZ, AND A.SCHRIJVER Geometric Algotithms and Combinatorial Optimization, volume 2 of Combinatorics. Springer-Verlag, 1988.

[4] C. BERGE Farbung von Graphen, deren samtiche bzv. deren ungerade Kreise starr sind (Zusammenfassung), Wissenschaftliche

Zeitschrift, Martin Luther Universitat Halle-Wittenberg, Mathematisch-Naturwissenschaftliche Reihe, pp.114-115, 1961.

[5] C. BERGE Sur une conjecture relative au probleme des codes optimaux, Communication, 13eme assemblee generake de l'URSI, Tokyo, 1962.

[6] L. LÓVASZ Normal hypergraphs and the perfect graph conjecture, Discrete Mathematics, 2:253-267, 1972.

[7] F.BARAHONA, M,GRÖTSHEL AND A.MAHJOUB Facets of the bipartite subgraph polytope. Mathematics of Operations Research, 10 (1985), 340-358.

[8] F.BARAHONA AND A.MAHJOUB On the cut polytope. Mathematics Programming, 36 (1986), 157-173.

[9] M.GRÖTSHEL AND W.PULLEYBLANK Weakly bipartite graphs and the max-cut problem. Operations Research letters. 1, 1981, 23-27.

[10] LASLO LÓVASZ On the Shannon-capacity of graph. IEEE Transactions on information theory. IT-25, 1, 1979, 1-7.

[11] SHANNON, C. The zero-error capacity of a noisy channel. IRe Transactions Information Theory. IT, 2, 1956, 8-19.

[12] SHRIJVER, A. A comparison of the Delsarte and Lovash bounds. IEEE Transactions on information theory. IT-25, 4, 1979.

[13] PANOS M. PARDALOS AND HENRY WOLKOWICZ (EDS), Topics in Semidefinite and Interior-Point Methods, Fields Institute Communications Series Vol. 18, American Mathematical Society (1998).

[14] MCELICE, R., RODEMICH, E., ROMSEY, JR. H. The Lovash bound and some generalization. Journal of combinatorics, Information and System Science. 3, No.3, 1978. 134-152.

[15] W.F. DONATH AND A.J. HOFFMAN Lower bounds for the partitioning of graphs, IBM Journal Res. Dev., 17, 1973.

[16] ANJOS M.F., WOLKOWICZ H. A strengthened SDP relaxation via second lifting for the max-cut problem, University of Waterloo, Department of Combinatorics and Optimization, Research Report CORR 99-55, 28 pp.

[17] DELORME C. AND POLJAK S. Laplacian eigenvalues and the maximum cut problem, Math. Programming 62(3), 557-574, 1993.

[18] GOEMANS M.X. Semidefinite programming in combinatorial optimization, Math. Programming, 79, 143-162, 1997.

[19] ALIZADEH F. Interior point methods in semidefinite programming with applications to combinatorial optimization. SIAM Journal on Optimization, 5 (1995), 13-51.

[20] LAWRENT M., POLJAK S., AND RENDL F. Connections between the semidefinite relaxations of the max-cut and stable set problems, Math. Programming, 77 (1997), 225-246.

[21] L. LÓVASZ, A. SHRIJVER Cones of matrices and set functions, and 0-1 optimization. SIAM Journal on Optimization, 1 (1991), 166-190.

[22] POLJAK S., AND RENDL F. Nonpolyhedral relaxations of graph-bisection problems. SIAM Journal on Optimization, 5 (1995), 467-487.

[23] M.F.ANJOS, H.WOLKOWICZ A Strengthened SDP Relaxation via a Second Lifting for the Max-Cut Problem. University of Waterloo, Canada. Research Report CORR 99-55, 1999, p.1-28.

Chapter 10

# GENERALIZED DUALITY IN VARIATIONAL ANALYSIS

Stephen M. Robinson
*Department of Industrial Engineering*
*University of Wisconsin–Madison*
*1513 University Avenue*
*Madison, WI 53706-1572, USA*
smrobins@facstaff.wisc.edu

**Abstract** This paper deals with extensions of the duality structure familiar from convex optimization to more general problems of variational analysis. We begin by showing how several duality structures, which have already proven to be useful in various areas of optimization and equilibrium analysis, can be obtained as special cases of a general approach involving inverses of (possibly set-valued) mappings associated with the problems involved. This approach leads to a general procedure that we have elsewhere called *composition duality.*

Duality schemes provide most of their real usefulness through the interplay between the general duality principles and the special structure of a particular problem to which they are applied. We review some examples of this kind of usefulness, using the general duality scheme just mentioned. Finally, we look at possibilities for extending these schemes to wider classes of problems within the general field of variational analysis.

**Keywords:** Complementarity, Composition duality, Maximal monotonicity, Preprocessing, Variational analysis.

## 1 INTRODUCTION

This expository paper describes a general duality structure for operator inclusions. The structure yields reformulation methodology that is tailored to available special structure appearing in the inclusion. This methodology includes a large number of useful duality forms and other



*N. Hadjisavvas and P.M. Pardalos (eds.), Advances in Convex Analysis and Global Optimization,* 205–219.

known techniques as special cases, and it can give useful theoretical insights. It can also yield computational savings when applied to reduce the number of variables in a variational inequality. We include here both examples of the theoretical applications and some information on results obtained by Ferris and Munson on computational applications.

The rest of this paper is organized in five sections. Section 2 explains the basic method and points out some important known special cases. Then Section 3 shows important connections with maximal monotonicity, especially the preservation of maximality under a certain transformation that the duality structure uses. Section 4 shows several ways in which this set of methods can be applied analytically to exploit special structure, while Section 5 complements this by reporting a significant computational application. Finally, Section 6 reviews what has been done in the paper and suggests some areas in which further work might pay dividends.

# 2 VARIATIONAL DUALITY STRUCTURES

In this paper we are concerned with *variational duality structures,* which for our purposes will mean correspondences that associate with a "primal" generalized equation $0 \in F(x)$ a "dual" inclusion, $0 \in G(y)$, where both $F$ and $G$ are multifunctions but possibly between different spaces. These two inclusions should be related in suitable ways; for us this will mean that one is solvable if and only if the other is, and that solutions of one inclusion have some relationship to solutions of the other. Depending on the context, given a primal solution it may be fairly easy to identify a dual solution, and vice versa. Also, it is desirable that the association go in both directions: that is, that the "dual of the dual" be the original primal inclusion.

Duality theory is very well known in convex programming. For a good basic treatment of this see, e.g., [20], and for a more general exposition, Chapter 11 of [21]. Several authors have extended ideas from duality to apply to generalized equations, especially variational inequalities. For example, the duality scheme of [14] applies to the sum of a subdifferential and a maximal monotone operator. Mosco's treatment was extended in 1983 by [9], and Fukushima gave in [8] a duality scheme for the sum of two maximal monotone operators. Several schemes were unified by the recent work of [1].

All of these duality formulations can be unified by a simple operator formalism that we called *composition duality* in [17]. In the rest of this section we explain the idea behind composition duality, and we indicate how the duality schemes just mentioned comprise special cases.

We start by considering an inclusion involving the composition of several operators, of the form

$$0 \in A(x) + QBP(x), \tag{10.1}$$

where $A : X \to Y$, $P : X \to U$, $B : U \to V$, $Q : V \to Y$, and where $X$, $Y$, $U$, $V$ are linear spaces. Here and in what follows we use the term "operator" interchangeably with "multifunction," so that in particular these operators need not be single-valued, and they may have empty values at certain points.

Simple arguments involving the graphs of these operators lead to the following duality theorem (Theorem 1 of [17]):

**Theorem 1.** *The inclusion* (10.1) *is solvable if and only if the inclusion*

$$0 \in (-P)A^{-1}(-Q)(y) + B^{-1}(y) \tag{10.2}$$

*is solvable.*

Note that the multifunction appearing in (10.1) goes from $X$ to $Y$, while that appearing in (10.2) goes from $V$ to $U$. This is in contrast to the setup of [1], in which the primal and dual inclusions involve multifunctions between the same pair of spaces. An important special case of Theorem 1 occurs when $A$ and $B$ are maximal monotone operators and $P$ and $Q$ are linear transformations.

Several well known duality setups appear as special cases of (10.1) and (10.2) for various choices of $A$, $B$, $P$, and $Q$. One of these is the scheme of [1], which we can recover by setting $X = U$, $Y = V$, and $P = I = Q$. If we take $X$ to be a locally convex Hausdorff topological vector space and $Y$ to be its dual $X^*$, with $U = X$, $V = Y$, and $P = I = Q$, and if we require $A$ to be an injective map and take $B$ to be $\partial f$ for a lower semicontinuous convex function $f$, we obtain the duality relation of [14]. This relation was extended by [9] to cases in which $A$ is a continuous linear operator.

As pointed out in [17], another very well known duality setup with extensive applications is the conjugate duality of Rockafellar; see e.g. [19, 20] and references therein. To represent it in the present framework, we take $X$ to be a real Hilbert space (identified with its dual, as we shall do with all such spaces that we use here) and $Y = X$; we introduce another real Hilbert space $J$ and let $U = X \times J$. Let $A$ be the zero operator on $X$, let $F$ be a lower semicontinuous proper convex function on $U$ and take $B = \partial F$ (thus $V = U$). Finally, define $S : X \to U$ by $S(x) = (x, 0)$ and let $P = S$ and $Q = S^*$. Then we can rewrite (10.1) and (10.2) as

$$0 \in 0(x) + S^*(\partial F)S(x) \tag{10.3}$$

and

$$0 \in (-S)(0^{-1})(-S^*)(y) + \partial F^*(y), \tag{10.4}$$

where $0^{-1}$ is the operator $X$ defined by taking its value at $z$ to be $X$ if $z = 0$ and is empty otherwise.

The following theorem shows the relationship between the general duality setup above and the conjugate duality construction. The theorem uses the *Lagrangian* function $\mathcal{L}$ defined on $X \times J$ by

$$\mathcal{L}(x, p^*) := \inf_p \{\langle p^*, p\rangle + F(x, p)\}.$$

**Theorem 2.** *([17], Theorem 2) Let $X$ and $J$ be real Hilbert spaces and let $K = X \times J$. Let $F$ be a lower semicontinuous convex function from $K$ to $\mathbb{R} \cup \{+\infty\}$, and define $S : X \to K$ by $S(x) = (x, 0)$. For a pair $(x_0, y_0) \in X \times K$ the following are equivalent:*

1. *$x_0$ satisfies (10.3) and $y_0$ satisfies (10.4).*
2. *$y_0$ is of the form $(0, -p_0^*)$ and $y_0 \in \partial F(x_0, 0)$.*
3. *$y_0$ is of the form $(0, -p_0^*)$ and $(x_0, p_0^*)$ is a saddle point of $\mathcal{L}$.*

*Moreover, these equivalent conditions imply that $F(x_0, 0) = \mathcal{L}(x_0, p_0^*)$, and that if there is some $\hat{x} \in X$ with $F(\hat{x}, 0) < +\infty$ then this value is finite.*

For a viewpoint about conjugate duality similar to this, see Chapter 5 of [2] and Chapter III, Section 3 of [5].

We have seen that several known duality structures fit into the setup presented here, and we will see more in the next section. It is worth noting here, however, that some well known devices in linear algebra also turn out to be closely related. One of these is the Sherman-Morrison-Woodbury formula for inversion of a perturbation of a nonsingular matrix by a matrix of relatively low rank, for which see [10, Sec. 5.1]. Another is Kron's "method of tearing" for electrical network analysis; see [11, 12, 13].

# 3 CONNECTIONS WITH MAXIMAL MONOTONICITY

One of the most interesting instances of the duality structure of Section 2 occurs when we specialize to the situation in which $X$ and $U$ are real Hilbert spaces and $A$ and $B$ are maximal monotone operators, with $A : X \to X$ and $B : U \to U$, and where $L$ is a continuous linear operator from $X \to U$ with $P = L$ and $Q = L^*$. Then the primal and dual inclusions become

$$0 \in A(x) + L^*BL(x), \qquad 0 \in (-L)A^{-1}(-L^*)(y) + B^{-1}(y), \qquad (10.5)$$

and a natural question to ask is: when are *all* terms in these two inclusions maximal monotone operators? This question is of special interest in the context of computational solution of these inclusions, because with maximal monotonicity we have a better chance of applying splitting methods to solve these inclusions.

Of course, the inverse of a maximal monotone operator is also maximal monotone, so that we can reformulate the essential part of this question as follows: if $L$ is a continuous linear transformation from $X$ to $U$ and $T$ is a maximal monotone operator from $U$ to $U$, then $L^*TL$ is clearly monotone; when is it *maximal* monotone?

Gabay considered this question ([9]), and showed that maximal monotonicity of $L^*TL$ holds if either $T^{-1}$ is strongly monotone, or $LL^*$ is an isomorphism of $U$. He was particularly interested in the case where the operator $B$ was the subdifferential of a closed proper convex function, so that the two inclusions were

$$0 \in A(x) + L^*(\partial f)L(x), \qquad 0 \in (-L)A^{-1}(-L^*)(y) + \partial f^*(y).$$

[8] also studied problems of this type, in the more general form (10.5), in the context of splitting algorithms.

In many problems we may want to avoid strong monotonicity, and the second condition (that $LL^*$ be an isomorphism of $U$) imposes strong restrictions on the problem. Therefore it is natural to ask whether there is a less restrictive condition.

In fact there is, and to find it we begin by transforming the question into a more familiar one by dealing with sums of operators instead of products. We then find the following very close correspondence, nearly a characterization. We write $I_C$ for the (convex) indicator of a convex set $C$.

**Theorem 3.** ([17], Theorem 3) *Let $X$ and $U$ be real Hilbert spaces; let $L$ be a continuous linear transformation from $X$ to $U$ and let $T$ be a monotone operator from $U$ to $U$. If $\operatorname{im} L^*$ is closed in $X$ and $T + \partial I_{\operatorname{im} L}$ is maximal monotone, then $L^*TL$ is maximal monotone. If $\operatorname{im} L$ is closed in $U$ and $L^*TL$ is maximal monotone, then $T + \partial I_{\operatorname{im} L}$ is maximal monotone.*

In particular this result shows that maximality of $L^*TL$ depends only on the images of $L$ and $L^*$, rather than on the operators themselves.

The sum formulation now yields a simple sufficient condition.

**Theorem 4.** ([17], Theorem 4) *Let $X$ and $U$ be real Hilbert spaces; let $L$ be a continuous linear transformation from $X$ to $U$ such that* $\operatorname{im} L$ *is closed in $U$ and* $\operatorname{im} L^*$ *is closed in $X$. Let $T$ be a maximal monotone operator from $U$ to $U$. If either* $\operatorname{im} L$ *meets* $\operatorname{int} \operatorname{dom} T$ *or* $\operatorname{im} L = U$*, then $L^*TL$ is maximal monotone.*

If the closedness conditions on the images of $L$ and $L^*$ in Theorem 4 hold, then the conclusion of that theorem covers the results of Gabay. In particular, if $T^{-1}$ is strongly monotone and maximal then $\operatorname{dom} T = U$, so $L^*TL$ is maximal by Theorem 4. For the other condition we note that if $LL^*$ is an isomorphism of $U$ then $\operatorname{im} L = U$, so again $L^*TL$ is maximal by Theorem 4.

Thus, we have established a simple condition for *both* primal and dual inclusions in (10.5) to involve maximal monoton operators. This can help in providing theoretical justification for the use of numerical solution methods that require maximal monotonicity for their successful use.

# 4 EXPLOITING SPECIAL STRUCTURE

To date, relatively few applications of the composition duality setup have been explored. Therefore the full scope of its possible applications is still unclear. In the remainder of this paper, we will consider ways in which it can help exploit special structure in *large, finite-dimensional* variational inequality problems. Accordingly, from here on we specialize the setup of Sections 2 and 3 to the case in which $X = \mathbb{R}^n = Y$ and $U = \mathbb{R}^m = V$. Accordingly, we take

$$A : \mathbb{R}^n \to \mathbb{R}^n, \qquad B : \mathbb{R}^m \to \mathbb{R}^m,$$

$$P : \mathbb{R}^n \to \mathbb{R}^m, \qquad Q : \mathbb{R}^m \to \mathbb{R}^n,$$

and thus the primal problem involves a sum of operators from $\mathbb{R}^n$ to $\mathbb{R}^n$ and the dual involves a sum of operators from $\mathbb{R}^m$ to $\mathbb{R}^m$.

In what follows we will try to illustrate how composition duality can contribute to both theory (principles for dealing with classes of problems) and application (analysis and processing of problems found in practice). The illustrations of principles will involve applications to complementarity problems, of two kinds: solving problems subject to a simple rank-$k$ perturbation, and reducing problem dimension by partial solution. The illustration of practice will involve a system for preprocessing very large complementarity problems before numerically solving them.

## 4.1 SIMPLE RANK-K PERTURBATION

The setting for the first application is the numerical solution of structured operator inclusions; in particular, solving problems that have been altered by a simple rank-$k$ perturbation. We will describe the problem first, then show how composition duality can sometimes help with its solution. We described a more specific case of this transformation in [18].

Suppose we have a (possibly very large) inclusion of the form

$$0 \in (M + UV^T)(x), \tag{10.6}$$

where $M$ is an operator from $\mathbb{R}^n$ to $\mathbb{R}^n$ and $U$ and $V$ are $n \times k$ matrices of rank $k$. Assume that this problem would be "easy" without the rank-$k$ perturbation $UV^T$. For example, in an extreme case $M$ might be of the form $M(x) = Dx - d + N_+(x)$, where $D$ is diagonal, $d \in \mathbb{R}^n$, and $N_+(x)$ denotes the normal cone to the nonnegative orthant $\mathbb{R}^n_+$. In that case, if $0 \in M(x)$ has a solution we can solve it componentwise by solving $n$ one-dimensional inclusions.

Suppose that we define $A(x) = M(x)$, $B(\xi) = \xi$ for $\xi \in \mathbb{R}^k$, $Q(\xi) = U\xi$, and $P(x) = V^T x$. Then our problem is

$$0 \in A(x) + QBP(x).$$

We can therefore apply composition duality to obtain the equivalent dual problem given by

$$0 \in (-P)A^{-1}(-Q)(\xi) + B^{-1}(\xi), \tag{10.7}$$

which is an equation in $\mathbb{R}^k$, no matter how large $n$ is. Therefore, if $k$ is much smaller than $n$ then there will be a substantial reduction in the dimensionality of the problem to be solved. This illustrates a conceptual contribution of the duality approach: we could in fact derive an equivalent method from a splitting argument, but that would express the problem as an equation in $\mathbb{R}^n$, and it would require some manipulation to obtain the equation (10.7) in $\mathbb{R}^k$. The duality approach guides us naturally to the equation in $\mathbb{R}^k$.

Of course, up to now we have not assumed anything that would guarantee that (10.6) can be solved, and therefore that (10.7) can be solved. To obtain any general results about solvability, we therefore have to make some additional assumptions.

To illustrate one set of assumptions that yields solvability, suppose that $M$ is strongly monotone with modulus $\rho$ and that $\rho^{-1}\|U\|\|V\| < 1$. This means that the perturbation matrices $U$ and $V$ are "sufficiently small" compared to the modulus $\rho$ of $M$. Now rewrite the dual (10.7) as

$$0 = \phi(\xi) = \xi - V^T M^{-1}(d - U\xi).$$

Our assumptions guarantee that $M^{-1}$ is single-valued and Lipschitzian with modulus $\rho^{-1}$, and we have assumed that we can numerically evaluate $M^{-1}(y)$ for any $y$ (as we assumed that solving $0 \in M(x)$ was "easy").

The function of $\xi$ given by $V^T M^{-1}(d - U\xi)$ is a strong contraction with modulus $\rho^{-1}\|U\|\|V\| < 1$, so we can solve $\phi(\xi) = 0$ by iteration, and the solution will be unique. Once we find $\bar{\xi}$ with $\phi(\bar{\xi}) = 0$ (that is, the solution of (10.7)), then the point $\bar{x} \in \mathbb{R}^n$ given by $\bar{x} = F(d - U\bar{\xi})$ solves the original problem (10.6). We have therefore solved the perturbed inclusion (10.6) by solving an equation in $\mathbb{R}^k$ together with "easy" problems in $\mathbb{R}^n$.

As we pointed out in [18], this reduction applies in particular to "weakly coupled" complementarity problems of the form

$$\begin{bmatrix} X & \cdots & X & & & \\ \vdots & & \vdots & & & \\ X & \cdots & X & \beta & & \\ & & \alpha & X & \cdots & X \\ & & & \vdots & & \vdots \\ & & & X & \cdots & X \end{bmatrix},$$

as this is the sum of a block-diagonal problem and

$$\begin{pmatrix} e_i & e_j \end{pmatrix} \begin{pmatrix} 0 & \beta \\ \alpha & 0 \end{pmatrix} \begin{pmatrix} e_i^T \\ e_j^T \end{pmatrix},$$

where the $e$'s are elementary vectors. In fact, in the context described here the block subproblems need not even be linear, though the illustration we just gave used linearity to make the structure easy to see.

If $M$ is of the form $M(x) = Tx - t$, where $T$ is an $n \times n$ nonsingular matrix and $t \in \mathbb{R}^n$, then this transformation recovers the Sherman-Morrison-Woodbury formula for inverting a matrix that has been perturbed by addition of a low-rank matrix. This case is also closely related to Kron's "method of tearing" for solving electrical network problems (see [11, 12, 13]).

## 4.2 REDUCING DIMENSION BY PARTIAL SOLUTION

In this section we examine a method for what might be called partial solution of an operator inclusion. It is a method for general operators that mimics the procedure of eliminating some of the variables in a linear system by solving a subset of the equations in the system; as is well known, that procedure leads to the Schur complement ([3]), so that the operator we describe here could be regarded as a nonlinear analogue of the Schur complement. We developed various aspects of this partial solution method in [16] and [18], and general properties of such methods have been developed in the studies of partial inversion in [22] and [15].

To describe the procedure, we start with an inclusion in $\mathbb{R}^{m+n}$ of the form

$$0 \in A(x) + S^* B S(x), \tag{10.8}$$

where $A$ and $B$ are operators from $\mathbb{R}^{m+n}$ to $\mathbb{R}^{m+n}$ and from $\mathbb{R}^m$ to $\mathbb{R}^m$ respectively, and where $S$ is the linear transformation from $\mathbb{R}^{m+n}$ to $\mathbb{R}^m$ given by $S(u,v) = u$. Notice that the effect of the composition $S^*BS$ is analogous to what in a linear system would amount to placing $B$ in the upper left corner of an $(m+n) \times (m+n)$ matrix and filling the rest of the matrix with zeros.

We can dualize (10.8) to obtain

$$0 \in (-S)A^{-1}(-S^*)(y) + B^{-1}(y), \tag{10.9}$$

an equation in $\mathbb{R}^m$. However, suppose that we want a form in which $A$ and $B$ appear, instead of their inverses. Apply Attouch-Théra duality to (10.9) in $\mathbb{R}^m$ to produce the equivalent inclusion

$$0 \in (S^*)^{-1} A S^{-1}(z) + B(z), \tag{10.10}$$

also in $\mathbb{R}^m$. What is this operator $(S^*)^{-1}AS^{-1}$ that appears in (10.10)?

The following proposition characterizes $(S^*)^{-1}AS^{-1}$ in terms of its graph, which makes visible the action of the operator on points. Recall that $A : \mathbb{R}^{m+n} \to \mathbb{R}^{m+n}$, so that points of the domain and range of $A$ can be written as pairs $(x, y)$ with $x \in \mathbb{R}^m$ and $y \in \mathbb{R}^n$.

**Proposition 5.** ([18], Proposition 2.) *Let $A$ be an operator from $\mathbb{R}^{m+n}$ to $\mathbb{R}^{m+n}$, and let $S$ be the linear transformation from $\mathbb{R}^{m+n}$ to $\mathbb{R}^m$ given by $S(u,v) = u$. The graph of $(S^*)^{-1}AS^{-1}$ is the set of pairs $(x, x^*)$ in $\mathbb{R}^{2m}$ for which there is some $y \in \mathbb{R}^n$ such that*

$$\begin{pmatrix} x^* \\ 0 \end{pmatrix} \in A \begin{pmatrix} x \\ y \end{pmatrix}.$$

So this is indeed a kind of "partial solution" operator, which solves the operator inclusion with respect to the second group of variables (here, the variables in $\mathbb{R}^n$) while maintaining those in the first group (here, the variables in $\mathbb{R}^m$) at a predetermined value $x$. By solving the inclusion with respect to the second group, we mean causing a zero to appear in that portion of a point in the image, corresponding to the second group of variables. As a result of this process, the first group of variables in the image point contains some point $x^*$, and this is then an image of $x$ under the operator $(S^*)^{-1}AS^{-1}$.

As an example of the application of this operator, consider the inclusion

$$0 \in \begin{pmatrix} g(x) \\ 0 \end{pmatrix} + A \begin{pmatrix} x \\ y \end{pmatrix},$$

where $g$ is a function, $A$ is a structured multifunction (*e.g.*, a linear transformation plus a normal-cone operator). We can rewrite this inclusion as

$$0 \in S^* g S \begin{pmatrix} x \\ y \end{pmatrix} + A \begin{pmatrix} x \\ y \end{pmatrix},$$

so we know from the above discussion that it is equivalent to

$$0 \in g(z) + (S^*)^{-1}AS^{-1}(z).$$

Several consequences follow from this. First consider the case in which

$$A \begin{pmatrix} x \\ y \end{pmatrix} = \begin{pmatrix} 0 & M \\ -M^T & 0 \end{pmatrix} \begin{pmatrix} x \\ y \end{pmatrix} + \begin{pmatrix} 0 \\ m \end{pmatrix} + \begin{pmatrix} N_P(x) \\ N_Q(y) \end{pmatrix},$$

with $P$ convex and $Q$ a polyhedral convex cone, for example $\mathbb{R}^n_+$. This form arises in Walrasian equilibrium problems in economics; see the example in [16] and references therein. For this setup,

$$(S^*)^{-1}AS^{-1} = N_{P\cap Z}, \tag{10.11}$$

where $Z$ is the polyhedral convex set defined by

$$Z = \{z \mid M^T z - m \in Q^\circ\},$$

and $Q^\circ$ is the polar cone of $Q$. Therefore we have reduced the original problem to

$$0 \in g(z) + N_{P\cap Z}(z),$$

a generalized equation in a smaller space. We shall use (10.11) again in the next section, in the context of preprocessing for numerical solution of certain variational inequalities.

For another example, suppose A is a matrix of block form,

$$A = \begin{pmatrix} A_{11} & A_{12} \\ A_{21} & A_{22} \end{pmatrix},$$

with $A_{11}$ and $A_{22}$ square and $A_{22}$ nonsingular. Then

$$(S^*)^{-1}AS^{-1} = (A/A_{22}),$$

where $(A/A_{22}) = A_{11} - A_{12}A_{22}^{-1}A_{21}$, the Schur complement of $A_{22}$ in $A$. This justifies our earlier remark about the connection with Schur complements.

We can take this a step further by introducing what might be called generalized Schur complements. Here is one such construction, which is studied in [18]. Let

$$A\begin{pmatrix} x \\ y \end{pmatrix} = \begin{pmatrix} A_{11} & A_{12} \\ A_{21} & A_{22} \end{pmatrix}\begin{pmatrix} x \\ y \end{pmatrix} + \begin{pmatrix} a_1 \\ a_2 \end{pmatrix} + \begin{pmatrix} N_P(x) \\ \{0\} \end{pmatrix},$$

where $P$ is a convex set in $\mathbb{R}^p$ and $A_{22}$ is nonsingular. Then

$$(S^*)^{-1}AS^{-1}(x) = (A/A_{22})(x) + b + N_P(x),$$

with $b = a_1 - A_{12}A_{22}^{-1}a_2$. This transformation occurs in local reduction of a complementarity problem by use of the critical cone.

# 5 APPLICATION TO PRACTICE: PREPROCESSING

Preprocessing techniques are widely used and very effective in the solution of large linear programming (LP) problems. The idea is to identify, before the actual solution of the problem, special structure that can be exploited. Then one adjusts the problem to take advantage of that structure, for example by reducing the problem's size, solves the reduced problem, and finally recovers a solution of the original problem from information stored during the preprocessing and solution phases. Preprocessing can contribute to increasing the effectiveness of solvers for optimization by enabling quicker solution, less storage use, and better accuracy, among other benefits.

Because of the benefits associated with the use of preprocessing, it would be very beneficial to be able to apply it also to the solution of problems more general than linear programming. This has recently been done in [7], and their implementation makes substantial use of composition duality: in particular, of the partial-solution operator described in Section 4.2. In this section we will give an overview of some of the work in [7], focusing particularly on how the methods described there use partial solution.

The work described in [7] applies to mixed complementarity problems (MCP). These are finite-dimensional variational inequalities over a "box" (that is, a product of intervals), and they have the form

$$0 \in g(x) + N_B(x),$$

where $B = \Pi_{i=1}^n [L_i, U_i]$, a product of intervals. Here $L_i$ may be $-\infty$ and $U_i$ may be $+\infty$; thus we are dealing with closed intervals, but they may be halflines in $\mathbb{R}$ or even all of $\mathbb{R}$.

The idea is to apply the formula in (10.11), namely

$$(S^*)^{-1} A S^{-1} = N_{P \cap Z},$$

where $Z$ is the polyhedral convex set defined by

$$Z = \{z \mid M^T z - m \in Q^\circ\},$$

to the case in which $P$ is most of $B$ and $Q$ is the rest of it. This means selecting a few rows/columns of the MCP that have the skew-symmetric linear structure required for (10.11), and for which at least one of the $L_i$ and $U_i$ is infinite.

Of the several kinds of reductions described in [7], we show only one, given in Section 2.1 of that paper, as an example of the use of composition duality. Suppose that we can express the problem that we are dealing with in the form

$$0 \in \begin{pmatrix} F(x) - A^T y \\ Ax - b \end{pmatrix} + \begin{pmatrix} N_X(x) \\ N_Y(y) \end{pmatrix}, \tag{10.12}$$

where $x \in \mathbb{R}^n$, $y \in \mathbb{R}^m$, $F$ is a continuously differentiable function from $\mathbb{R}^n$ to $\mathbb{R}^n$, $A$ is a linear transformation from $\mathbb{R}^n$ to $\mathbb{R}^m$, and $b \in \mathbb{R}^m$. The set $X$ is a Cartesian product of $n$ closed intervals, and $Y$ is a Cartesian product of $m$ intervals, each of which is $\mathbb{R}_-$, $\mathbb{R}$, or $\mathbb{R}_+$. The quantities $N_X(x)$ and $N_Y(y)$ are the normal cones of the sets $X$ and $Y$ at $x$ and $y$ respectively.

Note that the quantities $F$ and $A$ in (10.12) will differ at different stages of the preprocessing. Essentially, (10.12) says that we have been able to find certain variables (the elements of $y$) that enter the problem linearly, and that are related to the other variables in a way defined by the skew-symmetric structure shown. All of the other variables are lumped together in the vector $x$. It is quite possible that $y$ might have only one element at a particular stage of the process, for example if we are eliminating a single variable as described below. At the next stage $y$ will consist of other variables. Thus, (10.12) only expresses our view of the problem at a particular time.

For illustration, suppose that we choose $A$ to consist of a single row and that this row has just one nonzero element, say $r$, which appears in the $j$th column. Thus, $Ax = rx_j$. Suppose also that $Q$ (a single interval, in this case) is $\mathbb{R}_+$. Then $Z = \{z \mid Az - b \in Q^\circ\}$ is just the set of points satisfying a simple bound, either $x_j \leq r^{-1}b$ if $r > 0$ or $x_j \geq r^{-1}b$ if $r < 0$. Thus $Z$ is a product of intervals, each of which is $\mathbb{R}$ except for the $j$th, which is either $(-\infty, r^{-1}b]$ or $[r^{-1}b, +\infty)$. Let us call this interval $I$.

Recall that (10.11) involves $N_{P\cap Z}$, and $P$ is itself a product of intervals. Therefore $P \cap Z$ is a product of intervals, each of which except the $j$th is just the corresponding interval from $P$. The $j$th interval in $P \cap Z$ is the intersection of $I$ with the $j$th interval of $P$. If this intersection is empty then the problem has no solution and no further work need be done. Otherwise we can replace the $j$th interval of $P$ by the intersection and remove the variable that we labeled as $y$ from the problem. We have now reduced the number of variables by one, and we proceed to look for more preprocessing opportunities.

Of course, it is necessary not only to reduce the problem as shown here, but to have some means of recovering, from the solution of the reduced problem, a solution of the original unreduced problem. This is done by means of two propositions from [16], which are summarized as Theorem 1 of [7].

Ferris and Munson give extensive computational results. Here we just quote a few of these to illustrate the power of this preprocessor. First, on a number of problems from MCPLIB (see [4]), preprocessing significantly reduced the solution time. Reductions reported (in seconds) include

- `electric`: 1.3 to 0.5
- `golanmcp`: 80.9 to 25.0
- `merge`: 2254.6 to 1954.2

Even more significant reductions were obtained on a large equilibrium model of Chavas and Cox, designated as `hwayoung`, for simulating trade liberalization in the world dairy industry. For that model, preprocessing reduced the problem from 46,123 variables to 13,655, a reduction of more than 70%. The original version of the model could not be solved, but preprocessing rendered the model solvable ([6]).

# 6 CONCLUSIONS

In this expository paper, we have exhibited a general duality structure for operator inclusions and have described some of its applications. The duality technique yields reformulation methodology that is tailored to available special structure, and that can be implemented in practice (e.g. by preprocessing). It has already produced computationally useful results on practical models. It also includes a large number of useful duality forms and other known techniques as special cases.

Given the results obtained to date, it seems likely that more benefit could be gained by exploring further properties of this general methodology. Such benefits might include increasing the sophistication of computational preprocessors (e.g. by developing aggregation methods), finding out more about what the duality transformation does to particular problem structures, and exploiting that knowledge to increase computational effectiveness in solving variational models.

## Acknowledgments

The research reported here was sponsored by the U. S. Army Research Office under Grant DAAG55-97-1-0324.

## References

[1] Attouch, H. and Théra, M. (1996). A general duality principle for the sum of two operators. *Journal of Convex Analysis*, 3:1–24.

[2] Aubin, J.-P. (1984). *L'Analyse Non Linéaire et Ses Motivations Économiques*. Masson, Paris.

[3] Cottle, R. W. (1974). Manifestations of the Schur complement. *Linear Algebra and Its Applications*, 8:189–211.

[4] Dirkse, S. and Ferris, M. C. (1995). MCPLIB: A collection of nonlinear mixed complementarity problems. *Optimization Methods and Software*, 5:319–345.

[5] Ekeland, I. and Turnbull, T. (1983). *Infinite-Dimensional Optimization and Convexity*. Chicago Lectures in Mathematics. The University of Chicago Press, Chicago and London.

[6] Ferris, M. C. (2000). Private communication to the author.

[7] Ferris, M. C. and Munson, T. S. (2001). Preprocessing complementarity problems. In Ferris, M. C., Mangasarian, O. L., and Pang, J.-S., editors, *Complementarity: Algorithms, Applications and Extensions*, Dordrecht. Kluwer Academic Publishers.

[8] Fukushima, M. (1996). The primal Douglas-Rachford splitting algorithm for a class of monotone mappings with application to the traffic equilibrium problem. *Mathematical Programming*, 72:1–15.

[9] Gabay, D. (1983). Applications of the method of multipliers to variational inequalities. In Fortin, M. and Glowinski, R., editors, *Augmented Lagrangian Methods: Applications to the Numerical Solution*

*of Boundary-Value Problems*, number 15 in Studies in Mathematics and Its Applications, pages 299–331. North-Holland, Amsterdam.

[10] Householder, A. S. (1975). *The Theory of Matrices in Numerical Analysis.* Dover Publications, Inc., New York. Originally published 1964 by Blaisdell Publishing Co.

[11] Kron, G. (1939). *Tensor Analysis of Networks.* Wiley, London.

[12] Kron, G. (1959). *Tensors for Circuits, 2d Ed.* Dover, New York.

[13] Kron, G. (1963). *Diakoptics; the piecewise solution of large-scale systems.* Macdonald, London.

[14] Mosco, U. (1972). Dual variational inequalities. *Journal of Mathematical Analysis and Applications*, 40:202–206.

[15] Pennanen, T. (1998). Dualization of generalized equations of maximal monotone type. Department of Mathematics, University of Washington, Seattle, WA.

[16] Robinson, S. M. (1998). A reduction method for variational inequalities. *Mathematical Programming*, 80:161–169.

[17] Robinson, S. M. (1999). Composition duality and maximal monotonicity. *Mathematical Programming*, 85:1–13.

[18] Robinson, S. M. (2000). Structural methods in the solution of variational inequalities. In Di Pillo, G. and Giannessi, F., editors, *Nonlinear Optimization and Related Topics*, pages 369–380, Dordrecht. Kluwer Academic Publishers.

[19] Rockafellar, R. T. (1970). *Convex Analysis.* Princeton University Press, Princeton, NJ.

[20] Rockafellar, R. T. (1974). *Conjugate Duality and Optimization.* Number 16 in CBMS Regional Conference Series in Applied Mathematics. Society for Industrial and Applied Mathematics, Philadelphia, PA.

[21] Rockafellar, R. T. and Wets, R. J. (1998). *Variational Analysis.* Number 317 in Grundlehren der mathematischen Wissenschaften. Springer-Verlag, Berlin.

[22] Spingarn, J. E. (1983). Partial inverse of a monotone operator. *Applied Mathematics and Optimization*, 10:247–265.

# Chapter 11

# CLUSTERING VIA D. C. OPTIMIZATION

Hoang Tuy
*Institute of Mathematics*
*P. O. Box 631, Bo Ho, Hanoi, Vietnam*

A.M. Bagirov and A.M. Rubinov*
*School of Information Technology and Mathematical Sciences,*
*University of Ballarat, Ballarat, Victoria, 3353, Australia*

**Abstract** The cluster analysis problem is formulated as a problem of global minimization of a function represented as a difference of two convex functions over the unit simplex. The version of branch and bound method for the solution of this problem is studied. Computational testing of suggested algorithm was carried out on Wisconsin Diagnostic Breast Cancer database. We present the results of numerical experiments and discuss them.

**Keywords:** global optimization, d.c. optimization, clustering, data mining.

## 1 INTRODUCTION

The classification of objects into groups or clusters is important in many fields of operations research and applied sciences. During the last decades many approaches to this problem, including statistical ( [9]), machine learning ([7]), mathematical programming and in particular linear and integer programming approaches ([2, 10, 12]) have been suggested and studied. Approaches based on the mathematical programming techniques are most effective among them (see, for example, [10]).

*This research was supported by the Australian Research Counsil under grant No. A49906152. It has been completed when first author visited School of Information Technology and Mathematical Sciences, University of Ballarat, Australia.



*N. Hadjisavvas and P.M. Pardalos (eds.), Advances in Convex Analysis and Global Optimization,* 221–234.

But algorithms for the solution of cluster analysis problems developed in these papers are based on methods which allow to find only local minima points. Therefore, there is no guaranteed way to ensure global optimality of the solution obtained by these techniques. So different starting points can be used to initiate the algorithms. But even in this case the obtained solution is not necessarily a global one. Such heuristics strategies guarantee, at best, the finding of some "locally good" classification which might be quite far from the "best" one. From the theoretical point of view the global optimization methods provide the best solution determining centers of clusters. The global optimization approach to the clustering problem and some difficulties connected with it is discussed in [11]. Another approach based on so-called cutting angle method is studied in [3].

In this paper we study the d.c. programming approach to the clustering problem. To verify of effectiveness of this approach numerical experiments have been carried out. In numerical experiments as data set the Wisconsin Diagnostic Breast Cancer Database is used. We present and discuss the results of these experiments.

The problem of cluster analysis can be formulated as follows. For a given set of points $a^i, i = 1, \ldots, m$ in $R^n$, find cluster centers $x^l, l = 1, \ldots, p$ in $R^n$ such that the sum of the minima over $l \in \{1, \ldots, p\}$ of the distance between each point $a^i$ and the cluster centers $x^l, l = 1, \ldots, p$ is minimized. Using 1-norm for measuring distances, this problem can be formulated as (see e.g. [10]):

$$\begin{array}{ll} \text{minimize} & \sum_{i=1}^{m} \min_{l=1,\ldots,p} e^\top y^{il} \\ \text{subject to} & -y^{il} \le a^i - x^l \le y^{il} \quad i = 1, \ldots, m, \ l = 1, \ldots, p \end{array} \tag{11.1}$$

where $e = (1, \ldots, 1) \in R^{mp}$.

Since for each $i = 1, \ldots, m$ the function $\min_{l=1,\ldots,p} e^\top y^{il}$ is concave piecewise linear, this is a piecewise linear concave minimization problem, for which there are several available algorithms, see [8], [16]. The main difficulty with this approach is that in practice $m$ is too large for these algorihtms to perform efficiently. However, due to the special form of the objective function, problem (11.1) can be turned into an equivalent bilinear program. Specifically, as was shown in [4], (11.1) can be rewritten as

$$\begin{array}{ll} \min & \sum_{i=1}^{m} \sum_{l=1}^{p} e^\top y^{il} t_{il} \\ \text{s.t.} & -y^{il} \le a^i - x^l \le y^{il} \quad i = 1, \ldots, m, \ l = 1, \ldots, p \\ & \sum_{l=1}^{p} t_{il} = 1, \ t_{il} \ge 0 \quad i = 1, \ldots, m, \ l = 1, \ldots, p \end{array} \tag{11.2}$$

In [4] this bilinear program has been solved by an algorithm which alternates between solving a linear program in the variable $t$ and a linear program in the variables $(x, y)$. Although this method terminates after finitely many iterations at a stationary point satisfying the necessary optimality conditions, this stationary point is not guaranteed to be a global optimal solution.

On the other hand, when the square of the 2-norm is used instead of the 1-norm, the problem can no longer be formulated as a concave minimization problem. Instead, it becomes a considerably harder problem, known as the Weber multisource location problem (see e.g. [15]). In [10] the bilinear program (11.2) has been solved by a $p$-mean algorithm, which again does not guarantee a global optimal solution.

The purpose of this paper is to present an approach to the clustering problem based on d.c. optimization techniques developed in [15] for continuous location problems. In [1] results of solving the generalized Weber problem with $m = 10,000$, $p = 2$, and $m = 1,000$, $p = 3$ have been reported. We do not know any computational result for this problem with $p > 3$. However, from the results in [1] it appears that the d.c. approach might be useful for finding a global optimal solution to the clustering problem, although the complexity of the problem quickly grows with $p > 3$.

## 2 CLUSTERING AS A DC OPTIMIZATION PROBLEM

The problem we are concerned with is

$$\min \sum_{i=1}^{m} \min_{l=1,\ldots,p} \|x^l - a^i\| \quad \text{s.t.} \quad x^l \in S \subset R^n,\ l = 1,\ldots,p. \tag{11.3}$$

where $S$ is a compact convex set in $R^n$. Without loss of generality we can assume that $S$ is the unit simplex of $R^n_+$. It is well known (see e.g. [16]) that

$$\min_{l=1,\ldots,p} \|x^l - a^i\| = \sum_{l=1}^{p} \|x^l - a^i\| - \max_r \sum_{l \neq r} \|x^l - a^i\| \tag{11.4}$$

and since the functions

$$\sum_{l=1}^{p} \|x^l - a^i\|, \quad \max_r \sum_{l \neq r} \|x^l - a^i\|$$

are convex it follows that (11.3) is a *d.c. optimization problem,* namely

$$\min \sum_{i=1}^{m}\sum_{l=1}^{p} \|x^l - a^i\| - \sum_{i=1}^{m} \max_r \sum_{l\neq r} \|x^l - a^i\| \quad \text{s.t. } x^l \in S, l = 1, \ldots, p. \tag{11.5}$$

A common method for solving d.c. optimization problems is by branch and bound. A simplicial branch and bound algorithm for the Weber multisource location problem (11.3) was outlined in [15] and applied to location on the plane ($n = 2$) in [1]. Below we adapt this algorithm to the above formulated clustering problem, where usually $n > 2$.

The problem involves $p \times n$ variables, namely $x^l_j, j = 1, \ldots, n, l = 1, \ldots, p$. For the sake of clarity any vector in $X = \prod_{l=1}^p X_l,\ X_l \in R^n$, will be denoted by a boldface low case letter, e.g. $\mathbf{x} = (x^1, \ldots, x^p)$ where $x^l \in X_l$. We also agree that $\mathbf{x}_l = x^l$. Then the problem takes on the form

$$\text{minimize } f(\mathbf{x}) := g(\mathbf{x}) - h(\mathbf{x}) \quad \text{subject to } \mathbf{x} \in \mathbf{S} \tag{11.6}$$

where

$$g(\mathbf{x}) = \textstyle\sum_{i=1}^m \sum_{l=1}^p g_{il}(\mathbf{x}), \quad g_{il}(\mathbf{x}) = \|\mathbf{x}_l - a^i\|,$$
$$h(\mathbf{x}) = \textstyle\sum_{i=1}^m h_i(\mathbf{x}), \quad h_i(\mathbf{x}) = \max_r \sum_{l\neq r} \|\mathbf{x}_l - a^i\|$$

and $\mathbf{S}$ is a simplex in $X = \prod_{l=1}^p X_l,\ X_l = R^n$.

## 2.1 BOUNDING

Let $M$ be a subsimplex of $\mathbf{S}$, with vertex set $V(M)$ (note that $|V(M)| = pn + 1$). To compute a lower bound $\beta(M)$ for the value of $f(\mathbf{x})$ over $M$ we replace the problem by a linear relaxation constructed as follows.

For each $\mathbf{v} \in V(M)$ let $p_{il}(\mathbf{v}) \in \partial g_{il}(\mathbf{v})$, i.e. (see, for example, [15]):

$$p_{il}(\mathbf{v}) = \begin{cases} \frac{\mathbf{v}_l - a^i}{\|\mathbf{v}_l - a^i\|} & \text{if } \mathbf{v}_l \neq a^i, \\ 0 & \text{otherwise.} \end{cases} \tag{11.7}$$

Then an affine minorant of $g_{il}(\mathbf{x})$ is $\langle p_{il}(\mathbf{v}), \mathbf{x}_l - \mathbf{v}_l\rangle + \|\mathbf{v}_l - a^i\|$, so a minorant of $g(\mathbf{x})$ is

$$\varphi(\mathbf{x}) = \max_{\mathbf{v}\in V(M)} \sum_{i=1}^{m}\sum_{l=1}^{p} [\langle p_{il}(\mathbf{v}), \mathbf{x}_l - \mathbf{v}_l\rangle + \|\mathbf{v}_l - a^i\|].$$

Also let $\psi_i(\mathbf{x})$ be a minorant of $h_i(x)$ provided by the affine function which matches with $h_i(\mathbf{x})$ at every $\mathbf{v} \in V(M)$, i.e.

$$\begin{aligned} \psi_i(\mathbf{x}) = \textstyle\sum_{\mathbf{v}\in V(M)} \lambda_{\mathbf{v}} \max_r \sum_{l\neq r} \|\mathbf{v}_l - a^i\| \text{ for every} \\ \mathbf{x} = \textstyle\sum_{\mathbf{v}\in V(M)} \lambda_{\mathbf{v}} \mathbf{v},\ \sum_{\mathbf{v}\in V(M)} \lambda_{\mathbf{v}} = 1,\ \lambda_{\mathbf{v}} \geq 0. \end{aligned} \tag{11.8}$$

**Proposition 1** *A lower bound of $f(\mathbf{x})$ over $M$ is given by the value*

$$\beta(M) := \min_{t,\lambda}\{t - \sum_{i=1}^{m} \sum_{\mathbf{v}\in V(M)} \lambda_{\mathbf{v}} \max_{r} \sum_{l\neq r} \|\mathbf{v}_l - a^i\|\} \quad \textit{subject to}$$
$$\sum_{i=1}^{m} \sum_{l=1}^{p} [\langle p_{il}(\mathbf{v}), \mathbf{x}_l - \mathbf{v}_l\rangle + \|\mathbf{v}_l - a^i\|] \leq t \quad \forall \mathbf{v} \in V(M),$$
$$\mathbf{x}_l = \sum_{\mathbf{v}\in V(M)} \lambda_{\mathbf{v}} \mathbf{v}_l, \quad \sum_{\mathbf{v}\in V(M)} \lambda_{\mathbf{v}} = 1, \ \lambda_{\mathbf{v}} \geq 0. \tag{11.9}$$

*Proof.* A relaxation of the subproblem associated with $M$ is

$$\min\{\varphi(\mathbf{x}) - \sum_{i=1}^{m} \psi_i(\mathbf{x}) | \quad \mathbf{x} \in M\} \tag{11.10}$$

which can also be written as

$$\min\{t - \sum_{i=1}^{m} \psi_i(\mathbf{x}) | \quad \varphi(\mathbf{x}) \leq t, \ \mathbf{x} \in M\}.$$

The latter problem in turn is nothing but (11.9). □We shall refer to the subproblem (11.10), with optimal value $\beta(M)$, as the *bounding subproblem* associated with $M$. Note that if $(\bar{t}, \bar{\lambda})$ is an optimal solution of the linear problem (11.9) then an optimal solution of the bounding subproblem (11.10) is $\bar{\mathbf{x}}(M) = \sum_{\mathbf{v}\in V(M)} \bar{\lambda}_{\mathbf{v}} \mathbf{v}$.

**Proposition 2** *We have $\varphi(\mathbf{x}) = g(\mathbf{x})$ and $\psi_i(\mathbf{x}) = h_i(\mathbf{x})$ $\forall i = 1, \ldots, m$ at every vertex of $M$. (In particular, if $\bar{\mathbf{x}}(M)$ is a vertex of $M$, i.e. if all but one $\bar{\lambda}_{\mathbf{v}}$, $\mathbf{v} \in V(M)$, are zeros, then $\bar{\mathbf{x}}(M)$ is a minimiser of $f(\mathbf{x})$ over $M$).*

*Proof.* Obvious from the definition of $\varphi(\mathbf{x})$ and $\psi_i(\mathbf{x})$. □

**Remark 1** *If we only replace $h(\mathbf{x})$ by a concave majorant, then we have a convex relaxation of the problem:*

$$\min_{\lambda}\{g(\mathbf{x}) - \sum_{i=1}^{m} \sum_{\mathbf{v}\in V(M)} \lambda_{\mathbf{v}} \max_{r} \sum_{l\neq r} \|\mathbf{v}_l - a^i\|\} \quad \textit{subject to}$$
$$\mathbf{x} = \sum_{\mathbf{v}\in V(M)} \lambda_{\mathbf{v}} \mathbf{v}, \quad \sum_{\mathbf{v}\in V(M)} \lambda_{\mathbf{v}} = 1, \ \lambda_{\mathbf{v}} \geq 0. \tag{11.11}$$

*The lower bound computed by solving this convex relaxation is of course in general tighter than the previous one (but requires more computational effort). Proposition 2 is still valid for his bounding method.*

## 2.2 BRANCHING

At each iteration of a branch and bound procedure a partition set $M$ is chosen for further subdivision. A common subdivision method is *bisection*, which consists in partitionning $M$ into two subsimplices of equal volume by a hyperplane through the midpoint of the longest edge of $M$. Specifically, if the vertex set of $M$ is $V(M) = \{\mathbf{v}^1, \ldots, \mathbf{v}^{pn+1}\}$ and $[\mathbf{v}^1, \mathbf{v}^2]$ is a longest edge, with midpoint at $\mathbf{w}$, then $M$ is subdivided into two subsimplices $M_1, M_2$ such that the vertex set of $M_j$, $j = 1, 2$, is obtained from $V$ by replacing $\mathbf{v}^j$ with $\mathbf{w}$, i.e.

$$M_1 = [\mathbf{w}, \mathbf{v}^2, \ldots, \mathbf{v}^{pn+1}], \quad M_2 = [\mathbf{v}^1, \mathbf{w}, \ldots, \mathbf{v}^{pn+1}].$$

Another subdivision method, often referred to as *$\omega$-subdivision*, is defined as follows. Let $\bar{\mathbf{x}}(M) = \sum_{\mathbf{v} \in V(M)} \bar{\lambda}_{\mathbf{v}} \mathbf{v}$ be an optimal solution of the relaxed problem (11.10) and let $J(M) = \{\mathbf{v} \in V(M) |\ \bar{\lambda}_{\mathbf{v}} > 0\}$. Then $M$ is divided into $|J(M)|$ subsimplices $M_j, j \in J(M)$, such that the vertex set of each $M_j$ is obtained from $V(M)$ by replacing $\mathbf{v}^j$ with $\bar{\mathbf{x}}(M)$. (We also say that $M$ is subdivided via the point $\bar{\mathbf{x}}(M)$.)

It is known (see [8] or [16]) that if bisection is used at every iteration then the subdivision process (which will be called a pure bisection process) is *exhaustive*, meaning that diam $M_{k_\nu} \to 0$ for any generated infinite nested sequence of simplices $\{M_{k_\nu}, \nu = 1, 2, \ldots\}$. It can be proved that with bounding as above, exhaustiveness of the subdivision process will ensure convergence. However this convergence may be slow, as it may take many iterations for diam $M_{k_\nu}$ to become sufficiently small. Therefore, to obtain a faster convergence, the following hybrid subddivision rule is proposed.

*Normal Simplicial Subdivision*: Select an infinite sequence $\Delta \subset \{1, 2, \ldots\}$. Set $\sigma(M) = 1$ for every initial simplex $M$ and $\sigma(M') = \sigma(M) + 1$ whenever $M'$ is a child of $M$ (so $\sigma(M)$ is the generation index of $M$). If $\sigma(M) \in \Delta$ then bisect $M$ otherwise $\omega$-subdivide $M$. (i.e. subdivide $M$ via $\bar{\mathbf{x}}(M)$).

Here the choice of $\Delta$ is up to the user and can be made adaptively, from iteration to iteration, to enhance the efficiency of the algorithm. A simple choice is to take $\Delta$ to be the set of multiples of some fixed natural number $N$. Then the subdivision rule is to bisect $M$ if $\sigma(M)$ is a multiple of $N$, and $\omega$-subdivide $M$ if not.

## 2.3 ALGORITHM

Once the bounding and branching rules have been specified, the BB algorithm proceeds according to a standard scheme (see e.g. [8]). At any iteration $k$ of this scheme:

- The candidate for branching is $M_k \in \text{argmin}\{\beta(M)|\ M \in \mathcal{R}_k\}$, where $\mathcal{R}_k$ denotes the collection of all partition sets still of interest at this iteration.
- Let $\bar{\mathbf{x}}^k := \bar{\mathbf{x}}(M_k)$ be the optimal solution of the bounding subproblem associated with $M_k$. If $\bar{\mathbf{x}}^k$ is a vertex of $M_k$, then $\varphi(\bar{\mathbf{x}}^k) = h(\bar{\mathbf{x}}^k)$, $\psi(\bar{\mathbf{x}}^k) = h_i(\bar{\mathbf{x}}^k)$, $i = 1, \ldots, m$, hence $\beta(M_k) = f(\bar{\mathbf{x}}^k)$ and since $\beta(M_k) \leq f(\mathbf{x})\ \forall x \in \mathbf{S}$, it follows that $\bar{\mathbf{x}}^k$ solves (11.6), i.e. our problem, so the algorithm terminates.
- At any iteration $k$, compute $f(\bar{x}(M))$ for every current partition set. Record the incumbent, i.e. the vector $\tilde{\mathbf{x}}^k$ which is the best among all feasible solutions encountered up to this iteration. Remove all partition sets $M$ with $\beta(M) \geq f(\tilde{\mathbf{x}}^k) - \varepsilon$. ( Here $\varepsilon > 0$ is a given tolerance). The general termination criterion is: $f(\tilde{\mathbf{x}}^k) - \beta(M_k) \leq \varepsilon$ (then $\tilde{\mathbf{x}}^k$ is an $\varepsilon$-optimal solution).

**Theorem 1** *The above branch and bound algorithm terminates after finitely many iterations, yielding an $\varepsilon$-optimal solution.*

*Proof.* If the algorithm is infinite, it generates at least an infinite nested sequence $\{M_{k_\nu}\}$. By compactness, we may assume that $\bar{\mathbf{x}}^{k_\nu} \to \bar{\mathbf{x}}$. Furthermore, by normality of the subdivision, we may assume, see e.g. [16], that $\bar{\mathbf{x}}(M_{k_\nu}) \to \mathbf{x}^*$, such that $\mathbf{x}^*$ is a vertex of the simplex $M_\infty = \cap_\nu M_{k_\nu}$. It then easily follows from Proposition 2 that $\mathbf{x}^*$ is a minimizer of $f(\mathbf{x})$ over $M_\infty$. Since $\beta(M_{k_\nu}) \leq \beta(M)$ for all partition sets $M$ still of interest at iteration $k_\nu$ it follows that $f(\mathbf{x}(M_{k_\nu})) \leq f(\mathbf{x})$ for all $\mathbf{x} \in \mathbf{S}$ and hence, by continuity, $f(\mathbf{x}^*) \leq f(\mathbf{x})$ for all $\mathbf{x} \in \mathbf{S}$. This proves the optimality of $\mathbf{x}^*$. Therefore $f(\bar{\mathbf{x}}^{k_\nu}) - \beta(M_{k_\nu}) \leq \varepsilon$ for sufficiently large $\nu$. □

# 3 APPROXIMATE APPROACH

For $p = 2$ (2 cluster centers) the above approach leads to solving a d.c. optimization problem of $2n$ variables. An approximate approach permitting a reduction of the size of the problem is the following.

First solve the problem for $p = 1$, i.e. find a cluster center $c$. Then, assuming the two unknown cluster centers to be $x$ and $y = c + t(c - x)$ with $t \geq 0$, solve the problem:

$$\min \sum_{i=1}^{m} \min\{\|x - a^i\|, \|y - a^i\|\} \tag{11.12}$$

$$\text{s.t.} \quad y = x + t(c - x), \tag{11.13}$$

$$x \in S,\ t \geq 0. \tag{11.14}$$

A difficulty of this problem is that the functions $d_i(t,x) := \|x + t(c - x) - a^i\|$, $i = 1,\dots,m$ are nonconvex in $(t,x) \in R \times R^n$. Two alternative methods may be proposed for solving this problem.

## 3.1 SIMPLICIAL BB METHOD

The main issue here is how to compute a lower bound $\beta(M)$ for the subproblem on a simplex $M \subset R \times R^n$, i.e. the subproblem

$$\min \sum_{i=1}^{m} \min\{\|x - a^i\|, d_i(t,x))\} \quad \text{s.t. } (t,x) \in M. \tag{11.15}$$

Denote every vertex $\mathbf{v}$ of $M$ by $\mathbf{v} = (v_0, v) \in R \times R^n$ and let $V(M)$ be the vertex set of $M$. Clearly $d_i(t,x)$ is a Lipschitz function on $M$. Let $L$ be the Lipschitz constant of this function on $M$ (of course $L$ does not depend on $i$).

**Proposition 3** *Let $(t_M, x_M)$ be an interior point of $M$, (e.g. the barycentre of $M$), and let $\rho_i^M(t,x)$ be the affine function that matches with the following function at every vertex $(v_0, v)$ of $M$ :*

$$d_i(t_M, x_M) - L\|(t,x) - (t_M, x_M)\|. \tag{11.16}$$

*Then $\rho_i^M(t,x)$ is a* lower linearization *of $d_i(t,x)$, i.e. an affine minorant of $d_i(t,x)$ such that*

$$\sup_{(t,x)\in M} \{d_i(t,x) - \rho_i^M(t,x)\} \to 0 \qquad \text{as } diam M \to 0. \tag{11.17}$$

*Proof.* See [16], page 296. □Note that the function (11.16) is a concave minorant of $d_i(t,x)$. It follows from the above Proposition that

$$\min \sum_{i=1}^{m} \min\{\|x - a^i\|, \rho_i^M(t,x)\} \mid \quad (t,x) \in M\} \tag{11.18}$$

is a relaxation of the subproblem (11.15), which can also be written as

$$\min \sum_{i=1}^{m} [\|x - a^i\| + \rho_i^M(t,x) - \max\{\|x - a^i\|, \rho_i^M(t,x)\}] \quad \text{s.t. } (t,x) \in M. \tag{11.19}$$

Let $\psi_i^M(t,x)$ be the affine function that matches with $\max\{\|x - a^i\|, \rho_i^M(t,x)\}$ at every vertex $\mathbf{v}$ of $M$. Then a further relaxation of (11.15) is obtained by substituting $\psi_i^M(t,x)$ for $\max\{\|x - a^i\|, \rho_i^M(t,x)\}$ in (11.19), i.e.

$$\min \sum_{i=1}^{m} [\|x - a^i\| + \rho_i^M(t,x) - \psi_i^M(t,x)] \quad \text{s.t. } (t,x) \in M. \tag{11.20}$$

**Proposition 4** *A lower bound on $M$ is*

$$\beta(M) = \min\{\sum_{i=1}^{m}[\|x - a^i\| + \eta_i^M(t,x)] \mid \quad (t,x) \in M\}. \qquad (11.21)$$

*where $\eta_i^M(t,x) := \rho_i^M(t,x) - \psi_i^M(t,x)$ is the affine function that satisfies*

$$\eta_i^M(\mathbf{v}) = \min\{0,\ d_i(t_M, x_M) - L\|(v_0, v) - (t_M, x_M)\| - \|v - a^i\|\} \quad (11.22)$$

$$\forall \mathbf{v} = (v_0, v) \in V(M).$$

*Proof.* The function $\eta_i^M(t,x) := \rho_i^M(t,x) - \psi_i^M(t,x)$ is affine and satisfies

$$\eta_i^M(v_0, v) = \rho_i^M(v_0, v) - \max\{\|v - a^i\|, \rho_i^M(v_0, v)\} \quad \forall (v_0, v) \in V(M) \qquad (11.23)$$

hence the conclusion follows, by noting that

$$\rho_i^M(v_0, v) = d_i(t_M, x_M) - L\|(v_0, v) - (t_M, x_M)\|$$

and, on the other hand:

$$A - \max\{B, A\} = \begin{cases} 0 & \text{if } A \geq B \\ A - B & \text{if } A < B. \end{cases}$$

i.e. $A - \max\{B, A\} = \min\{0, A - B\}$. □

**Remark 2** *With the above bounding, if the subdivision rule for branching is (pure) bisection then, due to property (11.17), the convergence of the algorithm is guaranteed.*

## 3.2 POLYBLOCK APPROXIMATION METHOD

By rescaling if necessary one can assume that

$$t + \sum_{i=1}^{n} x_i \leq 1.$$

Then, introducing an additional variable $s = 1 - t - \sum_{i=1}^{n} x_i$ and setting $\mathbf{x} = (t, x_1, \ldots, x_n, s)$ one can write the problem as

$$\min \ f(t,x,s) := \sum_{i=1}^{n} \min\{\|x - a^i\|, d_i(t,x)\} \quad \text{s.t. } (t,x,s) \in \mathbf{S}$$

where $\mathbf{S} = \{\mathbf{x} = (t,x,s) \in R \times R^n \times R| \quad t + \sum_{i=1}^{n} x_i + s = 1,\ t, x, s \geq 0\}$. Noting then that the function $f(t,x,s)$ is Lipschitz on the simplex $\mathbf{S}$ one can extend this function to an increasing function on the orthant $t \geq 0, x \geq 0, s \geq 0$ and apply the polyblock approximation method ([14], [17]) to this monotonic optimization problem.

# 4 IMPLEMENTATION AND COMPUTATIONAL RESULTS

To verify the effectiveness of the suggested algorithm we applied it to the Wisconsin Diagnostic Breast Cancer database (see [18, 19, 20, 21]). This database consists of 569 vectors with known outcomes. It was created by W.H. Wolberg, General Surgery Department, University of Wisconsin, Clinical Sciences Center, W.N. Street and O.L. Mangasarian, Computer Sciences Department, University of Wisconsin.

Ten real-valued features are computed for each cell nucleus: radius (mean of distances from center to points on the perimeter), texture (standard deviation of gray-scale values), perimeter, area, smoothness (local variation in radius lengths), compactness, concavity (severity of concave portions of the contour), concave points (number of concave portions of the contour), symmetry, fractal dimension (of the boundary). The mean value , standard error and extreme value (i.e., largest or worst value: biggest size, most irregular shape) of each of these cellular features are computed for each image, resulting in a total of 30 real-valued features. Thus $n = 30$ in the case under consideration. The papers [18, 19, 20, 21] contain more detailed description of this database.

The direct application of global optimization methods is impossible because of the large number of parameters. Therefore, the finding of the most informative parameters for the database under consideration is very important. For this purpose we use the algorithm for the reduction of the dimensionality, which was described in [3]. The essence of the algorithm is the following. For each training set only one center of cluster is computed using methods of convex programming. Then all coordinates which differ slightly are excluded (see, for details, [3]). Our numerical experiments show that using all 30 parameters one can determine the set with almost 90% accuracy whereas the use of only 5 parameters: mean values of perimeter, area, compactness, concavity and concave points allows one to determine this set with almost 89% accuracy. The same result was obtained for the largest values (instead of means) of the same parameters. For further investigations we chose only mean values. Then we considered various groups of parameters which contained 3 or 4 parameters from the choosen 5. Various groups of 4 parameters describe the data with accuracy from 85% to 88% whereas various groups of 3 parameters describe them with accuracy from 82% to 87%. Thus we can conclude that for computing the centers of clusters it is sufficient to use only 3 parameters. The best results we obtained for the mean values of perimeter, area and concavity which allow the description of all data, using only one cluster for each set, with 87% accuracy.

We took 80% of all vectors as a training set and remaining 20% of vectors were used for testing of the obtained clusters. By applying the algorithm and using only the mentioned parameters we computed two clusters for both sets. Thus the dimensionality of global minimization problem is $n = 6$. Both 1-norm and 2-norm have been used in numerical experiments.

The problem (11.11) was used for the computing of lower bounds. Convex relaxation of the cluster analysis problem requires more computational efforts than the linear relaxation (11.9). However, it gives a more exact lower bound and the results of our numerical experiments show that the application of convex relaxation is more efficient. Computational efforts in problem (11.11) also depend on the chosen norm. For the 1-norm the objective function becomes nonsmooth. The objective function with this norm is more complicated than with the 2-norm. However numerical experiments show that the 1-norm allows us to obtain better results. Therefore, we chose the 1-norm. One of the effective methods for solving nonsmooth convex problems is the exact penalty method (see [6] and references therein). We use this method for the solution of the problem (11.11) with the 1-norm.

Often, solution of this problem requires a long CPU time and in a case of searching more than two clusters the algorithm becomes not sufficiently effective. On the other hand, results of our numerical experiments show that the suggested approach is quite effective for determining two clusters.

We took $\epsilon = 0.0001$ in our numerical experiments. More precise solution of the problem (11.11) is not essential for the cluster analysis problem under consideration. Indeed, it does not lead to the improvement of the description of the sets; on the other hand it sharply increases the execution time of the algorithm.

Note that the required CPU time weakly depends on the number of vectors which are contained in database. The effectiveness of the suggested approach also depends on the subdivision procedure. Numerical experiments show that the *normal simplicial subdivision* procedure is more effective.

The results of numerical experiments were as follows. The two clusters obtained for each set describe them with 96% accuracy. We also carried out numerical experiments with both 4 parameters and 3 clusters. Using 4 parameters and computing 2 clusters we did not obtain improvement of the results with respect to case with 3 parameters. The computation of 3 clusters (with both 3 and 4 parameters) did not also lead to improvement of the results. In this case the dimension of the problem is 9 or 12, respectively, which is complicated enough for the methods of global optimization.

## 5 CONCLUSIONS

In this paper we have studied the algorithm for the solution of the clustering problem based on the d.c. programming approach. This algorithm has been applied for the analysis of the Wisconsin Diagnostic Breast Cancer database. Analysing the results of numerical experiments we can draw the following conclusions.

For the database under consideration the suggested algorithm allows effectively to compute two clusters centers. This gives a good enough description of the sets. Computational efforts in present algorithm do not depend as strongly on the number of vectors which is the advantage of the global optimization approach in comparison with other approaches mentioned in the Introduction. However they strongly depend on the number of clusters and the dimensionality of vectors. Therefore the procedure for the reduction of the dimensionality of the vectors is the important part of the algorithm. The results also depend on the precision of the solution of problem (11.11), especially in the case of 1-norm. Since increasing of the number of clusters leads to increasing of the dimensionality of global optimization problem (in the case under consideration $n \geq 9$) so the present algorithm did not allow to compute more than 2 clusters. Therefore we can conclude that the present approach has its numerical limitations.

## Acknowledgments

The authors wish to thank the anonymous referees for their valuable comments, which improved the presentation of the paper.

## References

[1] Faiz A. Al-Khayyal, H. Tuy and F. Zhou (1997), D.C. Optimization Methods for Multisource Location Problems, Preprint, School of Industrial and Systems Engineering, Georgia Institute of Technology, Atlanta.

[2] K. Al-Sultan (1995), A Tabu search approach to the clustering problem. *Pattern Recognotion*, **28(9)**, 1443-1453.

[3] A.M. Bagirov, A.M. Rubinov, A. Stranieri and J. Yearwood (1999), The global optimization approach to the clustering analysis, Working Paper 45/99, University of Ballarat, Australia.

[4] P.S. Bradley, O.L. Mangasarian aand W.N. Street (1997), Clustering via concave minimization, Techincal Report 96-03, Computert Sciences Department, University of Wisconsin, Madison, Wisconsin, May 1996. Advances In Neural Information processing Systems 9, MIT Press, Cambridge, MA, 368-374, M.C. Mozer, M.I. Jordan and T. Petsche, editors. Available by ftp: //ftp.cs.wisc.edu/math-prog/tech-trports/96-03.ps.Z

[5] P.-C. Chen, P. Hansen, B. Jaumard and H. Tuy (1998), Solution of the multifacility Weber and conditional Weber problems by D.C. Programming, *Operations Research*, **46**, 548-562.

[6] Yu. G. Evtushenko (1982), *Solution Methods of Extremal Problems and Their Application to Optimization Systems*, Moscow, Nauka. (in Russian)

[7] D. Fisher (1987), Knowledge acquisition via incremental conceptual clustering. *Machine Learning*, **2**, 139-172.

[8] R. Horst and H. Tuy (1996), *Global Optimization* (Deterministic Approaches), Springer, third edition.

[9] A.K. Jain and R.C. Dubes (1988), *Algorithms for Clustering Data.* Prentice-Hall Inc., Englewood Cliffs, NJ.

[10] O.L. Mangasarian (1997), Mathematical Programming in Data Mining, *Data Mining and Knowledge Discovery* **1**, 183-201.

[11] J.D. Pinter (1996), *Global Optimization in Action. Continuous and Lipschitz Optimization: Algorithms, Implementation and Applications*, Kluwer Academic Publishers, Dordrecht.

[12] M.R. Rao (1971), Cluster analysis and mathematical programming. *Journal of the American Statistical Association,* **66**, 622-626.

[13] A. Rubinov and M. Andramonov (1999), Lipschitz programming via increasing convex-along-rays functions. *Optimization Methods and Software* **10**, 763-78.

[14] A. Rubinov, H. Tuy and H. Mays, Algorithm for a Monotonic Global Optimization Problem, *Optimization*, to appear.

[15] H. Tuy (1997), A General D.C. Approach to Location Problems. in *State of the Art in Global Optimization: Computational Methods and Applications*, C. Floudas and P. Pardalos, eds., Kluwer, 413-432.

[16] H. Tuy (1998), *Convex Analysis and Global Optimization*, Kluwer.

[17] H. Tuy (1999), Monotonic Optimization: Problems and Solution Approaches, Preprint, Institute of Mathematics, Hanoi.

[18] W.H. Wolberg, W.N. Street and O.L. Mangasarian (1994), Machine learning techniques to diagnose breast cancer from fine-needle aspirates. *Cancer Letters,* **77**, 163-171.

[19] W.H. Wolberg, W.N. Street and O.L. Mangasarian (1995a), Image analysis and machine learning applied to breast cancer diagnosis and prognosis. *Analytical and Quantitative Cytology and Histology,* **17** No. 2, 77-87.

[20] W.H. Wolberg, W.N. Street, D.M. Heisey and O.L. Mangasarian (1995b), Computerized breast cancer diagnosis and prognosis from fine-needle aspirates. *Archives of Surgery,* **130**, 511-516.

[21] W.H. Wolberg, W.N. Street, D.M. Heisey and O.L. Mangasarian (1995c), Computer - derived nuclear features distinguish malignant from benign breast cytology . *Human Pathology,* **26**, 792-796.

Chapter 12

# ALGORITHMS AND MERIT FUNCTIONS FOR THE PRINCIPAL EIGENVALUE

Giles Auchmuty
*Department of Mathematics*
*University of Houston*
*Houston, TX 77204-3476*
auchmuty@uh.edu

**Abstract** This paper describes an efficient algorithm for finding the principal eigenvalue and a corresponding positive eigenvector of a positive matrix. It is based on the use of a merit function for the problem. A separately quasi-convex function defined on the positive unit simplex cross the positive numbers is described whose minimizers yield the eigenvalue and a corresponding normalized eigenvector. The algorithm provides explicit formulae for descent directions at each stage which yield strict descent. A relative error estimate for the principal eigenvalue is described and it is used in the stopping condition for the algorithm.

**Keywords:** Principal eigenvalue and eigenvector, Positive matrix, Descent algorithm, Merit function.

## 1 INTRODUCTION

When $A$ is a matrix all of whose entries are positive then Perron's theorem (see Chapter 8 of [3]), says that the spectral radius of $A$ is a simple eigenvalue of $A$ and the associated eigenvector can be taken to have all positive entries. This eigenvalue is often called the principal eigenvalue, or Perron root, of $A$.

The problem to be treated here is that of developing accurate and efficient algorithms for for computing this eigenvalue and the associated eigenvector. The problem only requires that we find a particular eigenvalue of a matrix - not all of them. The matrix $A$, however, need not be symmetric, or even normal, so the usual variational principles for



*N. Hadjisavvas and P.M. Pardalos (eds.), Advances in Convex Analysis and Global Optimization*, 235–244.

eigenvalues are not directly applicable. Nevertheless, there are a number of saddle point characterizations of the principal eigenvalue that are described in [3], [4] and elsewhere.

Here we shall describe a merit function approach to determining this principal eigenvalue and an associated eigenvector. That is, we describe a function whose (global) minima occur precisely at the desired eigenpair $(v, \lambda)$. There are many possible choices for such merit functions. We will describe a particular function which is quasi-convex in each of $v, \lambda$ separately and for which we have extra information available for doing the minimization. In particular there are exact formulae for minimization at various steps; so many of the common issues associated with line searches are easily resolved.

As far as the author can determine the approach that will be described here is quite different to those in the current literature for this problem. Recent papers on these issues include those by [1] and by [5]. The first uses repeated inverse iteration - which requires the solution of a linear system at each stage. The second uses some nice inequalities to find upper and lower bounds on the eigenvalue. Quite often users just run general purpose eigenvalue routines such as those found in MATLAB or other packages. This may not be practical for very large systems. It appears that algorithms based on the optimization problem to be described here provide fast and efficient methods for computing the principal eigenvalue and an associated eigenvector and may be the most practical method if the system is very large.

This approach to be described here may be of particular interest to many participants in this conference as it provides a concrete example of many of the themes of convex analysis. In particular, the function is not differentiable at its minima in either $\lambda$ or $v$. So classical optimization theory and Newton or quasi-Newton methods are not appropriate tools for analyzing this problem and developing good algorithms. Instead convexity techniques are used to motivate the descent algorithm.

This paper will concentrate on the description and analysis of the merit function and provide a general algorithm for solving the eigenproblem. A detailed presentation of computational details and comparisons hopefully will be published elsewhere.

## 2 NOTATION AND DEFINITIONS

Let $\mathbb{R}^n$ be the usual vector space of all real $n$-tuples, $x = (x_1, x_2, \ldots, x_n)$, with the inner product

$$\langle x, y\rangle := \sum_{j=1}^{n} x_j y_j \qquad \text{for } x, y \in \mathbb{R}^n.$$

$\mathbb{P}^n$ is the subset of $\mathbb{R}^n$ of all vectors whose components are positive; that is $x_j > 0$ for $1 \le j \le n$. Such vectors will be called positive vectors and $\mathbb{P}^n$ is an open, convex cone in $\mathbb{R}^n$.

Let $A = (a_{jk})$ be a $n \times n$ real matrix. Then $A$ is said to be a positive matrix if $a_{jk} > 0$ for all $j, k$. Our problem is to find the spectral radius $r(A)$ of $A$. That is, we wish to find a pair $(v, \lambda)$ in $\mathbb{R}^n \times \mathbb{R}$ with $|\lambda|$ is maximal, $v \ne 0$ and

$$Av \ = \ \lambda v. \tag{12.1}$$

When $A$ is positive, then Perron's theorem (see [3]), Theorem 8.2.11) says that $r(A)$ is the unique positive real eigenvalue of $A$, it is a simple eigenvalue and $v$ may be chosen to be positive. This eigenvalue is called the principal eigenvalue of $A$ and an associated eigenvector is called a principal eigenvector.

Our interest is in describing an optimization problem whose solution will yield $r(A)$ and the principal eigenvector. The notation and definitions used here should be taken as in [2] unless indicated otherwise. When $C$ is a convex subset of $\mathbb{R}^n, \varphi : C \to \mathbb{R}$ is a function, then the synoptic, or sublevel, sets of $\varphi(x)$ are the sets

$$S_r(\varphi) \ := \ \{x \in C : \varphi(x) \le r\}. \tag{12.2}$$

$\varphi$ is said to be quasi-convex on $C$ if $S_r(\varphi)$ is convex for all $r$. When $\varphi$ is convex on $C$, then it will be quasi-convex. Many of the functions arising in this paper, will be quasi-convex - but not convex.

Let $\Delta_n = \{x \in \mathbb{P}^n : \sum_{j=1}^n x_j = 1\}$ be the (open) unit simplex in $\mathbb{R}^n$. For $x \in \mathbb{P}^n$, define

$$|x| \ := \ \sum_{j=1}^n x_j, \text{ and } \qquad Ux = x/|x|. \tag{12.3}$$

Then $|x| > 0$, $Ux$ is in $\Delta_n$ and the mapping of $\mathbb{P}^n$ to $\Delta_n \times (0, \infty)$ which associates $x$ to $(Ux, |x|)$ is a smooth diffeomorphism. The projective metric on $\Delta_n$ is the metric defined by

$$d(x, y) \ := \ \max_{1 \le j \le n} \ | \ln x_j - \ln y_j \ |. \tag{12.4}$$

Here $ln$ is the natural logarithm function and $\Delta_n$ is a complete metric space under this metric.

When $y^{(1)}, y^{(2)}$ are in $\Delta_n$, then $z = y^{(2)} - y^{(1)}$ obeys

$$\sum_{j=1}^n z_j \ = \ 0 \tag{12.5}$$

The subspace of $\mathbb{R}^n$ of vectors obeying (12.5) will be denoted $M_0$. When $w$ is in $\mathbb{R}^n$, its projection onto $M_0$ is

$$\mathcal{P}_0 w \;:=\; w - \bar{w}\, e$$

where $e = (1, 1, \ldots, 1)$ and $\bar{w} = n^{-1} \langle w, e \rangle$. The barycenter of $\Delta_n$ is the vector $\hat{e} = n^{-1}\, e$.

## 3 THE MERIT FUNCTION

Consider the function $\Phi : \Delta_n \times (0, \infty) \to [0, \infty)$ defined by

$$\Phi(y, \lambda) := \max_{1 \le j \le n} \left| \ln \left( \sum_{k=1}^{n} a_{jk} y_k \right) - \ln \lambda y_j \right|. \tag{12.6}$$

$\Phi$ is a continuous, nonnegative function and $\Phi(\tilde{y}, \tilde{\lambda}) = 0$ if and only if $(\tilde{y}, \tilde{\lambda})$ is a normalized solution of (12.1). Thus $\Phi$ is a merit function for the positive eigenproblem associated with $A$ and $\Phi(y, \lambda)$ is a projective distance between $Ay$ and $\lambda y$ in $\mathbb{P}^n$.

The optimization problem to be considered here is that of minimizing $\Phi$ on $\Delta_n \times (0, \infty)$. To analyze this problem, we shall repeatedly use properties of the function $\psi : \mathbb{P}^2 \to [0, \infty)$ defined by

$$\psi(s, t) \;:=\; |\ \ln(s/t)\ |. \tag{12.7}$$

$\psi$ is continuous and symmetric in $s$ and $t$. It is quasi-convex, but not convex, on $\mathbb{P}^2$, and the synoptic sets $S_r(\psi)$ are the (unbounded) convex sets of points $(s, t) \in \mathbb{P}^2$ obeying the inequalities

$$t e^{-r} \;\le\; s \;\le\; t e^{r} \tag{12.8}$$

Let the rows of $A$ be denoted $a^{(j)}$ with $1 \le j \le n$. Then each $a^{(j)}$ is in $\mathbb{P}^n$ and

$$(Ay)_j \;=\; \sum_{k=1}^{n} a_{jk} y_k \;=\; \langle a^{(j)}, y \rangle.$$

Define $H : \Delta_n \to \mathbb{P}^n$ by $H(y) \;:=\; (h_1(y), \ldots, h_n(y))$ with

$$h_j(y) \;:=\; y_j^{-1} \sum_{k=1}^{n} a_{jk}\, y_k. \tag{12.9}$$

Then (12.6) may be rewritten, using (12.7) and (12.9), as

$$\Phi(y; \lambda) \;=\; \max_{1 \le j \le n} \psi(h_j(y), \lambda). \tag{12.10}$$

**Proposition 3.1** *Given $y \in \Delta_n$, the function $\Phi(y; .)$ is quasi-convex on $(0, \infty)$.*

PROOF: We need to show that the set of $\lambda$'s for which $\Phi(y; \lambda) \leq r$ is convex for each $r \geq 0$. From (12.10), $\Phi(y, \lambda) \leq r$ if and only if

$$\psi(h_j(y), \lambda) \leq r \qquad \text{for all } 1 \leq j \leq n. \tag{12.11}$$

Thus $S_r(\Phi(y, .))$ is the intersection of $n$ closed subintervals of $(0, \infty)$, hence it is a bounded, convex subset of $(0, \infty)$ when $r \geq 0$. □

There are many possible choices of merit functions for this problem. One motivation for the selection of (12.6) is that there is an explicit formula for the optimal choice of $\lambda$, as will be described in the next theorem. To describe this result we introduce the quantities

$$\gamma_0(y) = \min_{1 \leq j \leq n} \ln h_j(y) \qquad \text{and } \gamma_1(y) = \max_{1 \leq j \leq n} \ln h_j(y). \tag{12.12}$$

**Theorem 3.2** *Given a positive matrix $A$ and $y$ in $\Delta_n$, $\Phi(y, .)$ is minimized in $\lambda$ at*

$$\hat{\lambda} = \exp((\gamma_0(y) + \gamma_1(y))/2), \tag{12.13}$$

*with the value*

$$\Phi(y, \hat{\lambda}) = (\gamma_1(y) - \gamma_0(y))/2. \tag{12.14}$$

PROOF: (12.10) may be rewritten as

$$\Phi(y, \lambda) = \max_{1 \leq j \leq n} | \ln h_j(y) - \ln \lambda | .$$

When $\lambda$ varies, $y$ is fixed, this is minimized where $\ln \lambda$ is the mean of the extreme values of $\ln h_j(y)$. Then (12.13)-(12.14) follow. □

It is worth noting that this minimizer $\hat{\lambda}$ is unique and

$$\Phi(y, \lambda) = \Phi(y, \hat{\lambda}) + \left| \ln(\lambda/\hat{\lambda}) \right| .$$

This equation provides an exact formula for the dependence of $\Phi(y, \lambda)$ on $\lambda$ and may be used to show that $\Phi(y, .)$ is not differentiable at $\hat{\lambda}$.

# 4 QUASI-CONVEXITY OF $\Phi(., \lambda)$ ON $\Delta_N$

When $C$ is a convex subset of $\mathbb{R}^n$ and $\varphi : C \to \mathbb{R}$ is a continuous function, then a vector $d \in \mathbb{R}^n$ will be called a descent direction for $\varphi$ at $x$ provided there is a $\delta > 0$ such that, for $0 < t < \delta$, $x + td$ is in $C$ and $\varphi(x + td) < \varphi(x)$.

In this section $\lambda$ will be regarded as fixed and the dependence of functions on $\lambda$ will not be indicated explicitly. First consider the problem of finding the descent directions for the functions $\psi_j : \Delta_n \to [0, \infty)$ defined by

$$\psi_j(y) := \psi(h_j(y), \lambda) = \left| \ln\left(\sum_{k=1}^{n} a_{jk} y_k\right) - \ln(\lambda y_j) \right| \tag{12.15}$$

When $h_j(y) \neq \lambda$, $\psi_j$ is Gateaux differentiable at $y$ with

$$\nabla \psi_j(y) = \pm \frac{n^{(j)}(y)}{\langle a^{(j)}, y \rangle}. \tag{12.16}$$

Here

$$n^{(j)}(y) := a^{(j)} - h_j(y) e^{(j)} \tag{12.17}$$

is a normal to the level set of $\psi_j$ at $y$ and $e^{(j)}$ is the $j$-th unit vector. The plus sign holds in (12.16) if $h_j(y) > \lambda$, while the negative sign holds when $h_j(y) < \lambda$. If $h_j(y) = \lambda$, then $y$ minimizes $\psi_j$ on $\Delta_n$ and $h_j$ is not differentiable at $y$.

**Proposition 4.1** *Suppose $y \in \Delta_n$ and $\psi_j(y) \neq 0$. A nonzero vector $d$ is a descent direction for $\psi_j$ at $y$ if and only if $d$ is in $M_0$ and*

$$\langle \nabla \psi_j(y), d \rangle < 0. \tag{12.18}$$

PROOF: When $y$ is in $\Delta_n$, then $y + td$ is in $\Delta_n$ for small positive $t$ if and only if $d$ is in $M_0$. Now

$$h_j(y + td) = \frac{\langle a^{(j)}, y + td \rangle}{y_j + td_j} = h_j(y) \frac{1 + c_j t}{1 + b_j t} \tag{12.19}$$

where

$$c_j = \frac{\langle a^{(j)}, d \rangle}{\langle a^{(j)}, y \rangle} \quad \text{and} \quad b_j = d_j / y_j. \tag{12.20}$$

From (12.17) and these definitions,

$$\langle n^{(j)}(y), d \rangle = \langle a^{(j)}, y \rangle (c_j - b_j) \tag{12.21}$$

so $\langle \nabla \psi_j(y), d \rangle = \pm(c_j - b_j)$ from (12.16). It is easy to demonstrate from (12.19), that $h_j(y + td)$ decreases for t small and positive when $c_j < b_j$

and it increases if $c_j > b_j$. When $b_j = c_j$, $h_j$ is constant along this line. Substitute in (12.15) and the result follows. □

It is worth nothing that for $d \neq 0$ and a given $y$, there is a finite value $\tau_m$ such that $y + td$ is in $\Delta_n$ for $0 \leq t < \tau_m$ and $y + \tau_m d \notin \Delta_n$. since $\Delta_n$ is bounded.

If $c_j < 0$ and $\tau_c := -{c_j}^{-1} < \tau_m$, then (12.19) shows that $h_j(y+\tau_c d) = 0$, so we have an explicit value for the optimal value of $t$ in this case. When $\tau_c \geq \tau_m$, then $\psi_j$ decreases for all $t$ such that $y + td$ is in $\Delta_n$.

The condition $c_j = b_j$ is, from (12.21), equivalent to

$$\langle \nabla \psi_j(y), d \rangle \;=\; \langle n^{(j)}(y), d \rangle \;=\; 0. \tag{12.22}$$

This defines a subspace $M_j$ of $M_0$ and (12.19) shows that when $d \in M_j$, and $y + td$ is in $\Delta_n$, then

$$h_j(y + td) \;=\; h_j(y) \tag{12.23}$$

**Corollary 4.2** *$\psi_j$ is quasi-convex on $\Delta_n$.*

PROOF: The subset of $\Delta_n$ where $\psi_j$ is zero is the set of all points $y \in \Delta_n$ obeying $\langle a^{(j)}, y \rangle \;=\; \lambda y_j$. This is a linear equation, so the synoptic set $S_0(\psi_j)$ is a convex subset of $\Delta_n$.

When $\psi_j(y) = r > 0$, let $S_r(\psi_j)$ be the synoptic set associated with $\psi_j$ on $\Delta_n$ and $z$ be another vector in $S_r(\psi_j)$. We wish to show that $y + t(z - y)$ is in $S_r(\psi)$ for $0 \leq t \leq 1$. Since $\Delta_n$ is convex, such vectors are in $\Delta_n$ and, from (12.17),

$$\langle n^{(j)}(y), z - y \rangle \;=\; z_j \, (h_j(z) - h_j(y)). \tag{12.24}$$

If $h_j(y) \neq \lambda$, $h_j(z) = h_j(y)$, then $d \;=\; z - y$ obeys (12.22) so, from (12.23), we have $\psi_j(y + td) \;=\; \psi_j(y) \;=\; r$ for $0 \;\leq\; t \;\leq\; 1$.

If $h_j(y) > \lambda$ and $h_j(z) < h_j(y)$ then, from (12.24) and (12.16), $(z-y)$ is a descent direction for $\psi_j$, so $c_j < b_j$ in (12.19) and $y + t(z - y)$ is in $S_r(\psi_j)$ for $0 \leq t \leq 1$. When $h_j(y) < \lambda$ and $h_j(z) > h_j(y)$, then again $y + t(z - y)$ is in $S_r(\psi_j)$ for $0 \leq t \leq 1$, so the corollary follows. □

With these results we can now prove that $\Phi(.,\lambda)$ is quasi-convex on $\Delta_n$ and describe its descent directions at a point.

**Theorem 4.3** *For each $\lambda > 0$, $\Phi(.,\lambda)$ is quasi-convex on $\Delta_n$. If $\tilde{y}$ is in $\Delta_n$, then either*

*(i) $\tilde{y}$ minimizes $\Phi(.,\lambda)$ on $\Delta_n$, or else*

*(ii) there is a descent direction $\tilde{d}$ for $\Phi(.,\lambda)$ at $\tilde{y}$.*

PROOF: From (12.15), (12.10) and the preceding corollary, $\Phi(.,\lambda)$ is the maximum of a family of quasi-convex functions, so it is quasi-convex.

Suppose (i) does not hold, then there is a $z$ in $\Delta_n$ with $\Phi(z,\lambda) < \Phi(\tilde{y},\lambda)$.

Let $\mathcal{J}(\tilde{y})$ be the set of indices for which $\psi_j(\tilde{y}) = \Phi(\tilde{y},\lambda)$. Then for $j \in \mathcal{J}(\tilde{y}), \psi_j(z) < \psi_j(\tilde{y})$, so $\tilde{d} = z - \tilde{y}$ is a descent direction for each $\psi_j$, and

$$\psi_j(\tilde{y} + t\tilde{d}) < \psi_j(\tilde{y}) \qquad \text{for all } 0 < t \leq 1.$$

Thus $\tilde{d}$ will be a descent direction for $\Phi$ as each $\psi_j$ is continuous and (ii) holds. □

Note that while each $\psi_j$ is differentiable except at its minima, this is not true for $\Phi(.,\lambda)$. It need not be differentiable at any point $y$ where $\psi_j(y) = \psi_l(y)$ and $j \neq \ell \in \{1, 2, \ldots, n\}$.

A similar argument leads to the result that if $\tilde{y}$ is not a minimizer of $\Phi(.,\lambda)$ on $\Delta_n$, then $d$ is a descent direction for $\Phi(.,\lambda)$ at $\tilde{y}$ provided it is a descent direction for each $\psi_j$ at $\tilde{y}$ with $j \in \mathcal{J}(\tilde{y})$. The set of all such descent directions is a non-empty convex cone in $M_0$.

It is also worthwhile to obtain some error estimates for approximate solutions. Suppose that $\Phi(\tilde{v},\tilde{\lambda}) = \epsilon > 0$ and that $\hat{\lambda} = r(A)$ is the principal eigenvalue of $A$.

**Theorem 4.4** *If $A$ is a positive matrix $(\tilde{y},\tilde{\lambda}) \in \Delta_n \times (0,\infty)$ with $\Phi(\tilde{y},\tilde{\lambda}) = \epsilon$, then*

$$\left|(\tilde{\lambda} - \hat{\lambda})/\hat{\lambda}\right| \leq e^{\epsilon} - 1. \tag{12.25}$$

PROOF: This value of $\Phi$ means that $(\tilde{y},\tilde{\lambda})$ satisfies

$$\lambda e^{-\epsilon}\, y_j \leq \langle a^{(j)}, y\rangle \leq \lambda e^{\epsilon}\, y_j$$

for each $1 \leq j \leq n$. From corollary 8.1.29 of [3]), this implies that $\lambda e^{-\epsilon} \leq \hat{\lambda} \leq \lambda e^{\epsilon}$ so (12.25) follows. □

## 5 MINIMIZATION ALGORITHM

When $A$ is a positive matrix, Perron's theorem guarantees that there is a unique minimizer $(\hat{v},\hat{\lambda})$ of $\Phi$ defined by (12.6). By using the convexity properties of $\Phi$ described in the preceding two sections, we can outline a descent algorithm for finding the principal eigenvalue and eigenvector of $A$.

As a consequence of theorem 4.4, the stopping criterion will be that $\Phi(\tilde{v},\tilde{\lambda}) < \epsilon$ where $\epsilon$ is preassigned. This guarantees that the relative error in $\hat{\lambda}$ will be less than $(e^{\epsilon} - 1)$.

The algorithm requires that an initial $y^{(0)}$ and the stopping criterion $\epsilon > 0$ be given. For $k \geq 0$,

(1): Evaluate $Ay^{(k)}$ and $h_j(y^{(k)})$ for $1 \leq j \leq n$

(2): Evaluate $\gamma_0^k := \gamma_0(y^{(k)})$ and $\gamma_1^k := \gamma_1(y^{(k)})$

(3): Compute $\lambda_{k+1} = \exp((\gamma_0^k + \gamma_1^k)/2)$, and $\Phi_k := \Phi(y^{(k)}, \lambda_{k+1})$.

(4): If $\Phi_k < \epsilon$, then $(y^{(k)}, \lambda_{k+1})$ is an approximate solution and stop, else continue.

(5): Determine the set $\mathcal{J}(y^{(k)})$ of indices $j$ such that $\psi_j(y^{(k)}) \cong \Phi_j$.

(6): Choose a direction $d^{(k)}$ in $M_0$ which is a descent direction for $\Phi(., \lambda_{k+1})$ at $y^{(k)}$.

(7): Choose a $\tau_k > 0$ and $y^{(k+1)} = y^{(k)} + \tau_k d^{(k)}$ so that $\Phi(y^{(k+1)}, \lambda_{k+1}) < \Phi(y^{(k)}, \lambda_k)$.

(8): Go to 1 and continue.

In step 6, we choose the descent direction for $\Phi(., \lambda_{k+1})$ at $y^{(k)}$ by ensuring that it is a descent direction for each $\psi_j$ at $(y^{(k)}, \lambda_{k+1})$ with $j$ in $\mathcal{J}(y^{(k)})$. This is relatively straightforward when $\mathcal{J}(y^{(k)})$ only has 1 or 2 elements. When $\mathcal{J}(y^k)$ has more than 2 indices, we have usually restricted attention to 2 at a time.

In step 7 we have generally required that $\tau_k$ is chosen so that a "sufficient decrease condition" holds. Such conditions enable proofs of the convergence of the numerical methods.

More detailed discussions of the implementation of this algorithm and comparison with other methods will be described elsewhere.

## References

[1] Elsner, L. M. Koltracht, M. Neumann and Xiao, D. (1993). On accurate computation of the Perron root. *SIAM J. Matrix Anal. Appns*, 14:456–467.

[2] Hiriart-Urruty, J.-B. and Lemarechal, C. (1993). *Convex Analysis and Minimization Algorithms.* Springer-Verlag, Berlin.

[3] Horn, R. and Johnson, C. (1985). *Matrix Analysis.* Cambridge U. Press, Cambridge.

[4] Seneta, E. (1981). *Nonnegative Matrices and Markov Chains, 2nd ed.* Springer Verlag, New York.

[5] Yeh, L. (1997). Inequalities for the Maximal Eigenvalue of a Nonnegative Matrix. *Glasgow Math. J.*, 39:275–284.

# Chapter 13

# MODIFIED VERSIONS OF THE CUTTING ANGLE METHOD

A. M. Bagirov and A.M. Rubinov*

*School of Information Technology and Mathematical Sciences*

*The University of Ballarat, Vic 3353, Australia*

abagirov@students.ballarat.edu.au

**Abstract** The cutting angle method for solving global optimization problems has recently been developed. We propose some modifications and improvements of this method and report results of numerical experiments which demonstrate that the cutting angle method becomes a powerful tool for solving many problems of global optimization.

**Keywords:** Global optimization, cutting angle method, Lipschitz programming, IPH function.

## 1 INTRODUCTION

Let $f$ be a Lipschitz function defined on the unit simplex $S$. Consider the following problem of global optimization

$$f(x) \longrightarrow \min \text{ subject to } x \in S. \tag{13.1}$$

Note that many problems of unconstrained and constrained global optimization can be reduced to this problem. (One example of unconstrained problem can be found in Section 9; many problems of constrained optimization can be reduced to (13.1) by a transformation of variables and penalization.) The cutting angle method for solving (13.1) has been proposed and studied in [1, 10]. A new version of this method was discussed in [4]. We propose some modifications and improvements of this version.

*This research was supported by the Australian Research Council



*N. Hadjisavvas and P.M. Pardalos (eds.), Advances in Convex Analysis and Global Optimization,* 245–268.

It was shown in [10] that the problem (13.1) can be reformulated as the global minimization of the so-called IPH function over the unit simplex

$$S = \{x = (x_1, \dots, x_n) \in \mathbb{R}^n : \sum_{i=1}^{n} x_i = 1, x_1 \geq 0, \dots, x_n \geq 0\} \quad (13.2)$$

(IPH is an abbreviation for an increasing positively homogeneous function of degree one.) Taking into account this result, we firstly study the following problem:

$$f(x) \longrightarrow \min \text{ subject to } x \in S, \quad (13.3)$$

where $f$ is an IPH function. The cutting angle method reduces the problem (13.3) to the sequence of auxiliary problems of the form:

$$h_j(x) \longrightarrow \min \text{ subject to } x \in S \quad (13.4)$$

where

$$h_j(x) = \max_{k \leq j} \min_{i \in I(l^k)} l_i^k x_i, \quad (13.5)$$

$$l^k = (l_1^k, \dots, l_n^k) \in \mathbb{R}^n, \ l_i^k \geq 0, \ i = 1, \dots, n, \ I(l^k) = \{i : l_i^k > 0\},$$
$$k = 1, \dots, j, \ j \geq n.$$

The auxiliary problem (13.4) is essentially of combinatorial nature.

Let $x^{j+1}$ be a solution of problem (13.4) and $\lambda_j = \min_{x \in S} h_j(x) = h_j(x^{j+1})$ be the value of this problem. It can be shown (see, for example, [8]) that $\lambda_j \longrightarrow f_* := \min_{x \in S} f(x)$ as $j \to +\infty$ and any limit point of the sequence $(x^j)$ is a global minimizer of the function $f$ over the simplex $S$.

An algorithm for solving auxiliary problem (13.4) was suggested in [4]. This algorithm is based on the calculation of all local minimizers of the auxiliary problem (13.4). Then a global minimizer can be found by sorting local minimizers. The results of numerical experiments presented in [4] demonstrate the effectiveness of this algorithm when $n$ is not very large. Certain parameters should be chosen for numerical implementation of the algorithm. The good choice of these parameters allowed the authors to solve some problems even with $n = 50$. This algorithm significantly extends the possibilities of the cutting angle method.

A number of numerical experiments were recently carried out by the authors. The results of these experiments clarified the following:

1) The auxiliary function $h_j$ defined by (13.5) has many local minimizers. The number of local minimizers depends on the objective function $f$ of the initial problem (13.3). For instance, the

number of local minimizers of the auxiliary function $h_j$ is significantly smaller for an IPH function $f(x) = \sqrt{[Ax, x]}$ where $A$ is a matrix with nonnegative entries, than for an IPH function $f(x) = \max_j \min_i [a^{ij}, x]$ with $a^{ij} \geq 0$ (here $[a, x]$ is the inner product of vectors $a$ and $x$). The number of local minimizers of the function $h_j$ increases as $j$ increases. For some IPH functions and for fairly large $j$ the function $h_j$ has several hundred local minimizers for $n = 10$ and several thousand local minimizers for $n = 20$. So the complete description of the set of all local minimizers may require a long CPU time.

2) Only one global minimizer of problem (13.4) is calculated on each iteration of the cutting angle method. However, often the auxiliary function $h_j$ has many global minimizers. In such a case we need to choose only one of them. As it turned out, the remaining global minimizers of $h_j$ can be global minimizers of the next auxiliary function $h_{j+1}$. In such a case the cutting angle method converges very slowly.

3) Results of numerical experiments show that the cutting angle method is suitable for the search of approximate global minimizers (with the relative precision $\delta = 0.05 \div 0.01$). To make an approximate solution more precise with the help of this method is very time-consuming. Different methods, in particular a local search, can be used for this purpose.

4) Results of numerical experiments demonstrate that the sequence $\lambda_j$ converges very slowly. Actually slow increasing of $\lambda_j$ is one of the main drawbacks of the cutting angle method. Very often the method computes the approximate global minimizer quickly enough, but slow convergence of the sequence $\lambda_j$ does not allow us to assert that exactly an approximate global minimizer has been computed. Thus, we have to pay a high price for the confirmation that a found point is close enough to a global minimizer.

We propose some modifications of the cutting angle method which allow us to surmount (at least partly) the aforementioned drawbacks of this method and improve its convergence. Firstly we describe the principal ideas of these modifications related to solving the auxiliary problems.

1) The auxiliary problems form a connected sequence. Indeed the objective function $h_j$ at the current iteration $j$ has a form

$$h_j(x) = \max(h_{j-1}(x), \min_{i \in I(l^j)} l_i^j x_i) \tag{13.6}$$

where $h_{j-1}$ is the objective function at the previous iteration $j-1$. Many local minimizers of the function $h_{j-1}$ remain local minimizers of the function $h_j$. Therefore it is beneficial to find a modification of the algorithm, where only new local minimizers of the function $h_j$ are calculated. (In some modifications of the cutting angle method the function $h_j$ can be considered as the maximum of $h_{j-1}$ and several new min-type functions, formed by vectors $l^{j_1}, \dots l^{j_k}$. The same question remains for these modifications as well.)

2) A modification of the method, where all global minimizers of the auxiliary function are taking into account, can be considered. This modification sometimes allows us to accelerate the convergence of the algorithm and to achieve the strict increase of the estimates $\lambda_j$.

3) Finally computing not only exact global minimizers but also all approximate global minimizers of auxiliary functions leads to guaranteed increase of $\lambda_j$.

Improvements of the algorithm for solving the auxiliary problems, proposed in this paper, substantially accelerate the convergence of the cutting angle method. Numerical experiments show that the new version is 10-20 times faster then the old one.

Assume now that the objective function $f$ of problem (13.1) is Lipschitz. It was proved in [10] that there exists a constant $c > 0$ such that the function $f_1(x) = f(x) + c$ can be considered as the restriction of an IPH function to $S$, and so the cutting angle method can be applied to the function $f_1$. Numerical experiments demonstrated that the method is rather sensitive to the choice of the constant $c$. Nevertheless we can find such a constant for some problems with the Lipschitz function $f$.

An auxiliary function $h_j$ defined by (13.5) is the maximum of a number of min-type functions. If this number is fairly large, the function $h_j$ has a large number of local minimizers and the global minimization of this function is very time-consuming. If we take into account all global minimizers and moreover, some approximate global minimizers, the number of min-type functions increases sharply. So the cutting angle method produces fairly quickly several first iterations and much more time is required for the next iterations. This observation leads to a fruitful idea to combine the cutting angle method with a local search. Using local search we can arrive to a stationary point, say $x$. If $x$ is not a global minimizer then by applying the cutting angle method to a certain new function, we can sufficiently quickly leave the point $x$ and find a new point $y$ such that $f(y) < f(x)$. Using a local search again we can find a new stationary point and so on.

We do not study combination of the cutting angle method with a local search in this paper. Note that modifications of the cutting angle method, which are considered here, lead to succesful applications of combinations with a local search.

The paper has the following structure. In Section 2 we provide some brief preliminary definitions and results related to IPH functions. One of the versions of the cutting angle method is decribed in Section 3. A basic algorithm for the computation of local minimizers of the auxiliary function and some of its modifications are discussed in Section 4. Section 5 contains an algorithm for computation of general maxmin functions. Some properties of the auxiliary functions are studied in Section 6 and new modifications of the basic algorithm based on these properties are given in Section 7. Some numerical experiments for comparison of various modifications of the basic algorithm were carried out. Section 8 contains description of the results of these experiments. A version of the cutting angle method for global minimization of a Lipschitz function can be found in Section 9.

## 2 PRELIMINARIES

Consider an $n$-dimensional vector space $\mathbb{R}^n$. We shall use the following notation:

- $I = \{1, \dots, n\}$;
- $x_i$ is the $i$-th coordinate of a vector $x \in \mathbb{R}^n$;
- $[l, x] = \sum_{i \in I} l_i x_i$ is the inner product of vectors $l$ and $x$;
- if $x, y \in \mathbb{R}^n$ then $x \geq y \iff x_i \geq y_i$ for all $i \in I$;
- if $x, y \in \mathbb{R}^n$ then $x \gg y \iff x_i > y_i$ for all $i \in I$;
- $\mathbb{R}^n_+ := \{x = (x_1, \dots, x_n) \in \mathbb{R}^n : \ x_i \geq 0 \text{ for all } i \in I\}$ (the nonnegative orthant);
- $S = \{x \in \mathbb{R}^n_+ : \ \sum_{i \in I} x_i = 1\}$ (the unit simplex).

Recall that a function $f$ defined on $\mathbb{R}^n_+$ is called increasing if $x \geq y$ implies $f(x) \geq f(y)$; the function $f$ is positively homogeneous of degree one if $f(\lambda x) = \lambda f(x)$ for all $x \in \mathbb{R}^n_+$ and $\lambda > 0$. We shall use the abbreviation IPH for increasing positively homogeneous functions of degree one.

The following functions may serve as examples of IPH functions defined on $\mathbb{R}^n_+$:

**Example 2.1** 1) $f(x) = \sum_{i\in I} a_i x_i$ with $a_i \geq 0$;

2) $p_k(x) = (\sum_{i\in I} x_i^k)^{\frac{1}{k}}$ $(k > 0)$;

3) $f(x) = \sqrt{[Ax, x]}$ where $A$ is a matrix with nonnegative entries;

4) $f(x) = \prod_{j\in J} x_j^{t_j}$ where $J \subset I$, $t_j > 0$, $\sum_{j\in J} t_j = 1$.

Let $l \in \mathbb{R}^n_+, l \neq 0$ and $I(l) = \{i \in I : l_i > 0\}$. The vector $l$ generates the function

$$x \mapsto \langle l, x\rangle := \min_{i\in I(l)} l_i x_i, \tag{13.7}$$

which will be denoted by the same symbol $l$. The function $l$ defined by (13.7) is called the min-type function generated by the vector $l$. Clearly a min-type function is IPH. We shall denote the set of all min-type functions $l$ with $l \in \mathbb{R}^n_+ \setminus \{0\}$ by $L$.

We shall use the following notation for $c \in \mathbb{R}$ and $l \in \mathbb{R}_n^+$:

$$\left(\frac{c}{l}\right)_i = \begin{cases} \frac{c}{l_i} & \text{if } i \in I(l), \\ 0 & \text{if } i \notin I(l). \end{cases} \tag{13.8}$$

The following result holds (see [8]).

**Theorem 2.1** *1) A finite function $f$ defined on $\mathbb{R}^n_+$ is IPH if and only if there exists a set $U \subset L$ such that*

$$f(x) = \max\{\langle l, x\rangle : l \in U\}, \qquad x \in \mathbb{R}^n_+.$$

*2) Let $x^0 \in \mathbb{R}^n_+$ be a vector such that $f(x^0) > 0$ and $l = f(x^0)/x^0$. Then $\langle l, x\rangle \leq f(x)$ for all $x \in \mathbb{R}^n_+$ and $\langle l, x^0\rangle = f(x^0)$.*

The vector $f(x^0)/x^0$ is called the support vector of a function $f$ at a point $x^0$.

Subsequently we will use the unit vectors $e^m = (0, \ldots, 0, 1, 0, \ldots, 0)$ with $I(e^m) = \{m\}$. Clearly the vector $l = f(e^m)/e^m$ can be represented in the form $l = f(e^m)e^m$. We also have $\langle f(e^m)e^m, x\rangle = f(e^m)x_m$.

## 3 THE CUTTING ANGLE METHOD

In this section we give a brief description of a version of the cutting angle method for solving problem (13.3). Note that an IPH function is nonnegative on $\mathbb{R}^n_+$, since $f(x) \geq f(0) = 0$. We assume that $f(x) > 0$ for all $x \in S$. It follows from positiveness of $f$ that $I(l) = I(x)$ for all $x \in S$ and $l = f(x)/x$.

**Algorithm 1**

*Step 0.* (Initialization) Take points $x^k = e^k$, $k = 1, \ldots, n$. Let $l^k = f(x^k)/x^k$, $k = 1, \ldots, n$. Define the function $h_n$ :

$$h_n(x) = \max_{k \leq n} \min_{i \in I(l^k)} l_i^k x_i = \max_{k \leq n} l_k^k x_k$$

and set $j = n$.

*Step 1.* Find a solution $x^*$ of the problem

$$h_j(x) \longrightarrow \min \quad \text{subject to} \quad x \in S. \tag{13.9}$$

*Step 2.* Set $j = j + 1$ and $x^j = x^*$.

*Step 3.* Compute $l^j = f(x^j)/x^j$, define the function

$$h_j(x) = \max\{h_{j-1}(x), \min_{i \in I} l_i^j x_i\} \equiv \max_{k \leq j} \min_{i \in I(l^k)} l_i^k x_i$$

and go to Step 1.

This algorithm can be considered as a version of the cutting angle method ( [1]). A more general version of this algorithm, known as the $\Phi$ - bundle method, has been discussed in [7]. Convergence of the $\Phi$-bundle method was proved in [7] under very mild assumptions.

The algorithm provides lower and upper estimates for the value $f_*$ of the global minimum of the function $f$.

Let

$$\lambda_j = \min_{x \in S} h_j(x) = h_j(x^{j+1}) \tag{13.10}$$

be the value of the problem (13.9). It follows from Theorem 2.1 that

$$\langle l^k, x \rangle \equiv \min_{i \in I(l^k)} l_i^k x_i \leq f(x) \text{ for all } x \in S \text{ and } k = 1, \ldots, j.$$

Hence $h_j(x) \leq f(x)$ for all $x \in S$ and

$$\lambda_j \equiv \min_{x \in S} h_j(x) \leq \min_{x \in S} f(x).$$

Thus $\lambda_j$ is a lower estimate of the global minimum $f_*$. Consider the number $\mu_j = f(x^j)$. Clearly $\mu_j \geq f_*$, so $\mu_j$ is an upper estimate of $f^*$. It can be shown (see, for example, [8]) that $\lambda_j$ is an increasing sequence. (Note that the sequence $\mu_j$ is not decreasing.) It can be shown (see, for example, [8]) that $\mu_j - \lambda_j \to 0$ as $j \to +\infty$. Thus, we have a stopping criterion, which enables us to obtain an approximate solution with an arbitrary given tolerance.

# 4 THE AUXILIARY PROBLEM

The *Step 1* (find the global minimum of the problem (13.9)) is the most difficult part of the algorithm. This problem can be represented in the following form:

$$h_j(x) \longrightarrow \min \text{ subject to } x \in S \tag{13.11}$$

where

$$h_j(x) = \max_{k \le j} \min_{x \in I(l^k)} l_i^k x_i \tag{13.12}$$

and $j \ge n$, $l^k = f(x^k)/x^k$ are given vectors $(k = 1, \ldots, j)$ . Note that $x^k = e^k$, $k = 1, \ldots, n$. Let

$$\phi_k(x) = \min_{i \in I(l^k)} l_i^k x_i, \tag{13.13}$$

then

$$h_j(x) = \max_{k \le j} \phi_k(x).$$

**Proposition 4.1** *(see [4]) Let $j \ge n$, $l^k = l_k^k e^k$, $k = 1, \ldots, n$, $l^k \gg 0$, $k = n+1, \ldots, j$. Then each local minimizer of the function $h_j$ over the simplex $S$ is a strictly positive vector.*

**Corollary 4.1** *Let $(x^j)$ be a sequence generated by the algorithm. Then $x^j \gg 0$ for all $j > n$, hence $l^j \gg 0$ for all $j > n$.*

Let $\operatorname{ri} S$ be the relative interior of the simplex $S$:

$$\operatorname{ri} S = \{x \in S : x_i > 0 \text{ for all } i \in I\}.$$

It follows from Proposition 4.1 and Corollary 4.1 that problem (13.11) can be solved by sorting all local minimizers of the function $h_j$ over the set $\operatorname{ri} S$. We now describe some properties of local minimizers of $h_j$ on $\operatorname{ri} S$. The following result has been established in [4].

**Proposition 4.2** *Let $x \gg 0$ be a local minimizer of the function $h_j$ over the set $\operatorname{ri} S$. Then there exists a subset $\{l^{k_1}, l^{k_2}, \ldots, l^{k_n}\}$ of the set $\{l^1, \ldots, l^j\}$ such that*

*1)*

$$x = \left(\frac{d}{l_1^{k_1}}, \ldots, \frac{d}{l_n^{k_n}}\right) \text{ where } d = \frac{1}{\sum_{i \in I} \frac{1}{l_i^{k_i}}}; \tag{13.14}$$

*2)*

$$\max_{k \le j} \min_{i \in I(l^k)} \frac{l_i^k}{l_i^{k_i}} = 1; \qquad (13.15)$$

*3) Either $k_i = \{i\}$ for all $i \in I$ or there exists $m \in I$ such that $k_m \ge n+1$; if $k_m \le n$ then $k_m = m$; if $k_m \ge n+1$ then $l_i^{k_m} > l_i^{k_i}$ for all $i \in I, i \ne m$.*

An algorithm for solving the auxiliary problem based on the results of Propositions 4.1 and 4.2 has been proposed in [4]. This algorithm consists of two parts. Firstly we sort out subsets consisting of $n$ different elements from the given set of $j$ vectors $\{l^1, \dots, l^j\}$, which provide a local minimum of the function $h_j$. Secondly we choose a subset with the least value of this function.

Let $\epsilon > 0$ be a given tolerance. Note that the choice of the tolerance $\epsilon > 0$ depends on the concrete problem. For a broad class of problems a suitable tolerance is $\epsilon = 10^{-4} \div 10^{-7}$.

We now present an algorithm for calculation of a local minimizer of the auxiliary function $h_j$ over the unit simplex.

**Algorithm 2**

*Step 0.* (Initialization) Set $t = 0$, $m = 0$, $p(k) = 0$, $k = 1, \dots, n$.

*Step 1.* Set $m = m + 1$. If $m > n$ go to Step 5.

*Step 2.* Set $p(m) = p(m) + 1$ and $i = p(m)$. If $i > j$ then go to Step 3. Otherwise take a vector $l^i$ and go to Step 4.

*Step 3.* Set $p(m) = 0$ and $m = m - 1$. If $m = 0$ then go to Step 8, otherwise go to Step 2.

*Step 4. (checking the vector $l^i$).* Set $q = 0$.

*Step 4a.* If $l_m^i < \epsilon$ go to Step 2.

*Step 4b.* Set $q = q + 1$. If $q > m - 1$ then go to Step 4g.

*Step 4c.* If $l_q^i < \epsilon$ then go to Step 4e.

*Step 4d.* If $l_q^{k_q} \ge l_q^i$ then go to Step 2.

*Step 4e.* If $l_m^{k_q} < \epsilon$ go to Step 4b.

*Step 4f.* If $l^i_m \geq l^{k_q}_m$ then go to Step 2, otherwise go to Step 4b.

*Step 4g.* Set $k_m = i$ and go to Step 1.

*Step 5.* If

$$\max_{k \leq j} \min_{i \in I(l^k)} \frac{l^k_i}{l^{k_i}_i} = 1$$

then go to Step 6, otherwise set $m = n$ and go to Step 2.

*Step 6.* Calculate the number

$$d = \left( \sum_{i \in I} \frac{1}{l^{k_i}_i} \right)^{-1}$$

and the point $x \in S$ with coordinates

$$x_i = \frac{d}{l^{k_i}_i}.$$

*Step 7.* Set $t = t+1, \; g(t) = x, \; v(t) = d, \; m = n$ and go to Step 2.

*Step 8.* Compute the global minimum of the auxiliary function by

$$h_* = \min\{v(r) : \; r = 1, \ldots, t\} = v(r^*)$$

and a global minimizer of this function by

$$x_* = g(r^*)$$

and the process terminates.

Algorithm 2 allows us to compute all local minimizers of the auxiliary problem. The conditions of item 3) of Proposition 4.2 are checked in Step 4 and the condition (13.15) is checked in Step 5. A global minimizer of the auxiliary problem is computed in Step 8.

Algorithm 1 constructs the following auxiliary function $h_{j+1}$ on the iteration $j$ with $j \geq n$:

$$h_{j+1}(x) = \max(h_j(x), \min_{i \in I} l^{j+1}_i x_i)$$

where $l^{j+1}_i = f(x^{j+1})/x^{j+1}_i$ and $x^{j+1}$ is a global minimizer of the function $h_j$, which is found by Algorithm 2.

Often the function $h_j$ has more than one global minimizer. In such a case we can consider the following modification of Algorithm 1.

**Algorithm 1′**

*Step 0.* This step coincides with *Step 0* of Algorithm 1.
*Step 1.* Find all solutions $x^*(1), \ldots, x^*(p)$ of the problem (13.9).
*Step 2.* Set $j = j+1$ and $x^j(s) = x^*(s)$, $s = 1, \ldots, p$.
*Step 3.* Compute $l^j(s) = f(x^j(s))/x^j(s)$ $s = 1, \ldots, p$, define the function

$$h_j(x) = \max(h_{j-1}(x), \min_{i \in I}(l^j(1))_i x_i, \ldots, \min_{i \in I}(l^j(p))_i x_i)$$

and go to Step 1.

For the implementation of Algorithm 1′ we need to consider a version of Algorithm 2, which allows us to calculate all global minimizers of the functions $h_j$. We need to change only Step 8 of Algorithm 2 and replace it by the following

*Step 8′.* Compute the global minimum of the auxiliary function by

$$h_* = \min\{v(r) : \ r = 1, \ldots, t\},$$

construct the set

$$R = \{r = 1, \ldots, t : v(r) = h_*\},$$

compute the set of global minimizers $V$ as follows:

$$V = \{x : \exists\, r \in R, x = g(r)\}$$

and the process terminates.

Numerical experiments show that Algorithm 1′ leads to a strict increase of estimates $\lambda_j$ whereas Algorithm 1 does not guarantee such an increase. However, even the computation of all global minimizers does not quicken a slow convergence of the sequence $(\lambda_j)$ in many instances. As a rule the auxiliary function $h_j$ has a large number of local minimizers $x_{(s)}$ and the value $h_j(x_{(s)})$ for some $s$ is very close to the global minimum $(h_j)_*$ of this function. We consider this point as one of the main reasons for the slow convergence of estimates $\lambda_j$. Hence, in order to accelerate the convergence we need to compute not only all exact global minimizers but also *approximate* global minimizers, that is, local minimizers $x_{(s)}$ such that $h_j(x_{(s)}) - (h_j)_* < \delta$, where $\delta > 0$ is a given tolerance. (We call such points $\delta$-global minimizers.) Taking into account all $\delta$-global minimizers accelerates the rate of increase of the sequence $\lambda_j$ in many instances.

A corresponding version of Algorithm 1′ can be constructed by the following modification of Step 8 of Algorithm 2:

*Step 8''*. Compute the global minimum of the auxiliary function by

$$h_* = \min\{v(r) : \ r = 1, \ldots, t\},$$

construct the set

$$R_\delta = \{r = 1, \ldots, t : v(r) \leq h_* + \delta\},$$

compute the set $V_\delta$ of $\delta$-global minimizers:

$$V_\delta = \{x : \exists \, r \in R_\delta, x = g(r)\}$$

and the process terminates.

The number of $\delta$-global minimizers may be very large, so the computation of new support vectors at all $\delta$-global minimizers can slow down the convergence process. Note, that we use all $\delta$-global minimizers in order to achieve a guaranteed increase of the sequence $\lambda_j$. However, sometimes we can achieve this increase without the calculation of support vectors at all $\delta$-global minimizers of the previous auxiliary function. This can be done by a certain modification of Step 3 of Algorithm 1', and correspondingly by the following modification of this algorithm.

**Algorithm 1''**

*Step 0.* This step coincides with *Step 0* of Algorithm 1.
*Step 1.* Find all $\delta$-global minimizers $x^*(1), \ldots, x^*(p)$ of the problem (13.9) and compute

$$h_\delta = \max\{h_j(x^*(i)), \ i = 1, \ldots, p\}.$$

*Step 2.* Set $j = j + 1$ and $x^j(s) = x^*(s)$, $s = 1, \ldots, p$.
*Step 3.* Set $s = 1$ and define the function $h_j^{s-1}(x) = 0$.
*Step 3a.* Compute $l^j(s) = f(x^j(s))/x^j(s)$ and define the function

$$h_j^s(x) = \max\{h_j^{s-1}(x), \min_{i \in I}(l^j(s))_i x_i\}.$$

*Step 3b.* Set $s = s + 1$. If $s > p$ then go to Step 3d.
*Step 3c.* Compute $h_j^{s-1}(x^*(s))$. If $h_j^{s-1}(x^*(s)) > h_\delta + \delta$ then go to Step 3b. Otherwise go to Step 3a.
*Step 3d.* Define the function

$$h_j(x) = \max\{h_{j-1}(x), h_j^p(x)\}$$

and go to Step 1.

Step 3c of this algorithm allows us to remove a $\delta$-global minimizer if we have already had a guaranteed increase of the auxiliary function at this minimizer.

# 5 AN ALGORITHM FOR THE COMPUTATION OF MAXMIN FUNCTIONS

The objective function of the auxiliary problem is a special kind of maxmin functions. This function is computed only in Step 5 of Algorithm 2 and its modifications in order to check the validity of (13.15). Each iteration requires a large number (from several thousands till several tens of thousands) of executions of Step 5, so this step occupies a considerable part of CPU time. We now describe an algorithm for the computation of general maxmin-type functions, which allows us to significantly decrease CPU time. This algorithm was suggested by Yu. G. Evtushenko (see [6]).

Let $X$ be a set and $f_{ij}$ ($k = 1, \dots, p$, $j = 1, \dots, m$) be functions defined on $X$. Let

$$h(y) = \max_{k=1,\dots,p} \min_{j=1,\dots,m} f_{kj}(y), \qquad (y \in X).$$

Consider also a point $x \in X$.

**Algorithm 3**

*Step 1.* Compute

$$\varphi_1(x) = \min_{j=1,\dots,m} f_{1j}(x),$$

set $f = \varphi_1(x)$ and $k = 1$.

*Step 2.* Set $k = k + 1$. If $k > p$ then go to Step 6. Otherwise set $j = 0$.

*Step 3.* Set $j = j + 1$. If $j > m$ go to Step 5, otherwise go to Step 4.

*Step 4.* Compute

$$f_{kj} = f_{kj}(x).$$

If $f_{kj} \leq f$ then go to Step 2, otherwise go to Step 3.

*Step 5.* Compute

$$f_k = \min_{j=1,\dots,m} f_{kj}, \quad f = \max\{f, f_k\}$$

and go to Step 2.

*Step 6.* Set $h(x) = f$ and the algorithm terminates.

## 6 SOME PROPERTIES OF THE AUXILIARY FUNCTION

In this section we study some properties of the auxiliary function $h_j(x)$ defined by (13.12). For the sake of simplicity we assume that Algorithm 1 is considered, that is, only one solution of the auxiliary problem is computed at each iteration. Let $x^{j+1}$ be a solution of the problem (13.11). Compute the vector $l^{j+1} = f(x^{j+1})/x^{j+1}$ and consider the new auxiliary function $h_{j+1}(x)$:

$$h_{j+1}(x) = \max_{k \leq j+1} \min_{i \in I(l^k)} l_i^k x_i = \max\{h_j(x),\ \varphi_{j+1}(x)\},$$

where $\varphi_{j+1}(x) = \min_{i \in I} l_i^{j+1} x_i$.

**Proposition 6.1** *The point $x^{j+1}$ is the unique strict global maximizer of the function $\varphi_{j+1}(x)$ over the simplex $S$.*

*Proof.* Since $x^{j+1} \gg 0$ and $l^{j+1} = \dfrac{f(x^{j+1})}{x^{j+1}}$, it follows that

$$l_i^{j+1} x_i^{j+1} = f(x^{j+1}) = \varphi_{j+1}(x^{j+1})$$

for all $i \in I$. Let $x \in S$. Then there exists $i \in I$ such that $x_i < x_i^{j+1}$. So, $l_i^{j+1} x_i^{j+1} > l_i^{j+1} x_i$ and consequently

$$\varphi_{j+1}(x^{j+1}) = l_i^{j+1} x_i^{j+1} > \min_{i'} l_{i'}^{j+1} x_{i'} = \varphi_{j+1}(x). \qquad \triangle$$

**Proposition 6.2** *If the point $x^{j+1}$ is not a global minimizer of the function $f$ over the simplex $S$ then $x^{j+1}$ is a strict local maximizer of the function $h_{j+1}(x)$.*

*Proof.* Since the point $x^{j+1}$ is a global minimizer of the function $h_j$ and is not a global minimizer of the function $f$ then

$$h_j(x^{j+1}) < f(x^{j+1}). \tag{13.16}$$

Indeed, $h_j(x) \leq f(x)$ for all $x$. Since $\lambda_j \leq \min_{x \in S} f(x)$, it follows that the equality $\lambda_j = h_j(x^{j+1}) = f(x^{j+1})$ is valid if and only if $x^{j+1}$ is a global minimizer of $f$, so (13.16) holds. On the other hand $\varphi_{j+1}(x^{j+1}) = f(x^{j+1})$, so applying (13.16) we conclude that

$$\begin{aligned} h_{j+1}(x^{j+1}) &= \max\{h_j(x^{j+1}), \varphi_{j+1}(x^{j+1}\} = \max\{h_j(x^{j+1}), f(x^{j+1})\} \\ &= f(x^{j+1}) = \varphi_{j+1}(x^{j+1}). \end{aligned} \tag{13.17}$$

The continuity of the functions $h_j$ and $\varphi_{j+1}$ and (13.16) implies the existence of $\epsilon > 0$ such that $\varphi_{j+1}(x) > h_j(x)$ for all $x \in S$ with $\|x - x^{j+1}\| < \epsilon$. It follows from Proposition 6.1 that $x^{j+1}$ is a global maximizer of the function $\varphi_{j+1}(x)$, so we have by applying (13.17):

$$h_{j+1}(x) = \varphi_{j+1}(x) < \varphi_{j+1}(x^{j+1}) = h_{j+1}(x^{j+1})$$

for all $x \in S$ with $\|x - x^{j+1}\| < \epsilon$. △

**Proposition 6.3** *If $x^{j+1}$ is not a global minimizer of the function $f$ over the simplex $S$ then $l_i^{j+1} > l_i^i$ for all $i = 1, ..., n$ and $j \geq n$ where $l^i = f(e_i)e_i$, $i = 1, \dots, n$, $l^{j+1} = f(x^{j+1})/x^{j+1}$.*

*Proof.* Since $x^{j+1}$ is a global minimizer of the function $h_j$ and is not a global minimizer of the function $f$ then (see (13.16))

$$h_j(x^{j+1}) < f(x^{j+1}).$$

Since also $\varphi_{j+1}(x^{j+1}) = f(x^{j+1})$, it follows that

$$\varphi_{j+1}(x^{j+1}) > h_j(x^{j+1}). \tag{13.18}$$

Assume that there exists $i \in I$ such that $l_i^{j+1} \leq l_i^i$. Then

$$\min_i l_i^{j+1} x_i^{j+1} \leq \max_i l_i^i x_i^{j+1}.$$

Consequently,

$$\begin{aligned} h_j(x^{j+1}) &= \max\{\max_{i \in I} l_i^i x_i^{j+1}, \max_{n+1 \leq k \leq j} \min_i l_i^k x_i^{j+1}\} \\ &\geq \max_i l_i^i x_i^{j+1} \geq \min_i l_i^{j+1} x_i^{j+1} = \varphi_{j+1}(x^{j+1}) \end{aligned}$$

which contradicts (13.18). Thus $l_i^{j+1} > l_i^i$ for all $i \in I$. △

**Proposition 6.4** *Let the point $x^{j+1}$ is not a global minimizer of the function $f$ over the simplex $S$. Then for any $k = n + 1, \dots, j$ an $i \in I$ can be found such that $l_i^{j+1} > l_i^k$.*

*Proof.* Since $x^{j+1}$ is not a global minimizer of the function $f$ it follows (see (13.16)) that

$$f(x^{j+1}) > h_j(x^{j+1}) := \max_{k \leq j} \min_{i \in I(l^k)} l_i^k x_i^{j+1}.$$

On the other hand

$$\min_i l_i^{j+1} x_i^{j+1} := \varphi_{j+1}(x^{j+1}) = f(x^{j+1}).$$

Consequently,

$$\min_i l_i^{j+1} x_i^{j+1} > \max_{k \le j} \min_{i \in I(l^k)} l_i^k x_i^{j+1}.$$

Thus for any $k \ge n+1$

$$\min_i l_i^{j+1} x_i^{j+1} > \min_i l_i^k x_i^{j+1}. \tag{13.19}$$

Assume that $l_i^{j+1} \le l_i^k$ for each $i = 1, \dots, n$. Since $x^{j+1} \gg 0$, we have

$$\min_i l_i^{j+1} x_i^{j+1} \le \min_i l_i^k x_i^{j+1}$$

which contradicts (13.19). △

# 7 A NEW VERSION OF ALGORITHM 2

The results of Section 6 allow us to propose a new version of the algorithm for solving the auxiliary problem. First, we describe the set of all local minimizers of the function $h_{j+1}(x)$ with $j \ge n$. Recall that

$$h_{j+1}(x) = \max\{h_j(x), \varphi_{j+1}(x)\}.$$

We consider the following three sets:

$$S_1 = \{x \in S \;:\; h_{j+1}(x) = \varphi_{j+1}(x) > h_j(x)\},$$

$$S_2 = \{x \in S \;:\; h_{j+1}(x) = h_j(x) > \varphi_{j+1}(x)\},$$

$$S_3 = \{x \in S \;:\; h_{j+1}(x) = \varphi_{j+1}(x) = h_j(x)\}.$$

Let $j \ge n$. Since the function $\varphi_{j+1}$ is concave, $\varphi_{j+1}(x) > 0$ for all $x \in \mathrm{ri} S$ and $\varphi_{j+1}(x) = 0$ for all $x \in \mathrm{bd}\, S$ it follows that the set of local (and also global) minimizers of $\varphi_{j+1}$ coincides with the set of boundary points of the simplex $S$.

Since functions $h_j, h_{j+1}$ and $\varphi_{j+1}$ are continuous, it follows that the set of the local minimizers of the function $h_j$ on the set $S_2$ and the set of local minimizers of the function $\varphi_{j+1}$ on the set $S_1$ coincide with the corresponding sets for $h_{j+1}$ on $S_2$ and $S_1$, respectively. Since all local minimizers of $\varphi_{j+1}$ are boundary points of the simplex and $h_{j+1}$ does not attain its local minimum on the boundary of $S$ it follows that there are no local minimizers of $h_{j+1}$ on the set $S_1$.

The set of the local minimizers of $h_j$ and consequently, of $h_{j+1}$ over the set $S_2$ is known from the previous iteration. So it remains to describe the local minimizers of the function $h_{j+1}$ on the set $S_3$. Let $x^* \in S_3$ be a local minimizer of $h_{j+1}$. Then there exists a subset $\{k_1, \dots, k_n\}$ such that

$$x_i^* = \frac{d}{l_i^{k_i}}, \quad i = 1, \dots, n, \quad \text{where } d = h_{j+1}(x^*). \tag{13.20}$$

We have

$$h_{j+1}(x^*) = \max\{h_j(x^*), \varphi_{j+1}(x^*)\} = \varphi_{j+1}(x^*).$$

Then

$$h_{j+1}(x^*) = \min_i l_i^{j+1} x_i^* = h_{j+1}(x^*) \min_i \frac{l_i^{j+1}}{l_i^{k_i}}.$$

Thus for each local minimizer $x^* \in S_3$ with the coordinates given by (13.20), the following holds:

$$\min_i \frac{l_i^{j+1}}{l_i^{k_i}} = 1.$$

On other words

$$l_i^{k_i} \le l_i^{j+1}, \; i = 1, ..., n. \tag{13.21}$$

Consequently, for the computation of the local minimizers of $h_{j+1}$ over the set $S_3$ we need to add (13.21) to the conditions 1)–3) in Proposition 4.2.

Let $M_j$ be the set of all local minimizers of the function $h_j$ on $S$. The discussion above shows that $M_{j+1}$ can be presented as follows:

$$M_{j+1} = M_{j+1}^1 \cup M_{j+1}^2$$

where

$$M_{j+1}^1 = \{x \in M_j : h_j(x) > \varphi_{j+1}(x)\},$$

and $M_{j+1}^2$ is the set of all points $x \in S_3$ such that $x$ satisfies the conditions 1) – 3) in Proposition 4.2 and (13.21). We now can propose a new version of Algorithm 2, which calculates global minimizers of $h_{j+1}$ assuming that the set of all local minimizers of $h_j$ is known. This new version differs from Algorithm 2 by Steps 0, 4a, and 5. We substitute these steps for the following Step 0′, Step 4a′ and Step 5′, respectively.

*Step 0′*. (Initialization and checking the local minimizers, which were computed at previous iteration). Set $t = 0, m = 0, p(k) = 0, k = 1, \ldots, n$. Take a point $x \in M_j$. If $h_j(x) > \varphi_{j+1}(x)$ then set $t = t + 1, \; g(t) = x, \; v(t) = h_j(x)$, otherwise exclude this point.

*Step 4a′*. If $l_m^i < \epsilon$ or $l_m^i > l_m^{j+1}$ then go to Step 2.

*Step 5′*. If

$$\min_i \frac{l_i^{j+1}}{l_i^{k_i}} = 1 \tag{13.22}$$

and

$$\max_{k \le j} \min_{i \in I(l^k)} \frac{l_i^k}{l_i^{k_i}} = 1 \tag{13.23}$$

then go to Step 6, otherwise set $m = n$ and go to Step 2.

Steps 1–4 of Algorithm 2 and all its modifications lead to construction of subsets $\{l^{k_1}, \ldots, l^{k_n}\}$ of the set $\{l^1, \ldots, l^n\}$. Then subsets, for which (13.15) holds, are selected, see Step 5. (For the new version (13.22) and (13.23) need to be checked.) As a rule the number of selected subsets is much larger than the number of the local minimizers. We have already mentioned that the computation of the auxiliary function in Step 5 occupies a considerable part of the CPU time. Therefore it is beneficial to modify Algorithm 2 and its versions so that to exclude selected subsets, which certainly do not provide local minima. We can do it, for example, using the known global minimum of the auxiliary function at the previous iteration. Recall that the value of the auxiliary function at a local minimizer $x$, which is defined by vectors $l^{k_1}, \ldots, l^{k_n}$, has the form:

$$h(x) = \left( \sum_{i \in I} \frac{1}{l_i^{k_i}} \right)^{-1}.$$

It is clear that

$$\left( \sum_{i=1}^{i_2} \frac{1}{l_i^{k_i}} \right)^{-1} < \left( \sum_{i=1}^{i_1} \frac{1}{l_i^{k_i}} \right)^{-1}.$$

for any $i_1, i_2 \in I$, $i_1 < i_2$. Let $h_j^*$ be the global minimum of the auxiliary function at $j$-th iteration. Since $h_j^* \ge h_{j-1}^*$, it follows that a subset $\{l^{k_1}, \ldots, l^{k_n}\}$, for which there exists $i_1 \in I$ such that

$$\left( \sum_{i=1}^{i_1} \frac{1}{l_i^{k_i}} \right)^{-1} < h_{j-1}^*,$$

does not provide local (and so global) minimum at $j$-th iteration. This observation leads to the following modification of Algorithm 2: between Steps 4f and 4g we consider new Step 4g′:

*Step 4g′.* Compute

$$d_0 = \left( \sum_{i=1}^{m} \frac{1}{l_i^{k_i}} \right)^{-1}.$$

If $d_0 < h_{j-1}^*$ then go to Step 2.

## 8 THE COMPARISON OF DIFFERENT VERSIONS OF ALGORITIIM 2

In this section we give a comparison of different versions of Algorithm 2 for solving the auxiliary problem, which were discussed in previous sections. This comparison is based on application of Algorithm 1 (Algorithm 1′) for minimizing some functions, by assuming that Step 2 (Step 2′, respectively) of this Algorithm is accomplished by several versions of Algorithm 2. We consider seven different versions of this algorithm. Version 1 is the initial version of Algorithm 2, Version 2 differs from Version 1 only by using Algorithm 3 for computation of the auxiliary function. Version 3 (Version 4) is obtained from Version 2 by substitution of Step 8 for Step 8′ (Step 8″, respectively). Version 5 corresponds to Algorithm 1″. Version 6 is obtained from Version 5 by substitution of Steps 0, 4a and 5 for Steps 0′, 4a′ and 5′, respectively. If we add Step 4g′ to Version 6, we get Version 7.

We consider the minimization of three IPH functions over the unit simplex for comparison of proposed versions of the algorithm. The following objective functions have been chosen:

**Problem 8.1**

$$f(x) = \left( \sum_{i=1}^{n} \sum_{j=1}^{n} a_{ij} x_i x_j \right)^{1/2},$$

$$a_{ij} = \begin{cases} 12 + n/i & \text{if } i = j, \\ 0 & \text{if } i = j+1, \\ 0 & \text{if } j = i+2, \\ 15/(i+0.1j) & \text{otherwise.} \end{cases}$$

**Problem 8.2**

$$f(x) = \max\{[a^i,\ x] : i = 1, \dots, 40\} + \min\{[b^j, x] : j = 1, \dots, 20\},$$

$$a^i_k = \frac{20i}{k(1+|i-k|)},\ k = 1, \dots, n;\ i = 1, \dots, 40,$$

$$b^j_k = 5|\sin(j)\ \sin(k)|,\ k = 1, \dots, n,\ j = 1, \dots, 20.$$

**Problem 8.3**

$$f(x) = \max_{1 \le i \le 20} \min_{1 \le j \le n} [a^{ij}, x],$$

$$a^{ij}_k = \frac{10j}{k(1+|k-j|)} |\cos(i-1)|,\ i = 1, \dots, 20,\ j = 1, \dots, n,\ k = 1, \dots, n.$$

We consider Problems 8.1 - 8.3 for $n = 3, 5, 10$. Thus we have $3\times 3 = 9$ problems. First we performed a certain number of iterations for each of these problems by using Version 1 (for example, we made 100 iterations for Problem 8.1 with $n = 3$ and 45 iterations for problem 8.3 with $n = 10$). The obtained results have been fixed. Then we tried to get the same results by using the other versions of the algorithm. Therefore, the main index for the comparison of various versions is the computational time. Numerical experriments have been curried out on PC Pentium-S with Main Processor 150 MHz.

Results of numerical experiments are presented in Tables 13.1, 13.2 and 13.3. We use the following notation: $j$ is the number of iterations and $t$ is the computational time in seconds.

Tables 13.1 – 13.3 confirm that the last version of Algorithm 2 is much more efficient than the first version. The last version almost 26 times faster than Version 1 in the best case and it 10.5 times faster than this version in the worst case.

*Table 13.1* Results of numerical experiments: Problem 8.1

| Version | *n =3* | | *n=5* | | *n=10* | |
|---|---|---|---|---|---|---|
| | $j$ | $t$ | $j$ | $t$ | $j$ | $t$ |
| 1 | 100 | 342.40 | 80 | 743.14 | 70 | 1503.26 |
| 2 | 100 | 279.73 | 80 | 623.90 | 70 | 1378.08 |
| 3 | 100 | 281.93 | 80 | 623.68 | 70 | 1386.32 |
| 4 | 82 | 177.08 | 47 | 380.52 | 18 | 249.53 |
| 5 | 82 | 164.34 | 47 | 227.94 | 18 | 215.03 |
| 6 | 82 | 56.08 | 47 | 109.63 | 18 | 131.44 |
| 7 | 82 | 17.09 | 47 | 70.68 | 18 | 117.55 |

## 9 GLOBAL MINIMIZATION OF LIPSCHITZ FUNCTIONS

It is known (see [10, 8]) that the global minimization of Lipschitz function over the simplex $S$ can be reduced to the global minimization of a certain IPH function over $S$. We now recall the relevant results.

**Theorem 9.1** *Let $f : S \to \mathbb{R}$ be a Lipschitz function and let*

$$L = \sup_{x,y\in S, x\neq y} \frac{f(x) - f(y)}{\|x - y\|_1} \qquad (13.24)$$

*Table 13.2* Results of numerical experiments: Problem 8.2

| Version | $n$ =3 | | $n$=5 | | $n$=10 | |
|---|---|---|---|---|---|---|
| | $j$ | $t$ | $j$ | $t$ | $j$ | $t$ |
| 1 | 80 | 180.27 | 60 | 416.66 | 50 | 1768.49 |
| 2 | 80 | 183.23 | 60 | 404.26 | 50 | 1705.60 |
| 3 | 80 | 185.65 | 60 | 406.12 | 50 | 1707.02 |
| 4 | 52 | 122.87 | 35 | 147.04 | 18 | 225.44 |
| 5 | 56 | 125.89 | 36 | 101.01 | 18 | 108.08 |
| 6 | 56 | 29.17 | 36 | 53.88 | 18 | 93.59 |
| 7 | 56 | 9.06 | 36 | 35.93 | 18 | 68.38 |

*Table 13.3* Results of numerical experiments: Problem 8.3

| Version | $n$ =3 | | $n$=5 | | $n$=10 | |
|---|---|---|---|---|---|---|
| | $j$ | $t$ | $j$ | $t$ | $j$ | $t$ |
| 1 | 80 | 212.23 | 60 | 500.10 | 45 | 1530.00 |
| 2 | 80 | 224.64 | 60 | 495.60 | 45 | 1441.74 |
| 3 | 80 | 223.00 | 60 | 498.01 | 45 | 1448.60 |
| 4 | 37 | 78.49 | 20 | 172.85 | 13 | 130.88 |
| 5 | 37 | 76.51 | 21 | 58.88 | 13 | 129.18 |
| 6 | 37 | 25.76 | 21 | 30.04 | 13 | 96.84 |
| 7 | 37 | 13.56 | 21 | 26.97 | 13 | 71.90 |

*be the least Lipschitz constant of $f$ in $\|\cdot\|_1$-norm, where $\|x\|_1 = \sum_{i\in I} |x_i|$. Assume that $\min_{x\in S} f(x) \geq 2L$. Then there exists an IPH function $g : \mathbb{R}^n_+ \to \mathbb{R}$ such that $g(x) = f(x)$ for all $x \in S$.*

Let $f : S \to \mathbb{R}$ be a Lipschitz function and let

$$c \geq 2L - \min_{x\in S} f(x), \tag{13.25}$$

where $L$ is defined by (13.24). Let $f_1(x) = f(x) + c$. It follows from Theorem 9.1 that the function $f_1$ can be extended to an IPH function $g$. Consider the problem

$$g(x) \longrightarrow \min \quad \text{subject to} \quad x \in S. \tag{13.26}$$

Clearly this problem is equivalent to the problem

$$f_1(x) \longrightarrow \min \quad \text{subject to} \quad x \in S. \tag{13.27}$$

Solving problem (13.26) by the cutting angle method, we use the values $g(x)$ of $g$ only for points $x \in S$, in particular we can apply the cutting angle method for solution of the problem (13.27). Clearly functions $f$ and $f_1$ have the same minimizers on the simplex $S$. Thus if the constant $c$ of (13.25) is known, the cutting angle method can be applied for the minimization of a Lipschitz function $f$ over $S$. We can assume that $c$ is a sufficiently large number, however numerical experiments show that the cutting angle method is rather sensitive to the choice of $c$, in particular, when $c$ is very large, the method works badly. In order to estimate $c$ we need to know an upper estimation of the least Lipschitz constant $L$ and a lower estimation of the desired global minimum of the function $f$.

Assume now that we consider the following problem:

$$f(x) \longrightarrow \min \quad \text{subject to} \quad x \in D = \{x \in \mathbb{R}^n : x_i \geq a_i \, (i \in I) \ \sum_{i \in I} x_i \leq b\} \tag{13.28}$$

where $f$ is a Lipschitz function, $a_i, (i \in I)$ and $b$ are given numbers. Assume that $b > \sum_{i \in I} a_i$. In order to apply the cutting angle method for the solution of the problem (13.28) we need to transform the set $D$ to the unit simplex. First we consider the following set:

$$D_1 = D - a = \{y \in \mathbb{R}^n : y_i = x_i - a_i, \ x \in D, \ i \in I\}$$

$$\equiv \{y \in \mathbb{R}^n : y_i \geq 0, \ i \in I, \ \sum_{i \in I} y_i \leq b - \sum_{i \in I} a_i\}.$$

Let $t = b - \sum_{i=1}^n a_i$. Clearly $t > 0$. Adding a new variable $y_{n+1}$ and replacing $y_i/t$ by $z_i$ we transform the set $D_1$ to the unit simplex $S_* \subset \mathbb{R}^{n+1}$:

$$S_* = \{z \in \mathbb{R}_+^{n+1} : \sum_{i=1}^{n+1} z_i = 1\}.$$

Thus we transform the problem (13.28) to the following problem:

$$f(tz^n + a) \longrightarrow \min \quad \text{subject to} \quad z \in S_* \tag{13.29}$$

where $z^n = (z_1, \ldots, z_n), a = (a_1, \ldots, a_n)$.

Some problems of Lipschitz programming were successfully solved by the cutting angle method.

Consider the following question which arises in the study of some problems of cluster analysis (see [5]). Let $Q$ be a finite set of $n$-dimensional vectors. Consider a set $X = \{x^1, \ldots, x^k\} \subset \mathbb{R}^n$. The distance $d(q, X)$ between a point $q \in Q$ and the set $X$ is $\min_{1 \leq j \leq k} \|x^j - q\|$. The quantity

$$d(Q, X) = \sum_{q \in N} d(q, X) = \sum_{q \in N} \min_{1 \leq j \leq k} \|x_j - q\|$$

is the total distance between the set $Q$ and the set $X$. Consider the following problem:

$$\sum_{q \in N} \min_{j=1,\dots,k} \|x_j - q\| \longrightarrow \min \quad \text{subject to } x_1, \dots, x_k \in \mathbb{R}^n. \qquad (13.30)$$

This is a unconstrained problem of Lipschitz minimization. Let

$$a_i = \min_{q \in N} q_i \, (i \in I), \; b_i = \max_{q \in N} q_i \, (i \in I).$$

Clearly the inequalities $a_i \leq \bar{x}_i^k \, (i \in I)$ and $\sum_{1<j<k} \sum_{i \in I} \bar{x}_i^k \leq k \sum_{i \in I} b_i$ hold for a solution $\bar{x}^1, \dots, \bar{x}^k$ of the problem (13.30). Thus we can consider (13.30) as a special case of (13.28) If $k = 1$ then the objective function of problem (13.30) is convex, so this problem can be solved by well-known methods of convex optimization. If $k > 1$ this problem becomes a very difficult problem of global optimization. We consider now a more simple problem. Assume that the solution $\bar{x}^1$ of the problem (13.30) with $k = 1$ is known and consider the problem

$$\sum_{q \in Q} \min\{\|x - q\|, \|x^1 - q\|\} \longrightarrow \min \quad \text{subject to } \; x \in \mathbb{R}^n.$$

This problem was solved by the cutting angle method with the set $Q$ containing 357 three-dimensional vectors (see [5] for details). The problem (13.30) can be solved by a combination of the cutting angle method and a local search for not fairly small numbers $n$ and $k$. We used the derivative-free discrete gradient method for a local search (see [2, 3]). The cardinality of the set $Q$ is not very important for a combination of the cutting angle method and discrete gradient method.

## Acknowledgments

The authors wish to thank the anonymous referees for their valuable comments, which improved the presentation of the paper.

## References

[1] Andramonov, M. Yu., Rubinov, A.M. and Glover, B. M. (1999), Cutting angle method in global optimization, *Applied Mathematics Letters,* Vol. 12, 95-100.

[2] Bagirov, A.M. (1999a), Minimization methods for one class of nonsmooth functions and calculation of semi-equilibrium prices, in Ser. *Applied Optimization,* Vol. 30: *Progress in Optimization: Contributions from Australasia,* Eberhard, A. et al. (eds.), Kluwer Academic Publishers, 147-175.

[3] Bagirov, A.M. (1999b), Derivative-free methods for unconstrained nonsmooth optimization and its numerical analysis. *Investigacao Operacional*, Vol. 19, 75-93.

[4] Bagirov, A.M. and Rubinov, A.M. (2001), Global minimization of increasing positively homogeneous functions over the unit simplex, *Annals Oper. Research,* to appear. ( Research Report, 37/99, University of Ballarat, 1999.)

[5] Bagirov, A.M., Rubinov, A.M., Stranieri, A. and Yearwood, A. (1999), The global optimization approach to the clustering analysis. Research Report, 45/99, University of Ballarat, Australia, 1999.

[6] Evtushenko, Yu. G. (1972), A numerical method for finding best guaranteed estimates, *U.S.S.R. Computational Mathematics and Mathematical Physics,* Vol. 12, 109-128.

[7] Pallaschke, D. and Rolewicz, S. (1997), *Foundations of Mathematical Optimization (Convex Analysis without Linearity)*, Kluwer Academic Publishers, Dordrecht, 1997.

[8] Rubinov, A.M. (2000), *Abstract Convexity and Global Optimization*, Kluwer Academic Publishers, 2000.

[9] Rubinov, A.M. and Andramonov, M.Yu. (1999a), Minimizing increasing star-shaped functions based on abstract convexity, *Journal of Global Optimization*, Vol. 15, 19-39.

[10] Rubinov, A. and Andramonov, M. (1999b), Lipschitz programming via increasing convex-along-rays functions, *Optimization Methods and Software*, Vol. 10, 763-781.

Chapter 14

# THEORETICAL AND COMPUTATIONAL RESULTS FOR A LINEAR BILEVEL PROBLEM

Manoel Campêlo
*Dep. Estatística e Mat. Aplicada, Universidade Federal do Ceará - UFC*
mcampelo@ufc.br

Susana Scheimberg
*DCC/IM, COPPE/Sistemas, Universidade Federal do Rio de Janeiro - UFRJ*
susana@cos.ufrj.br

**Abstract** In this paper a linear bilevel programming problem (LBP) is considered. Two related problems $(P)$ and $P(M)$ are examined. The three problems are shown to be simultaneously infeasible or unbounded for all $M \geq 0$ or else they have equivalent global solution sets for some finite $M^*$ and all $M > M^*$. The concept of equilibrium point of $P(M)$ is introduced and its properties are analysed. An algorithm to obtain an $\varepsilon$-solution to (LBP) is developed. It is based on equilibrium point properties and a modified version of outer approximation. Infeasible and unbounded cases are also identified by the method. Computational results are given.

**Keywords:** Bilevel linear programming, Global optimization, Exact penalty methods.

## 1 INTRODUCTION

In this article we focus our attention on the following linear bilevel program:

$$
\begin{array}{llll}
\text{(LBP)} & \max_{x,y} & c_1^T x + c_2^T y & \\
 & \text{s.t.} & x \geq 0, y \text{ solves} & \\
 & & \max_y & b^T y \\
 & & \text{s.t.} & A_1 x + A_2 y \leq a \\
 & & & y \geq 0
\end{array}
$$



*N. Hadjisavvas and P.M. Pardalos (eds.), Advances in Convex Analysis and Global Optimization*, 269–281.

where $c_1, x \in \mathcal{R}^{n_1}$, $c_2, b, y \in \mathcal{R}^{n_2}$, $a \in \mathcal{R}^m$, $A_1 \in \mathcal{R}^{m \times n_1}$ and $A_2 \in \mathcal{R}^{m \times n_2}$. Bilevel Programming has applicability in many domains, e.g. in network design, transportation, game theory, engineering and economics [13].

Let us note that (LBP) is a nonconvex problem. Actually it is strongly NP-hard [8]. This problem has been extensively studied in the literature, see e.g. [1, 2, 11, 17]. We refer to [16] for a bibliography survey.

The aim of this paper is to develop an algorithm for finding an $\varepsilon$-global solution to problem (LBP). It is based on the new notion of equilibrium point and a modified version of the outer approximation method. In the procedure we can identify two parts. The first one achieves a local solution of a problem related to (LBP). The second part introduces a cut that separates the feasible set containing the global solution from those points just obtained. Thus, we consider problem (LBP) with a linear constraint in the first level. This problem is denoted by (LBP1). We observe that the method recognizes infeasible and unbounded cases.

The paper is organized as follows. Section 2 contains theoretical background supporting the proposed method. In section 3 we present the algorithm. We prove that it is well defined and finishes in a finite number of steps. Finally, we give some computational results.

# 2 THEORETICAL PRELIMINARIES

## 2.1 AUXILIARY PROBLEMS

In this subsection we define two auxiliary problems of (LBP). They will be the basis of our approach. We also introduce a simplified notation. Finally we give a property showing a way to find feasible points of (LBP) and related problems, which is computationally attractive.

The first auxiliary problem $(P)$ is obtained by replacing the inner linear problem of (LBP) by its KKT conditions. The second one $P(M)$ comes from a penalization of the complementarity constraints in $(P)$ with a parameter $M \geq 0$. Thus, we have the following models:

$(P)$

$$\begin{array}{ll} \max & c_1^T x + c_2^T y \\ \text{s.t.} & A_1 x + A_2 y + w = a, \\ & x \geq 0, y \geq 0, w \geq 0, \\ & A_2^T u - v = b, \\ & u \geq 0, v \geq 0, \\ & u^T w + v^T y = 0 \end{array}$$

$P(M)$

$$\begin{array}{ll} \max & c_1^T x + c_2^T y - M(u^T w + v^T y) \\ \text{s.t.} & A_1 x + A_2 y + w = a, \\ & x \geq 0, y \geq 0, w \geq 0, \\ & A_2^T u - v = b, \\ & u \geq 0, v \geq 0 \end{array}$$

where $w \in \mathcal{R}^m$, $u \in \mathcal{R}^m$ and $v \in \mathcal{R}^{n_2}$.

Formulation $(P)$ is often used instead of model (LBP), see e.g. [3, 11, 14, 17]. Indeed, they are globally equivalent, in the sense that a global solution of $(P)$, if it exists, corresponds to a global solution of (LBP) and vice-versa. The correspondence also holds in the cases of infeasibility and unboundedness. However, a local solution of $(P)$ may not be associated with a local solution of (LBP). In [7] we give a counterexample.

The penalty function has also been considered in the literature [1, 14, 17] as an approach to identify global optima of (LBP) under restrictive assumptions. Actually, under compactness hypothesis on the set defined by the primal constraints, Bard [3] shows that extreme global solutions of $P(M^\star)$ yield extreme global optima of $(P)$ and vice-versa, for some finite $M^\star \geq 0$. In [6] we prove that it is sufficient to consider the objective function of $(P)$ bounded above on that set. Now, we show that this penalty is still useful in a wider context involving unbounded and infeasible situations. We point out that the penalty parameter can be implicitly considered, which is computationally attractive.

In order to simplify the notation, we consider the block matrices $A = [A_1 A_2 I_m] \in \mathcal{R}^{m\times n}$, $B = [0 - I_{n_2} A_2^T] \in \mathcal{R}^{n_2\times n}$ and $c^T = (c_1^T, c_2^T, 0) \in \mathcal{R}^n$, where $n = n_1 + n_2 + m$, $I_k$ is the $(k \times k)$-identity matrix and 0 is a null matrix of appropriate dimension for each case. We define the sets $Z = \{z \in \mathcal{R}^n : Az = a, z^T = (x^T, y^T, w^T) \geq 0\}$ and $S = \{s \in \mathcal{R}^n : Bs = b, s^T = (0, v^T, u^T) \geq 0\}$. Thus, the auxiliary problems are rewritten as:

$$(P) \quad \begin{array}{ll} \max & F(z,s) = c^T z \\ \text{s.t.} & z \in Z, s \in S \\ & s^T z = 0 \end{array} \qquad P(M) \quad \begin{array}{ll} \max & F_M(z,s) = c^T z - M s^T z \\ \text{s.t.} & z \in Z, s \in S \end{array}$$

We denote by $X_v$ the vertex set of a polyhedron $X$ and by $V[\mathcal{P}]$ the optimal value of a maximization problem $(\mathcal{P})$, regarding that $V[\mathcal{P}]$ is $-\infty$ or $+\infty$ if $(\mathcal{P})$ is infeasible or unbounded respectively. Let us recall that the infinity norm of $\nu = (\nu_1, \nu_2, \ldots, \nu_p)$ is $\| \nu \|_\infty = \max\{|\nu_i| : 1 \leq i \leq p\}$ and $B_\varepsilon(\bar{\nu}) = \{\nu \in \mathcal{R}^p : \| \nu - \bar{\nu} \|_\infty \leq \varepsilon\}$ is an $\varepsilon$-neighborhood of $\bar{\nu} \in \mathcal{R}^p$.

**Proposition 1** *If $Z$ and $S$ are nonempty sets then $(P)$ is feasible. Moreover, for each $z \in Z$ and $s \in S$, it holds:*

*(1) If $\hat{s} \in \arg\min\{z^T s : s \in S_v\}$ and $\hat{z} \in \arg\min\{\hat{s}^T z : s \in Z\}$ then $(\hat{z}, \hat{s})$ is feasible to $(P)$.*

*(2) If $\hat{z} \in \arg\min\{s^T z : z \in Z_v\}$ and $\hat{s} \in \arg\min\{\hat{z}^T s : s \in S\}$ then $(\hat{z}, \hat{s})$ is feasible to $(P)$.*

*Proof:* **(1)** Let $z \in Z$ and $\hat{s} \in \arg\min\{z^T s : s \in S_v\}$. Consider $B = [E\, R]$ where $E$ is a basis of $B$ associated to $\hat{s}^T = ((E^{-1}b)^T, 0)$. Let be $z^T = (z_E^T, z_R^T)$. Consider the optimal reduced cost vetor $\tilde{z}^T = (0, z_R^T - z_E^T E^{-1} R) = z^T - z_E^T E^{-1} B \geq 0$. Since $BA^T = 0$ we obtain that

$A\tilde{z} = Az = a$. Hence, $\tilde{z} \in Z$. The partitions of $\tilde{z}$ and $\hat{s}$ results $\hat{s}^T\tilde{z} = 0$. So it must be $\hat{s}^T\hat{z} = 0$ for each $\hat{z} \in \arg\min\{\hat{s}^T z : z \in Z\}$.
**(2)** It is proved in a similar way. ■

## 2.2 GLOBAL EQUIVALENCE

Here we characterize the existence or nonexistence of a global solution to problem (LBP) by considering the auxiliary problems. We start with the following property:

**Proposition 2** *If problem $(P)$ is feasible and problem $P(M_0)$ has solution for some $M_0 \geq 0$, then there exists $M^\star \geq M_0$ such that:*

*(1) There exists $(z^\star, s^\star) \in Z_v \times S_v$ satisfying $(s^\star)^T z^\star = 0$ and $V[P] = V[P(M)] = F(z^\star, s^\star)$ for all $M \geq M^\star$.*

*(2) Moreover, $(P)$ and $P(M)$ have the same nonempty global solution set for all $M > M^\star$.*

*Proof:* **(1)** As $P(M)$ is less restricted than $(P)$ and $s^T z \geq 0$ for all $(z, s) \in Z \times S$, $P(M)$ has solution for all $M \geq M_0$. Moreover, at least one of its solutions is attained at a vertex, since it is a bilinear program [10, proposition IX.1]. Then, it follows that

$$V[P(M)] = \max\{F(z, s) - Ms^T z : z \in Z_v, s \in S_v\} \quad \forall M \geq M_0. \tag{13.1}$$

Let us split $Z_v \times S_v$ into the finite subsets, $C_0 = \{(z, s) \in Z_v \times S_v : z^T s = 0\}$ and $C_1 = (Z_v \times S_v) \setminus C_0$. As $(P)$ is feasible, it must be $C_0 \neq \emptyset$. Let $(z^\star, s^\star) \in C_0$ satisfying $F(z^\star, s^\star) = \max\{F(z, s) : (z, s) \in C_0\}$. If $C_1 \neq \emptyset$, let us take

$$M^\star = \max\left\{M_0, \max\{(F(z, s) - F(z^\star, s^\star))/s^T z : (z, s) \in C_1\}\right\}. \tag{13.2}$$

Thus, for all $M \geq M^\star$ we have

$$L(M) = \max\{F(z, s) - Ms^T z : (z, s) \in C_1\} \leq F(z^\star, s^\star) \tag{13.3}$$

By (13.1)-(13.3) and taking into account that $M^\star \geq M_0$, it follows that

$$V[P(M)] = \max\{F(z^\star, s^\star), L(M)\} = F(z^\star, s^\star) \quad \forall M \geq M^\star. \tag{13.4}$$

If $C_1 = \emptyset$, we trivially have $V[P(M)] = F(z^\star, s^\star)$ for all $M \geq M^\star = M_0$. So, it results

$$(s^\star)^T z^\star = 0 \quad \text{and} \quad V[P] = V[P(M)] = F(z^\star, s^\star) \quad \forall M \geq M^\star. \tag{13.5}$$

**(2)** Let $\bar{M} > M^\star$. Let us denote by $O(P)$ and $O(P(\bar{M}))$ the optimal sets of problems $(P)$ and $P(\bar{M})$. By (13.5), we have that $O(P) \neq \emptyset$

and $O(P(\bar{M})) \neq \emptyset$, with $V[P] = V[P(\bar{M})]$. Moreover, since any feasible solution of $(P)$ is feasible to $P(\bar{M})$, it follows that $O(P) \subseteq O(P(\bar{M}))$.

Conversely, let $(z, s) \in O(P(\bar{M}))$. Using again (13.5) together with $\bar{M} > M^\star$, we have that $V[P] = V[P(\bar{M})] = F(z,s) - \bar{M}s^T z \leq F(z,s) - M^\star s^T z \leq V[P(M^\star)] = V[P]$. Thus, equality holds among the relations. It yields that $(\bar{M} - M^\star)s^T z = 0$, which implies $s^T z = 0$. Therefore, $(z, s)$ is feasible to $(P)$ and $V[P] = F(z,s)$, that is, $(z,s) \in O(P)$. Hence, $O(P(\bar{M})) \subseteq O(P)$. So, we obtain the desired conclusion. ■

Let us note that the definition of $M^\star$ coincides with the expression given in [3], when $C_1 \neq \emptyset$ and $P(M)$ has solution for all $M$. Also, we observe that item (2) of proposition 2 may not hold for $M = M^\star$ [7]. Finally, it is worth stressing that the hypotheses of the above proposition are weaker than those ones considered in [3, 6].

Now we address our attention to the situations not covered by proposition 2, that is, unboundedness and infeasibility. For the first case, we are able to present the following equivalence:

**Proposition 3** *Problem $(P)$ is unbounded if, and only if, problem $P(M)$ is unbounded for all $M \geq 0$.*

*Proof:* If $(P)$ is unbounded, $P(M)$ is unbounded for all $M \geq 0$, since $V[P(M)] \geq V[P] = +\infty$ for all $M \geq 0$. Now, in order to prove the converse claim we consider the recession cone of $Z$ restricted to the unitary hypercube $Z_r = \{h \in \mathcal{R}^n : Ah = 0, h \geq 0, \| h \|_\infty \leq 1\}$. Let $P(M)$ be unbounded for all $M \geq 0$. It means that there exists $(h^\star, s^\star) \in (Z_r)_v \times S_v$ such that $(s^\star)^T h^\star = 0$ and $c^T h^\star > 0$. In addition, since $Z \neq \emptyset$ and $(s^\star)^T z \geq 0$ for all $z \in Z$, we can take $z^\star \in \arg\min\{(s^\star)^T z : z \in Z_v\}$. We are going to prove that there exists $\hat{s} \in S$, such that $(z^\star + \alpha h^\star, \hat{s})$ is feasible to $(P)$ for all $\alpha \geq 0$. Let $A = [D\ N]$ be a partition of $A$, where $D$ is a basis associated with $(z^\star)^T = ((D^{-1}a)^T, 0)$. Let $(s^\star)^T = ((s^\star_D)^T, (s^\star_N)^T)$ and consider the optimal reduced cost vector $\hat{s}_o^T = \left(0, (s^\star_N)^T - (s^\star_D)^T D^{-1} N\right) = (s^\star)^T - (s^\star_D)^T D^{-1} A \geq 0$. By the partitions of $\hat{s}_o$ and $z^\star$, we have $\hat{s}_o^T z^\star = 0$. Moreover, as $Ah^\star = 0$ and $AB^T = 0$, the second expression for $\hat{s}_o$ yields $\hat{s}_o^T h^\star = (s^\star)^T h^\star = 0$ and $B\hat{s}_o = Bs^\star = b$. Now, let us partition $\hat{s}_o^T = (\hat{r}^T, \hat{v}^T, \hat{u}^T)$ according to the block matrix $B$ and define $\hat{s}^T = (0, \hat{v}^T, \hat{u}^T)$. It is clear that $\hat{s} \geq 0$ and $B\hat{s} = B\hat{s}_o = b$ that is, $\hat{s} \in S$. Furthermore, since $\hat{s}_o \geq 0$, $z^\star \geq 0$ and $h^\star \geq 0$, we have $\hat{s}^T z^\star = \hat{s}_o^T z^\star = 0$ and $\hat{s}^T h^\star = \hat{s}_o^T h^\star = 0$. Thus, $(z^\star + \alpha h^\star, \hat{s})$ is feasible to $(P)$ for all $\alpha \geq 0$. Since $c^T h^\star > 0$, it follows that problem $(P)$ is unbounded. ■

For the infeasible case we have the next property:

**Proposition 4** *Let $\bar{M} \geq 0$. Problem $(P)$ is infeasible if, and only if, problem $P(\bar{M})$ is infeasible.*

*Proof:* Let $\bar{M} \geq 0$. If $P(\bar{M})$ is infeasible, clearly $(P)$ is infeasible. Conversely, assume that $P(\bar{M})$ is feasible. Then, $Z$ and $S$ are nonempty. By proposition 1, $(P)$ is feasible. ■

Now we summarize the relations between the auxiliary problems as follows:

**Theorem 5** *Exactly one of the following assertions must hold:*

*(1) problems $(P)$ and $P(M)$ are infeasible for all $M \geq 0$;*

*(2) problems $(P)$ and $P(M)$ are unbounded for all $M \geq 0$;*

*(3) problems $(P)$ and $P(M)$ have the same nonempty global solution set for all $M > M^\star$ for some $M^\star \geq 0$.*

Finally, we express theorem 5 in terms of the original problem:

**Theorem 6** *Problems $(LBP)$ and $P(M)$ are both infeasible or unbounded for all $M \geq 0$, or else they have solutions for some $M_0$. In this case, $z^\star$ is a global solution of $(LBP)$ if, and only if, there is $s^\star$ such that $(z^\star, s^\star)$ is a global solution of $P(M)$ for all $M > M^\star$, for some $M^\star \geq M_0$.*

**Remark:** All the above results remain valid when we add to the first level linear constraints involving only variable $x$. Besides, even if these constraints depend on variable $y$, proposition 2 still holds and we can guarantee as well the last part of theorem 6. However, propositions 3 and 4 cannot be extended to this more general (LBP). See [7] for details.

## 2.3 EQUILIBRIUM POINTS

In this part we consider the linear bilevel problem with linear constraints in the first level, (LBP1). Let us denote its related problems by $(\tilde{P})$ and $\tilde{P}(M)$. We introduce the concept of equilibrium point of the parametric problem. We establish some properties used in the algorithmic section. We also characterize local solutions of $(\tilde{P})$ in terms of this new concept.

Problems $(\tilde{P})$ and $\tilde{P}(M)$ can be obtained from $(P)$ and $P(M)$ by replacing $Z$ by $\tilde{Z} = Z \cap \{z \in \mathcal{R}^n : D_1 x + D_2 y \leq d\}$, where $D_1 \in \mathcal{R}^{p \times n_1}$, $D_2 \in \mathcal{R}^{p \times n_2}$ and $d \in \mathcal{R}^p$. We introduce the following notion.

**Definition 7** *A point $(\bar{z}, \bar{s}) \in \tilde{Z} \times S$ is an* **equilibrium point** *of the penalized problem if there is $\bar{M} \geq 0$ such that, for each $M \geq \bar{M}$, it holds*

$$\max\{F_M(\bar{z}, s) : s \in S\} = F_M(\bar{z}, \bar{s}) = \max\{F_M(z, \bar{s}) : z \in \tilde{Z}\}. \quad (13.6)$$

The equilibrium required in the above definition extends that one considered by the mountain climbing algorithm given for a bilinear program [12]. Indeed, in our case, equality (13.6) must be fulfilled for all $M \geq \bar{M}$, which means that the equilibrium has to be held by a family of parametric bilinear problems $P(M)$. This yields the following property related to the complementarity term of the penalty function.

**Lemma 8** *If $(\bar{z}, \bar{s})$ is an equilibrium point of the penalized problem then*

$$\min\{\bar{z}^T s : s \in S\} = \bar{s}^T \bar{z} = \min\{\bar{s}^T z : z \in \tilde{Z}\}. \tag{13.7}$$

*Proof:* Let $(\bar{z}, \bar{s})$ be an equilibrium point of the penalized problem. The left equality in (13.7) results from the first one given in (13.6). Now, suppose, by contradiction, that $\bar{s}^T \bar{z} > \bar{s}^T \hat{z} = \min\{\bar{s}^T z : z \in \tilde{Z}\}$. Then, $\bar{s}^T(\bar{z} - \hat{z}) > 0$. Take $M > \max\{\bar{M}, c^T(\bar{z} - \hat{z})/\bar{s}^T(\bar{z} - \hat{z})\}$. Thus, $c^T \hat{z} - M\bar{s}^T \hat{z} > c^T \bar{z} - M\bar{s}^T \bar{z}$, which contradicts the second equality in (13.6), since $M > \bar{M}$. Therefore, $\bar{s}^T \bar{z} = \min\{\bar{s}^T z : z \in Z\}$, and the proof is complete. ■

When $\tilde{Z} = Z$ we can show that an equilibrium point $(\bar{z}, \bar{s})$ satisfies $\bar{z}^T \bar{s} = 0$. This is essential to establish the equivalence between an equilibrium point and a local solution of $(P)$ [7]. In the general case, we have the following characterization of a local solution of $(\tilde{P})$:

**Theorem 9** *Let $(\bar{z}, \bar{s})$ be feasible to $(\tilde{P})$. The point $(\bar{z}, \bar{s})$ is a local solution of $(\tilde{P})$, if and only if, $(\bar{z}, \bar{s})$ is an equilibrium point of the corresponding penalized problem.*

*Proof:* Assume that $(\bar{z}, \bar{s})$ is a local solution to $(\tilde{P})$. So, there is $\varepsilon > 0$ such that $c^T \bar{z} \geq c^T z$ for all $(z, s) \in B_\varepsilon(\bar{z}, \bar{s})$. If $(\bar{z}, \bar{s})$ is not an equilibrium point, for all $\bar{M} \geq 0$ there are $M > \bar{M}$ and $z \in \tilde{Z}$ such that $F_M(z, \bar{s}) > F_M(\bar{z}, \bar{s}) = c^T \bar{z}$. Define $z_\alpha = \alpha z + (1 - \alpha)\bar{z}$ such that $\alpha \in (0, 1)$ and $(z_\alpha, \bar{s}) \in B_\varepsilon(\bar{z}, \bar{s})$. The convexity of $\tilde{Z}$ assures that $z_\alpha \in \tilde{Z}$. Hence, $c^T z_\alpha \geq F_M(z_\alpha, \bar{s}) = \alpha F_M(z, \bar{s}) + (1 - \alpha)c^T \bar{z} > c^T \bar{z}$. This yields is a contradiction.

Conversely, let $(\bar{z}, \bar{s})$ be feasible to $(\tilde{P})$ and an equilibrium point of the penalized problem. Then, $\bar{z}^T \bar{s} = 0$ and there is $\bar{M} \geq 0$ such that

$$c^T z - \bar{M}\bar{s}^T z \leq c^T \bar{z} \quad \forall z \in \tilde{Z}. \tag{13.8}$$

If $\bar{s} = 0$, the inequality above implies that $c^T z \leq c^T \bar{z}$ for all $(z, s)$ feasible to $(\tilde{P})$. Now suppose that $\bar{s} \neq 0$. Let us split the index set $J = \{1, 2, \ldots, n\}$ into the subsets $J_0 = \{j \in J : \bar{s}_j = 0\}$ and $J_1 = J \setminus J_0$. Since $\bar{s} \neq 0$, it must be $J_1 \neq \emptyset$. For each $j \in J_1$, let us define the set $C_j = \{s \in \mathcal{R}^n : s_j = 0\}$. Then, set $C_j$ is closed, convex and nonempty $(0 \in C_j)$. As $\bar{s} \notin C_j$, there is an hyperplane separating $\bar{s}$ and

$C_j$ strongly. So it is known that $d(\bar{s}, C_j) = \min\{\| s - \bar{s} \|_\infty : s \in C_j\} > 0$. Hence, there exists $\varepsilon \in \mathcal{R}$ such that $0 < \varepsilon < \min\{d(\bar{s}, C_j) : j \in J_1\}$. Let $s \in S \cap B_\varepsilon(\bar{s}) \neq \emptyset$. Then, $s \notin C_j$ for all $j \in J_1$, that is, $s_j > 0$ for all $j \in J_1$. Then,

$$0 = s^T z = \bar{s}^T z \quad \forall (z, s) \in B_\varepsilon(\bar{z}, \bar{s}), \text{ feasible to } (\tilde{P}). \tag{13.9}$$

Using (13.8)-(13.9), we conclude that $c^T z \leq c^T \bar{z}$ for all $(z, s) \in B_\varepsilon(\bar{z}, \bar{s})$, feasible to $(\tilde{P})$. Therefore, $(\bar{z}, \bar{s})$ is a local solution of $(\tilde{P})$. ■

# 3 ALGORITHMS

## 3.1 EQUILIBRIUM POINT ALGORITHM

In this subsection we describe an algorithm for finding an equilibrium point. Such a procedure will be used in the global algorithm.

Let $(\bar{z}, \bar{s})$ be an equilibrium point of the penalized problem. According to definition 7, $\bar{z}$ and $\bar{s}$ are respectively solutions of the following parametric linear problems for all $M \geq \tilde{M}$.

$$\tilde{P}(M; \bar{s}) \quad \begin{array}{ll} \max & F_M(z, \bar{s}) \\ \text{s.t.} & z \in \tilde{Z} \end{array} \qquad\qquad \tilde{P}(M; \bar{z}) \quad \begin{array}{ll} \max & F_M(\bar{z}, s) \\ \text{s.t.} & s \in S \end{array}$$

We can solve problems $\tilde{P}(M; \bar{s})$ and $\tilde{P}(M; \bar{z})$ considering $M$ implicitily as a large dominating value like in the big-M simplex method [4]. Particularly, the value of $M$ does not affect the solution point of $\tilde{P}(M; \bar{z})$.

We describe below a procedure to find an equilibrium point starting from a point $z^0 \in \tilde{Z}$. A similar algorihtm can be designed from a point $s^0 \in S$. We assume *a priori* that $\tilde{Z} \times S \neq \emptyset$.

**Algorithm 1** (Equilibrium Point Algorithm)

0. Let $z^0 \in \tilde{Z}$. Set $k = 0$.
1. Solve $\tilde{P}(M; z^k)$ using simplex method to obtain a solution $s^k$.
2. Attempt to solve $\tilde{P}(M; s^k)$ using the big-M simplex method. If the parametric problem is unbounded, conclude that $\tilde{P}(M)$ is inbounded for all $M \geq 0$. Otherwise obtain a solution $z^{k+1}$. If $F_M(z^k, s^k) = F_M(z^{k+1}, s^k)$, stop: $(z^k, s^k)$ is an equilibrium point.
3. Solve $\tilde{P}(M; z^{k+1})$ using simplex method to obtain a solution $s^{k+1}$. If $F_M(z^{k+1}, s^k) = F_M(z^{k+1}, s^{k+1})$, stop: $(z^{k+1}, s^k)$ is an equilibrium point. Otherwise set $k = k + 1$ and return to step 2.

We note that problems $\tilde{P}(M; z^k)$ and $\tilde{P}(M; s^k)$ cannot be infeasible because $S \times \tilde{Z} \neq \emptyset$. Also, $\tilde{P}(M; z^k)$ is not unbounded since $(z^k)^T s \geq 0$ for all $s \in S$. However, it is possible to have $P(M; s^k)$ unbounded for all $M \geq 0$ meaning that the parametric problem is unbounded.

**Proposition 10** *Algorithm 1 halts after a finite number of steps.*

*Proof:* Let be $(z^k, s^k)$ and $(z^{k+1}, s^{k+1})$ two consecutive points of the sequence generated by the algorithm. By step 2, $z^{k+1}$ solves $\tilde{P}(M; s^k)$. Thus,

$$(s^k)^T z^{k+1} \leq (s^k)^T z^k. \tag{13.10}$$

If the algorithm does not stop at step 3, we have that $F_M(z^{k+1}, s^{k+1}) \neq F_M(z^{k+1}, s^k)$ for some $M \geq 0$. As $s^{k+1}$ solves $\tilde{P}(M; z^{k+1})$, it must be $F_M(z^{k+1}, s^{k+1}) > F_M(z^{k+1}, s^k)$. So, it follows that

$$(s^{k+1})^T z^{k+1} < (s^k)^T z^{k+1}. \tag{13.11}$$

Therefore, by (13.11)-(13.10), we deduce that the points cannot be repeated. Since $\tilde{Z}_v \times S_v$ is a finite set, we conclude that the sequence $\{(z^k, s^k)\}$ is finite. ■

**Proposition 11** *Algorithm 1 finds an equilibrium point of the penalized problem or it verifies that $\tilde{P}(M)$ is unbounded for all $M \geq 0$.*

*Proof:* By proposition 10, the algorithm halts. Suppose, at first, that it stops at step 2, because $\tilde{P}(M; s^k)$ is unbounded for all $M \geq 0$. So, $\tilde{P}(M)$ is unbounded for all $M \geq 0$. Now suppose that the algorithm stops at step 2 due to $F_M(z^k, s^k) = F_M(z^{k+1}, s^k)$. Note that $s^k$ solves $\tilde{P}(M; z^k)$ and $z^{k+1}$ solves $\tilde{P}(M; s^k)$. Then, $F_M(z^k, s^k) \geq F_M(z^k, s)$ for all $s \in S$ and $F_M(z^k, s^k) = F_M(z^{k+1}, s^k) \geq F_M(z, s^k)$ for all $z \in \tilde{Z}$. Therefore, $(z^k, s^k)$ is an equilibrium point. Finally, the last stopping rule is $F_M(z^{k+1}, s^k) = F_M(z^{k+1}, s^{k+1})$ (step 3). Similarly to the previous case, we can conclude that $(z^{k+1}, s^k)$ is an equilibrium. ■

A simpler version of algorithm 1 can be used to find an equilibrium point of the penalized problem associated to $(P)$. In fact, by proposition 1, we deduce that $(s^k)^T z^{k+1} = 0$ for $k = 0$. Hence, algorithm 1 could stop at the initial iteration of step 2 returning $(z^1, s^0)$ as an equilibrium point or concluding that $P(M)$ is unbounded for all $M \geq 0$.

## 3.2 GLOBAL ALGORITHM

In this subsection we propose an iterative algorithm for finding an $\varepsilon$-global solution to problem (LBP). At iteration $k = 0$, we consider problem $(P)$. The procedure finds an equilibrium point $(z^0, s^0)$ of the penalized problem $P(M)$, which is a local solution of $(P)$. Then, a cut is added. At each iteration $k \geq 1$, we consider the problem $(\tilde{P})$ and the penalized problem $\tilde{P}(M)$, where $\tilde{Z} = Z \cap \{z \in \mathcal{R}^n : c^T z \geq c^T z^{k-1} + \varepsilon\}$. We note that this kind of cut is also used in [1, 11, 14, 15]. Then, a new equilibrium point $(z^k, s^k)$ is obtained. If $(z^k)^T s^k = 0$ the cut is updated. Otherwise we apply an outer approximation procedure [10, section IX.1.5] to the following bilinear problem:

$$(PC) \;\; \min s^T z \;\; \text{s.t.} \;\; (z, s) \in \tilde{Z} \times S.$$

Note that $(PC)$ cannot be unbounded. Moreover, if $(PC)$ is infeasible or its optimal value is nonzero, then $(\tilde{P})$ is infeasible. Otherwise, any optimal solution to $(PC)$ is feasible to $(\tilde{P})$.

We can stop the outer approximation procedure as long as it gets a point $(\hat{z}, \hat{s})$ such that $\hat{z}^T \hat{s} < \bar{z}^T \bar{s}$. Then, we restart the equilibrium point algorithm from $\hat{z}$ or $\hat{s}$. Actually, at each iteration of the outer approximation method we determine an upper bound for $(PC)$ given by a feasible point $(\hat{z}, \hat{s})$ (see [9, section 3.5.2]). If $\hat{z}^T \hat{s} < \bar{z}^T \bar{s}$, the procedure stops and returns $(\hat{z}, \hat{s})$. If this is never the case, the procedure finds out that $(\bar{z}^T \bar{s})$ is a solution to $(PC)$ and returns this point.

**Algorithm 2** (Global Algorithm)

1. If $S = \emptyset$, go to step 6. Otherwise, set $\tilde{Z} = Z$ and $k = 0$.
2. If $\tilde{Z} = \emptyset$, go to step 6. Otherwise, find $\hat{z} \in \tilde{Z}$.
3. Apply algorithm 1 from $\hat{z}$. If $\tilde{P}(M)$ is umbounded for all $M \geq 0$, stop: (LBP) is unbounded. Otherwise, let $(\bar{z}, \bar{s})$ be the returned equilibrium point.
4. If $\bar{s}^T \bar{z} = 0$, then $(\bar{z}, \bar{s})$ is a local solution to $(P)$; set $(z^\star, s^\star) = (\bar{z}, \bar{s})$; apply a linear cut defining $\tilde{Z} = Z \cap \{z \in \mathcal{R}^n : c^T z \geq c^T z^\star + \varepsilon\}$, with $\varepsilon > 0$; set $k = k + 1$ and return to step 2.
5. If $\bar{s}^T \bar{z} > 0$, apply the modified *outer approximation* procedure to obtain a point $(\hat{z}, \hat{s})$. If $\hat{s}^T \hat{z} \geq \bar{s}^T \bar{z}$, go to step 6. Otherwise, go to step 3.
6. Stop. If $k = 0$, (LBP) is infeasible. Otherwise, $z^\star$ is an $\varepsilon$-global solution to (LBP).

Let us make some remarks about algorithm 2. At iteration $k = 0$, phase I of the simplex method may be used to check whether $\tilde{Z}$ or $S$ are empty sets as well as to obtain the initial point $\hat{z} \in \tilde{Z}$. At iteration $k \geq 1$, it is better to apply the dual-simplex method at step 2. Finally, we note that it is more interesting to use a version of algorithm 1 that starts from a point $\hat{s} \in S$ imediately after returning from step 5 to step 3.

**Proposition 12** *The global algorithm halts after a finite number of steps.*

*Proof:* Suppose, by contradiction, that an infinite loop occurs. Since each new cut discards at least one more vertex of $Z$, we deduce that step 4 is no longer executed after some iteration. Then, the loop runs between steps 3 and 5. So, an infinite sequence of vertices of $\tilde{Z} \times S$ is generated by step 3. Thus, some points are repeated.

Let $(\bar{z}, \bar{s})$ and $(\hat{z}, \hat{s})$ in $\tilde{Z} \times S$ be the equilibrium point generated by step 3 and the point returned by the outer approximation at a certain iteration of the infinite loop. At step 5, $\hat{s}^T\hat{z} < \bar{s}^T\bar{z}$ and algorithm 1 is applied from $z^0 = \hat{z}$. Then, the point $s^0$, obtained by step 1 of algorithm 1, satisfies $(s^0)^T\hat{z} \leq \hat{s}^T\hat{z}$. Let $(\bar{z}', \bar{s}')$ be the next equilibrium point generated. By (13.10)-(13.11) we conclude that $(\bar{s}')^T\bar{z}' \leq (s^0)^T\hat{z}$. By the last three inequalities it follows that the sequence cannot have repeated points, which is a contradiction. ■

**Proposition 13** *The global algorithm finds an $\varepsilon$-global solution to (LBP) or it verifies that the problem is unbounded or infeasible.*

*Proof:* By proposition 12, the algorithm stops. According to proposition 11, the stopping rule of step 3 guarantees that $\tilde{P}(M)$ is unbounded for all $M \geq 0$ and so is $P(M)$. Then, by theorem 5, $(P)$ is unbounded. If the algorithm halts at step 6 with $k = 0$, we clearly have that $(P)$ is infeasible. In these two cases, the same conclusions hold for (LPB) due to its equivalence with $(P)$. Finally, the algorithm may stop when there is no more feasible points to $(P)$ in the set $\tilde{Z} \times S$, where $\tilde{Z} = Z \cap \{z \in \mathcal{R}^n : c^T z \geq c^T z^\star + \varepsilon\}$. This case means that $z^\star$ is an $\varepsilon$-global solution to (LBP). ■

## 3.3 COMPUTATIONAL RESULTS

The global algorithm was implemented in $C$ and was run on a Pentium III, 600MHz. The value of $\varepsilon$ was fixed at 0.013. Computational results for some random test problems are presented in the following.

The test problems were obtained with a method developed in [5]. The method allow us to control the density and the conditioning number of the constraint matrix and the number of local and global optima of the problem. Actually, it first generates a separable linear bilevel problem with the desired dimensions and number of optima. Then, it disguises the separability with a matrix transformation $HD^{-1}H$ while keeping the number of optima. The invertible diagonal matrix $D$ and the block-diagonal Householder matrix $H$ are randomly generated so that $10^\delta$ is the conditioning number of $D$ and $\tau$ is the percentual of nonzero entries in the basis vector of $H$, where $\delta$ and $\tau$ are given parameters (see [5] for details). Following suggestions of the authors, we introduced some modifications in the method to reduce the number of constraints. We also added nonnegativeness constraints for variables.

Table 13.1 presents computational results for 60 test problems. Each number in the last two columns is the average elapsed time in seconds for

*Table 13.1* Computational results for random test problems

| $n_1 + n_2$ | $n_2$ | $m$ | $\tau = 0.1$ | $\tau = 0.8$ |
|---|---|---|---|---|
| 50 | 12 | 15,17,19 | 0.13 | 0.04 |
| 50 | 38 | 15,17,19 | 4.36 | 1.87 |
| 60 | 15 | 18,20,22 | 0.26 | 0.02 |
| 60 | 45 | 18,20,22 | 31.80 | 10.51 |
| 70 | 17 | 20,22,24 | 0.40 | < 0.01 |
| 70 | 53 | 20,22,24 | 37.70 | 10.49* |
| 80 | 20 | 23,25,27 | 0.59 | 0.09 |
| 80 | 60 | 23,25,27 | 19.56 | 18.11 |
| 90 | 22 | 25,27,29 | 1.19 | 1.96 |
| 90 | 68 | 25,27,29 | 45.11 | 23.80* |

three problems with $\tau = 0.1$ or $\tau = 0.8$. The $i$-th problem ($i = 1, 2, 3$) is related to the $i$-th number in column $m$. It has $2^i$ local solutions (one of them is a global one) before considering $x \geq 0$ and $y \geq 0$. We used $\delta = 1$.

These results show that the algorithm had a good behavior in general. It was slightly affected by the density of the constraint matrix. On the other hand, we can observe that times increase very much when more variables are assigned to the second level. We remark that (*) indicates that one of the problems could not be solved in a limit time of 900s. This difficulty is due to the outer approximation procedure. For a future work it is interesting to consider alternatives to the outer approximation procedure.

## Acknowledgments

We would like to thank CAPES and CNPq grant 302393/85-4(RN).

## References

[1] M. Amouzegar and K. Moshirvazari. A penalty method for linear bilevel programming problems. Nonconvex Optimization and its Applications 20: 251–271, Kluwer, 1998.

[2] J. Bard. An efficient point algorithm for a linear two-stage optimization problem. *Operations Research*, 31:670–684, 1983.

[3] J. Bard. An investigation of the linear three level programming problem. *IEEE Trans. Systems, Man and Cybernetics*, 14:711–717, 1984.

[4] M. Bazaraa, J. Jarvis, and H. Sherali. *Linear programming and network flows.* Wiley, New York, 2nd edition, 1990.

[5] P. Calamai and L. Vicente. Generating linear and linear-quadratic bilevel programming problems. *SIAM J. Scientific and Stat. Computing*, 14:770–782, 1993.

[6] M. Campêlo, S. Dantas, and S. Scheimberg. A note on a penalty function approach for solving bilevel linear programs. *J. Global Optimization*, 16:245–255, 2000.

[7] M. Campêlo and S. Scheimberg. Characterizing local and global solutions of a linear bilevel problem. Technical Report ES-515/99, Univ. Federal do Rio de Janeiro, COPPE/Engenharia de Sistemas e Computação, 1999.

[8] P. Hansen, B. Jaumard, and G. Savard. New branch-and-bound rules for linear bilevel programming. *SIAM J. Scientific and Stat. Computing*, 13:1194–1217, 1992.

[9] R. Horst, P. Pardalos, and N. Thoai. *Introduction to global optimization*, volume 3 of *Nonconvex Optimization and Its Aplications.* Kluwer, 1995.

[10] R. Horst and H. Tuy. *Global optimization: deterministic approaches.* Spring-Verlag, Berlin, 1990.

[11] J. Júdice and A. Faustino. A sequential LCP method for bilevel linear programming. *Annals Operations Research*, 34:89–106, 1992.

[12] H. Konno. A cutting plane algorithm for solving bilinear programs. *Math. Programming*, 11:14–27, 1976.

[13] A. Migdalas, P. Pardalos, and P. Värbrand, editors. *Multilevel Optimization: algorithms and applications*, volume 20 of *Nonconvex Optimization and Its Aplications.* Kluwer, 1998.

[14] H. Önal. A modified simplex approach for solving bilevel linear programming problems. *Eur. J. Operational Research*, 67:126–135, 1993.

[15] H. Tuy, A. Migdalas, and P. Värbrand. A global optimization approach for the linear two-level program. *J. Global Optimization*, 3:1–23, 1993.

[16] L. Vicente and P. Calamai. Bilevel and multilevel programming: a bibliography review. *J. Global Optimization*, 5:291–306, 1994.

[17] D. White and G. Anandalingam. A penalty function approach for solving bi-level linear programs. *J. Global Optimization*, 3:397–419, 1993.

Chapter 15

# THE LAGRANGIAN SEARCH METHOD

Pavlos S. Efraimidis*
*Computer Technology Institute*
*Comp. Eng. and Informatics Dept., Univ. of Patras*
efraimid@cti.gr

Paul G. Spirakis
*Computer Technology Institute*
*Comp. Eng. and Informatics Dept., Univ. of Patras*
spirakis@cti.gr

**Abstract** In this work, the Lagrangian Search Method (LSM), an algorithmic framework based on lagrangian relaxation, relaxed decision procedures and binary search, is presented. LSM is used to efficiently extend the class of positive linear programs (PLP), a submodel of linear programming that admits very efficient approximations in sequential, parallel and distributed settings. Theoretical guarantees for the efficiency and the approximation ratio of LSM for extended PLP, are derived. The extended PLP model is used for fast parallel (and sequential) approximations to a class of hard combinatorial optimization problems.

**Keywords:** Lagrangian relaxation, Approximate linear programming, Parallel algorithms, Combinatorial optimization.

## 1 INTRODUCTION

Positive linear programming (PLP) is the class of linear programs in packing or covering form where all coefficients of matrix A and vectors b and c are non-negative:

*Financial support from the Bodosaki Foundation to perform doctoral studies is gratefully announced. Bodosaki Foundation, Leoforos Amalias 20, 10557 Athina, Greece



*N. Hadjisavvas and P.M. Pardalos (eds.), Advances in Convex Analysis and Global Optimization*, 283–294.

| PLP in packing form | PLP in covering form |
|---|---|
| $\max \sum_{i=1}^{m} c_j x_j$ subject to | $\min \sum_{j=1}^{n} b_i y_i$ subject to |
| $\forall i: \sum_{j=1}^{n} a_{ij} x_j \leq b_i$ | $\forall j: \sum_{i=1}^{m} a_{ij} y_i \geq c_j$ |
| $\forall j: x_j \geq 0$ | $\forall i: y_i \geq 0$ |

*Table 15.1* The Positive Linear Programming (PLP) model. All entries of matrix A and vectors b and c are non-negative. The two forms of PLP correspond to the primal and the dual of the same problem.

PLP admits very efficient approximations in sequential ([7]), parallel ([7]) and distributed settings ([1]). However the syntax of PLP is rather weak and hence only a very limited number of combinatorial optimization problems (including matching in bipartite graphs and set covering, [7]) admit native PLP formulations. We present the Lagrangian Search Method (LSM), an algorithmic framework that is based on lagrangian relaxation, relaxed decision procedures and binary search and use it to extend PLP in a natural way so that the extended PLP model

- can model a larger number of combinatorial optimization problems, and
- at the same time it admits efficient sequential and parallel approximations.

## 2 EXTENDED PLP

Extended PLP (ePLP) are positive linear programs (PLP) with a number of PLP-Violations that are equality constraints, covering constraints for PLP in packing form or the opposite, and variables with negative coefficients. While PLP admits efficient parallel approximations, even simple extensions of PLP with one equality constraint are hard to parallelize ([3]). We show, that if a bounded violation of certain constraints is admitted, then efficient sequential and parallel approximations can be obtained for the decision version of an ePLP. The decision problem is then used within a two-level binary search procedure to approximately solve the optimization version of the problem.

Important concepts for handling ePLP are relaxed constraints and relaxed decision procedures (RDP). An $\epsilon$-relaxed equality constraint is an equality constraint that has been relaxed to a pair of one packing and one covering constraint that allow at most relative gap $\epsilon$ between them. An $\epsilon$-relaxed packing (covering) constraint is a packing (covering) constraint that can violate the right-hand side by at most a factor $\epsilon$.

| original constraint | $\epsilon$-relaxed constraint |
|---|---|
| $\sum_j a_{i,j}x_j = b_i$ | $\sum_j a_{i,j}x_j \leq b_i$ *and* $\sum_j a_{i,j}x_j \geq b_i(1-\epsilon)$ |
| $\sum_j a_{i,j}x_j \leq b_i$ | $\sum_j a_{i,j}x_j \leq b_i(1+\epsilon)$ |
| $\sum_j a_{i,j}x_j \geq b_i$ | $\sum_j a_{i,j}x_j \geq b_i(1-\epsilon)$ |

An $\epsilon$-relaxed solution to a ePLP is a solution that is feasible when specific problem constraints of the ePLP are $\epsilon$-relaxed. An $\epsilon$-relaxed $(1-\rho)$-approximate solution to a ePLP is an $\epsilon$-relaxed solution of objective value at least $1-\rho$ times the optimum of the (non-relaxed) ePLP.

An $\epsilon$-relaxed decision procedure (RDP) ([8]) for a problem P, is an algorithm that, given an instance of P and a constant $\epsilon > 0$ either returns a $\epsilon$-relaxed solution to P or decides that no exact solution exists.

# 3 THE LAGRANGIAN SEARCH METHOD

The Lagrangian Search Method (LSM), uses the general strategy of lagrangian relaxation, that is to relax specific problem constraints by bringing them into the objective function of the linear program. The main idea is given an ePLP to build a corresponding PLP, and then to approximate the PLP with the approximation algorithm LN of [7]. However certain issues arising from the limitations of the PLP model and the fact that PLP is only approximately solved, have to be circumvented.

First LSM transforms the ePLP at hand to a PLP in the following way:

1. All violations are transformed to appropriate combinations of packing and equality constraints so that the only kind of violations that remain are equality constraints.

2. The equality constraints are relaxed to packing constraints and a corresponding term is added to the objective function.

It is easy to show that the resulting PLP is equivalent to the original ePLP problem, if both problems are solved optimally. However a simple approximation to the PLP does not always imply an equivalent approximation to the original ePLP, because the existence of an approximation error in the PLP admits individual terms of the objective function of the PLP to decline significantly from their respective optimal values.

The novel approach of LSM to the above shortcoming is the addition of Z, a new parameter, to the PLP. The modified PLP is now called PLP(Z) and instead of solving the optimization version of ePLP, PLP(Z) aims at solving the decision version of ePLP. The role of parameter Z is critical:

- Z is an estimation of the optimal objective value of the ePLP.

- A constraint assures that in PLP(Z) the value of the original objective function of ePLP does not exceed Z.

- Knowing upper bounds for all terms of the objective function, permits the algorithm to force equidistribution of the objective function to all its terms.

It will be shown that approximating PLP(Z) corresponds to solving a relaxed decision procedure for the original ePLP problem with the condition that specific problem constraints of ePLP can be violated by at most an arbitrary small constant factor.

The complexity of the LSM method depends on the number of PLP-Violations. For extended PLP problems with up to a $O(polylogN)$ number of PLP-Violations the complexity of the overall methodology does not significantly increase compared to the complexity of the LN algorithm.

The LSM methodology is illustrated on ePLP with equality constraints, but it can easily be adapted to any kind and any combination of at most polylog PLP-Violations. For more details the reader is referred to [2]. Given an ePLP with a $k_e$ equality constraints,

$$\begin{array}{ll} \max & c^T x \\ \text{s. t.} & A_p x \leq b_p \\ & A_e x = b_e \\ & x \geq 0, \end{array}$$

first the linear program PLP(Z), a normal PLP, is constructed.

$$PLP(Z) \left\{ \begin{array}{ll} \max & c^T x + w_e A_e x \\ \text{s. t.} & A_p x \leq b_p \\ & A_e x \leq b_e \\ & c^T x \leq Z \\ & x \geq 0 \text{ and } w_e^i = \frac{Z}{k_e b_e^i} \end{array} \right.$$

In PLP(Z) all the equality constraints of the ePLP problem are replaced by packing constraints. For each equality constraint, that is replaced, a lagrangian-like term is introduced into the objective function of the PLP. This term forces the packing constraint, that corresponds to the equality constraint, to become tight or almost tight in the solution of PLP(Z). The coefficients $w_e^i$ of the lagrangian-like terms are chosen so that the respective terms have the same weight in the problem formulation.

Being a valid PLP admits PLP(Z) to be approximated efficiently with appropriate algorithms, like LN of [7]. The following results show that LSM achieves an efficient sequential and parallel RDP for the relaxed ePLP problem.

**Theorem 1**

**(A):** *If the value of Z is feasible for the original ePLP problem then the approximation to PLP(Z) has objective value at least* $2Z(1-\epsilon)$ *(for PLP(Z)).*

**Proof:** *For simplicity, the slightly inaccurate equalities* $(1+\epsilon_1)(1+\epsilon_2) = (1+\epsilon_1+\epsilon_2)$ *and* $(1/(1+\epsilon)) = (1-\epsilon)$ *are used. This is done only a constant number of times.*

*It is sufficient to show that any feasible solution of the ePLP of objective value Z corresponds to a feasible solution of PLP(Z) with objective value 2Z. Let* $x$ *be a feasible solution to the (exact) ePLP problem of objective value* $c^T x = Z$. *Then solution* $x$ *satisfies the following constraints:*

$$A_p x \leq b_p$$
$$A_e x = b_e$$
$$c^T x = Z$$

*Hence it satisfies all the constraints of PLP(Z). So the vector* $x$ *corresponds to a feasible solution for PLP(Z). The objective value of this feasible solution is:* $(c^T x + w_e A_e x) = Z + Z = 2Z$.

*Now, since the objective value of PLP(Z) is at least 2Z, the approximation algorithm with approximation ratio* $1+\epsilon$ *will return a solution for PLP(Z) of objective value at least* $2Z(1-\epsilon)$. ■

**(B):** *The solution of (A) implies a relaxed approximate solution to ePLP of objective value at least* $Z(1-2\epsilon)$. *The value of the lagrangian like terms is also at least* $Z(1-2\epsilon)$.

**Proof:** *Since Z is feasible for the ePLP, the optimal objective of the corresponding PLP(Z) is 2Z (from (A)). Hence the* $(1+\epsilon_1)$-*approximate solution to PLP(Z) will have objective value at least* $(1-\epsilon_1)2Z$. *In the objective function of PLP(Z), let* $Z_1 = c^T x$ *be the value of the original objective function of the ePLP and* $Z_2 = w_e A_e x$ *be the sum of the lagrangian-terms. Then* $Z = Z_1 + Z_2 \geq (1-\epsilon)2Z$, $0 \leq Z_1 \leq Z$, *and* $0 \leq Z_2 \leq Z$. *This gives* $Z(1-2\epsilon) \leq Z_1 \leq Z$ *and* $Z(1-2\epsilon) \leq Z_2 \leq Z$. ■

**(C):** *Approximating PLP(Z) gives a RDP for the $\epsilon$-relaxed ePLP.*

**Proof:** *From (A) and (B) it is clear that if Z is feasible then the $(1+\epsilon)$-approximate solution x to PLP(Z) satisfies:*

$$Z(1-2\epsilon) \leq c^T x \leq Z$$
$$Z(1-2\epsilon) \leq w_e A_e x \leq Z$$

*The bound on $c^T x$ shows that the objective value of ePLP for x is $(1+2\epsilon)$-approximation and the bound on $w_e A_e x$ that the relaxed constraints are violated by at most a factor $(1+2k_e\epsilon)$. Hence x is a relaxed approximate solution to ePLP. Hence approximating PLP(Z) gives a RDP for the $\epsilon$-relaxed ePLP.* ■

**(D):** *The complexity of approximating PLP(Z) is O(polylog) on a linear number of processors.*

**Proof:** *The complexity for approximating PLP(Z) is determined by the complexity of algorithm LN of [7], which runs in $O(log(nm)log(n)log(m/\epsilon_1)/(\epsilon_1)^4)$ on $O(N)$ processors, where n is the number of variables, m the number of constraints, and N is the number of non-zero elements of the matrix.*

*Note that since algorithm LN is used to solve PLP(Z) and not the ePLP, the actual size of the problem is the size of the PLP. However, the difference is within a constant factor and hence it can be assumed that the parameters n, m and N refer to the ePLP.*
■

We have shown how to construct a positive linear program PLP(Z) form the extended problem ePLP and than to approximately solve the decision procedure for a specific value of parameter Z. The decision procedure will be used to build an approximation algorithm for the extended problem ePLP.

## 3.1 SEARCHING WITH DECISION PROBLEMS

A relaxed decision procedure RDP for a CO problem Π can be used within a binary search to produce an approximation algorithm for Π. The running time of the whole scheme depends on the complexity of the RDP and the number of binary search steps. In sequential algorithms the reduction of a search or optimization problem to its decision version is called self-reducibility. In parallel algorithms where the same reduction is called parallel self-reducibility, it does not always produce an efficient solution ([5]).

We examine the concept of self-reducibility for efficient sequential and parallel approximation algorithms, and show that for the problems of this work, efficient relaxed decision procedures can be used for efficient approximations in both sequential and parallel settings. For the applications In LSM, the relaxed decision problem is repeatedly solved within a binary search procedure until the largest possible value of parameter Z is found. Due to an interesting property of the combinatorial optimization problems of this work, called the poly-bottleneck property, the number of binary search steps is at most poly-logarithmic.

**Definition 1** *We say that a combinatorial optimization problem has the poly-bottleneck property if its optimal objective value is always within a polynomial factor of one of its input weights.*

The poly-bottleneck property holds for all problems considered in this work and interestingly, it seems to be valid for a large number of combinatorial optimization problems including the k-median, the traveling salesman and many others. For each case the proof is usually a simple combinatorial argument.

Given a hard problem with a relaxed decision procedure and the poly-bottleneck property, a 2-level binary search procedure can find an approximate solution in at most a poly-logarithmic number of steps. At each step LSM is used to solve a relaxed decision procedure for the problem and in this way to decide if the current value of parameter Z is feasible for the original ePLP. By exploiting the poly-bottleneck property, in the first level the binary search procedure detects in $O(\log n)$ steps a weight such that the optimal value is within a polynomial factor of that weight. At the second level, the procedure approximates the optimal objective value in $O(polylog(n/\epsilon))$ steps in the interval defined by the selected weight and the given polynomial factor of the poly-bottleneck property. At the end the binary search selects the largest value of Z that is feasible for the relaxed problem.

**Input:** A poly-bottleneck ePLP problem, the polynomial factor $f(n)$ given by the poly-bottleneck property, and constants $\epsilon > 0$ for the relaxation of certain constraints and $\rho > 0$ for the approximation ratio.

**Output:** An $\epsilon$-relaxed $(1+\rho)$-approximate solution to ePLP or ePLP is infeasible.

**Step 1:** The positive linear program PLP(Z) is constructed from the ePLP.

**Step 2:** Let $w_1, w_2, \ldots, w_n$ be the input weights of the problem specification. Sort the values $w_1, w_1 f(n), w_2, w_2 f(n), \ldots w_n f(n)$

**Step 3:** Let $l_1 > l_2 > \ldots > l_{n'}$ be the sorted sequence of the input weights, where $n' \leq 2n$ (equal weight values are discarded).

**Step 4:** In $O(\log(2n))$ indexed binary search steps find an $i$ such that $l_i$ is feasible and $l_{i+1}$ is infeasible for ePLP.

**Step 5:** In $O(\log(f(n)/\epsilon))$ binary search steps find the largest value $Z^*$ in $[l_i, l_{i+1})$ such that $Z^*$ is feasible for ePLP.

At each binary search step (of lines 3 and 4) $PLP(Z)$ is solved for a value $Z$. The solution for the $PLP(Z^*)$ gives an $\epsilon$-relaxed $\rho$-approximate solution to ePLP. If there is no feasible Z then ePLP is infeasible. The above above algorithm proves that:

**Theorem 2** *Given a poly-bottleneck problem $\Pi$ and a RDP for $\Pi$, a two-level binary search procedure that calls the RDP at each step, finds in $O(polylog(n/\epsilon))$ steps a (relaxed) approximate solution to $\Pi$.*

# 4 APPLICATIONS

The PLP extensions and the LSM algorithm are used for efficient sequential and parallel approximations to a number of hard combinatorial optimization problems. The problems are Global Routing in Gate Arrays, Scheduling Unrelated Parallel Machines, Generalized Assignment, and extensions of Simple k-Matching in Hypergraphs. The algorithms run in poly-logarithmic time on a polynomial number of processors and achieve logarithmic or poly-logarithmic approximation guarantees. These are, to our knowledge, the best known parallel approximation results for the corresponding problems.

## 4.1 GLOBAL ROUTING IN GATE ARRAYS

Global Routing in Gate Arrays (GRGA) is a routing problem that arises in the design of a certain class of VLSI circuits. Raghavan and Thompson showed that it is NP-Complete and provided a sequential randomized approximation algorithm based on Linear Programming (LP) and Randomized Rounding ([10]). In this section, the ePLP model is used for an efficient parallel approximation algorithm to GRGA. Nets $\alpha_i$ have each to be assigned to a Steiner tree $b_{ij}$ with the objective to minimize the width of the solution, which is the maximum usage of any edge. In the integer program formulation of GRGA the binary variables $x_{ij}$ denote whether or not the tree $b_{ij}$ is chosen for net $\alpha_i$. More details on GRGA can be found in [10].

$$\min W \; s.t. \begin{cases} \forall i, \sum_j x_{i,j} = 1 \\ \forall e, \sum_{b_{i,j} \ni e} x_{i,j} \leq W \\ \forall i,j : x_{ij} \in \{0,1\} \end{cases}$$

Trivial upper and lower bounds on the optimal objective value $W^*$ of GRGA are $U = n$ and $L = 1$ and hence GRGA is trivially a poly-bottleneck problem. Relaxing the integrality constraints on variables $x_{ij}$ to $x_{ij} \geq 0$ gives the LP relaxation for GRGA. The linear program LP–GRGA is an ePLP, but with a large number of PLP-Violations: the n equality constraints. LSM cannot directly solve this problem efficiently since it can satisfy at most a polylog number of relaxed equalities. To handle this case the n equality constraints are treated all together as one violation: they are replaced by $n$ packing constraints and 1 equality constraint. This gives the linear program ePLP-GRGA:

$$\min W \; s.t. \begin{cases} \sum_i \sum_j x_{i,j} = n \\ \forall i, \sum_j x_{i,j} \leq 1 \\ \forall e, \sum_{b_{i,j} \ni e} x_{i,j} \leq W \\ \forall i,j : x_{ij} \in [0,1] \end{cases}$$

The corresponding PLP(Z) for LSM is:

$$\max \sum_i \sum_j x_{ij} \; s.t. \begin{cases} \forall i, \sum_j x_{i,j} \leq 1 \\ \forall e, \sum_{b_{i,j} \ni e} x_{i,j} \leq Z \\ \forall i,j : x_{ij} \in [0,1] \end{cases}$$

PLP(Z) is a valid PLP and can be approximated with algorithm LN of [7] for any fixed parameter $\epsilon > 0$. If $Z$ is a feasible value for the objective value $W$ of GRGA, i.e. $Z$ is not less than the minimum objective value $W^*$,then algorithm LN returns a relaxed approximate vector $x$:

$$\begin{array}{l} \sum_i \sum_j x_{ij} \geq n(1-\epsilon) \\ \forall i, \sum_j x_{ij} \leq 1 \\ \forall e, \sum_{b_{ij} \ni e} x_{ij} \leq Z \\ \forall i,j : x_{ij} \in [0,1] \end{array}$$

In the above solution, at least half of the nodes $i$ are covered by $\sum_j x_{ij} \geq 1 - 2\epsilon$, because else $\sum_i \sum_j x_{ij} < n(1-\epsilon)$ a contradiction. Each vector $i$ of the the solution $x_{ij}$ is scaled by $\min\{1+2\epsilon, 1/(1-\sum_i x_{ij})\}$. Now the width of the solution is upper bounded by $Z(1+\epsilon)(1+2\epsilon)$ and at least the half nodes $i$ are totally covered ($\sum_j x_{ij} = 1$).

**Satisfying all equality constraints.** The solution found so far covers at least a constant fraction $\frac{1}{2}$ of the nets. This fraction can be increased to any constant less than 1 by using an appropriate approximation ratio for PLP(Z). However, in order to satisfy all nodes the above procedure is repeated (each time on the remaining nets) until they are all covered, and all partial solution vectors are summed up to give the final solution. By covering every time at least the half nodes, at most $\lceil \log n + 1 \rceil$ repetitions are needed to cover all nodes. The width of the final fractional solution is bounded by $W' = Z(1 + 2\epsilon)\lceil 1 + \log n \rceil$.

**Binary Search.** A two-level binary search procedure selects after a total number of at most $O(\log(n/\epsilon))$ steps the smallest $Z^*$ such that algorithm LSM gives a fractional solution that covers all nodes and has width at most $W'$. At each binary search step either a fractional solution (covering all nodes) is calculated for a value $Z$, or $Z$ is considered infeasible, that is the fractional solution has width larger than $W'$. Let $Z_1$ be the smallest feasible $Z$ found, then $Z_1 \leq W^*(1 + \epsilon)$. The corresponding $W$ found for PLP$(Z_1)$ satisfies $W \leq W'' = W'(1+\epsilon) = W^*(1+3\epsilon)\lceil 1+\log n \rceil$.

**Rounding.** The best fractional solution is selected and rounded randomly to an integer solution, in the following way: For each equality constraint at most one $x_{ij}$ is randomly chosen and set to 1. All the other variables of the equality constraint are set to 0. The probability for each variable $x_{ij}$ to get the value 1 is equal to its fractional value $x_{ij}$. This is the (exclusive) randomized rounding procedure of [10].

$$x_i = \begin{cases} 0, & \text{with probability } 1 - x_i \\ 1, & \text{with probability } x_i \end{cases}$$

The performance guarantee on the rounded solution is given by an adaptation of [10, Theorem 2.1].

**Theorem 3** *([10]). Let $\delta$ be an appropriate positive real such that $0 < \delta < 1$. Provided that $W \geq 3\ln(2n(n-1)/\delta)$, the width of the solution produced by the above procedure does not exceed $W'' + \left(3W'' \ln \frac{2n(n-1)}{\delta}\right)^{1/2}$ with probability at least $1 - \delta$, for $W'' = W^*(1 + 3\epsilon)\lceil 1 + \log n \rceil$.*

The complexity of the complete algorithm is within a *polylog*$(n/\epsilon)$ factor of the complexity of algorithm LN and hence it runs in *polylog*$(n/\epsilon)$ time on $O(N)$ processors, where $N$ is the number of non-zero elements of the problem description.

## 4.2 SCHEDULING UNRELATED PARALLEL MACHINES

Scheduling $n$ independent jobs on $m$ Unrelated Parallel Machines SCHED is the problem of assigning $n$ jobs $j = \{1, \ldots, n\}$ to $m$ machines $i = \{1, \ldots, m\}$ so that each job is processed without interruption on one of the machines, and at any time, every machine processes at most one job. The processing time for job $j$ on machine $i$ is $p_{ij}$. For each schedule, the makespan is the maximum load on any machine. The objective is to find a schedule of minimum makespan. The integer program formulation is:

$$\min T \ s.t. \left\{ \begin{array}{l} \forall j, \sum_i x_{ij} = 1 \\ \forall i, \sum_j p_{ij} x_{ij} \leq T \\ \forall i, j : x_{ij} \in \{0, 1\} \end{array} \right.$$

If all coefficients $p_{ij}$ in SCHED are equal to 1 then the problem is equivalent to problem GRGA. An algorithm for SCHED similar to the algorithm for GRGA can be obtained by applying the techniques of the previous section.

## 4.3 THE GENERALIZED ASSIGNMENT PROBLEM

The Generalized Assignment Problem (GAP) is also known as the generalization of SCHED to jobs with costs. Its definition is the same as for SCHED, but in GAP the assignment of each job $j$ to a machine $i$ has, besides the processing time $p_{ij}$, a cost $c_{ij}$, and the objective is to find a schedule of minimum cost $C$ and makespan at most a given value $T$. LSM and ePLP can be used to give an approximation algorithm for GAP.

## 4.4 SIMPLE K-MATCHING IN HYPERGRAPHS

A hypergraph is a finite collection of non-empty finite sets called edges. The elements of edges are called vertices. The set of vertices of a hypergraph H is denoted by V(H). The degree of $x \in V(H)$ is the number of edges containing x. A k-matching of H is a collection of edges such that each point belongs to at most k of them. A k-matching is simple if no edge occurs in it more then once. Matching in 3-uniform (all edges have cardinality 3) hypergraphs is NP-hard [4] and hence the general problem of SKMH is also NP-hard.

LSM can be used for fast sequential and parallel approximations to SKMH and to the generalization of SKMH that admits a small number of additional linear constraints. An extended positive linear program in covering form can be used to model SKMH. The additional constraints of SKMH can be handled as PLP-Violations. The overall methodology is very similar to the approach in the previous applications.

## References


[1] Bartal, Y., Byers, J., and Raz, D. (1997). Global optimization using local information with applications to flow control. In *38th IEEE FOCS*, pages 303–312.

[2] Efraimidis, P. and Spirakis, P. (1999). Fast parallel approximations to positive linear programs with a small number of constraint violations. Technical Report TR99.04.02, Computer Technology Institute.

[3] Efraimidis, P. and Spirakis, P. (2000). Positive linear programming extensions: Complexity and applications. In *European Conference on Parallel Computing (Euro-PAR 2000)*, Technical Univeristy of Muenich, Germany.

[4] Garey, M. and Johnson, D. (1979). *Computers and Intractability: A Guide to the Theory of NP-Completeness.* W.H. Freeman and Company.

[5] Greenlaw, R., Hoover, H. J., and Ruzzo, W. L. (1995). *Limits to Parallel Computation: P-Completeness Theory.* Oxford University Press.

[6] Lovász, L. (1975). On the ratio of optimal and fractional covers. *Discrete Mathematics*, 13:383–390.

[7] Luby, M. and Nisan, N. (1993). A parallel approximation algorithm for positive linear programming. In *25th ACM Symp. on Theory of Computing*, pages 448–457.

[8] Plotkin, S., Shmoys, D., and Tardos, E. (1995). Fast approximation algorithms for fractional packing and covering problems. *Mathematics of Operations Research*, 20:257–301.

[9] Raghavan, P. (1986). *Randomized Rounding and Discrete Ham-Sandwich Theorems: Provably Good Algorithms for Routing and Packing Problems.* PhD thesis, Computer Science Division, UC Berkeley.

[10] Raghavan, P. and Thompson, C. (1987). Randomized rounding: A technique for provably good algorithms and algorithmic proofs. *Combinatorica*, 7:365–374.

Chapter 16

# AN $\varepsilon$-MAXIMUM PRINCIPLE FOR GENERALIZED CONTROL SYSTEMS

Andreas H. Hamel
*Department of Mathematics and Computer Science*
*University Halle-Wittenberg, D-06099 Halle.*
hamel@mathematik.uni-halle.de

**Abstract** A suboptimality theorem, i.e. an $\varepsilon$-optimality condition for generalized control systems is proven which also applies in cases if there is no optimal solution. We apply the result to obtain the first suboptimality theorem for differential inclusion problems.

**Keywords:** Differential inclusion, Near-optimal control.

## 1 INTRODUCTION

There are easy-to-construct-examples of problems without any classical solution in optimal control theory, i. e. there is no measurable control function which gives the infimal value of the cost functional. Therefore, it might be reasonable to look for a substitute for necessary conditions which always applies if the problem has admissible solutions and, in case of minimization, is bounded from below. We call a theorem about the existence of an $\varepsilon$-optimal solution satisfying $\varepsilon$-optimality conditions of first order a *suboptimality theorem.*

The first suboptimality theorem in optimal control theory is theorem 7.1 in [2] for problems with free end points and smooth data. This results has been extended to optimal control problems in parametrized form with general end point constraints under assumption used by Clarke [1], see [3].

In this paper we consider generalized control systems introduced by Kaskosz and Lojasiewicz in [5]. The main result is a suboptimality



*N. Hadjisavvas and P.M. Pardalos (eds.), Advances in Convex Analysis and Global Optimization,* 295–301.

theorem for this type of problems. As an important application we obtain the first suboptimality theorem for differential inclusion problems.

After introducing some notation we state and prove the main result. The last section is devoted to differential inclusion problems.

Let us denote by $|\cdot|$ the Euclidean norm and by $B$ the corresponding closed unit ball in any $\mathbb{R}^k$, $k$ a natural number. The symbol $\chi_M(\cdot)$ is used for the indicator function of a set $M$.

## 2 MAIN THEOREM

We consider the following problem (P):

$$\begin{array}{ll} \text{minimize} & l\left(x\left(0\right),x\left(1\right)\right) \\ \text{subject to} & \left\{ \begin{array}{l} \dot{x}\left(t\right)=f\left(t,x\left(t\right)\right),\ f\in S \\ \left(x\left(0\right),x\left(1\right)\right)\in C \end{array} \right. \end{array}$$

The family $S$ of vector fields is a collection of right hand sides for the differential equation above. We impose the following hypotheses.
(H1) The function $l:\mathbb{R}^n\times\mathbb{R}^n\to\mathbb{R}$ is locally Lipschitz.
(H2) The set $C\subset\mathbb{R}^n\times\mathbb{R}^n$ is closed.
(H3) If $f_i\in S$, $T_i\subset[0,1]$ Lesbegue-measurable, $i=1,2,\ldots$, $T_i\cap T_j=\emptyset$ for $i\neq j$, $\cup_{i=1}^{\infty}T_i=[0,1]$ then

$$\sum_{i=1}^{\infty}\chi_{T_i}\left(t\right)f_i\left(t,x\right)\in S.$$

(H4) Any $f\in S$ is Carathéodory; for each $f\in S$ and for each bounded set $A\subset\mathbb{R}^n$ there is $k_f\in L^1$ such that

$$\left|f\left(t,x\right)-f\left(t,y\right)\right|\leq k_f\left(t\right)\left|x-y\right| \text{ for } x,y\in A.$$

(H5) For every bounded set $A\subset\mathbb{R}^n$ there is a function $c_A\in L^1$ such that for $f\in S$:

$$\left|f\left(t,x\right)\right|\leq c_A\left(t\right) \text{ for a.e. } t\in[0,1],\ x\in A.$$

(H6) $-\infty<\inf(P)<+\infty$ holds.

Of course, there is no guarantee for the existence of a solution of the optimal control problem under these assumptions. But, if the problem is reasonable (cf. assumption (H6)), i.e. there are admissible trajectories and the value of the problem is bounded below, we are able to give a substitute for necessary conditions.

**Theorem 1** *Let (H1)-(H6) be satisfied. Then, for each $\varepsilon>0$ there are $f_\varepsilon\in S$, $x_\varepsilon,p_\varepsilon\in W^{1,1}$, $\lambda_\varepsilon\geq 0$ such that we have*

*(1) ε-minimality*

$$l\left(x_{\varepsilon}(0), x_{\varepsilon}(1)\right) \leq \inf(P)+\varepsilon^{2}$$

*(2) the adjoint inclusion*

$$-\dot{p}_{\varepsilon}(t) \in \partial_{x}\left(p_{\varepsilon}(t) \cdot f_{\varepsilon}\left(t, x_{\varepsilon}(t)\right)\right)$$

*(3) the transversality condition*

$$\left(p_{\varepsilon}(0),-p_{\varepsilon}(1)\right) \in \lambda_{\varepsilon}\left[\partial l\left(x_{\varepsilon}(0), x_{\varepsilon}(1)\right)+(\varepsilon B \times\{0\})\right]+\mathcal{N}_{C}\left(x_{\varepsilon}(0), x_{\varepsilon}(1)\right)$$

*(4) the ε-maximum condition*

$$p_{\varepsilon}(t) f_{\varepsilon}\left(t, x_{\varepsilon}(t)\right) \geq p_{\varepsilon}(t) f\left(t, x_{\varepsilon}(t)\right)-\varepsilon \lambda_{\varepsilon} \ \forall f \in \mathcal{F} \text{ a.e.}$$

*(5) nontriviality of the multipliers*

$$\left\|p_{\varepsilon}\right\|+\left|\lambda_{\varepsilon}\right|>0.$$

The symbol $\partial_x$ in (2) stands for Clarke's subdifferential; in (3) the limiting subdifferential and the limiting normal cone are used, see for example [6] for definitions and related results.

# 3 PROOF OF THE THEOREM

*Step 1:* In the first place we shall consider the following problem:

$$\text{minimize } g(x(0), x(1))$$

$$\text{subject to } \dot{x}(t)=f(t, x(t)), \ f \in S$$

assuming (H3)-(H6) and
(H1') the function $g: \mathbb{R}^{n} \times \mathbb{R}^{n} \rightarrow \mathbb{R} \cup\{+\infty\}$ is lower semicontinuous.
We write $x_{\xi, f}$ for the unique $W^{1,1}$-solution of

$$x(t)=\xi+\int_{0}^{t} f(s, x(s)) d s, \ \xi \in \mathbb{R}^{n}, \ f \in S$$

and call it the trajectory corresponding to $\xi, f$. Define

$$\mathcal{W}:=\left\{(\xi, f) \in \mathbb{R}^{n} \times S: \ x_{\xi, f} \text{ exists on the whole interval }[0,1]\right\}.$$

We shall introduce a metric on $\mathcal{W}$ by defining ($\mu$ denotes the Lebesgue measure.)

$$\tilde{d}\left(f_{1}, f_{2}\right):=\mu\left\{t \in[0,1]: \sup _{x \in \mathbb{R}^{n}}\left|f_{1}(t, x)-f_{2}(t, x)\right|>0\right\},$$

$$d\left(\left(\xi_1, f_1\right), \left(\xi_2, f_2\right)\right) := \left|\xi_1 - \xi_2\right| + \tilde{d}\left(f_1, f_2\right)$$

for $\xi_1, \xi_2 \in \mathbb{R}^n$ and $f_1, f_2 \in S$.

**Lemma 1** *$(\mathcal{W}, d)$ is a complete metric space; the functional*

$$(\xi, f) \longrightarrow g\left(x_{\xi,f}(0), x_{\xi,f}(1)\right)$$

*is lower semicontinuous with respect to d.*

*Proof:* Kaskosz and Lojasiewicz [5] introduced the metric $d$ and gave the essentials of the proof of the completeness of $(\mathcal{W}, d)$. Let $\{(\xi_n, f_n)\}$ be a convergent sequence in $\mathcal{W}$, $(\xi, f)$ its limit. Let $x_n := x_{\xi_n, f_n}$ and $x := x_{\xi,f}$ be the corresponding trajectories. Then

$$|x_n(t) - x(t)| \leq$$

$$|\xi_n - \xi| + \int_0^t |f_n(s, x_n(s)) - f_n(s, x(s))|\, ds + \int_0^t |f_n(s, x(s)) - f(s, x(s))|\, ds.$$

Suppose that $\tilde{d}(f_n, f) =: \mu(B_n) \to 0$ where $B_n = \{t \in [0,1] : \sup_{x \in \mathbb{R}^n} |f_n(t,x) - f(t,x)| > 0\}$. The definition of $\tilde{d}$ and assumption (H5) imply

$$\int_0^t |f_n(s, x(s)) - f(s, x(s))|\, ds \leq 2 \int_{B_n} c_A(t)\, dt \to 0$$

where $A = \{x(t) : t \in [0,1]\}$ which is bounded since $x$ is a continuous function. The Lipschitz continuity of $f_n$ and Gronwall's inequality imply $x_n \to x$ uniformly. The lower semicontinuity of $g$ gives the result. ■

We apply Ekeland's principle to get $\xi_\varepsilon, f_\varepsilon$ and a corresponding $x_\varepsilon := x_{\xi_\varepsilon, f_\varepsilon}$ such that the latter, together with the vector field $f_\varepsilon$, is a solution of the following problem.

$$\text{minimize } g(x(0), x(1)) + \varepsilon |x(0) - x_\varepsilon(0)| + \varepsilon y(1)$$

$$\text{subject to} \begin{cases} \dot{x}(t) = f(t, x(t)), \ f \in S \\ \dot{y}(t) = \phi_{\varepsilon,f}(t), \ y(0) = 0 \end{cases}$$

The function $\phi_{\varepsilon,f}$ is defined by

$$\phi_{\varepsilon,f}(t) = \begin{cases} 1 & : \ \sup_{x \in \mathbb{R}^n} |f(t,x) - f_\varepsilon(t,x)| > 0 \\ 0 & : \ \text{otherwise} \end{cases}$$

*Step 2:* We wish to apply the necessary conditions of theorem 5 of [4]. To this aim the problem has to be transformed into a problem satisfying

(H1) as well. Therefore, using an idea of [7] we introduce the following problem $(P_\varepsilon)$:

$$\text{minimize } z(1)$$

$$\text{subject to} \begin{cases} \dot{x}(t) = f(t, x(t)), \; f \in S \\ \dot{y}(t) = \phi_{\varepsilon,f}(t) \\ \dot{z}(t) = 0 \\ (x(0), x(1), z(1)) \in \operatorname{epi}\left\{\tilde{h}_\varepsilon\right\}, \; y(0) = 0 \end{cases}$$

where $\tilde{h}_\varepsilon(\xi, \xi', \eta, \eta') := g(\xi, \xi') + \varepsilon|\xi - x_\varepsilon(0)| + \varepsilon\eta' + \chi_{\{0\}}(\eta)$.

All assumptions of theorem 5 in [4] are satisfied. Therefore, there exists a nontrivial set of multipliers $\lambda_\varepsilon \geq 0$, $p_\varepsilon, q_\varepsilon \in W^{1,1}$ such that we have the adjoint equations

$$\begin{cases} -\dot{p}_\varepsilon(t) & \in & \partial_x(p_\varepsilon(t) \cdot f_\varepsilon(t, x_\varepsilon(t))) \\ -\dot{q}_\varepsilon(t) & = & 0 \end{cases}$$

the maximum condition

$$p_\varepsilon(t) f_\varepsilon(t, x_\varepsilon(t)) \geq p_\varepsilon(t) f(t, x_\varepsilon(t)) + q_\varepsilon(t)\phi_{\varepsilon,f}(t)$$

for all $f \in S$ and almost all $t \in [0, 1]$ and the transversality conditions

$$(p_\varepsilon(0), q_\varepsilon(0), -p_\varepsilon(1), -q_\varepsilon(1), -\lambda_\varepsilon) \in \mathcal{N}_{\operatorname{epi}\tilde{h}_\varepsilon}\left(x_\varepsilon(0), y_\varepsilon(0), x_\varepsilon(1), y_\varepsilon(1), \tilde{h}_\varepsilon(x_\varepsilon(0), y_\varepsilon(0), x_\varepsilon(1), y_\varepsilon(1))\right).$$

Since $\tilde{h}_\varepsilon$ is separated in $(x, z)$ and $y$ the transversality condition implies

$$q_\varepsilon(1) = -\lambda_\varepsilon \cdot \varepsilon$$

and

$$(p_\varepsilon(0), -p_\varepsilon(1), -\lambda_\varepsilon) \in \mathcal{N}_{\operatorname{epi} h_\varepsilon}(x_\varepsilon(0), x_\varepsilon(1), h_\varepsilon(x_\varepsilon(0), x_\varepsilon(1)))$$

where $h_\varepsilon(\xi, \xi') := g(\xi, \xi') + \varepsilon|\xi - x_\varepsilon(0)|$.

*Step 3:* Using the function

$$g(\xi, \xi') = l(\xi, \xi') + \chi_C(\xi, \xi')$$

of "Lipschitz-plus-indicator"-type we may deduce the assertions of the theorem because the "bad case" of proposition 5A.4 of [6] (sum rule for the limiting subdifferential) can not occur and $\partial^\infty h_\varepsilon(\xi, \xi') = \mathcal{N}_C(\xi, \xi')$. This completes the proof. ■

# 4 APPLICATION TO DIFFERENTIAL INCLUSION PROBLEMS

There are several potential applications of the theorem, for example to optimal control problems in parametrized form. Compare [3] for corresponding results. Here let us consider differential inclusion problems as a special case.

$$\begin{array}{ll} \text{minimize} & l(x(0), x(1)) \\ \text{subject to} & \left\{ \begin{array}{l} \dot{x}(t) \in F(t, x(t)) \\ (x(0), x(1)) \in C \end{array} \right. \end{array}$$

We posit the following assumptions:
(D1) $F(t,x)$ is non-empty, compact and convex for all $t \in [0,1]$, $x \in \mathbb{R}^n$.
(D2) $F(\cdot,x)$ is a measurable multifunction for each fixed $x \in \mathbb{R}^n$; i.e. for each closed subset $A$ of $\mathbb{R}^n$ the set $\{t \in [0,1] : F(t,x) \cap A \neq \emptyset\}$ is measurable.
(D3) For each bounded subset $A$ of $\mathbb{R}^n$ there exists a function $k_A \in L^1$ such that

$$\rho_H\left(F(t,x), F(t,x')\right) \leq k_A(t)\left|x - x'\right| \text{ for } x, x' \in A,\ t \in [0,1],$$

where $\rho_H$ denotes the Hausdorff distance.
(D4) For every bounded set $A \subset \mathbb{R}^n$ there is a function $c_A \in L^1$ such that

$$F(t,x) \subset c_A(t)\, B, \text{ for } t \in [0,1],\ x \in A.$$

Let $S$ be the set of all functions $f : [0,1] \times \mathbb{R}^n \to \mathbb{R}^n$ such that $f(t,x) \in F(t,x)$ for all $t \in [0,1]$, $x \in \mathbb{R}^n$ satisfying (H4). Then, $S$ is a generalized control system with properties (H3)-(H5) and the trajectories of $S$ and of the differential inclusion are the same, compare [5]. From theorem 1 we obtain

**Corollary 1** *Let (H1), (H2), (D1)-(D4), (H6) be satisfied for the system $S$ generated by the differential inclusion above. Then, there exists for each $\varepsilon > 0$ a selection $f_\varepsilon$ of $F$ and $x_\varepsilon, p_\varepsilon \in W^{1,1}$, $\lambda_\varepsilon \geq 0$ such that (1)-(3), (5) of theorem 1 and the maximum condition*

$$p_\varepsilon(t)\,\dot{x}_\varepsilon(t) = p_\varepsilon(t)\, f_\varepsilon(t, x_\varepsilon(t)) \geq \max\{p_\varepsilon(t)\, w : w \in F(t, x_\varepsilon(t))\} - \lambda_\varepsilon \varepsilon$$

*for a.e. $t \in [0,1]$ hold true.*

To the knowlegde of the author, this is the first suboptimality theorem for differential inclusion problems.

## References


[1] CLARKE, F. H. . The maximum principle under minimum hypotheses. *SIAM Journal Control Optimization*, 14(6):1078–1091, 1976.

[2] EKELAND, I. . On the variational principle. *Journal Mathematical Analysis Application*, 47:324–353, 1974.

[3] HAMEL, A. H. . Suboptimality theorems for control problems with general boundary conditions. Reports of the Institut of Optimization and Stochastics 27, Martin-Luther-University, Department of Mathematics and Computer Science, 1999.

[4] IOFFE, A. D. . Euler-Lagrange and Hamiltonian formalism in dynamic optimization. *Transaction of American Mathematical Society*, 349(7):2871–2900, 1997.

[5] KASKOSZ, B. and LOJASIEWICZ, S. . A maximum principle for generalized control systems. *Nonlinear Analysis. Theory, Methods & Applications*, 9(2):109–130, 1985.

[6] LOEWEN, P. D. . *Optimal control via nonsmooth analysis*, volume 2 of *CMR Proceedings and Lecture Notes*. American Mathematical Society, Providence, Rhode Island USA, 1993.

[7] VINTER, R. B. and ZHENG, H. . The extended Euler-Lagrange condition for nonconvex variational problems. *SIAM Journal Control Optimization*, 35(1):56–77, 1997.

Chapter 17

# D.C. OPTIMIZATION APPROACHES VIA MARKOV MODELS FOR RESTORATION OF SIGNAL (1-D) AND (2-D)

Le Thi Hoai An, Pham Dinh Tao

*Mathematical Modelling and Applied Optimization Group*

*Laboratory of Mathematics*

*National Institute for Applied Sciences-Rouen*

*BP 8, F 76 131 Mont Saint Aignan Cedex, France*

lethi@insa-rouen.fr, pham@insa-rouen.fr

**Abstract** A new approach based on d.c. (difference of convex functions) is presented for signal (1-D) and (2-D) restoration problems. It consists in applying d.c. algorithms (DCA) to suitable d.c. programs. Some choices of starting points for DCA have been discussed. Finally computational results are reported which prove the globality of sought solutions, the robustness and the efficiency of our method with respect to the Graduated NonConvexity algorithm (GNC).

**Keywords:** D.C. programming, D.C. duality, Local optimality, DCA, GNC.

## 1 INTRODUCTION

Nonconvex programming constitutes successful approaches to solving inverse problems in signal (1-D) and (2-D) processing by the way of a suitable formulation of a cost function whose global minimum is the desired solution. In this paper we address the problem of restoring signal and image affected by additive white Gaussian noise. Such problems play a crucial role in low-level vision problems. The cost functions in these problems are usually derived from discretization of functionals obtained from regularization principles or stochastic estimation techniques using Markov random field models ([9], [10], [6],[7]). We then are faced with



*N. Hadjisavvas and P.M. Pardalos (eds.), Advances in Convex Analysis and Global Optimization,* 303–317.

the following nonconvex program denoted (P)

$$\begin{array}{ll} \min \quad E(u,x) = & \frac{1}{2}\sum_{i=1}^{n}(u_i - d_i)^2 + \frac{\lambda^2}{2}\sum_{i=1}^{n-1}(u_i - u_{i-1})^2(1 - x_i) \\ & +\frac{\alpha}{2}\sum_{i=1}^{n-1} x_i \\ s.t. & u \in \mathbb{R}^n, x \in \{0,1\}^n. \end{array}$$

That is a mixed 0-1 optimization problem in which u is the estimated height, d is the noisy data and the cost function, called the Gibbs energy function, is not convex. If $(u^*, x^*)$ is a global solution, then $u^*$ is the restored signal (i.e. the original signal) and the values $x_i^* = 1$ represent the points of discontinuity of this signal. $\alpha$ is a penalty per unit length and $\lambda$ is a characteristic length for smoothing the continuous portions of the data d.

In the case of signal (2-D) restoration, the cost function will have the same form in which the vectors $u$ and $x$ become the matrices $U$ and $X$. For simplicity of presentation we restrict our attention to our d.c. approach and DCA for solving this nonconvex program in case of (1-D). Problem (P) has been solved by several methods among them simulated annealing in the stochastic approaches ([9], [10]) and the graduated nonconvexity algorithm (GNC) in the deterministic approaches ( [6], [7]), have been the most used. In [6], [7] the authors tackled the following problem equivalent to (P) in the continuous variables u:

$$\min\{F(u) : u \in \mathbb{R}^n\}$$

where $F(u) = \inf\{E(u,x) : x \in [0,1]^n\}$

To find a global minimum of the cost function $F$, the GNC algorithm due to Blake ([6], [7]) starts with a convex approximation $\overline{F}$ to $F$. Then a whole sequence of cost functions $F^{(p)}, 0 \le p \le 1$ is constructed so that $F^{(0)} = F$ and $F^{(1)} = \overline{F}$. In between, $F^{(p)}$ changes continuously, between $F^{(1)}$ and $F^{(0)}$. The GNC algorithm optimizes a whole subsequence of $F^{(p)}$, for example $\{F^{(1)}, F^{(1/2)}, F^{(1/4)}, F^{(1/8)}, F^{(1/16)}, ...\}$, one after the other, using the optimal solution of previous optimization step as the starting point for the next. The optimization step based is then performed using the sequential over relaxation (SOR) method in ([6],[7]). In [7], a comparison of the efficiency of deterministic and stochastic algorithms for visual reconstruction proved that the GNC is by far better than simulated annealing for solving Problem (P).

The major advantage of Blake's continuation approach relies on the quite fitting choice of the approximations $F^{(p)}$ to $F$ while improvements could be done in the use of local algorithms for computing global minimum of $F^{(p)}$. Extensive computations on test problems ([8]) proved that the GNC with local optimizer SOR might not converge to an optimal solution.

In convex analysis approach to nondifferentiable nonconvex programming, the d.c. (difference of convex functions) optimization and its solution algorithms (DCA) developed by Pham Dinh Tao and Le Thi H. An ([1], [2], [18] - [20] and references therein) constitute a natural and logical extension of Pham Dinh Tao's earlier works concerning convex maximization and its subgradient algorithms ([14] - [17] and references therein). Our method in this paper, based on the d.c. optimization approach, aims at solving two d.c. programs which are equivalent to the original problem (P), (see Sections 2, 3).

The DCA is a primal-dual subgradient method for solving a general d.c. program that consists in the minimization of difference of convex functions on the whole space. It is at the present time one of a few algorithms in the local approach which has been successfully applied to many large-scale d.c. optimization problems and proved to be more robust and efficient than related standard methods. Due to its local character it cannot guarantee the globality of computed solutions for general d.c. programs. However, we observe that with a suitable starting point it converges quite often to a global one (see e.g. [1], [2], [3], [21]).

The paper is composed of four sections. Section 2 is devoted to the outline of d.c. programming with the description of the DCA for solving general d.c. program. The study of the two d.c. programs equivalent to the original problem (P) and the description of the DCA for solving these d.c. programs are given in Section 3. Experimental results are provided in the last section to illustrate the robustness and the efficiency of our method.

# 2 D.C. PROGRAMMING AND DCA

Here we summarize the material needed for an easy understanding of d.c. programming and DCA which will be used to solve d.c. programs . Our working space is $X = \mathbb{R}^n$ equipped with the canonical inner product $\langle \cdot, \cdot \rangle$ and the corresponding Euclidean norm $\|\cdot\|$, thus the dual space $Y$ of $X$ can be identified with $X$ itself. We follow [22] for definitions of usual tools of modern convex analysis where functions could take the infinite values $\pm\infty$. A function $\theta : X \to \mathbb{R} \cup \{\pm\infty\}$ is said to be proper if it takes nowhere the value $-\infty$ and is not identically equal to $+\infty$. The effective domain of $\theta$, denoted by *dom* $\theta$, is *dom* $\theta = \{x \in X : \theta(x) < \infty\}$.The set of all lower semicontinuous proper convex functions on $X$ is denoted

defined by $y^k \in \partial h(x^k)$ (resp. $x^k \in \partial g^*(y^{k-1})$), the DCA yields the next scheme: $y^k \in \partial h(x^k)$; $x^{k+1} \in \partial g^*(y^k)$. It corresponds actually to the *simplified* DCA (which will be in short called DCA through the paper for simplicity) where $x^{k+1}$ (resp. $y^k$) is arbitrarily chosen in $\partial g^*(y^k)$ (resp. $\partial h(x^k)$. In the *complete* form of DCA, we impose the following natural choice: $x^{k+1} \in \arg\min\{g(x) - h(x) : x \in \partial g^*(y^k)\}$ and $y^k \in \arg\min\{h^*(y) - g^*(y) : y \in \partial h(x^k)\}$. These problems are equivalent to convex maximization problems : $x^{k+1} \in \arg\min\{\langle x, y^k\rangle - h(x) : x \in \partial g^*(y^k)\}$ and $y^k \in \arg\min\{\langle x^k, y\rangle - g^*(y) : y \in \partial h(x^k)\}$ respectively. The complete DCA ensures that $(x^\infty, y^\infty) \in \mathcal{P}_l \times \mathcal{D}_l$. From a practical point of view, although the latter convex maximization problems are simpler than $(P_{dc})$, $(D_{dc})$, they remain nonconvex programs and thus are still hard to solve. In practice one generally uses the simplified DCA to solve d.c. programs.

The DCA was introduced as an extension of the aforementioned subgradient algorithms (for convex maximization programming) to d.c. programming by Pham in 1986. But this field has been really developed from 1994 only with joint works by Le Thi and Pham ([1] - [3]), solving nonsmooth nonconvex optimization problems. To our knowledge, DCA is actually one of a few algorithms (in the convex analysis approach to d.c. programming) which allows to solve large-scale d.c. programs.

It had been proved in Pham and Le Thi [21], [20] that, for the simplified DCA, we have

(i) The sequences $\{g(x^k) - h(x^k)\}$ and $\{h^*(y^k) - g^*(y^k)\}$ are decreasing and

- $g(x^{k+1}) - h(x^{k+1}) = g(x^k) - h(x^k)$ if and only if $y^k \in \partial g(x^k) \cap \partial h(x^k)$, $y^k \in \partial g(x^{k+1}) \cap \partial h(x^{k+1})$ and $[\rho(g) + \rho(h)]\|x^{k+1} - x^k\| = 0$.
- $h^*(y^{k+1}) - g^*(y^{k+1}) = h^*(y^k) - g^*(y^k)$ if and only if $x^{k+1} \in \partial g^*(y^k) \cap \partial h^*(y^k)$, $x^{k+1} \in \partial g^*(y^{k+1}) \cap \partial h^*(y^{k+1})$ and $[\rho(g^*) + \rho(h^*)]\|y^{k+1} - y^k\| = 0$.

  In such a case DCA terminates at the $k^{th}$ iteration.

(ii) If $\rho(g) + \rho(h) > 0$ (resp. $\rho(g^*) + \rho(h^*) > 0$), then the series $\{\|x^{k+1} - x^k\|^2\}$ (resp. $\{\|y^{k+1} - y^k\|^2\}$) converges.

(iii) If the optimal value $\alpha$ of problem $(P_{dc})$ is finite and the sequences $\{x^k\}$ and $\{y^k\}$ are bounded then every limit point $x^\infty$ (resp. $y^\infty$) of the sequence $\{x^k\}$ (resp. $\{y^k\}$) is a critical point of $g - h$ (resp. $h^* - g^*$).

(iv) DCA has a linear convergence for general d.c. programs.

(v) In polyhedral d.c. programs, the sequences DCA $\{x^k\}$ and $\{y^k\}$ contain finitely many elements and DCA has a finite convergence.

We have the same results for the complete DCA, except that in (i) (resp. (iii)) we must add the following property:
$\partial h(x^k) \subset \partial g(x^k)$ and $\partial g^*(y^k) \subset \partial h^*(y^k)$ (resp. $\partial h(x^\infty) \subset \partial g(x^\infty)$ and $\partial g^*(y^\infty) \subset \partial h^*(y^\infty)$).

**Remark 1** *(i) So there are as many DCA as there are d.c. decompositions and it is of particular interest to study various equivalent d.c. forms for the primal and dual d.c. problems. It is worth mentioning, for instance, that by using conjointly suitable d.c. decompositions of convex functions and proximal regularization techniques we can obtain the proximal point algorithm and the Goldstein-Levitin-Polyak subgradient method (in convex programming) as special cases of DCA.*

*(ii) The above description of DCA does not really reveal the main features of this approach which could partly explain the qualities (running time, robustness, stability, rate of convergence and globality of sought solutions) of DCA from the computational viewpoint. We have recently introduced a deeper interpretation of DCA that has the merit of offering a valuable insight into DCA ([4]).*

*(iii) The choice of the d.c. decomposition of the objective function in a d.c. program and the initial point for DCA are open questions to be studied for the specific structure of the problem being considered. In practice, for solving a given d.c. program, we try to choose g and h such that sequences $\{x^k\}$ and $\{y^k\}$ can be easily calculated, i.e. either they are in explicit form or their computations are inexpensive.*

## 3 D.C. FORMULATIONS OF PROBLEM ($P$) AND DCA FOR ITS SOLUTION

To apply d.c. programming to the mixed 0-1 optimization problem (P), we first set the two d.c. programs equivalent to (P). We have $\sum_{i=1}^{n} (u_i - d_i)^2 \leq E(u,x)$, for all $u \in \mathbb{R}^n$ and $x \in \{0,1\}^{n-1}$. Hence $\lim_{\|u\| \to +\infty} E(u,x) = +\infty, \forall x \in \{0,1\}^{n-1}$. It follows that the energy function $E(u,x)$ is coercive on its feasible domain and the solution set of (P), denoted by $\mathcal{P}$, is nonempty. For each $x \in ]-\infty, 1]^{n-1}$, $A(x)$ is the $(n-1) \times n$ matrix defined by $A(x) = D(x)A$, where $D(x)$ is the $(n-1) \times (n-1)$ diagonal matrix function whose entries are $D(x)_{i,i} = (1-x_i)^{1/2}$, for $i = 1, ..., n-1$, and A is the $(n-1) \times n$ bidiagonal matrix whose nonzero entries are $A_{i,i} = -1, A_{i,i+1} = 1$, for $i = 1, ..., n-1$. We also consider the $n \times n$ bidiagonal symmetric matrix function $B(x)$ defined by: $B(x)_{11} = a_1 = -B(x)_{12}$, $B(x)_{ii} = a_{i-1} + a_i$ and $B(x)_{i,i+1} = -a_i$ for $i = 2, ..., n-1$, and $B(x)_{nn} = a_{n-1}$, where $a_i = 1 - x_i$ for $i = 1, ..., n-1$.

We note that $B(x) = A(x)^T A(x)$ for $x \in ]-\infty, 1]^{n-1}$ and the matrix expression of the function $E(u, x)$ for all $u \in \mathbb{R}^n$ and $x \in \{0,1\}^{n-1}$ : $E(u,x) = \frac{1}{2}\langle (I + \lambda^2 A(x)^T A(x))u, u\rangle - \langle u, d\rangle + \frac{1}{2}\|d\|^2 + \frac{\alpha}{2}\sum_{i=1}^{n-1} x_i$. Now let $F : \mathbb{R}^n \to \mathbb{R}$ be defined by : $F(u) = \inf\{E(u,x) : x \in \{0,1\}^n\}$ $(\overline{P}_u)$

It is clear that the solution set of $(\overline{P}_u)$ is nonempty and since for fixed $u$, the function $E(u,x)$ is affine with respect to the variables $x$, we can write: $F(u) = \inf\{E(u,x) : x \in [0,1]^n\}$ $(P_u)$. Let the solution sets of $(P_u)$ and $(\overline{P}_u)$ be denoted by $\mathcal{P}_u$ and $\overline{\mathcal{P}}_u$ respectively. The next result is straightforward

**Proposition 1** *For all $u \in \mathbb{R}^n, \mathcal{P}_u$ is the set of $x$ whose components are explicitly given by $x_i = 1$ if $\lambda^2(u_{i+1} - u_i)^2 > \alpha$, $0$ if $\lambda^2(u_{i+1} - u_i)^2 < \alpha, c \in [0,1]$ if $\lambda^2(u_{i+1} - u_i)^2 = \alpha$.*

*Moreover we have $\overline{\mathcal{P}}_u = \{x \in \mathcal{P}_u : x_i \in \{0,1\}$ if $\lambda^2(u_{i+1} - u_i)^2 = \alpha\}$*

It follows that $F(u) = \frac{1}{2}\sum_{i=1}^{n} (u_i - d_i)^2 + \frac{1}{2}\sum_{i=1}^{n-1} f(u_{i+1} - u_i)$ with $f(t) = \min\{\alpha^2, \lambda^2 t^2\}$.

From the d.c. decomposition of the function $f(t) : f(t) = \alpha + \lambda^2 t^2 - \max\{\alpha, \lambda^2 t^2\}$ we deduce that $F$ is a d.c. function on $\mathbb{R}^n$ : $F(u) = G(u) - H(u)$ where the convex d.c. components are defined by

$$G(u) := \frac{1}{2}\sum_{i=1}^{n} (u_i - d_i)^2 + \frac{\lambda^2}{2}\sum_{i=1}^{n-1} (u_{i+1} - u_i)^2 + \frac{\alpha(n-1)}{2},$$

$$H(u) = \sum_{i=1}^{n-1} H_i(u) \text{ with } H_i(u) = \frac{1}{2}\max\{\alpha, \lambda^2 (u_{i+1} - u_i)^2\}.$$

It is worth noting that $G(u)$ is a positive definite quadratic form $G(u) = \frac{1}{2}\langle (I + \lambda^2 A^T A)u, u\rangle - \langle u, d\rangle + \frac{1}{2}\|d\|^2 + \frac{\alpha(n-1)}{2}$.

## 3.1 THE FIRST D.C. PROGRAM $(P_1)$ EQUIVALENT TO $(P)$

Our first d.c. program is of the form: $\min\{G(u) - H(u) : u \in \mathbb{R}^n\}$ $(P_1)$ whose solution set is denoted by $\mathcal{P}_1$ and the equivalence between (P) and (P$_1$) is stated in the next result whose proof is immediate

**Proposition 2** $(u^*, x^*) \in \mathcal{P}$ *if and only if* $u^* \in \mathcal{P}_1$ *and* $x^* \in \overline{\mathcal{P}}_{u^*}$.

**Corollary 1** *(P) admits a unique solution $(u^*, x^*)$ if and only if $u^*$ is a unique solution to $(P_1)$ and $\lambda^2(u_{i+1} - u_i)^2 \neq \alpha$, $\forall i = 1, ..., n-1$.*

**Proof:** Immediate consequence of Proposition 1 and (2). □

## 3.2 THE SECOND D.C. PROGRAM ($Q_3$) EQUIVALENT TO ($P$)

We are now stating the other d.c. program equivalent to $(P)$. To this end, let K be the function defined on $\mathbb{R}^{n-1}$ by

$$K(x) := \inf\{E(u,x) : u \in \mathbb{R}^n\} \quad (Q_x)$$

and let $\mathcal{Q}_x$ be the solution set of $(Q_x)$. For $u$ fixed, $E(u,x)$ is affine function in $x$, so $-K(x)$ is lower semicontinuous proper convex function on $\mathbb{R}^{n-1}$ ([22]) and $domK = \{x \in \mathbb{R}^{n-1} : K(x) > -\infty\} \supset ]-\infty, 1]^{n-1}$. More precisely $domK \supset ]-\infty, 1+\varepsilon]^{n-1}$ with any positive number $\varepsilon < \frac{1}{2\lambda^2}$. Indeed, for $x \in ]-\infty, 1+\varepsilon]^{n-1}$, the symmetric matrix $I + \lambda^2 B(x)$, being strictly diagonally dominant with positive diagonal entries, is then positive definite. Hence Problem $(Q_x)$ has a unique solution $\overline{u}$ given by $(I + \lambda^2 B(x))\overline{u} = d$. Consider now the following problem: $\min\{K(x) : x \in \{0,1\}^{n-1}\}$ ($Q_1$) whose solution set is denoted by $\mathcal{Q}_1$.

Similar to Proposition 2 we have:

**Proposition 3** *$(u^*, x^*) \in \mathcal{P}$ if and only if $x^* \in \mathcal{Q}_1$ and $u^* \in \mathcal{Q}_{x^*}$.*

Since K is a concave function on $\mathbb{R}^{n-1}$, $(Q_2)$ is equivalent to the problem $\min\{K(x) : x \in [0,1]^{n-1}\}$, i.e., a d.c. program of the form:

$$\min\{\chi_T(x) - (-K)(x) : x \in \mathbb{R}^{n-1}\} \; (Q_3)$$

where $\chi_T$ denotes the (polyhedral convex) indicator function of $T = [0,1]^{n-1}$. Denote by $\mathcal{Q}_3$ the solution set of ($Q_3$). It is clear that $\mathcal{Q}_3 = \{x \in \mathcal{Q}_2 : x \in \{0,1\}^{n-1}\}$ and ($Q_3$) is a polyhedral d.c. program. In the sequel we are dealing with the two d.c. programs ($P_1$) and ($Q_3$) to which DCA will be applied.

# 4 DCA FOR SOLVING ($P_1$) AND ($Q_3$)

In this section we are concerned with the solution of these two d.c. programs by DCA.

## 4.1 SOLVING ($P_1$) BY DCA: DCA1

DCA applied to ($P_1$) consists of constructing the two sequences $\{u^k\}$ and $\{v^k\}$ such that $v^k \in \partial H(u^k)$, $u^{k+1} \in \partial G^*(v^k)$ with a starting point $u^0 \in dom\ \partial H$. Since $dom\ H_i = \mathbb{R}^n$ for $i = 1, ..., n$ and $H = \sum_{i=1}^{n-1} H_i$, we have ([22]): $\partial H(u) = \sum_{i=1}^{n-1} \partial H_i(u)$, for all $u \in \mathbb{R}^n$. Using the Valadier's Lemma ([11]), we obtain: $\partial H_i(u) = 0$ if $|u_{i+1} - u_i| < \frac{\sqrt{\alpha}}{\lambda}$,

$\lambda^2(u_{i+1}-u_i)(e_{i+1}-e_i)$ if $|u_{i+1}-u_i| > \frac{\sqrt{\alpha}}{\lambda}$, $[0, \lambda^2(u_{i+1}-u_i)(e_{i+1}-e_i)]$ if $|u_{i+1}-u_i| = \frac{\sqrt{\alpha}}{\lambda}$ where $e_i$ is the $i$-th unit vector of $\mathbb{R}^n$. Since $u^{k+1} \in \partial G^*(v^k)$ if and only if it is a solution of the next problem $\min\{G(u) - \langle u, v^k\rangle : u \in \mathbb{R}^n\}$, computing $u^{k+1}$ amounts to solving the following linear system

$$(I + \lambda^2 A^T A)u = d + v^k. \tag{17.1}$$

The sequence $\{u^k\}$ is well defined because the symmetric matrix $I + \lambda^2 A^T A$ is positive definite.

**Algorithm DCA1** (DCA applied to $(P_1)$)
Given $\varepsilon > 0$ be sufficiently small and $u^0 \in \mathbb{R}^n$, $k := 0$.
**Do**
Determine $v^k$ according to the preceding computation of $\partial H(u)$
Solve the linear system (17.1) to obtain $u^{k+1}$.
$k + 1 \longleftarrow k$
**While** $\|u^{k+1} - u^k\| > \varepsilon$.

## 4.2 SOLVING $(Q_3)$ BY DCA: DCA2

The sequences $\{x^k\}$ and $\{y^k\}$ generated by DCA for solving $(Q_3)$ are given by $y^k \in \partial(-K)(x^k), x^{k+1} \in \partial\chi_T^*(y^k)$ where $x^0$ is arbitrarily chosen in dom $\partial(-K)$. According to Valadier's Lemma [11] we have

$$\partial(-K)(x) \supset \{y : y_i = \lambda^2(\widetilde{u}_{i+1} - \widetilde{u}_i) - \alpha,\ i = 1, ...n-1\}$$

where $\widetilde{u}$ is the unique solution of the linear system $(I+\lambda^2 A(x)^T A(x))u = d$. More precisely we obtain the result

**Lemma 1** *The function $K$ is differentiable on* $]-\infty, 1]^{n-1}$.

**Proof.** One can, for example, demonstrate that any $s \in\ ]-\infty, 1]^{n-1}$ admits an open neighborhood $U$ and a compact set $C$ such that for all $x \in U$ one has:

$$K(x) = \inf\Big\{E(u,x) \mid \quad E(u,x) = \tfrac{1}{2}\langle (I + \lambda^2 B(x))u, u\rangle - \langle u, d\rangle + \tfrac{1}{2}\|d\|^2 + \tfrac{\alpha}{2}\textstyle\sum_{i=1}^{n-1} x_i,\ u \in C\Big\}.$$

This and Valadier's Lemma then imply Lemma 1. The proof is left to the reader □.

By definition, computing $x^{k+1}$ amounts to solving the problem $\min\{\chi_T(x) - \langle x, y^k\rangle : x \in \mathbb{R}^{n-1}\}$ or $\max\Big\{\langle x, y^k\rangle : x \in [0,1]^{n-1}\Big\}$ whose a solution is explicitly defined by:

$$x_i^{k+1} = 1 \text{ if } y_i^k > 0,\ 0 \text{ if } y_i^k < 0, a \in [0,1] \text{ if } y_i^k = 0. \tag{17.2}$$

**Algorithm DCA2** (DCA applied to ($Q_3$)):
Let $\varepsilon$ is sufficiently small, let $x^0 \in [0,1]^{n-1}$ , $k := 0$.
**Do**

Solve the linear system $((I + \lambda^2 A(x^k)^T A(x^k))u = d$ to obtain $\widetilde{u}$ and then compute $y^k$ by $y_i^k = \lambda^2(\widetilde{u}_{i+1} - \widetilde{u}_i) - \alpha, i = 1, ...n-1$.

Let $x^{k+1}$ be defined by (17.2)

$k + 1 \longleftarrow k$

**While** $\|x^{k+1} - x^k\| > \varepsilon$

It is worth noting ($Q_3$) is a polyhedral d.c.program for which DCA has a finite convergence (Section 2).

## 4.3 COMPUTATIONAL EXPERIMENTS

We have presented in Section 3 the two versions of DCA for solving the nonconvex mixed 0-1 optimization problem $(P)$ : DCA1 is the DCA applied to Problem ($P_1$) and DCA2 is the DCA applied to Problem ($Q_3$).

For these methods the following choices of starting points have been used: $u^0 = \frac{1}{2}d$ in DCA1 and $x^0 \in \mathcal{P}_{u^0}$ in DCA2. Different combinations between DCA and GNC have been considered. We present below numerical results through DCA1, DCA2 and the four combinations:

1. DCA1-2: DCA2 with starting point $x^0 \in \mathcal{P}_{\overline{u}}$, where $\overline{u}$ is the solution computed by DCA1
2. DCA2-1: DCA1 with starting point $u^0 \in Q_{\overline{x}}$, where $\overline{x}$ is the solution computed by DCA2
3. G-DCA1: DCA1 with starting point $u^0 = \overline{u}$ where $\overline{u}$ is the solution computed by GNC
4. G-DCA2: DCA2 with starting point $x^0 \in \mathcal{P}_{\overline{u}}$, where $\overline{u}$ is the solution computed by GNC.

In Table 1 we present our computational results on 5 examples where the original signal is affected with the level of noise is equal to five. The informations about the original signal that are the dimension (n) and the number of discontinuity (dp) are indicated in the first column. In the next columns we compare the performance of our variant algorithms and GNC (optimal value $(E(u^*, x^*))$, the number of discontinuity detected (discont), CPU time in second (CPU)) with the same choice of $(\lambda, \alpha)$) $((20, 80000)$ for the three first examples and $(5, 5000)$ for the last one). We illustrate the results by an example in Figure 1 (the original signal and the affected signal (Fig. 1a), the signal restored by DCA1-2 (Fig. 1b) and by GNC (Fig. 1c) in case $\lambda = 5, \alpha = 5000$, the signal restored by DCA1-2 (Fig. 1d) and by GNC (Fig. 1e) in case $\lambda = 20, \alpha = 80000$. From preliminary experiments we see that DCA is efficient to solve this type of problems, and among variant versions of DCA, DCA2 is the best (this corresponds to the polyhedral d.c. program to which DCA has a finite convergence). On the other hand, the DCA is not very depending upon the initial point and the choice of $\lambda$ and $\alpha$ while this choice is very important to GNC.

*Table 17.1* The performance of DCA and GNC

| n | | DCA1 | DCA2-1 | G-DCA1 | DCA2 | DCA1-2 | G-DCA2 | GNC |
|---|---|---|---|---|---|---|---|---|
| 128 | $E(u^*,x^*)$ | 83540 | 38956 | 83540 | 83540 | 83540 | 83540 | 563542 |
| dp=2 | discont | 1 | 0 | 1 | 1 | 1 | 1 | 7 |
| | CPU | 1.27 | 0.05 | 0.27 | 0.00 | 1.26 | 0.05 | 0.39 |
| 128 | $E(u^*,x^*)$ | 583593 | 64381 | 483593 | 439542 | 466497 | 403812 | 484037 |
| dp=5 | discont | 6 | 0 | 6 | 5 | 5 | 5 | 6 |
| | CPU | 1.48 | 0.16 | 1.43 | 0.00 | 1.48 | 0.00 | 0.39 |
| 128 | $E(u^*,x^*)$ | 403868 | 89893 | 403868 | 405833 | 403867 | 405833 | 647528 |
| dp=4 | discont | 5 | 0 | 5 | 5 | 5 | 5 | 8 |
| | CPU | 1.92 | 0.11 | 1.92 | 0.00 | 1.92 | 0.00 | 0.38 |
| 512 | $E(u^*,x^*)$ | 46058 | 15058 | 46058 | 32111 | 36239 | 36239 | 31736 |
| dp=6 | 8 | 0 | 8 | 5 | 6 | 6 | 5 | |
| | CPU | 4.01 | 3.36 | 3.68 | 0.00 | 3.68 | 0.00 | 0.05 |

## References


[1] Le Thi Hoai An. *Contribution à l'optimisation non convexe et l'optimisation globale: Théorie, Algorithmes et Applications.* Habilitation à Diriger des Recherches, Université de Rouen, Juin 1997.

[2] Le Thi Hoai An, Pham Dinh Tao. Solving a class of linearly constrained indefinite quadratic problems by D.c. algorithms. *Journal of Global Optimization,* 11:253-282, 1997.

[3] Le Thi Hoai An, Pham Dinh Tao, Le Dung Muu. Numerical solution for Optimization over the efficient set by D.c. Optimization Algorithm. *Operations Research Letters,* 19:117-128, 1996.

[4] Le Thi Hoai An, Pham Dinh Tao. D.C. programming approach for large-scale molecular optimization via the general distance geometry problem. In *Optimization in Computational Chemistry and Molecular Biology: Local and Global Approaches, C.A. Floudas and*

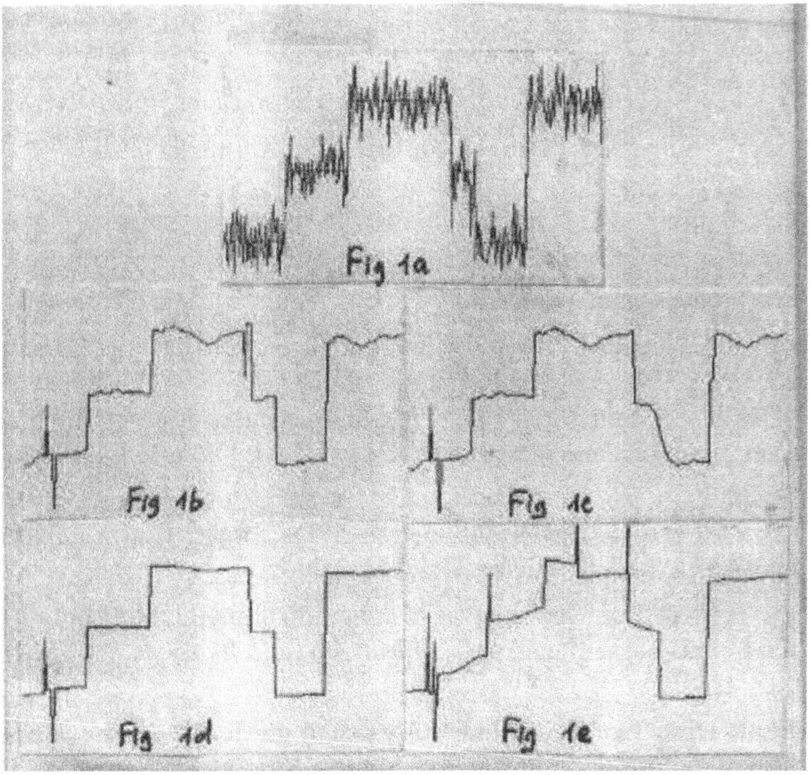


*Figure 17.1* An illustrated example with n=512.

*P.M. Pardalos (Eds.)*, pages 301-339, Kluwer Academic Publishers, 2000.

[5] Le Thi Hoai An. An efficient algorithm for globally minimizing a quadratic function over convex quadratic constraints. *Math. Program., Ser. A* 87:401-426, 2000.

[6] A. Blake, A. Zisserman. *Visual Reconstruction.* MIT Press, Cambridge, MA, 1987.

[7] A. Blake. A comparison of the efficiency of deterministic and stochastic algorithms for visual reconstruction. *IEEE Trans. Pattern Anal. Machine Intell.* Vol. 2, 1:2-3, 1989.

[8] M. Douimi. *Modélisation Markovienne et Optimisation Numérique pour la restauration des signaux en (1D) et (2D).* PhD Thesis, University of Rouen, France, October 1995.

[9] S. Geman, D. Geman. Stochastic relaxation, Gibbs distribution and the bayesian restoration of images. *IEEE Trans. Pattern Anal. Machine Intell.*, Vol. 6, 6:721-741, 1984.

[10] D. Geman, G. Reynolds. Constrained Restoration and the Recovery of Discontinuities. *IEEE Trans. Pattern Anal. Machine Intell.*, Vol. 14, 3: 367-383, 1992.

[11] P.J. Laurent. *Approximation et Optimisation.* Hermann, Paris, 1972.

[12] P. Mahey, Pham Dinh Tao. Partial regularization of the sum of two maximal monotone operators. *Math. Modell. Numer. Anal. ($M^2AN$)*, Vol. 27:375-395, 1993.

[13] P. Mahey, Pham Dinh Tao. Proximal decomposition of the graph of maximal monotone operator. *SIAM J. Optim.* 5:454-468, 1995.

[14] Pham Dinh Tao. *Contribution à la théorie de normes et ses applications à l'analyse numérique.* Thèse de Doctorat d'Etat Es Science, Université Joseph Fourier- Grenoble, 1981.

[15] Pham Dinh Tao. Convergence of subgradient method for computing the bound-norm of matrices. *Linear Alg. and Its Appls.*, Vol 62:163-182, 1984.

[16] Pham Dinh Tao. Algorithmes de calcul d'une forme quadratique sur la boule unité de la norme maximum. *Numer. Math.*, Vol 45:377-440, 1985.

[17] Pham Dinh Tao. Algorithms for solving a class of non convex optimization problems. Methods of subgradients. In *Fermat days 85. Mathematics for Optimization*, Elsevier Science Publishers B.V. North-Holland, 1986.

[18] Pham Dinh Tao. Duality in d.c. (difference of convex functions) optimization. Subgradient methods. *Trends in Mathematical Optimization, International Series of Numer Math.* Birkhauser, 84:277-293, 1988.

[19] Pham Dinh Tao, Le Thi Hoai An. Stabilité de la dualité lagrangienne en optimisation d.c. (différence de deux fonctions convexes). *C.R. Acad. Paris*, t.318, Série I:379-384, 1994.

[20] Pham Dinh Tao, Le Thi Hoai An. Convex analysis approach to d.c. programming: Theory, Algorithms and Applications (dedicated to Professor Hoang Tuy on the occasion of his 70th birthday). *Acta Mathematica Vietnamica*, Vol. 22, 1:289-355, 1997.

[21] Pham Dinh Tao, Le Thi Hoai An. D.c. optimization algorithms for trust region problem. *SIAM J. Optim.*, Vol. 8, 2:476-505, 1998.

[22] R.T. Rockafellar. *Convex Analysis.* Princeton University, Princeton, 1970.

[23] T. Simchony, R. Cjellapa, Z. Lichtenstein. The graduated non convexity algorithm for image estimation using compound Gauss-Markov field models. In *Glasgow, Editor, Proc. ICASSP,* May 1989.

[24] T. Simchony, R. Cjellapa, Z. Lichtenstein. Pyramid implementation of optimal-step conjugate search algorithms for some low-level vision problems. *IEEE Transactions on Systems, Man, and Cybernetics,* Vol. 19, No 6, November/December 1989.

[25] J.F. Toland. On subdifferential calculus and duality in nonconvex optimization. *Bull. Soc. Math. France,* Mémoire 60:177-183, 1979.

Chapter 18

# NEW POSITIVE SEMIDEFINITE RELAXATIONS FOR NONCONVEX QUADRATIC PROGRAMS

JEAN B. LASSERRE
*LAAS-CNRS*
*7 Avenue du Colonel Roche*
*31077 Toulouse Cédex 4, FRANCE*
lasserre@laas.fr

**Abstract** We consider the general nonconvex quadratic programming problem and provide a series of positive semidefinite (or LMI) relaxations whose sequence of optimal values is monotone and converges to the optimal value of the original problem. It improves and includes as a special case the well-known Shor's LMI relaxation.

**Keywords:** Semidefinite programming, Global optimization, Moment problem.

## 1 INTRODUCTION

In this paper we consider the general nonconvex quadratic programming problem

$$\mathbb{P} \mapsto p^* := \min_{x \in \mathbb{R}^n} \{g_0(x) \,|\, g_i(x) \geq 0,\ i = 1, 2, \ldots, m\}. \tag{18.1}$$

where the $g_i(x)$ are all quadratic polynomials, $i = 0, \ldots m$, and where neither the criterion nor the constraint set are assumed to be convex. We also allow equality constraints since they can be written with two opposite inequalities. In fact, any optimization problem with polynomials can be put in this form in an augmented space (see [11] and [4]). This problem is very difficult in general, for several NP-hard problems can be put in this form. However, there is a well-known relaxation of this problem, known as Shor's relaxation, which replaces $\mathbb{P}$ by a convex



*N. Hadjisavvas and P.M. Pardalos (eds.), Advances in Convex Analysis and Global Optimization,* 319–331.

LMI problem (positive semidefinite (psd) program). This relaxation has been proved very useful for computing estimates of global minima, even in combinatorial optimization. For instance, for the well-known MAX-CUT problem (a special case of (18.1)), very good approximations of a global minimum have been obtained in [5] using Shor's relaxation, followed by a randomized rounding procedure. For related topics, see e.g. [8]

In this paper, we propose a family $\{\mathbb{Q}_i\}$ of convex, linear LMI relaxations (or psd programs) that contains Shor's relaxation as its first member $\mathbb{Q}_1$ and with an associated *increasing* sequence of lower bounds $\{\inf \mathbb{Q}_i\}$.

When the constraint set is compact, the nondecreasing sequence of these lower bounds converges to the global optimal value $p^*$ in (18.1). In fact, in many cases, the global optimal value is reached exactly in a few steps. The approach is based on the theory of moments and recent results on the representation of polynomials that are strictly positive on a compact set. For results on the theory of moments and the representation of positive polynomials, the reader is referred to [3], [2], [10], [9], [6], [7]. It turns out that this theory is a natural and appropriate tool for global optimization since $g_0(x) - p^*$ is precisely a (non strictly) positive polynomial, a feature that distinguishes $p^*$ from the other (local) minima. The LMI relaxations are well-suited since both primal and dual psd programs perfectly match both sides of the same theory (moments and positive polynomials). Indeed, when the optimal value $p^*$ is obtained at some relaxation, say $\mathbb{Q}_i$, the primal psd program provides a global minimizer whereas the dual psd program $\mathbb{Q}_i^*$ provides the coefficients of the polynomials in the decomposition of $g_0(x) - p^*$ into a weighted sum of squares.

Of course, there is a price to pay for these refined relaxations. The number of variables in the LMI relaxation $\mathbb{Q}_i$ is the dimension of the vector space of real-valued polynomials in $n$ variables of degree $2i$, that is, $O(n^{2i})$. However, the $\mathbb{Q}_2$ relaxation is already interesting. On a sample of 50 randomly generated (elementary) MAX-CUT problems in $\mathbb{R}^4$ and $\mathbb{R}^5$, the $\mathbb{Q}_2$ relaxation always provided the optimal solution, in contrast to only 60% of the cases for the $\mathbb{Q}_1$ relaxation.

## 2 NOTATION AND PRELIMINARIES

For all $i = 0, 1, \ldots, m$, let $A_i$ be a real-valued symmetric $(n, n)$-matrix, $c_i \in \mathbb{R}^n$, and consider the general nonconvex quadratic programming problem (18.1) with $g_i(x) = x'A_ix + c_i'x + b_i$, $i = 1, \ldots m$ and $g_0(x) =$

$x'A_0x + c_0'x$. Let $\mathbb{K}$ be the feasible set of $\mathbb{P}$, that is, $\mathbb{K} = \{g_i(x) \geq 0,\ i = 1, \ldots m\}$.

Given any two real-valued symmetric matrices $A, B$ let $\langle A, B\rangle$ denote the usual scalar product trace$(AB)$ and let $A \succeq B$ (resp. $A \succ B$) stand for $A - B$ psd (resp. $A - B$ positive definite). Let

$$1, x_1, x_2, \ldots x_n, x_1^2, x_1x_2, \ldots, x_1x_n, x_2x_3, \ldots, x_n^2, \ldots, x_1^r, \ldots, x_n^r, \quad (18.2)$$

be a basis for the $r$-degree real-valued polynomials and let $s(r)$ be its dimension. Therefore, a $r$-degree polynomial $p(x) : \mathbb{R}^n \to \mathbb{R}$ is written

$$p(x) = \sum_{\alpha \leq r} p_\alpha x^\alpha, \qquad x \in \mathbb{R}^n,$$

where $x^\alpha = x_1^{\alpha_1} x_2^{\alpha_2} \ldots x_n^{\alpha_n}$, with $\sum_{i=1}^n \alpha_i = k$, is a monomial of degree $k$ with coefficient $p_\alpha$. Denote by $p = \{p_\alpha\} \in \mathbb{R}^{s(r)}$ the coefficients of the polynomial $p(x)$ in the basis (18.2).

Hence, the respective vectors of coefficients of the polynomials $g_i(x)$, $i = 0, 1, \ldots m$, are denoted $\{(g_i)_\alpha\} = g_i \in \mathbb{R}^{s(2)}$, $i = 0, 1, \ldots m$.

Given a $s(2r)$-sequence $(1, y_1, \ldots, )$, let $M_r(y)$ be the **moment** matrix of dimension $s(r)$, with rows and columns labelled by (18.2). For instance, for illustration purposes, and for clarity of exposition, consider the 2-dimensional case. The moment matrix $M_r(y)$ is the block matrix $\{M_{i,j}(y)\}_{0 \leq i,j \leq r}$ defined by

$$M_{i,j}(y) = \begin{bmatrix} y_{i+j,0} & y_{i+j-1,1} & \cdots & y_{i,j} \\ y_{i+j-1,1} & y_{i+j-2,2} & \cdots & y_{i-1,j+1} \\ \cdots & \cdots & \cdots & \cdots \\ y_{j,i} & y_{i+j-1,1} & \cdots & y_{0,i+j} \end{bmatrix}. \quad (18.3)$$

To fix ideas, when $n = 2$ and $r = 2$ one obtains

$$M_2(y) = \left[\begin{array}{c|cc|ccc} 1 & y_{10} & y_{0,1} & y_{20} & y_{11} & y_{0,2} \\ \hline y_{10} & y_{20} & y_{11} & y_{30} & y_{21} & y_{12} \\ y_{01} & y_{11} & y_{02} & y_{21} & y_{12} & y_{03} \\ \hline y_{20} & y_{30} & y_{21} & y_{40} & y_{31} & y_{22} \\ y_{11} & y_{21} & y_{12} & y_{3,1} & y_{22} & y_{13} \\ y_{02} & y_{12} & y_{03} & y_{22} & y_{13} & y_{04} \end{array}\right]$$

For the 3-dimensional case, $M_r(y)$ is defined via blocks $\{M_{i,j,k}(y)\}$, $0 \leq i, j, l \leq r$ in a similar fashion, and so on.

If the entry $(i,j)$ of the matrix $M_r(y)$ is $y_\beta$, let $\beta(ij)$ denotes the subscript $\beta$ of $y_\beta$. Next, for a $t$-degree polynomial $\theta(x) : \mathbb{R}^n \to \mathbb{R}$, let $M_r(\theta y)$ be the matrix defined by

$$M_r(\theta y)(i,j) = \sum_\alpha \theta_\alpha y_{\{\beta(i,j)+\alpha\}}. \tag{18.4}$$

For instance, with

$$M_1(y) = \begin{bmatrix} 1 & y_{10} & y_{01} \\ y_{10} & y_{20} & y_{11} \\ y_{01} & y_{11} & y_{02} \end{bmatrix} \quad \text{and } x \mapsto \theta(x) = a - x_1^2 - x_2^2,$$

we obtain

$$M_1(\theta y) = \begin{bmatrix} a - y_{20} - y_{02}, & ay_{10} - y_{30} - y_{12}, & ay_{01} - y_{21} - y_{03} \\ ay_{10} - y_{30} - y_{12}, & ay_{20} - y_{40} - y_{22}, & ay_{11} - y_{31} - y_{13} \\ ay_{01} - y_{21} - y_{03}, & ay_{11} - y_{31} - y_{13}, & ay_{02} - y_{22} - y_{04} \end{bmatrix}.$$

$M_r(y)$ define a bilinear form $\langle .,. \rangle_y$ on the space $\mathcal{A}_r$ of $r$-degree real-valued polynomials by

$$\langle q(x), v(x) \rangle_y := \langle q, M_r(y)v \rangle, \quad q(x), v(x) \in \mathcal{A}_r,$$

and if $y$ is a sequence of moments of some measure $\mu_y$, then

$$\langle q, M_r(y)q \rangle = \int q(x)^2 \, \mu_y(dx) \geq 0, \tag{18.5}$$

so that $M_r(y) \succeq 0$, and similarly,

$$\langle q, M_r(\theta y)q \rangle = \int \theta(x) q(x)^2 \, \mu_y(dx), \tag{18.6}$$

so that $M_r(\theta y) \succeq 0$ whenever $\mu_y$ has its support contained in $\{\theta(x) \geq 0\}$. The theory of *moments* identifies those sequences $y$ with $M_r(y) \succeq 0$, which are moment-sequences. For details and recent results, the interested reader is referred to [2], [3], [12], [10] and the many references therein.

## 3 A FAMILY OF LMI RELAXATIONS

Consider the problem $\mathbb{P}$ with feasible set $\mathbb{K} = \{g_i(x) \geq 0, i = 1, \ldots m\}$ and criterion $g_0(x)$. When needed below, the vectors $g_i \in \mathbb{R}^{s(2)}$ are extended to vectors of $\mathbb{R}^{s(2i)}$ by completing with zeros, i.e., the quadratic polynomials polynomials $g_i(x)$, $i = 0, \ldots m$, are considered as

$2i$ degree polynomials with null coefficients for terms of degree larger than 2. For $i = 1, 2, \ldots$, consider the following family of convex, linear LMI problems

$$\mathbb{Q}_i \begin{cases} \min\limits_y \sum\limits_\alpha (g_0)_\alpha y_\alpha & \\ \qquad M_i(y) \succeq 0 & \\ \qquad M_{i-1}(g_k y) \succeq 0, & k = 1, \ldots m. \end{cases} \tag{18.7}$$

with respective dual problems

$$\mathbb{Q}_i^* \begin{cases} \max\limits_{X, Z_k \succeq 0} -X(1,1) - \sum\limits_{k=1}^m g_k(0) Z_k(1,1) & \\ \qquad \langle X, B_\alpha \rangle + \sum\limits_{k=1}^m \langle Z_k, C_\alpha^k \rangle = (g_0)_\alpha, \ \forall \alpha \neq 0, & \end{cases} \tag{18.8}$$

where we have written

$$M_i(y) = \sum_\alpha B_\alpha y_\alpha; \ M_{i-1}(g_k y) = \sum_\alpha C_\alpha^k y_\alpha, \ k = 1, \ldots m,$$

(with $y_0 = 1$) for appropriate real-valued symmetric matrices $B_\alpha, C_\alpha^k$, $k = 1, \ldots m$. At this stage, it is worthy to write down the LMI program $\mathbb{Q}_1$, that is, when $i = 1$,

$$\mathbb{Q}_1 \begin{cases} \min\limits_y \sum\limits_\alpha (g_0)_\alpha y_\alpha & \\ \qquad M_1(y) \succeq 0 & \\ \qquad \sum\limits_\alpha (g_k)_\alpha y_\alpha \geq 0 & k = 1, \ldots, m \end{cases} \tag{18.9}$$

Observe that, if we write

$$M_1(y) = \begin{bmatrix} 1 & y' \\ y & Y \end{bmatrix},$$

and remembering that $g_k(x) = x' A_k x + c_k' x + b_k$, $k = 0, 1, \ldots m$, then $\mathbb{Q}_1$ has the equivalent form

$$\mathbb{Q}_1 \begin{cases} \min\limits_{Y,y} \langle A_0, Y \rangle + c_0' y & \\ \qquad \langle A_k, Y \rangle + c_k' y \geq -b_k, & k = 1, \ldots m \\ \qquad \begin{bmatrix} 1 & y' \\ y & Y \end{bmatrix} \succeq 0 & \end{cases} \tag{18.10}$$

which is the dual of the well-known Shor's relaxation for quadratic programs. Therefore, Shor's relaxation is the first LMI program $\mathbb{Q}_1$ in the hierarchy of relaxations $\{\mathbb{Q}_i\}$. We first have:

**Proposition 3.1.** *The family of convex LMI problems $\{\mathbb{Q}_i\}$ satisfies*

$$\inf \mathbb{Q}_i \leq \inf \mathbb{Q}_{i+1} \leq \inf \mathbb{P}, \qquad i = 1, 2, \ldots. \tag{18.11}$$

*Proof.* That $\inf \mathbb{Q}_i \leq \inf \mathbb{P}$ is immediate because to every admissible point $x$ of $\mathbb{P}$, corresponds a solution

$$y^x := (x_1, x_2, \ldots x_1^{2i}, \ldots x_n^{2i}),$$

admissible for $\mathbb{Q}_i$, and thus $\inf \mathbb{Q}_i \leq \inf \mathbb{P}$ for every $i = 1, \ldots$.

Next, consider an LMI problem $\mathbb{Q}_i$, with $i > 1$, and let $y = \{y_\alpha\}$ a feasible sequence for $\mathbb{Q}_i$, that is, $(1, y)$ is a vector of dimension $s(2i)$, the dimension of the vector space of real-valued polynomials $p(x) : \mathbb{R}^n \to \mathbb{R}$, of degree $2i$. If we write $y = (y_1, y_2)$ with $(1, y_1)$ of dimension $s(2(i-1))$, we have

$$M_i(y) = \begin{bmatrix} M_{i-1}(y_1) & M \\ M' & N \end{bmatrix}; \; M_{i-1}(g_k y) = \begin{bmatrix} M_{i-2}(g_k y_1) & V \\ V' & NS \end{bmatrix},$$

for appropriate matrices $M, N, V, S$. Therefore,

$$M_i(y) \succeq 0 \Rightarrow M_{i-1}(y_1) \succeq 0; \; M_{i-1}(g_k y) \succeq 0 \Rightarrow M_{i-2}(g_k y_1) \succeq 0,$$

so that $y_1$ is admissible for $\mathbb{Q}_{i-1}$. Moreover, the value of $y$ in $\mathbb{Q}_i$ is the same as the value of $y_1$ in $\mathbb{Q}_{i-1}$, and the result follows. □

Proposition 3.1 ensures that better and better lower bounds on $\mathbb{P}$ can be obtained by solving the relaxations $\mathbb{Q}_i$, $i = 1, \ldots$. The next result shows that in fact, whenever $\mathbb{K}$ is compact, one may approach as closely as desired, the optimal value $p^* = \inf \mathbb{P}$. We will use the fact that under a certain condition on the feasible set $\mathbb{K}$, every polynomial $p(x)$, strictly positive on $\mathbb{K}$, has the following representation:

$$p(x) = \sum_{j=1}^{r_0} q_j(x)^2 + \sum_{k=1}^{m} g_k(x) \left[ \sum_{j=1}^{r_k} q_{kj}(x)^2 \right] \tag{18.12}$$

for a finite family of polynomials $\{q_j(x)\}$, $j = 1, \ldots r_0$, and $\{q_{kj}(x)\}$, $j = 1, \ldots r_k$, $k = 1, \ldots m$. In fact, a necessary and sufficient condition for the representation (18.12) to exist is that there exists a polynomial $u(x)$ of the form (18.12) such that $\{u(x) \geq 0\}$ is compact (see [9] and [6]).

For instance, the representation (18.12) holds whenever $\{g_k(x) \geq 0\}$ is compact for some $k$, or when all the $g_k(x)$ are linear and $\mathbb{K}$ is compact. In particular, it holds for every 0-1 program. Indeed, write the integral constraints as $x_i^2 - x_i \geq 0$ and $x_i - x_i^2 \geq 0$, for all $i = 1, \ldots n$. Then, consider the polynomial $u(x) := \sum_{i=1}^n (x_i - x_i^2)$. Its level set $\{u(x) \geq 0\}$ is compact. Moreover, if one knows that a global minimizer is contained in some ball $\|x\|^2 \leq M$, for some $M$ large enough, then one may add the redundant constraint $g_{m+1}(x) := M - \sum_{i=1}^n x_i^2 \geq 0$, and the set $\mathbb{K}$ (defined as previously with the latter additional constraint $g_{m+1}(x) \geq 0$) has the required property. And, we have:

**Theorem 3.2.** *Assume that there is some polynomial $u(x) : \mathbb{R}^n \to \mathbb{R}$ of the form (18.12) and with $\{u(x) \geq 0\}$ compact. Then,*

$$\inf \mathbb{Q}_i \uparrow \min \mathbb{P}. \tag{18.13}$$

*If $\mathbb{K}$ has a nonempty interior, then*

$$\max \mathbb{Q}_i^* = \inf \mathbb{Q}_i \uparrow \min \mathbb{P}. \tag{18.14}$$

*The equality* $\max \mathbb{Q}_i^* = \min \mathbb{P}$ *occurs for all $i \geq i_0$ (for some index $i_0$) if and only if $g_0(x) - p^*$ is of the form (18.12).*

*Proof.* Let $\epsilon > 0$ be fixed arbitrary. Then, the polynomial $x \mapsto G_0(x) := g_0(x) - p^* + \epsilon$ is strictly positive on $\mathbb{K}$. From the assumption on $\mathbb{K}$, it follows that $G_0(x)$ has the representation

$$G_0(x) = \sum_{j=1}^{r_0} q_j(x)^2 + \sum_{k=1}^{m} g_k(x) \left[ \sum_{j=1}^{r_k} q_{kj}(x)^2 \right] \tag{18.15}$$

for some polynomials $q_j(x)$, $j = 1, \ldots r_0$, and $q_{kj}(x)$, $j = 1, \ldots r_k$, $k = 1, \ldots m$ (see e.g. [9], [6]). Now, let $i_1(\epsilon)$ (that for notational convenience we simply write $i_1$) be the maximum degree of $q_j(x)$, $j = 1, \ldots r_0$, and $i_2(\epsilon)$ (noted $i_2$) be the maximum degree of the polynomials $\{q_{kj}(x)\}$, so that the polynomials $\{g_k(x) q_{kj}(x)^2\}$ have maximum degree $2i_2 + 2$ (as the $g_k(x)$'s are quadratic polynomials). Let $i := \max[i_1, i_2 + 1]$. Let $q_j \in \mathbb{R}^{s(i)}$ and $q_{kj} \in \mathbb{R}^{s(i-1)}$ be the respective vectors of coefficients of the polynomials $\{q_j(x)\}$ and $\{q_{kj}(x)\}$, and write

$$X := \sum_{j=1}^{r_0} q_j q_j', \quad Z_k := \sum_{j=1}^{r_k} q_{kj} q_{kj}', \ k = 1, \ldots m. \tag{18.16}$$

Observe that with $y^x = (x_1, \ldots, x_1^{2i} \ldots x_n^{2i})$, one has

$$g_k(x) q_{kj}(x)^2 = \langle q_{kj}, M_{i-1}(g_k y^x) q_{kj} \rangle = \langle q_{kj} q_{kj}', M_{i-1}(g_k y^x) \rangle,$$

so that

$$g_k(x)\sum_{j=1}^{r_k} q_{kj}(x)^2 = \langle Z_k, M_{i-1}(g_k y^x)\rangle = \sum_\alpha x^\alpha \langle Z_k, C_\alpha^k\rangle, \qquad (18.17)$$

and similarly,

$$\sum_{j=1}^{r_0} q_j(x)^2 = \sum_\alpha x^\alpha \langle X, B_\alpha\rangle, \qquad (18.18)$$

with $B_\alpha, C_\alpha^k$ as in $\mathbb{Q}_i^*$. Therefore, from the representation (18.15),

$$\sum_\alpha x^\alpha \left[\langle X, B_\alpha\rangle + \sum_{k=1}^m \langle Z_k, C_\alpha^k\rangle\right] = \sum_\alpha x^\alpha (g_0)_\alpha - p^* + \epsilon. \qquad (18.19)$$

Identifying terms of same power yields

$$\langle X, B_\alpha\rangle + \sum_{k=1}^m \langle Z_k, C_\alpha^k\rangle = (g_0)_\alpha \;\; \forall \alpha \neq 0,$$

and, for the constant term, $X(1,1) + \sum_{k=1}^m Z_k(1,1)g_k(0) = -p^* + \epsilon$, which proves that $(X, Z_1 \ldots Z_m)$ is admissible for $\mathbb{Q}_i^*$ with value $p^* - \epsilon$. Therefore, $p^* - \epsilon \leq \inf \mathbb{Q}_i \leq p^*$, and from the monotonicity and the fact that $\epsilon$ was arbitrary, $\inf \mathbb{Q}_i \uparrow p^* = \min \mathbb{P}$.

The statement (18.14) follows from a standard result in convexity. Indeed, we just proved that $\mathbb{Q}_i^*$ has a feasible solution $(X, Z_1, \ldots Z_m)$. Moreover, if $\mathbb{K}$ has a nonempty interior, let $\mu$ be a probability measure with support $\mathbb{K}$ and with a uniform distribution. Denote by $y^\mu$ the vector of all the (well-defined) moments of $\mu$. From (18.5)-(18.6) and the fact that $\mu$ has a uniform distribution on $\mathbb{K}$, it follows easily that $M_i(y^\mu) \succ 0$ and $M_{i-1}(g_k y^\mu) \succ 0$ for all $k = 1, \ldots m$, so that $y^\mu$ is a strictly admissible solution. Therefore, there is no duality gap between $\mathbb{Q}_i^*$ and $\mathbb{Q}_i$, and (18.14) follows.

Finally, consider the *if part* of the final statement. From the representation (18.12) of $g_0(x) - p^*$ and using the same above arguments, the corresponding matrices $X, Z_1, \ldots Z_m$ form an admissible solution for $\mathbb{Q}_i^*$ with value $p^* = \min \mathbb{P}$, hence proving $\max \mathbb{Q}_i^* = \min \mathbb{P}$.

Conversely, the *only if part* of the final statement also follows easily from the above proof. Let $(X, Z_1, \ldots Z_m)$ be an optimal solution of $\mathbb{Q}_i^*$ with $\max \mathbb{Q}_i^* = p^* = \min \mathbb{P}$. As $X, Z_1, \ldots Z_m \succeq 0$, write them as in (18.16). The admissiblity of $(X, Z_1, \ldots Z_m)$ implies (18.19) with $p^*$ in lieu of $p^* - \epsilon$, which in turn, using (18.17)-(18.18) yields the desired result. □

The representation (18.12) has a nice interpretation as a **global optimality condition** à la Karush-Kuhn-Tucker. Indeed,

**Proposition 3.3.** *Assume that $g_0(x)-p^*$ has the representation (18.12). Then, at a global minimizer $x^*$, we have*

$$\nabla g_0(x^*) = \sum_{k=1}^{m} \nabla g_k(x^*) \left[\sum_{j=1}^{r_k} q_{kj}(x^*)^2\right] \tag{18.20}$$

$$g_k(x^*) \left[\sum_{j=1}^{r_k} q_{kj}(x^*)^2\right] = 0, \quad k = 1, \ldots m. \tag{18.21}$$

*Proof.* The proof is immediate. As $g_0(x^*) - p^* = 0$, the representation (18.12) leads

$$\sum_{j=1}^{r_0} q_j(x^*)^2 = 0; \; \sum_{k=1}^{m} g_k(x^*) \left[\sum_{j=1}^{r_k} q_{kj}(x^*)^2\right] = 0,$$

so that (18.21) follows. Differentiating in (18.12) and using the above relationship also yields (18.20). □

Hence, the theory of representation of polynomials, positive on the feasible set $\mathbb{K}$, can be viewed as a global optimality condition. The polynomials (sums of squares) $\sum_j q_{kj}(x)^2$ in the representation (18.12) of $g_0(x) - p^*$ (when it holds) are nothing less than *Lagrange Karush-Kuhn-Tucker* "multipliers". In contrast to the usual Kuhn-Tucker necessary optimality condition with "scalar multipliers" $\lambda_k$, a non-active constraint $g_k(x)$ at a global minimizer $x^*$ may have an nontrivial associated "polynomial" multiplier when this constraint plays a role to eliminate some better (non feasible) solutions. However, this polynomial multiplier vanishes at $x^*$, as in the usual Kuhn-Tucker conditions where $\lambda_k = 0$ at a non active constraint.

# 4 EXAMPLES

For illustration purposes we will consider the MAX-CUT problem and some elementary nonconvex quadratic problems.

## 4.1 NONCONVEX QUADRATIC PROBLEMS

Let $g_0(x) : \mathbb{R}^2 \to \mathbb{R}$ be the concave polynomial $x \mapsto p(x) := -(x_1 - 1)^2 - (x_1 - x_2)^2 - (x_2 - 3)^2$ and

$$\mathbb{K} := \{1 - (x_1 - 1)^2 \geq 0;\ 1 - (x_1 - x_2)^2 \geq 0;\ 1 - (x_2 - 3)^2 \geq 0\}.$$

The point $(1,2)$ is a global minimizer with optimal value $-2$. Solving $\mathbb{Q}_1$ with $\tilde{g}_0(x) = g_0(x) + 10$ (since we eliminate the constant term $-10$) yields an optimal value of 7 instead of the desired value 8. On the other hand, solving $\mathbb{Q}_2$ yields an optimal value 8.00017 and an approximate global minimizer $(1.0043, 2.0006)$ (the error 0.00017 is likely due to the use of an interior point method in MATLAB). Observe that there exist $\lambda_i = 1 \geq 0$, $i = 1, \ldots m$, such that $g_0(x) + 3 = 0 + \sum_{i=1}^m \lambda_i g_i(x)$, but $g_0(x) - p^*$ $(= g_0(x) + 2)$ cannot be written that way.

## 4.2 LINEAR-QUADRATIC PROBLEMS

We next consider the case where $g_0(x)$ is a nonconvex quadratic polynomial and the $g_i(x)$ are all linear, $i = 1, 2, \ldots m$, and $\mathbb{K}$ is compact. Clearly, in this case, $\mathbb{Q}_1$ provides a useless lower bound $-\infty$ as $Q_0$ has at least one negative eigenvalue. An alternative proposed in [11] is to add redundant quadratic inequalites $g_i(x)g_j(x) \geq 0$ for all $i, j = 1, \ldots m$ and then consider the $\mathbb{Q}_1$ relaxation of this new equivalent problem. When the constraint set $\mathbb{K}$ is compact with a nonempty interior $\mathbb{Q}_1$ will provide a finite lower bound.

Let $g_0(x) : \mathbb{R}^2 \to \mathbb{R}$ be the polynomial $x \mapsto g_0(x) := -a_1 x_1^2 - a_2 x_2^2$ and let $\mathbb{K}$ be the compact set

$$\mathbb{K} := \{x \in \mathbb{R}^2 \,|\, x_1 + x_2 \leq b_1; ax_1 + y \leq b_2;\ x_1, x_2 \geq 0\}.$$

As $\mathbb{K}$ is compact and all the constraints are linear, $\mathbb{K}$ satisfies the property required for the representation (18.12) of polynomials, strictly positive on $\mathbb{K}$. Whenever, $a_i > 0$, $g_0(x)$ is concave, so that we have a concave minimization problem and thus, some vertex of the polytope $\mathbb{K}$ is a global minimizer. We have solved $\mathbb{Q}_2$ for several values of $a_i > 0, b_i$, $i = 1, 2$ and $a < 0$, each time providing a global minimizer exactly. This is because $g_0(x) - p^*$ has the representation (18.12).

## 4.3 THE MAX-CUT PROBLEM

Roughly speaking, the MAX-CUT problem is a special case of (18.1) with

- $g_0(x) = x' A_0 x$ (and $\mathrm{diag}(A_0) = 0$)
- and the constraints $x_i^2 = 1$, $\forall i = 1, \ldots n$.

As for 0-1 programs, the feasible set $\mathbb{K}$ satisfies the condition required in Theorem 3.2. Indeed, write the integral constraint $x_i^2 = 1$ as $x_i^2 - 1 \geq 0$ and $1 - x_i^2 \geq 0$ for all $i = 1, \ldots n$, and consider the polynomial $u(x) := \sum_{i=1}^n (1 - x_i^2)$. Its level set $\{u(x) \geq 0\}$ is compact. Shor's relaxation (equivalently, $\mathbb{Q}_1$), is the convex LMI problem:

$$\mathbb{Q}_1 \left\{ \begin{array}{l} \min \langle Y, A_0 \rangle \\ \\ \left[ \begin{array}{ccc} 1 & | & y' \\ - & - & - \\ y & | & Y \end{array} \right] \succeq 0; \ \operatorname{diag}(Y) = e \end{array} \right. \tag{18.22}$$

where $e$ is a vector of ones. To visualize the $\mathbb{Q}_2$ relaxation, take an example in $\mathbb{R}^3$, in which case $\mathbb{Q}_2$ reads

$$\min q_{110}y_{110} + q_{101}y_{101} + q_{011}y_{011}$$

$$\left[ \begin{array}{cccccccccc}
1 & y_{100} & y_{010} & y_{001} & 1 & y_{110} & y_{101} & 1 & y_{011} & 1 \\
y_{100} & 1 & y_{110} & y_{101} & y_{100} & y_{010} & y_{001} & y_{100} & y_{111} & y_{100} \\
y_{010} & y_{110} & 1 & y_{011} & y_{010} & y_{100} & y_{111} & y_{110} & y_{001} & y_{010} \\
y_{001} & y_{101} & y_{011} & 1 & y_{001} & y_{111} & y_{100} & y_{001} & y_{010} & y_{001} \\
1 & y_{100} & y_{010} & y_{001} & 1 & y_{110} & y_{101} & 1 & y_{011} & 1 \\
y_{110} & y_{010} & y_{100} & y_{111} & y_{110} & 1 & y_{011} & y_{110} & y_{101} & y_{110} \\
y_{101} & y_{001} & y_{111} & y_{100} & y_{101} & y_{011} & 1 & y_{101} & y_{110} & y_{101} \\
1 & y_{100} & y_{110} & y_{001} & 1 & y_{110} & y_{101} & 1 & y_{011} & 1 \\
y_{011} & y_{111} & y_{001} & y_{010} & y_{011} & y_{101} & y_{110} & y_{011} & 1 & y_{011} \\
1 & y_{100} & y_{010} & y_{001} & 1 & y_{110} & y_{101} & 1 & y_{011} & 1
\end{array} \right] \succeq 0$$

where we have used the fact that the equality $x_{100}^2 = 1$ translates into $y_{200} = 1$ and therefore, a term like $y_{400}$ is replaced by 1, a term like $y_{210}$ is replaced by $y_{010}$, $y_{211}$ by $y_{011}$, etc. so that only the variables $y_\alpha$ with $\alpha_i \leq 1$ and $\sum_i \alpha_i \leq 4$ are present.

Hence, the $\mathbb{Q}_2$ relaxation of a MAX-CUT problem of dimension $n$ is an LMI problem with $\binom{n+1}{4} + \binom{n+1}{2}$ variables and one LMI constraint of dimension $(n+1)(n+2)/2$ in comparison with $\binom{n+1}{2}$ variables and one LMI constraint of dimension $n+1$ for the $\mathbb{Q}_1$ relaxation. Of course, as $i$ increases, then so does the computational burden, a price to pay to obtain better and better lower bounds. For a sample of 50 randomly generated problems in $\mathbb{R}^5$ with positive and negative weights, the $\mathbb{Q}_2$ relaxation provided the optimal solution in all cases whereas the $\mathbb{Q}_1$ relaxation in less than 50% of the cases only. For the case of equal weights, both $\mathbb{Q}_1$ and $\mathbb{Q}_2$ give the same lower bound as the relaxations proposed in [1], whereas $\mathbb{Q}_3$ provides the optimal value.

We have to mention that in using the MATLAB LMI toolbox, the running time for $\mathbb{Q}_2$ was surprisingly high compared to $\mathbb{Q}_1$, despite the fact that $\mathbb{Q}_2$ contains relatively few variables (15 for $\mathbb{R}^4$ and 30 for $\mathbb{R}^5$).

# 5 CONCLUSION

We have proposed a sequence of LMI relaxations $\{\mathbb{Q}_i\}$ whose sequence of optimal values converges to the global minimum $p^*$, and in many cases, $p^*$ is obtained at a particular relaxation. The primal and dual psd programs perfectly match both sides of the same theory (moments and positive polynomials). Moreover, the representation of polynomials, positive on the feasible set $\mathbb{K}$, is interpreted as a Karush-Kuhn-Tucker global optimality condition with "polynomial" Lagrange multipliers instead of scalar multipliers. However, high order relaxations require a lot of variables. It is hoped that for a large class of problems, the $\mathbb{Q}_2$ and $\mathbb{Q}_3$ relaxations will provide a global minimizer, or at least, a good lower bound that could be exploited in other optimization methods.

## References

[1] Anjos, M. and Wolkowicz, H. (2000). A tight semidefinite relaxation of the cut polytope. Technical Report CORR 2000-19, University of Waterloo, Waterloo, Canada.

[2] Berg, C. (1980). The multidimensional problem and semi-groups. In *Proceedings of Symposia in Applied Mathematics*, volume 37, pages 110–124.

[3] Curto, R. and Fialkow, L. (1991). Recursiveness, positivity, and truncated moment problems. *Houston J. Math*, 17:603–635.

[4] Ferrier, C. (1998). Hilbert's 17th problem and best dual bounds in quadratic minimization. *Cybern. Syst. Anal.*, 34:696–709.

[5] Goemans, M. and Williamson, D. (1995). Improved approximation algorithms for maximum cut and satisfiability problems using semidefinite programming. *J. ACM*, 42:1115–1145.

[6] Jacobi, T. (2000). A representation theorem for certain partially ordered commutative rings. *Mathematische Zeitschrift.*

[7] Jacobi, T. and Prestel, A. (2000). On special representations of strictly positive polynomials. *Technical Report.*

[8] Pardalos, P. P. and Wolkowicz, H., editors (1998). *Topics in Semidefinite and Interior-Point Methods.* Fields Institute Communications Series Vol. 18, American Mathematical Society.

[9] Putinar, M. (1993). Positive polynomials on compact semi-algebraic sets. *Ind. Univ. Math. J.*, 42:969–984.

[10] Schmudgen, K. (1991). The $K$-moment problem for compact semi-algebraic sets. *Math. Ann.*, 289:203–206.

[11] Shor, N. (1987). Quadratic optimization problems. *Tekhnicheskaya Kibernetika*, 1:128–139.

[12] Simon, B. (1998). The classical moment problem as a self-adjoint finite difference operator. *Adv. Math.*, 137:82–203.

Chapter 19

# INTERVAL ANALYSIS APPLIED TO GLOBAL MINIMIZATION OF POTENTIAL ENERGY FUNCTIONS

Carlile Lavor
*Federal University of Rio de Janeiro*
*P.O. Box 68511, Rio de Janeiro - RJ, Brazil*
carlile@cos.ufrj.br

Nelson Maculan
*Federal University of Rio de Janeiro*
*P.O. Box 68511, Rio de Janeiro - RJ, Brazil*
maculan@cos.ufrj.br

**Abstract** Computing the global minimum of a potential energy function is very difficult because it typically has a very large number of local minima which may grow exponentially with problem size. In this work we use a deterministic algorithm for finding the global minimum of this function. The algorithm is based on a branch and bound method that uses techniques of interval analysis. Using the Lennard-Jones potential function, the proposed approach was successfully applied to two example problems.

**Keywords:** Molecular conformations, Global optimization, Branch and bound, Interval analysis.

## 1 INTRODUCTION

The problem of minimizing potential energy functions is an instance of the general global optimization problem: given $f : \Re^n \to \Re$ a continuous function and $S \subset \Re^n$, find its global minimum $f^* = \min\{f(x) : x \in S\}$, and the set $X^*$ of all global minimizers $X^*(f) = \{x^* \in S : f(x^*) = f^*\}$.



*N. Hadjisavvas and P.M. Pardalos (eds.), Advances in Convex Analysis and Global Optimization,* 333–344.

Most of the nonlinear programming methods that have been developed find only a local optimum. In general, however, several local optima may exist and the corresponding function values may differ substantially.

A natural approach to solve the global optimization problem is through an appropriate generalization of branch and bound method, a deterministic technique that is well known from the area of combinatorial optimization ([3], [11], [14]).

A branch and bound algorithm generally alternate between two main steps: *branching*, which is a recursive subdivision of the set $S$; and *bounding*, which is the computation of lower and upper bounds for the global minimum of $f$ in a subregion of $S$. Subregions where the lower bound for $f$ is greater than the current upper bound for the global minimum of $f$ can be discarded because they cannot contain a global minimizer. Subregions that cannot be discarded in this way are kept in a list $L$ to be further processed. The algorithm stops based on the sizes of the subregions in $L$ and on the estimated range for $f$.

The lower bounding operations play a central role in branch and bound method. They are typically established by minorant functions of $f$, which are required to be increasingly more accurate.

A frequently applied technique to provide lower bounds for $f$ is to use suitable convex minorant functions, for example, convex envelopes ([1], [6], [10], [16]). Other possibility is to use techniques of *interval analysis*, which is the approach of this work.

The remainder of the paper is organized as follows. Section 2 describes the potential energy model used to represent the interactions between the atoms of a given molecule. Section 3 presents some definitions and results of interval analysis and the algorithm applied to the problem. In section 4 we present the computational results and section 5 gives some conclusions and the research in progress.

# 2 POTENTIAL ENERGY OF A MOLECULE

The number of possible conformations of a molecule depends on the number of the freely rotating bonds. The conformation of a molecule describes the spacial arrangement of groups of atoms about these rotating bonds. Changing the conformation of a molecule does not make a new molecule, but can change its properties. One of these conformations, the most stable one, is of particular importance because it dictates most of the properties of the molecule. Experimental evidence shows that in the great majority of cases the most stable conformation corresponds to the one involving the global minimum potential energy of the molecule.

The more accurate representation of the potential energy of a molecule is the *ab initio* quantum mechanical approach. However, quantum mechanical calculations are not feasible because of the large associated computational effort.

We use the results of *molecular mechanics*, where potential energy functions are derived empirically, as opposed to *ab initio* derivations from quantum mechanics.

A description of the total potential energy in a molecular system must include the intermolecular interactions among molecules and the intramolecular interactions among atoms within the molecule. The potential energy of a single molecule depends only on the intramolecular interactions. The total intramolecular potential energy $V$ is the sum of two types of interactions, the bonding and the nonbonding interactions. The bonding interactions are the covalent bonds that hold the atoms together, while the nonbonding interactions include electrostatic, dipolar, and steric interactions. To calculate the total potential, it is often sufficient to add together the bonded and nonbonded interactions between all pairs of atoms.

The total potential energy expression for a molecule can then be given by

$$V = \sum_{ij \in B} v_{ij} + \sum_{ij \in NB} v_{ij},$$

where $B$ is the set of bonded atoms, $NB$ is the set of nonbonded atoms, and $v_{ij}$ is the pairwise potential expression.

Each pairwise potential term $v_{ij}$ is a function of only the Euclidean distances $r_{ij}$ between the interacting atoms. In most conformational calculations, all covalent bond lengths and covalent bond angles are assumed to be fixed at their equilibrium values. Thus, in the Cartesian coordinate space, the problem can be formulated as follows:

$$\min \sum_{ij \in NB} v_{ij}(r_{ij})$$

subject to

$$\sqrt{(x_i - x_j)^2 + (y_i - y_j)^2 + (z_i - z_j)^2} = r^0_{ij}, \quad \forall ij \in B$$

and

$$-\frac{(x_i - x_j)(x_j - x_k) + (y_i - y_j)(y_j - y_k) + (z_i - z_j)(z_j - z_k)}{r^0_{ij} r^0_{jk}} = \cos(\theta^0_{ijk}),$$

$\forall ijk$ such that $ij, jk \in B$ and

$$x_1 = y_1 = z_1 = y_2 = z_2 = z_3 = 0,$$

where $v_{ij}(r_{ij})$ is the pairwise potential expression; $r_{ij}$ is the Euclidean distance between atom i and atom j; $r^0_{ij}$ and $\theta^0_{ijk}$ are the fixed covalent bond lengths and angles respectively; and $(x_i, y_i, z_i)$ are the Cartesian coordinates of the atom $i$. By imposing $x_1 = y_1 = z_1 = y_2 = z_2 = z_3 = 0$ we eliminate the translational and rotational degrees of freedom of the molecule.

By employing Cartesian coordinates we introduce a number of non-linear equality constraints, making the problem very difficult. On the other hand, because the covalent bond lengths and angles are assumed to be constant, the set of internal coordinates (bond lengths, covalent bond angles, and torsion angles) reduces to a set of torsion angles (in a chain $i - j - k - l$ of atoms, a torsion angle is defined as the angle between the normals through the planes determined by the atoms $i, j, k$ and $j, k, l$, respectively). In principle, this is a considerable advantage, because we can reduce the number of variables of the problem.

Then, if we use the set of internal coordinates, the problem can be written as follows:

$$\min \sum_{ij \in NB} v_{ij}\left(r_{ij}(w_k)\right)$$

$$\text{subject to } -\pi \leq w_k \leq \pi,$$

where $w_k$ is a torsion angle, $k = 1, ..., m$.

However, the relation between $r_{ij}$ and $w_k$ has not been established yet. Explicit expressions for the Cartesian coordinates $(x_m, y_m, z_m)$ of the $m^{th}$ atom in a m-atom chain can be obtained as function of the bond lengths, covalent bond angles, and torsion angles [9]. For example, the Cartesian coordinates $(x_4, y_4, z_4)$ of the fourth atom in a atom chain with four atoms can be given by ($t_i$ is the angle between the covalent bonds $(i-2, i-1)$ and $(i-1, i)$):

$$\begin{aligned}
x_4 &= -r_{1,2} + r_{2,3}\cos(t_3) - r_{3,4}\cos(t_3)\cos(t_4) + \\
&\quad r_{3,4}\sin(t_3)\sin(t_4)\cos(w_1), \\
y_4 &= r_{2,3}\sin(t_3) - r_{3,4}\sin(t_3)\cos(t_4) - r_{3,4}\cos(t_3)\sin(t_4)\cos(w_1), \\
z_4 &= -r_{3,4}\sin(t_4)\sin(w_1).
\end{aligned}$$

## 3 INTERVAL ANALYSIS AND THE ALGORITHM

Interval analysis, also known as interval arithmetic, is a model for numerical computation where each real quantity $x$ is represented by an interval $X$ of floating-point numbers. The arithmetic of interval analysis is defined on sets of intervals, rather than sets of real numbers. Those intervals are added, subtracted, multiplied, divided, etc., in such a way that each computed interval $X$ is guaranteed to contain the (unknown) value of the corresponding real quantity $x$.

Interval analysis has been used in many varied fields, such as numerical analysis, computer graphics, optimization, etc. The principal use of interval analysis is to bound rounding errors in numerical computations. However, it has had a much more profound impact on optimization than just providing a means for bounding rounding errors. It makes it possible to solve a global optimization problem and guarantee that a global optimum has been found.

Traditional global optimization algorithms can sample the objective function at only a finite number of distinct points. Hence, there is no way of knowing whether a function to be minimized drops to some unexpectedly small value between sample points. On the other hand, interval methods do not sample at points. They obtain bounds for a function over a continuum of points, including ones that are not finitely representable.

Modern development of interval analysis began with Moore's thesis [12], in 1962, then a Ph.D. student at Stanford University.

Despite the value of interval analysis for bounding rounding errors, interval mathematics is less used in practice than one might expect. However, there has been a resurgence of interest in interval analysis. Practitioners and researches in the most varied fields have found that interval analysis provides a simple method to solve computational problems that were intractable under the "classical" approach. This revival of interval analysis was greatly helped by the publishing and universal acceptance of the IEEE floating-point standard.

In the following, a real number $x$ will be considered equivalent to an interval $[x, x]$.

If $X = [\underline{x}, \overline{x}]$ and $Y = [\underline{y}, \overline{y}]$ are intervals of real numbers, the elementary operations for interval arithmetic are defined as

$$X \circ Y := [\min_{x \in X, y \in Y} (x \circ y), \max_{x \in X, y \in Y} (x \circ y)],$$

where $\circ$ can be $+$, $-$, $\times$ or $\div$.

Using this definition, we find that

$$\begin{aligned} X + Y &= [\underline{x} + \underline{y}, \overline{x} + \overline{y}], \\ X - Y &= [\underline{x} - \overline{y}, \overline{x} - \underline{y}], \\ X * Y &= [a, b] \ \ (a = \min\{\underline{x}\underline{y}, \overline{x}\underline{y}, \underline{x}\overline{y}, \overline{x}\overline{y}\} \text{ and } b = \max\{\underline{x}\underline{y}, \overline{x}\underline{y}, \underline{x}\overline{y}, \overline{x}\overline{y}\}), \\ \frac{1}{Y} &= \left[\frac{1}{\overline{y}}, \frac{1}{\underline{y}}\right] \ \ (0 \notin Y). \end{aligned}$$

We also define $(n = 0, 1, 2, ...)$

$$\frac{X}{Y} := X * \frac{1}{Y} \quad (0 \notin Y),$$

$$X^n := \begin{cases} [1,1] & \text{if } n = 0, \\ [\underline{x}^n, \overline{x}^n] & \text{if } \underline{x} \geq 0 \text{ or if } n \text{ is odd,} \\ [\overline{x}^n, \underline{x}^n] & \text{if } \overline{x} \leq 0 \text{ and } n \text{ is even,} \\ [0, \max(\underline{x}^n, \overline{x}^n)] & \text{if } \underline{x} \leq 0 \leq \overline{x} \text{ and } n \text{ is even.} \end{cases} \tag{19.1}$$

In the above rules of interval arithmetic, we excluded division by an interval containing zero. But it is possible to remove this restriction using *extended interval arithmetic* [7]. The IEEE standard for floating-point arithmetic makes it possible to do extended interval arithmetic on a computer. Chips with floating-point arithmetic having this facility have been available since the Intel 8087 was introduced.

The power of interval arithmetic lies in its implementation on computers. Much modern computers allows the result of an arithmetic operation to be rounded down to the nearest machine number less than or equal to the mathematically correct result and rounded up to the nearest machine number greater than or equal to the mathematically correct result. This is called *outward rounding*. It allows rigorous enclosures for the ranges of operations and functions, since the results are now intervals in which the exact result must lie. For example, take $X + Y = [\underline{x} + \underline{y}, \overline{x} + \overline{y}]$. If $\underline{x} + \underline{y}$ is rounded down after computation and $\overline{x} + \overline{y}$ is rounded up after computation, then the resulting interval that is represented in the machine must contain the exact range of $x + y$ for $x \in X$ and $y \in Y$.

Some programming languages, such as Ada, Fortran-90 and C++ make programming with intervals convenient by allowing operator overloading.

Expressions that are equivalent in real arithmetic can differ in interval arithmetic. Suppose we subtract the interval $X = [a, b]$ from itself. Using the definition, we obtain $[a - b, b - a]$. Unless $b = a$, we do not obtain $[0, 0]$ , as we might expect. In general, when a given variable occurs more than once in an interval computation, it is treated as a different variable in each occurrence. This causes widening of computed intervals and makes it more difficult to obtain sharp results in calculations. Thus, computations should be arranged so that overestimation of ranges is minimized. This problem is called the *dependency problem*. Definition for the $n$th power of an interval is included to overcome the dependency problem in multiplication. For example,

$$[-1, 2]^2 = [0, 4]$$

rather than

$$[-1,2]*[-1,2]=[-2,4].$$

For every real-valued function $f(x_1, x_2, ..., x_n)$ with $n$ real variables, interval arithmetic defines a corresponding *interval function* $F(X_1, X_2, ..., X_n)$ that returns some interval – preferably the smallest one – that contains all values of $f(x_1, x_2, ..., x_n)$ where the variables $x_i \in X_i$ ( $i = 1, 2, ..., n$). An interval function is an interval-valued function of one or more interval arguments.

If a particular interval variable occurs only once in a given form of a function, then it cannot give rise to dependency. Thus, dependency can occur in evaluating an interval function $F(X, Y)$ of the form

$$(X-Y)/(X+Y),$$

but not if it is rewritten as

$$1-2/(1+X/Y).$$

An interval function $F(X_1, X_2, ..., X_n)$ is said to be an *interval extension* of a real-valued function $f(x_1, x_2, ..., x_n)$ if

$$F(x_1, x_2, ..., x_n) = f(x_1, x_2, ..., x_n) \;\; \forall x_i \in X_i, \;\; i = 1, 2, ..., n.$$

This definition presupposes the use of exact interval arithmetic. In practice, when rounding occurs, we shall have

$$f(x_1, x_2, ..., x_n) \in F(x_1, x_2, ..., x_n) \;\; \forall x_i \in X_i, \;\; i = 1, 2, ..., n.$$

An interval function $F(X_1, X_2, ..., X_n)$ is said to be *inclusion monotonic* if

$$X_i \subset Y_i \;\; (i = 1, 2, ..., n) \Rightarrow F(X_1, X_2, ..., X_n) \subset F(Y_1, Y_2, ..., Y_n).$$

The following theorem shows that, for rational functions, inclusion monotonicity is easily assured (for proof, see [4]).

**Theorem 1** *Let $F(X_1, X_2, ..., X_n)$ be an interval function. Assume that $F$ is evaluated using a fixed form with a fixed sequence of operations involving only interval addition, subtraction, multiplication, and division. Then $F$ is inclusion monotonic.*

The following theorem is the most important one in interval analysis. It is known as *the fundamental theorem of interval analysis* [15] (for proof, see [4]).

**Theorem 2** *Let $F(X_1, X_2, ..., X_n)$ be an inclusion monotonic interval extension of a real function $f(x_1, x_2, ..., x_n)$. Then*

$$f(x_1, x_2, ..., x_n) \in F(X_1, X_2, ..., X_n) \text{ for all } x_i \in X_i \quad (i = 1, 2, ..., n).$$

As we have seen, the interval obtained when evaluating an interval function depends on the form of the function. Thus, while

$$\begin{aligned} F_1(X) &= X^2 - X \text{ and} \\ F_2(X) &= (X - 1/2)^2 - 1/4 \end{aligned}$$

are both interval extensions of

$$f(x) = x^2 - x,$$

they may not yield the same result when evaluated. For example,

$$\begin{aligned} F_1([0,2]) &= [-2, 4] \text{ and} \\ F_2([0,2]) &= [-1/4, 2]. \end{aligned}$$

Both interval results contain the range of $f$ over $[0, 2]$, which is $[-1/4, 2]$. This is assured by the fundamental theorem of interval analysis. The result for $F_2$ is precisely the range of $f$ over $[0, 2]$ because $X$ occurs only once in the expression for $F_2$. Thus, there is no dependency problem.

Considerable effort has been expended in attempting to produce systematic methods for representing an interval function to most sharply bound the range of a given real function over an interval. See, for example, [13] and [17].

We shall assume that the region where we seek the point(s) at which the global minimum of $f$ occurs is a box $X$ in $\Re^n$, i.e. $l_i \leq x_i \leq u_i$ $(i = 1, 2, ..., n)$.

In the following, we outline a branch and bound algorithm with interval arithmetic that we use to calculate the global minimum of a potential energy function. The algorithm admits endless variations, depending on how the branching and bounding steps are implemented. It proceeds roughly as follows.

**algorithm** {

**Initialization.** Set list $L$ to $X$, the box in which the global minimum is sought.

**Step 1.** Choose a box $B$ from the list $L$.

**Step 2.** The current box $B$ is partitioned in a given way. Choose one box from the partition, for example $B'$, and put others in $L$ to be further processed.

**Step 3.** Evaluate the function at some point in box $B'$. If that function value is less than any known function value, record this function value as the best upper bound $UB$ for the global minimum of $f$. Using interval arithmetic, calculate a lower bound for $f$ over box $B'$. If that value is greater than $UB$, that box can be eliminated.

**Step 4.** The algorithm stops based on the sizes of the boxes in $L$ and on the difference between the best values of the lower and upper bounds for the global minimum of $f$. }

We can also discard boxes in step 3 using information obtained from the first and second order derivatives of the function.

When the box volume tends to zero, interval arithmetic guarantees that the width of the range estimation of $f$ also tends to zero. From this we obtain the convergence proof [4].

# 4 COMPUTATIONAL EXPERIMENTS

We use the Lennard-Jones potential function to represent the pairwise potential interactions between atoms i and j:

$$v_{ij}(r_{ij}) = \frac{B_{ij}}{r_{ij}^{12}} - \frac{A_{ij}}{r_{ij}^{6}},$$

where $A_{ij}$ and $B_{ij}$ are parameters related to atoms $i$ and $j$, and $r_{ij}$ is the distance between atoms $i$ and $j$.

The Lennard-Jones potential has an exponential number of local minima on the numbers of atoms of the molecule [5], thus making the computation of the global minimum very difficult. The complexity of determining the global minimum energy of a system of atoms interacting via simple two-body forces belongs to the class NP-hard problems ([2], [18]).

We have successfully applied the algorithm to an example presented in [9] and another one presented in [8]. The program was coded in C++ and executed on a SUN Sparc Station Ultra 1 running SunOS 5.7.

## 4.1 EXAMPLE 1

The first example is the pseudoethane molecule (Fig.19.1). Such a molecule does not exist. It is a prototype for testing methods that predict molecular conformations. This molecule involves 9 chains composed of four covalently bonded atoms: C-C-C-C, C-C-C-N, C-C-C-O, O-C-C-C, O-C-C-N, O-C-C-O, N-C-C-C, N-C-C-N and N-C-C-O. Using the set of internal coordinates and considering all covalent bond lengths and covalent bond angles fixed at their equilibrium values, the resulting optimization problem has one degree of freedom, as shown in [9].

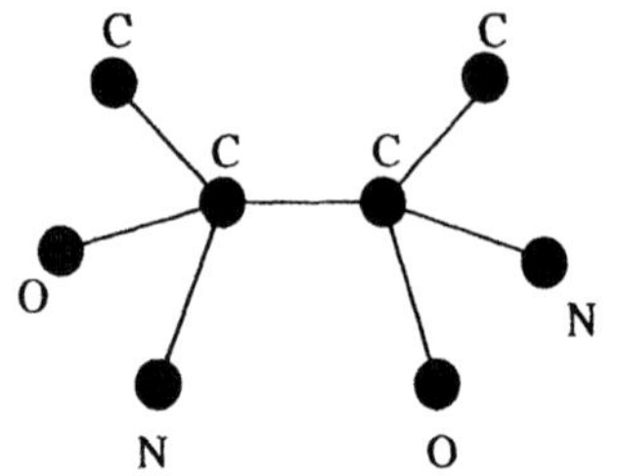


*Figure 19.1* The pseudoethane molecule.

Applying the algorithm, the global minimum solution presented in [9] is obtained. Below, we show some details about the performance of the algorithm applied to this example:

| | |
|---|---|
| Tolerance for the sizes of the boxes: | 0.00001 |
| Average number of real function evaluations: | 8 |
| Average number of interval function evaluations: | 4 |
| Average number of iterations: | 2 |
| Average CPU time (s): | 0.5 |

## 4.2 EXAMPLE 2

The second example is the 1,2,3-trichloro-1-fluoro-propane (CHClF-CHCl-$CH_2$Cl) molecule (Fig.19.2). This molecule involves 18 chains composed of four covalently bonded atoms (F-C-C-Cl, F-C-C-C, F-C-C-H, H-C-C-Cl, H-C-C-C, H-C-C-H, Cl-C-C-Cl, Cl-C-C-C, Cl-C-C-H, C-C-C-C, C-C-C-H, C-C-C-H, H-C-C-C, H-C-C-H, H-C-C-H, Cl-C-C-C, Cl-C-C-H, Cl-C-C-H) and 9 chains of five covalently bonded atoms (F-C-C-C-C, F-C-C-C-H, F-C-C-C-H, H-C-C-C-C, H-C-C-C-H, H-C-C-C-H, Cl-C-C-C-C, Cl-C-C-C-H, Cl-C-C-C-H). Using the set of internal coordinates and considering all covalent bond lengths and covalent bond angles fixed at their equilibrium values, the resulting optimization problem has two degrees of freedom, as shown in [8].

After applying the algorithm, the global minimum solution presented in [8] is obtained. In the following, we present some details about the

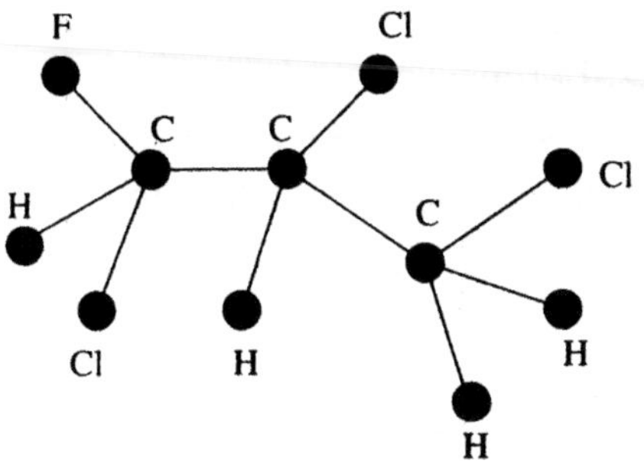


*Figure 19.2* 1,2,3-trichloro-1-fluoro-propane.

performance of the algorithm applied to this example:

| | |
|---|---|
| Tolerance for the sizes of the boxes: | 0.00001 |
| Average number of real function evaluations: | 500 |
| Average number of interval function evaluations: | 1200 |
| Average number of iterations: | 3600 |
| Average CPU time (s): | 90 |

## 5 CONCLUSIONS AND RESEARCH IN PROGRESS

In this paper we use a deterministic global optimization algorithm for identifying the global minimum potential energy conformation of small molecules based on the Lennard-Jones potential function. The algorithm is based on a branch and bound scheme that uses interval analysis techniques. The method was applied successfully to two example problems. The presented algorithm is not restricted to the specifics of the problems at hand. It can be applied to other classes of problems involving more complex interaction terms. Specifically, we are in the process of applying the proposed approach on polyalanines.

## References

[1] Falk, J. and Soland, R. (1969). An algorithm for separable nonconvex programming problems. *Management Science*, 15:550–569.

[2] Garey, M. and Johnson, D. (1979). *Computers and Intractability: A Guide to the Theory of NP-Completeness.* Freeman, San Francisco.

[3] Garfinkel, R. and Nemhauser, G. (1972). *Integer Programming.* Wiley, New York.

[4] Hansen, E. (1993). *Global Optimization using Interval Analysis.* Springer-Verlag, Berlin.

[5] Hoare, M. (1979). Structure and dynamics of simple microclusters. *Advances in Chemical Physics*, 40:49–135.

[6] Horst, R. (1984). On the convexification of nonconvex programming problems. *European Journal of Operational Research*, 15:382–392.

[7] Kahan, W. (1968). A more complete interval arithmetic. Lecture notes for a summer course, University of Michigan.

[8] Maranas, C. and Floudas, C. (1994a). A deterministic global optimization approach for molecular structure determination. *J. Chem. Phys.*, 100:1247–1261.

[9] Maranas, C. and Floudas, C. (1994b). Global minimum potential energy conformations of small molecules. *J. Global Opt.*, 4:135–170.

[10] McCormick, G. (1983). *Nonlinear Programming: Theory, Algorithms and Applications.* Wiley, New York.

[11] Mitten, L. (1970). Branch and bound method: General formulation and properties. *Operations Research*, 18:24–34.

[12] Moore, R. (1962). *Interval arithmetic and automatic error analysis in digital computation.* PhD thesis, Stanford University.

[13] Neumaier, A. (1990). *Interval Methods for Systems of Equations.* Cambridge University Press, London.

[14] Pardalos, P. and Rosen, J. (1987). Constrained global optimization: Algorithms and applications. *Lecture Notes in Computer Science*, 268.

[15] Rall, L. (1969). *Computational Solution of Nonlinear Operator Equations.* Wiley, New York.

[16] Rockafellar, R. (1970). *Convex Analysis.* Princeton University Press.

[17] Rokne, J. (1986). Low complexity $k$-dimensional centered forms. *Computing*, 37:247–253.

[18] Wille, L. and Vennik, J. (1985). Computational complexity of the ground-state determination of atomic clusters. *J. Phys. A*, 18:419–422.

Chapter 20

# APPROXIMATE ANALYTIC CENTER QUADRATIC CUT METHOD FOR STRONGLY MONOTONE VARIATIONAL INEQUALITIES

Hans-Jakob Lüthi and Benno Büeler
*Institute for Operations Research*
*Swiss Federal Institue of Technolgy, Zurich*
luethi@ifor.math.ethz.ch

**Abstract** For strongly monotone variational inequality problems (VIP) convergence of an algorithm is investigated which, at each iteration $k$, adds a quadratic cut through an approximate analytic center $x_k$ of the subsequently shrinking convex body. First it is shown that the sequence of $x_k$ converges to the unique solution $x^*$ of the VIP at $\mathcal{O}(1/\sqrt{k})$. As an interesting detail note that — for increasingly accurate analytic centers — the complexity constants converge to the quantities obtained for ACQCM with exact centers. Secondly we show that the arithmetic complexity to update from $x_k$ to $x_{k+1}$ after inserting a quadratic cut through $x_k$ is bounded by a constant number of Newton iterations plus $\mathcal{O}(n \cdot \ln \ln[\zeta \cdot \frac{k^2}{\varepsilon^4}])$, where $n$ is the space-dimension, $\varepsilon$ is the final solution accuracy $\|x_k - x^*\|$, and $\zeta$ depends on some problem-specific constants only.

**Keywords:** Interior-point methods, Approximate analytic center, Quadratic cuts, Variational inequalities.

## 1 INTRODUCTION

This paper extends the analysis in [4]. Here we replace the exact analytic centers by approximate analytic centers, turning the conceptual algorithm from [4] into a practical algorithm. To keep the paper short, we omit some proofs based on techniques similar to those used in [4]. We want to point to the parallel work reported in [7] where a homogeneous analytic center cutting plane method (HACCPM) with approxi-



*N. Hadjisavvas and P.M. Pardalos (eds.), Advances in Convex Analysis and Global Optimization,* 345–360.

mate centers is analyzed. Due to the construction of their iterates by using a weighted average of analytic centers, the final accuracy determines the accuracy of all previous approximate centers. In our analysis we can avoid this severe restriction. As a disadvantage of the quadratic cuts, however, we have to accept an increase in the computational complexity for updating the analytic center when the iteration number grows. This increase, however, grows very slowly at $\mathcal{O}(\ln \ln k^2)$. In case of HACCPM linearity of the cuts allows to keep the update complexity constant.
The following notation is used. Let $A$, $B \in \mathbb{R}^{n \times n}$ be symmetric. We then call $A$ p.s.d. if it is positive semidefinite, and p.d. if $A$ is positive definite. By $A \geq B$ we mean that $A - B$ is p.s.d., and by $A > B$ that $A - B$ is p.d. Furthermore, let $A$ be p.d.; we then define $\|x\|_A := \langle x, Ax \rangle^{1/2}$ for all $x \in \mathbb{R}^n$. Finally, by $\mathbb{1}$ we denote the unit matrix of appropriate dimension.

## 2 DESCRIPTION OF THE ACQCM

Given a closed, convex set $P \subset \mathbb{R}^n$ and a continuous map $f : P \to \mathbb{R}^n$ we want to solve the variational inequality problem (VIP)

$$\text{find } x^* \in P, \text{ such that } \langle f(x^*), x - x^* \rangle \geq 0 \quad \forall x \in P. \tag{20.1}$$

Throughout the paper we make the following assumptions.

**Assumption 2.1**

1. *$P$ has a non-empty interior and it can be described by an $m$-self-concordant barrier function $F_0$; furthermore, w.l.o.g. $P$ can be assumed to be compact.*
   *Hence $f$ is bounded on $P$, i.e. $\max_{x \in P} \|f(x)\| \leq L < \infty$.*

2. *$f$ is strongly monotone, i.e. there exists $\alpha > 0$ such that*

$$\langle f(y) - f(x), y - x \rangle \geq \alpha \|y - x\|^2 \quad \forall x,\, y \in P. \tag{20.2}$$

3. *$F_0''(x) > 0$ for all $x \in \operatorname{int} P$.*

Note that (i) as discussed in [4] can easily be met and are in this context not at all restrictive.
In addition, it is well known ([3]) that for strongly monotone VIPs continuity of $f$ implies the existence of a unique solution $x^* \in P$ of (20.1). Indeed, the existence of a solution $x^*$ and hence continuity of $f$ will not be needed for the algorithm nor for the convergence analysis as will be discussed in § 3.

For a definition and an in-depth discussion of self-concordant functions see [6].
It is known (cf. [6]) that a self-concordant barrier $F$ has a unique minimizer called (exact) analytic center of the corresponding domain of the barrier. We call $x$ an $\eta$-approximate center of $F$ if

$$\|F'(x)\|_{[F''(x)]^{-1}} \leq \eta. \tag{20.3}$$

For $\eta = 0$ we regain the analytic center. In order to apply the self-concordant theory we make in the rest of the paper the following assumption about $\eta$ in (20.3):

**Assumption 2.2**
$$0 \leq \eta \leq \frac{1}{3}.$$

Some properties of self-concordant barriers, which will be used in the sequel, are stated in the following lemma.

**Lemma 2.1** *Let $F$ be an $m$-self-concordant barrier with $F'' > 0$ on $\operatorname{dom} F$, and let $x, y \in \operatorname{dom} F$.*

*(i) Then we know from [8] (Lemma 9 ) that*

$$\langle F'(y) - F'(x), y - x\rangle \geq \frac{\|y-x\|^2_{F''(x)}}{1+\|y-x\|_{F''(x)}}, \tag{20.4}$$

*(ii) and from [6] ((2.2.1), Definition 2.3.1 and Proposition 2.3.2) we have*

$$\|F'(x)\|_{(F''(x))^{-1}} \leq \sqrt{m} \quad \textit{and} \quad \langle F'(x)\,, y - x\,\rangle \leq m\,, \tag{20.5}$$

*(iii) let $x^*$ be the analytic center; if $x$ is an $\eta$-approximate center, $\eta \leq \frac{1}{3}$, then*

$$\|x-x^*\|_{F''(x)} \leq \frac{\eta}{1-\eta} \leq \frac{1}{2}, \quad \textit{and} \tag{20.6}$$

*(iv)*

$$\|x-x^*\|_{F''(x^*)} \leq \beta(\eta) \leq \frac{1}{2}, \tag{20.7}$$

*where $\beta(\eta) := \frac{\eta}{2(1-\eta)}\left(\eta + \sqrt{\eta^2+4(1-\eta)}\right)$.*

**Proof**
Only item (iii) is proved here. Denote by $\gamma := \|x-x^*\|_{F''(x)}$, then we derive from (20.4) $\langle F'(x) - F'(x^*), x - x^*\rangle \geq \frac{\gamma^2}{1+\gamma}$. Since $F'(x^*) = 0$

we conclude $\|F'(x)\|_{(F''(x))^{-1}} \cdot \gamma \geq \frac{\gamma^2}{1+\gamma}$ and therefore $\eta\gamma(1+\gamma) \geq \gamma^2$. That is, $\gamma \leq \gamma(\eta) := \frac{\eta}{1-\eta} \leq \frac{1}{2}$. ■

Returning to VIP, strong monotonicity implies that the solution $x^*$ to (20.1) is not only unique but that

$$\begin{aligned} -\langle f(x), x^* - x\rangle - \alpha\|x - x^*\|^2 &\geq -\langle f(x^*), x^* - x\rangle \\ &= \langle f(x^*), x - x^*\rangle \geq 0 \quad \forall x \in P. \end{aligned} \tag{20.8}$$

If we denote by

$$t_k(x) := -\langle f(x_k), x - x_k\rangle - \alpha\|x_k - x\|^2 \tag{20.9}$$

the slack with respect to a quadratic cut through an arbitrary $x_k \in P$, then from (20.8) it immediately follows that the corresponding quadratic cut set

$$C_{x_k} := \{x \in \mathbb{R}^n \,|\, t_k(x) \geq 0\}$$

contains the solution $x^*$. Note that $C_{x_k}$ is a ball with center $z_k := x_k - \frac{f(x_k)}{2\alpha}$ and radius $r_k := \frac{\|f(x_k)\|}{2\alpha}$.

Let $x_0$ denote an $\eta$-approximate center of $F_0$. For $k \geq 1$ consider the logarithmic barrier

$$F_k(x) := F_0(x) - \sum_{i=0}^{k-1} \ln t_i(x),$$

where $x_k$, needed for the definition of $t_k(x)$, is an $\eta$-approximate center of $F_k(x)$ over $P_k := P \cap_{i=0}^{k-1} C_{x_i}$. By $x_k^*$ we denote the (exact) analytic center of $F_k$, whereas $x^*$ denotes the solution to the VIP. Based on this notation, the ACQCM presented below can be used to solve strongly monotone VIPs.

**ACQCM: Analytic Center Quadratic Cut Method**

(i) Let $k = 0$, $F_0$ as in the Assumption 2.1, and choose $\varepsilon > 0$.

(ii) Compute an $\eta$-approximate center $x_k$ of $F_k(x)$,
let $F_{k+1}(x) = F_k(x) - \ln(t_k(x))$ where $t_k$ is defined in(20.9) .

(iii) Stop if $\|x_k - x^*\| \leq \varepsilon$, otherwise set $k := k+1$ and return to (ii).

In the next section we bound the required number of iterations of ACQCM for an $\varepsilon$-solution. Finally, in the last section the arithmetic complexity of computing $x_{k+1}$ after inserting a quadratic cut through $x_k$ is discussed.

## 3 ITERATION COMPLEXITY OF ACQCM

In this section we investigate the number of iterations needed to achieve $\|x_k - x^*\| \leq \varepsilon$. The analysis shows that $\|x_k - x^*\| = \mathcal{O}(k^{-1/2})$, where the constants depend on $L$, $\alpha$, $\eta$ and $m$ only.

**Theorem 1** *Let* $\beta(\eta) := \frac{\eta}{2(1-\eta)}(\eta + \sqrt{\eta^2 + 4(1-\eta)})$ *and for all* $k \geq 1, S_k^* := \sum_{i=0}^{k-1}(t_i(x_k^*)^{-1}$, *where* $x_k^*$ *denotes the exact center in iteration* $k$.
*Then, we have for all* $k \geq 1$*:*

$$\|x_k - x^*\| \leq \sqrt{\frac{k+m}{\alpha S_k^*}} + \frac{\beta(\eta)}{\sqrt{2\alpha S_k^*}} \tag{20.10}$$

**Proof**
Since $x_k^*$ is by definition the minimizer of the barrier $F_k$ the first order optimality condition yields $F_k'(x_k^*) = 0$. In view of the definition of the barrier this implies for $k \geq 1$ that

$$F_0'(x_k^*) - \sum_{i=0}^{k-1} \frac{t_i'(x_k^*)}{t_i(x_k^*)} = 0. \tag{20.11}$$

Multiplying (20.11) by $(x_k^* - x^*)$ yields

$$\langle F_0'(x_k^*), (x_k^* - x^*)\rangle - \sum_{i=0}^{k-1} \frac{\langle t_i'(x_k^*), (x_k^* - x^*)\rangle}{t_i(x_k^*)} = 0. \tag{20.12}$$

Since $t_i(x^*) \geq 0*$ we derive from its quadratic form

$$\langle t_i'(x_k^*), (x_k^* - x^*)\rangle \leq t_i(x_k^*) - \alpha\|x^* - x_k^*\|^2,$$

and conclude from (20.12) and the definition of $S_k^*$

$$\langle F_0'(x_k^*), (x_k^* - x^*)\rangle - k + \alpha S_k^*\|x^* - x_k^*\|^2 \leq 0 \tag{20.13}$$

or, by rearranging the terms,

$$\alpha S_k^*\|x^* - x_k^*\|^2 \leq k + \langle F_0'(x_k^*), (x^* - x_k^*)\rangle \leq k + m, \tag{20.14}$$

where the last inequality is a consequence of property (iii) in Lemma 2.1. Hence for exact centers we obtain the bound

$$\|x_k^* * - x^*\| \leq \sqrt{\frac{k+m}{\alpha S_k^*}} \tag{20.15}$$

which corresponds to the first term in (20.10). Now

$$\|x^* - x_k\| \leq \|x^* - x_k^*\| + \|x_k^* - x_k\| \tag{20.16}$$

The first term in (20.16) is bounded by (20.15). To bound the second term we use (20.7) in Lemma 2.1 as follows

$$\beta(\eta) \geq \|x_k - x_k^*\|_{F''(x_k^*)} \tag{20.17}$$

$$\geq \sqrt{2\alpha S_k^*}\,\|x_k^* - x_k\| \tag{20.18}$$

where the last inequality follows directly from

$$\begin{aligned} F_k''(x_k^* *) &= F_0''(x_k^* *) - \sum_{i=0}^{k-1} \frac{t_i''(x_k^* *)}{t_i(x_k^* *)}\mathbb{1} + \sum_{i=0}^{k-1} \frac{t_i'(x_k^* *)(t_i'(x_k^* *))^T}{t_i^2(x_k^* *)} \\ &\geq \sum_{i=0}^{k-1} \frac{2\alpha}{t_i(x_k^* *)}\mathbb{1}, \end{aligned}$$

since both $F_0''(x_k^* *)$ and $t_i'(x_k^* *)(t_i'(x_k^* *))^T$ are positive (semi)-definit and $t_i(x_k^* *) > 0$. Hence

$$\|x_k^* - x_k\| \leq \frac{\beta(\eta)}{\sqrt{2\alpha S_k^*}}$$

which corresponds to the second term in (20.10). ■

As is seen in the proof of the above theorem, $x^*$ can be replaced by any feasible point $x \in P_k$. Hence, Theorem 1 simply states that at iteration $k$, we can bound the distance from the inexact analytic center $x_k$ to any feasible point $x \in P_k$ by (20.10).

The rest of this section is dedicated to derive a lower bound for $S_k^*$ leading to the claimed rate of convergence $\|x_k - x^*\| = \mathcal{O}(k^{-1/2})$ to the unique solution of (20.1). Towards that end we will prove the following theorem.

**Theorem 2** *Let* $\theta(\eta) := \frac{1+\eta+\sqrt{(1+\eta)^2+4(1-\eta)}}{2(1-\eta)}$. *For* $S_k^*$ *we obtain the lower bound*

$$S_k^* \geq k^2 e^{-2a}$$

*where* $a := \ln\left(L\sqrt{\frac{(1+\theta(\eta))(1+\eta\theta(\eta))}{2\alpha}}\right) + \sqrt{m}\left(\theta(\eta) + \frac{\eta}{1-\eta}\right) + 1.$

We remark first that for $\eta \to 0$ the complexity constants converge smoothly to the quantities obtained for ACQCM with exact centers, see [4].

The proof of theorem 2 is based on three lemmas which are very similar to the ones given in [4], hence we omit the proofs.

**Lemma 3.1** *Let $\theta(\eta)$ be as defined in theorem 2. For $k \geq 0$*

$$\|x_k - x^*_{k+1}\|_{F''_k(x_k)} \leq \theta(\eta), \tag{20.19}$$

$$\|x_k - x^*_{k+1}\|_{F''_k(x^*_{k+1})} \leq \theta(\eta), \tag{20.20}$$

**Lemma 3.2** *Let $\theta(\eta)$ be as defined in theorem 2. We have for all $k \geq 1$*

$$\|x^*_k - x_{k-1}\| \leq \sqrt{\frac{(1+\theta(\eta))(1+\eta\theta(\eta))}{2\alpha S^*_k}} \tag{20.21}$$

$$\leq \frac{L}{\alpha}\sqrt{\frac{(1+\theta(\eta))(1+\eta\theta(\eta))}{8k}}. \tag{20.22}$$

**Lemma 3.3** *Let $\theta_1 := \sqrt{m}(\theta(\eta) + \frac{\eta}{1-\eta})$; we then have for all $k \geq 1$*

$$-\sum_{i=0}^{k-1} \ln t_i(x^*_k*) \geq -\sum_{i=0}^{k-1} \ln t_i(x^*_{i+1}) - k \cdot \theta_1 \tag{20.23}$$

Based on the above lemmas we establish now a lower bound on the growth of $-\sum_{i=0}^{k-1} \ln t_i(x^*_{i+1})$ which depends on $k$ and some constants only.

**Lemma 3.4** *For $k \geq 1$, we have*

$$\sum_{i=0}^{k-1} \ln t_i(x^*_{i+1}) \leq -k \ln k + ck, \tag{20.24}$$

*where $c = 2\left[\ln(L\sqrt{\frac{(1+\theta(\eta))(1+\eta\theta(\eta))}{2\alpha}}) + \frac{\theta_1}{2} + 1\right]$, and $\theta_1$ is defined in Lemma 3.3*

**Proof**
Since the proof uses the same arguments as in [4] (Lemma 3.4) we will not give it here. ■

Now, lemma 3.4 leads immediately to a proof of theorem 2.
*Proof of theorem 2.*
Note that for all $k \geq 1$ (20.24) togehther with Lemma 3.3 implies

$$S^*_k \geq k^2 e^{-2a}. \tag{20.25}$$

where $a$ is given by

$$a = \ln(L\sqrt{\frac{(1+\theta(\eta))(1+\eta\theta(\eta))}{2\alpha}}) + \sqrt{m}(\theta(\eta) + \frac{\eta}{1-\eta}) + 1$$

which corresponds to the constant $a$ as defined in theorem 2. ■

# 4 COMPLEXITY OF UPDATING THE ANALYTIC CENTER

Let $k \geq 0$ and the approximation measure $\eta$ be in $(0, \frac{1}{3}]$. We then ask in this section how many Newton steps are required to compute $x_{k+1}$ given $x_k$. Together with the iteration bound from Theorem 1 we can derive an arithmetic complexity of ACQCM. By its nature such an analysis neglects the computational burden of evaluating $f$.

To determine $x_{k+1}$ we have to minimize the barrier $F_{k+1}$; if we want to apply Newton's method, however, we first have to find a starting point $x_k + \tau d$ which is 'reasonably' interior. Here $\tau$ is the steplength and $d$ the direction. With such a good starting point we then want—as usual in this kind of analysis—to bound $F_{k+1}(x_k + \tau d) - F_{k+1}(x^*_{k+1})$ from above, because this quantity bounds the number of Newton steps required to compute $x_{k+1}$ when starting from $x_k + \tau d$, see Nesterov and Nemirovsky [6, Theorem 2.2.3].

To enhance readability we abbreviate in this section $F''_k := F''_k(x_k)$ and $t'_k := t'_k(x_k)$. Furthermore, we can assume $\|d\|_{F''_k} = 1$, i.e. the direction $d$ is scaled to unit length with respect to the $F''_k$-metric, that is, $x_k + d$ lies on the border of the Dikins ellipsoid centered at $x_k$. Before discussing the choice of $\tau$ and $d$ more in depth we make the following observation.

**Lemma 4.1** *Let $\|d\|_{F''_k} = 1$ and $\tau \in (0,1)$ such that $t_k(x_k + \tau d) > 0$; we then have for all $k \geq 0$*

$$F_{k+1}(x_k + \tau d) - F_{k+1}(x^*_{k+1}) \\ \leq (\eta - 1)\tau - \ln(1-\tau) + \eta \cdot \theta(\eta) + \ln t_k(x^*_{k+1}) - \ln t_k(x_k + \tau d). \quad (20.26)$$

**Proof**

First note that $x_k + \tau d \in \mathrm{dom}(F_{k+1})$. To bound $F_{k+1}(x^*_{k+1})$ from below we observe

$$\begin{aligned} F_{k+1}(x^*_{k+1}) &= F_k(x^*_{k+1}) - \ln t_k(x^*_{k+1}) \\ &\geq F_k(x_k) + \langle F'_k(x_k), x^*_{k+1} - x_k \rangle - \ln t_k(x^*_{k+1}) \\ &\geq F_k(x_k) - \eta \cdot \theta(\eta) - \ln t_k(x^*_{k+1}). \end{aligned} \quad (20.27)$$

As for $F_{k+1}(x_k + \tau d)$ we conclude from [8, Lemma 10]

$$\begin{aligned} F_{k+1}(x_k + \tau d) &= F_k(x_k + \tau d) - \ln t_k(x_k + \tau d) \\ &\leq F_k(x_k) + \langle F'_k(x_k), \tau d \rangle - \|\tau d\|_{F''_k} - \ln(1 - \|\tau d\|_{F''_k}) - \ln t_k(x_k + \tau d) \\ &\leq F_k(x_k) + \|F'_k(x_k)\|_{[F''_k]^{-1}} \cdot \tau - \tau - \ln(1-\tau) - \ln t_k(x_k + \tau d) \\ &\leq F_k(x_k) + (\eta - 1)\tau - \ln(1-\tau) - \ln t_k(x_k + \tau d), \end{aligned}$$

where the third relation is a consequence of $\|\tau d\|_{F_k''} = \tau$. Hence, by (20.27)

$$\begin{aligned} &F_{k+1}(x_k + \tau d) - F_{k+1}(x_{k+1}^*) \\ &\leq (\eta - 1)\tau - \ln(1-\tau) - \ln t_k(x_k + \tau d) + \eta \cdot \theta(\eta) + \ln t_k(x_{k+1}^*) \end{aligned}$$

which proves the lemma. ∎

In view of the above lemma we want to bound

$$\ln t_k(x_{k+1}^*) - \ln t_k(x_k + \tau d) = \ln \left( \frac{t_k(x_{k+1}^*)}{t_k(x_k + \tau d)} \right)$$

from above, where $\|\tau d\|_{F_k''} < 1$.
Let

$$\delta^* := \arg\max\{t_k(x_k + \delta) \,|\, \delta \in W_k\}$$

and

$$\bar{\delta}^* := \arg\max\{t_k(x_k + \delta) \,|\, \delta \in \theta^2 \cdot W_k\}.$$

where $W_k := \{h \,|\, \langle h, F_k'' h\rangle \leq 1\}$ denotes the Dikins-Ellipsoid centered at $x_k$ and $\theta := \frac{1}{2}(1 + \sqrt{5})$ .
The use of the blown up factor $\theta^2$ is motivated by $\|x_k - x_{k+1}^*\|_{F_k''} \leq \theta(\eta) < \theta^2$ for $\eta \leq \frac{1}{3}$ (Lemma 3.1), and will be used below. Because of technical details involved in the complexity analysis of the approximate computation of $\delta^*$, however, we have chosen $\theta^2$ instead of $\theta(\eta)$. $\bar{\delta}^*$, on the other hand, is only needed in the arguments below and may not be computed:

**Lemma 4.2** $t_k(x_k + \bar{\delta}^*) \leq \theta^2 \cdot t_k(x_k + \delta^*)$.

**Proof**
From concavity of $t_k(x)$ and $t_k(x_k) = 0$ we conclude for all $\tau \in [0, 1]$

$$t_k(x_k + \tau\bar{\delta}^*) \geq \tau \cdot t_k(x_k + \bar{\delta}^*). \qquad (20.28)$$

Assume now to the contrary

$$t_k(x_k + \bar{\delta}^*) > \theta^2 \cdot t_k(x_k + \delta^*).$$

Because $1/\theta^2 \in (0, 1)$ we then have

$$t_k\big(x_k + \tfrac{1}{\theta^2}\bar{\delta}^*\big) \geq \tfrac{1}{\theta^2} \cdot t_k(x_k + \bar{\delta}^*) > \tfrac{1}{\theta^2}\theta^2 \cdot t_k(x_k + \delta^*).$$

Now $\frac{1}{\theta^2}\bar{\delta}^* \in W_k$ and so the above relation contradicts maximality of $\delta^*$. ∎

From (20.19) which implies $x^*_{k+1} \in x_k + \theta(\eta) \cdot W_k \subset x_k + \theta^2 \cdot W_k$ we conclude from the above lemma

$$t_k(x^*_{k+1}) \le t_k(x_k + \bar{\delta}^*) \le \theta^2 \cdot t_k(x_k + \delta^*).$$

Together this yields

$$\frac{t_k(x^*_{k+1})}{t_k(x_k + \tau\delta^*)} \le \frac{\theta^2 \cdot t_k(x_k + \delta^*)}{t_k(x_k + \tau\delta^*)} \le \frac{\theta^2 \cdot t_k(x_k + \delta^*)}{\tau \cdot t_k(x_k + \delta^*)} = \frac{\theta^2}{\tau}, \tag{20.29}$$

where the second relation is justified similarly to (20.28).

Finally we remark that as long as $\|\frac{t'_k}{2\alpha}\|_{F''_k} \le \theta^2$ the computation of $\delta^*$ can be omitted. In this case the direction $t'_k(x_k)/\|t'_k\|_{F''_k}$ leads directly to the bound

$$\frac{t_k(x^*_{k+1})}{t_k(x_k + \tau \cdot t'_k(x_k)/\|t'_k\|_{F''_k})} \le \frac{\theta^2}{\tau}.$$

Note that in case of linear cuts $\delta^* = [F''_k]^{-1} t'_k / \|t'_k\|_{[F''_k]^{-1}}$ is given analytically (see e.g. [7]). In view of the above theorem the knowledge of $\delta^*$ allows to keep the arithmetic complexity for upgrading $x_k \to x_{k+1}$ constant.

In case of quadratic constraints, however, their is no analytic solution for $\delta^*$ and hence an approximation $\delta_a$ of $\delta^*$ must be computed: For $a \in (0,1)$ we call $\delta_a \in W_k$ an $a$-approximation of $\delta^*$ if

$$\frac{t_k(x_k + \delta^*) - t_k(x_k + \delta_a)}{t_k(x_k + \delta^*)} \le a, \tag{20.30}$$

that is, we have

$$t_k(x_k + \delta_a) \ge (1 - a) \cdot t_k(x_k + \delta^*).$$

In such a situation (20.29) yields

$$\frac{t_k(x^*_{k+1})}{t_k(x_k + \tau\delta_a)} \le \frac{\theta^2 \cdot t_k(x_k + \delta^*)}{\tau \cdot t_k(x_k + \delta_a)} \le \frac{\theta^2 \cdot t_k(x_k + \delta^*)}{\tau(1-a) \cdot t_k(x_k + \delta^*)} = \frac{\theta^2}{\tau \cdot (1-a)}.$$

Hence, if $\delta^*$ is replaced by $\delta_a$ we find an additional summand :

$$\begin{aligned} & F_{k+1}(x_k + \tau\delta_a) - F_{k+1}(x^*_{k+1}) \\ & \le (\eta - 1)\tau - \ln[\tau(1-\tau)] + \eta \cdot \theta(\eta) + 2\ln\theta - \ln(1-a). \end{aligned} \tag{20.31}$$

Thus, we are left with the problem of computing $\delta_a$ for the case $\|\frac{t'_k}{2\alpha}\|_{F''_k} > \theta^2$ .

To apply the complexity bounds from Ye *et al* [1] and [9] we transform the ellipsoid-constrained problem

$$\max_{\langle \delta, F_k''\delta\rangle \leq 1} t_k(x_k + \delta) \quad \equiv \quad \max_{\langle \delta, F_k''\delta\rangle \leq 1} \langle t_k', \delta\rangle - \alpha\|\delta\|^2 \tag{20.32}$$

into the canonical ball-constrained form

$$\min_{\|\bar\delta\|^2 \leq 1} \frac{1}{2}\langle \bar\delta, 2\alpha D_k^{-1}\bar\delta\rangle - \langle c, \bar\delta\rangle \tag{20.33}$$

by using the decomposition $F_k'' = U_k D_k U_k^T$ and by defining

$$\bar\delta := D_k^{1/2} U_k^T \delta, \qquad c := D_k^{-1/2} U_k^T t_k'.$$

Because in the case to be treated the maximum of $t_k*$ is attained outside the Dikins ellipsoid the optimality conditions of (20.32) are

$$(2\alpha \mathbb{1} + \lambda^* F_k'')\delta^* = t_k', \qquad \|\delta^*\|_{F_k''}^2 = 1, \qquad \text{and } \lambda^* > 0.$$

Note that $\delta(\lambda) := (2\alpha\mathbb{1} + \lambda F_k'')^{-1} t_k'$ solves the first optimality condition. For convenience we write $t_k(\lambda) := t_k(x_k + \delta(\lambda))$. Similarly, the optimality conditions of (20.33) are

$$(2\alpha D_k^{-1} + \bar\lambda^* \mathbb{1})\bar\delta^* = c, \qquad \|\bar\delta^*\|^2 = 1, \qquad \text{and } \bar\lambda^* > 0.$$

Again $\bar\delta(\bar\lambda) := (2\alpha D_k^{-1} + \bar\lambda\mathbb{1})^{-1} c$ solves the first optimality condition, and hence we write $q(\bar\lambda) := \frac{1}{2}\langle \bar\delta(\bar\lambda), 2\alpha D_k^{-1}\bar\delta(\bar\lambda)\rangle - \langle c, \bar\delta(\bar\lambda)\rangle$. By construction, the optimality conditions of the two systems (20.32) and (20.33) are closely related; not only we have $\lambda^* = \bar\lambda^*$, but, moreover

$$t_k(\lambda) = -q(\lambda) \quad \text{for all} \quad \lambda \geq 0.$$

For this reason we write in the sequel $q(\lambda)$ and $\bar\delta(\lambda)$ instead of $q(\bar\lambda)$ and $\bar\delta(\bar\lambda)$ respectively. Assume $\bar\delta_a := \bar\delta(\lambda_a)$ for some $a \in (0,1)$ is an $a$-approximation of $\bar\delta^*$, i.e. (20.30) holds where $t_k(\lambda_a)$ is replaced by $-q(\lambda_a)$ and $t_k(\lambda^*)$ by $-q(\lambda^*)$. Then, by the previous relation, $\delta_a := \delta(\lambda_a)$ is an $a$-approximation of $\delta^*$. Thus, in the rest of this section we restrict the discussion on finding an $a$-approximation of $\bar\delta^*$ as a function of $\lambda > 0$. From the optimality conditions we know that $\bar\delta(\lambda) = \bar\delta^*$ if and only if $\|\bar\delta(\lambda)\| = 1$. To construct $\lambda^*$ the algorithmic concept we use here proceeds as follows. Assume we have a lower and an upper bound for $\lambda^*$. By a combination of binary search and Newton steps [9], abbreviated BSN, we then construct subsequently decreasing iterates $\lambda$ converging to $\lambda^*$, where always $\lambda \geq \lambda^*$. Because for $\lambda = \lambda^*$ we have already found the true solution $\bar\delta^*$ we assume in the sequel $\lambda > \lambda^*$.

Let $\lambda_k^{\max}$ and $\lambda_k^{\min}$ be the maximal and minimal Eigenvalue of $F_k''$ respectively. Then the following lemma establishes lower and upper bounds for $\lambda^*$ which are used in the sequel.

**Lemma 4.3** *If $\|t_k'/2\alpha\|_{F_k''} > \theta^2$ holds, then*

$$\frac{2\alpha\theta}{\lambda_k^{\max}} \leq \lambda^* \leq \frac{L}{\sqrt{\lambda_k^{\min}}}.$$

**Proof**
Because $\lambda^* > 0$ and $D_k^{-1} > 0$ the upper bound is a consequence of

$$1 = \|\bar{\delta}^*\| = \|\bar{\delta}(\lambda^*)\|^2 = c^T(2\alpha D_k^{-1} + \lambda^*\mathbb{1})^{-2}c \leq \|c\|^2/\lambda^{*2}$$

and $\|c\| = \|D_k^{-1/2}U_k^T t_k'\| \leq L/\sqrt{\lambda_k^{\min}}$. To verify the lower bound we start from the optimality condition $(2\alpha D_k^{-1} + \lambda^*\mathbb{1})\bar{\delta}^* = c$ which is equivalent to

$$D_k^{-1/2}U_k^T\frac{t_k'}{2\alpha} = \left(D_k^{-1} + \frac{\lambda^*}{2\alpha}\right)\bar{\delta}^*.$$

From the assumption we have $\theta^2 \leq \|D_k^{1/2}U_k^T\frac{t_k'}{2\alpha}\|$ and so

$$\theta^2 \leq 1 + \frac{\lambda_k^{\max}}{2\alpha}\lambda^*$$

which, by $\theta^2 - 1 = \theta$, completes the proof. ∎

The following lemma given without proof is an adaptation of [1, Case I p. 5]. It establishes a linear relation between the relative measure $(\lambda - \lambda^*)/\lambda^*$ and $a$.

**Lemma 4.4** *Assume $\|t_k'/2\alpha\|_{F_k''} > \theta^2$ and let $a \in (0,1)$; if*

$$0 \leq \lambda - \lambda^* \leq \lambda^* \cdot a \cdot \frac{2}{3} \cdot \frac{\theta - 1}{\theta}$$

*then*

$$0 \leq \frac{q(\lambda^*) - q(\lambda)}{q(\lambda^*)} \leq a.$$

Based on the above lemma a first complexity bound can be established.

**Lemma 4.5** *Let $C(n)$ denote the number of arithmetic operations required to compute the $U_kD_kU_k^T$-decomposition of $F_k''$. Then an $a$-approximation $\delta_a$ of $\delta^*$ can be computed in at most*

$$C(n)+\mathcal{O}\left(n \cdot \ln\ln\left(\frac{3}{4}\frac{1}{\theta-1}\frac{L}{\alpha} \cdot \frac{\lambda_k^{\max}}{\sqrt{\lambda_k^{\min}}} \cdot \frac{1}{a}\right)\ln\left(\frac{3}{4}\frac{1}{\theta-1}\frac{L}{\alpha}\right)\left(\frac{\lambda_k^{\max}}{\sqrt{\lambda_k^{\min}}}\right)\right)$$

*arithmetic operations.*

**Proof**

The $U_k D_k U_k^T$-decomposition of $F_k''$ must only be done once at the beginning of the $\bar{\delta}_a$ computation. The algorithm 'BSN' (Binary Search and Newton's Method, [9]), which then computes $\bar{\delta}_a$, requires in each iteration the construction of $(2\alpha D_k^{-1} + \lambda \mathbb{1})^{-1}$, which, by its diagonal structure, imposes $\mathcal{O}(n)$ arithmetic operations. The same applies also to the matrix-vector multiplications involved. We are thus left with estimating the number of iterations of BSN.

Let $\lambda_u := L/\sqrt{\lambda_k^{\min}}$ denote the upper bound of $\lambda^*$, and let $\varepsilon' := \lambda^* \cdot a \cdot \frac{2}{3} \cdot \frac{\theta-1}{\theta}$ be the absolute error tolerance for computing $\lambda$, i.e. $\lambda - \lambda^* \leq \varepsilon'$. Then from Ye [9] we know that BSN requires

$$\mathcal{O}\Big(\ln \ln \Big(\frac{\lambda_u}{\varepsilon'}\Big)\Big)$$

iterations to construct $\lambda$ such that $\lambda - \lambda^* \leq \varepsilon'$, that is, to compute $\bar{\delta}_a$. Using $2\alpha\theta/\lambda_k^{\max} \leq \lambda^*$ we find

$$\frac{\lambda_u}{\varepsilon'} \leq \frac{L}{\alpha} \cdot \frac{\lambda_k^{\max}}{\sqrt{\lambda_k^{\min}}} \cdot \frac{3}{4} \frac{1}{\theta - 1} \cdot \frac{1}{a}$$

which proves the lemma. ∎

Completing the complexity analysis demands for bounding $\lambda_k^{\max}$ from above and $\lambda_k^{\min}$ from below. To construct an upper bound for $\lambda_k^{\max}$ we first relate the stopping criterion $\|x_k - x^*\| \leq \varepsilon$ to an inscribed ball with a minimal radius. This requires bounding the condition number of the starting feasible set $P =: P_0$. If $\rho_0^i$ denotes the maximal radius of a sphere inscribed in $P_0$, and $\rho_0^c$ the minimal radius of a sphere circumscribing $P_0$, we know by item 1 of Assumption 2.1

$$\kappa := \frac{\rho_0^c}{\rho_0^i} < \infty.$$

**Lemma 4.6** *Let $k \geq 0$. If $\|x_k - x^*\| > \varepsilon$ then there exists a ball inscribed in $P_k$ with radius*

$$\rho^i = \min\left\{\frac{1}{\kappa}, \frac{\alpha\varepsilon}{4L}\right\} \cdot \frac{\varepsilon}{2}. \tag{20.34}$$

**Proof**

For $k = 0$ the claim holds trivially. For $k \geq 1$ there exists at least one quadratic constraint. In general, if the condition number of $P_k$ is bounded from above by $\kappa'$ it follows $\rho^i \geq \frac{\varepsilon}{2\kappa'}$. Next, $\kappa'$ is bounded by the maximum of $\kappa$ and the maximal condition number produced by a quadratic cut. A quadratic cut in turn produces a maximal condition number if it has minimal curvature and is intersected by a linear constraint. Considering this case, we assume the radius of the quadratic constraint to be $r^{\max} = \frac{L}{2\alpha}$, i.e. to be maximal. Denote by $\rho^i$ a lower bound for the radius of an inscribed sphere with maximal radius, and by $\rho^a$ an upper bound for the radius of an circumscribed sphere with minimal radius. Using the Pythagorean law $\rho^i$ is bounded from below by

$$(\rho^a)^2 \leq 2\rho^i(2r^{\max} - 2\rho^i) < 4\rho^i r^{\max} = \frac{2L}{\alpha}\rho^i.$$

Because $\rho^a > \frac{\varepsilon}{2}$ we obtain

$$\rho^i > \frac{(\rho^a)^2\alpha}{2L} \geq \frac{\varepsilon^2\alpha}{8L}$$

and hence conclude further that the condition number of a set constrained by a quadratic and a linear cut is bounded by

$$\frac{\rho^a}{\rho^i} \leq \frac{\sqrt{\frac{2L}{\alpha}\rho^i}}{\rho^i} \leq \frac{4L}{\alpha\varepsilon}.$$

Hence, $\kappa'$, the final condition number of $P_k$ is bounded by $\max\{\kappa, \frac{4L}{\alpha\varepsilon}\}$, implying

$$\rho^i \geq \frac{\rho^a}{\kappa'} \geq \min\left\{\frac{1}{\kappa}, \frac{\alpha\varepsilon}{4L}\right\}\frac{\varepsilon}{2}.$$

■

**Lemma 4.7** *Let $k \geq 0$. If $\|x_k - x^*\| > \varepsilon$ then*

$$F_k(x_k^*)'' \leq \max\left\{\frac{8L}{\varepsilon^2\alpha}, \frac{2\kappa}{\varepsilon}\right\}^2 \cdot \left(m + k + 2\sqrt{m+k}\right)^2 \cdot \mathbb{1}. \tag{20.35}$$

**Proof**

Let $f_k := m + k + 2\sqrt{m+k}$. By Lemma 4.6 there exists an inscribed sphere with radius $\rho^i := \min\left\{\frac{\varepsilon^2\alpha}{8L}, \frac{\varepsilon}{2\kappa}\right\}$; let $x^i$ be the corresponding center. From Jarre [2] then follows

$$x^i + \{h \,|\, \langle h, (1/\rho^i)^2 \cdot \mathbb{1}\, h\rangle \leq 1\} \subset P_k \subset x_k^* + f_k \cdot \{h \,|\, \langle h, F_k''(x_k^*)h\rangle \leq 1\}$$

and so

$$\frac{F_k''(x_k^*)}{f_k^2} \leq \left(\frac{1}{\rho^i}\right)^2 \mathbb{1}$$

which proves the claim. ∎

Based on the above discussion we can bound the arithmetic complexity of computing $\delta_a$ by

$$\mathcal{O}\left(n^3 + n \cdot \ln\ln\left(\zeta \cdot \frac{k^2}{\varepsilon^4}\right)\right) \tag{20.36}$$

where $\zeta$ depends on $L/\alpha, \theta, \kappa, \varepsilon, a, \lambda_0^{\min}$ and $\eta$. Here $\eta$ comes in because we need an Eigenvalue-bound for $F_k''(x_k)$ instead of $F_k''(x_k^*)$: In ([5]) it is shown that this factor is at most $(1-\beta(\eta))^{-2}$.

There is evidence that in the above complexity estimate $k^2$ can be reduced to $k$; indeed, in case of exact centers [4] we have

$$\lambda_k^{\min} \geq 2\alpha \sum_{i=0}^{k-1} \frac{1}{t_i(x_k^*)} \geq \mathcal{O}(k^2).$$

Given $\varepsilon$, the number $k$ of quadratic cuts is bounded, i.e. $\sum_{i=0}^{k-1}[t_i(x_k)]^{-1}$ has a bounded, finite number of terms only. Hence, by continuity of $t_i(x_k)$, i.e. $t_i(x_k) \to t_i(x_k^*)$ for $\eta \to 0$, we obtain

$$\lambda_k^{\min} \geq 2\alpha \sum_{i=0}^{k-1} \frac{1}{t_i(x_k)} \overset{\eta \to 0}{\longrightarrow} 2\alpha \sum_{i=0}^{k-1} \frac{1}{t_i(x_k^*)} \geq \mathcal{O}(k^2).$$

This can reduce the complexity in (20.36) from $k^2$ down to $k$ at the expense of increased constants.

But the question remains: Is there an update schema, which is independent of $k$?

# References


[1] M. Fu, Z.-Q. Luo, and Y. Ye. Approximation algorithms for quadratic programming. Technical report, University of Iowa, August 1996.

[2] F. Jarre. Optimal ellipsoidal approximations around the analytic center. *Applied Mathematics of Optimization*, 30:15–19, 1994.

[3] D. Kinderlehrer and G. Stampacchia. *An Introduction to Variational Inequalities and their Applications.* Academic Press, 1980.

[4] H.-J. Lüthi and B. Büeler. The analytic center quadratic cut algorithm for strongly monotone variational inequality problems. *SIAM Journal on Optimizatoion*, 10(2):415–426, 2000.

[5] Y. Nesterov. Complexity estimates of some cutting plane methods based on the analytic barrier. *Mathematical Programming*, 69:149–176, 1995.

[6] Y. Nesterov and A. S. Nemirovsky. *Interior-Point Polynomial Algorithms in Convex Programming.* SIAM, Studies in Applied Mathematics, 1994.

[7] Y. Nesterov, O. Péton, and J.-P. Vial. Homogeneous analytic center cutting plane methods with approximate centers. *To appear in Optimization Methods and Software*, March 1998.

[8] Y. Nesterov and J.-P. Vial. Homogeneous analytic center cutting plane methods for convex problems and variational inequalities. *SIAM Journal on Optimization*, 9:707–728, 1999.

[9] Y. Ye. Combining binary search and Newton's method to compute real roots for a class of real functions. *Journal of Complexity*, 10:271–280, 1994.

Chapter 21

# GENERATING CONVEX FUNCTIONS


Pierre Maréchal

*Laboratoire ACSIOM*

*Département de Mathématiques*

*Université de Montpellier 2, CC 051*

*Place Eugène Bataillon*

*34 095 Montpellier CEDEX 5*

*France*

marechal@darboux.math.univ-montp2.fr



**Abstract** A functional operation that generates convex functions on $\mathbb{R}^{n+m}$ from convex functions on $\mathbb{R}^n$ and $\mathbb{R}^m$ originated from the study of the multiplicative potential function. We review some of the properties of this functional operation, including associativity, right distributivity with respect to addition and left distributivity with respect to infimal convolution. We also show how it can be used to build a generalized version of the multiplicative potential function.



**Keywords:** Convexity, Conjugation, multiplicative potential function.


## 1 INTRODUCTION

The multiplicative potential and penalty functions are used as barrier functions for solving linear programming problems [4, 5, 10]. Their convexity results from the convexity of the function

$$k(z, \mathbf{x}) := \frac{1}{\gamma} \frac{|z|^\gamma}{x_1^{\alpha_1} \cdot \ldots \cdot x_n^{\alpha_n}}, \quad x_1, \ldots, x_n > 0$$

in either of the following cases [1, 2, 3]:

(a) $\alpha_1, \ldots, \alpha_n > 0$ and $\gamma \geq 1 + \sum_{k=1}^n \alpha_k$;

(b) $\alpha_1 = \ldots = \alpha_n = 1$, $\gamma \geq n$ and $\mathbf{x}$ is restricted to a simplex, that is, a set of the form $\{\mathbf{x} | \mathbf{x} \geq \mathbf{0}, \langle \mathbf{x}, \mathbf{s} \rangle = \sigma\}$ where $\sigma > 0$ and $\mathrm{s} > \mathbf{0}$.




*N. Hadjisavvas and P.M. Pardalos (eds.), Advances in Convex Analysis and Global Optimization,* 361–369.

In [7], the convexity of $k$ was also proved in the case where

(c) $\alpha_1, \ldots, \alpha_n > 0$, $\gamma \geq 1 + (n-1)\max\{\alpha_1, \ldots \alpha_n\}$ and $\mathbf{x}$ is restricted to a simplex.

In [7] and [8], it has been shown that the multiplicative potential function can be built from elementary convex functions by means of a new functional operation. The purpose of this note is to give an overview of some of the properties of this functional operation. The proofs of most results stated below can be found in [7] and [8].

Given a closed proper convex function $f$ on $\mathbb{R}^n$, we denote by $\tilde{f}$ the function defined on $\mathbb{R}^{n+1}$ by

$$\tilde{f}(\mathbf{x}, y) = \begin{cases} y\, f\left(\dfrac{\mathbf{x}}{y}\right) & \text{if } y > 0, \\ f0^+(\mathbf{x}) & \text{if } y = 0, \\ +\infty & \text{if } y < 0, \end{cases}$$

in which $f0^+$ is the recession function of $f$. The function $\tilde{f}$ differs from the *perspective* of $f$ (see [6], Section IV.2.2) only for $y = 0$. It is the support function of the *hypograph* of $-f^\star$ (see [9], Corollary 13.5.1). In other words, $\tilde{f}$ and the indicator function of $\operatorname{hypo}(-f^\star)$ are convex conjugate to each other:

$$\tilde{f}^\star = \delta(\cdot \,|\, \operatorname{hypo}(-f^\star)) \quad \text{and} \quad \tilde{f} = \delta^\star(\cdot \,|\, \operatorname{hypo}(-f^\star)).$$

In particular, $\tilde{f}$ is closed, proper, convex and positively homogeneous. If $f(\mathbf{0}) \leq 0$ then $\tilde{f}(\mathbf{x}, \cdot)$ is decreasing for all fixed $\mathbf{x} \in \mathbb{R}^d$. Consequently, since $\tilde{f}$ is lower semi-continuous, $\tilde{f}(\mathbf{x}, \cdot)$ is also right continuous.

Similar statements can be made when $f$ is a closed proper concave function. By symmetry, $f0^+$ is then the function whose hypograph is the recession cone of the hypograph of $f$, and $\tilde{f}$ is defined likewise, except for negative $y$ in which case it is set equal to $-\infty$. For clarity, we shall denote by $f_\star$ the concave conjugate of $f$.

**Definition 1** *Let $f$ and $g$ be closed, proper, convex or concave functions on $\mathbb{R}^n$ and $\mathbb{R}^m$, respectively. Assuming that $g(\operatorname{dom} g) \subset [0, \infty)$, we define the function $g \rhd f$ on $\mathbb{R}^{n+m}$ by*

$$(g \rhd f)(\mathbf{x}, \mathbf{y}) := \tilde{f}(\mathbf{x}, g(\mathbf{y})).$$

*When $\mathbf{y} \notin \operatorname{dom} g$ the above expression is understood to be $+\infty$ if $f$ is convex and $-\infty$ if $f$ is concave.*

## 2 MAIN RESULTS

When the epigraph of $f$ contains the origin, the following conjugation result holds.

**Theorem 1** *Let $f$ be a closed proper convex function such that $f(\mathbf{0}) \leq 0$, and let $g$ be a closed proper concave function such that $g(\operatorname{dom} g) \subset [0, \infty)$. Then*

$$(g \rhd f)^{\star}(\boldsymbol{\xi}, \boldsymbol{\eta}) = (f^{\star} \rhd (-g)^{\star})(\boldsymbol{\eta}, \boldsymbol{\xi})$$
$$\text{and} \quad (f^{\star} \rhd (-g)^{\star})^{\star}(\mathbf{y}, \mathbf{x}) = (g \rhd f)(\mathbf{x}, \mathbf{y}).$$

*In other words, $g \rhd f$ and $f^{\star} \rhd (-g)^{\star}$ are convex conjugate to each other, up to a change of the order of the variables.*

The proof of Theorem 1 relies essentially on the fact that conjugation changes *right scalar multiplication* into *left scalar multiplication* and vice versa [7].

An immediate consequence of Theorem 1 is that both $g \rhd f$ and $f^{\star} \rhd (-g)^{\star}$ are closed proper convex functions. As can be easily checked, $f(\mathbf{0}) \leq 0$ if and only if $f^{\star} \geq 0$, and $g(\operatorname{dom} g) \subset [0, \infty)$ if and only if $(-g)^{\star}$ is above its recession function. We thus have the following corollary:

**Corollary 1** *The function $g \rhd f$ is closed proper convex in either of the following cases:*

(1) *$f$ is closed proper convex, with $f(\mathbf{0}) \leq 0$, and $g$ is closed proper concave, with $g(\operatorname{dom} g) \subset [0, \infty)$;*

(2) *$f$ is closed proper convex, with $f \geq f0^{+}$, and $g$ is closed proper convex, with $g \geq 0$.*

One can also generate concave functions by taking $f$ concave:

**Theorem 2** *Let $f$ be a closed proper concave function such that $f(\mathbf{0}) \geq 0$, and let $g$ be a closed proper concave function such that $g(\operatorname{dom} g) \subset [0, \infty)$. Then*

$$(g \rhd f)_{\star}(\boldsymbol{\xi}, \boldsymbol{\eta}) = ((-f_{\star}) \rhd g_{\star})(\boldsymbol{\eta}, \boldsymbol{\xi})$$
$$\text{and} \quad ((-f_{\star}) \rhd g_{\star})_{\star}(\mathbf{y}, \mathbf{x}) = (g \rhd f)(\mathbf{x}, \mathbf{y}).$$

*In other words, $g \rhd f$ and $(-f_{\star}) \rhd g_{\star}$ are concave conjugate to each other, up to a change of the order of the variables.*

**Corollary 2** *The function $g \rhd f$ is closed proper concave in either of the following cases:*

(1') *$f$ is closed proper concave, with $f \leq f0^+$, and $g$ is closed proper convex, with $g \geq 0$;*

(2') *$f$ is closed proper concave, with $f(\mathbf{0}) \geq 0$, and $g$ is closed proper concave, with $g(\operatorname{dom} g) \subset [0, \infty)$.*

Now, having generated a convex function by means of Corollary 1(1), say, it may happen that the domain of $g$ contains the origin, so that $g \triangleright f$ itself is less than or equal to zero at the origin. Taking a closed proper concave function $h$ on $\mathbb{R}^l$ such that $h(\operatorname{dom} h) \subset [0, \infty)$ then gives rise to the closed proper convex function $h \triangleright [g \triangleright f]$, whose convex conjugate is, by Theorem 1, the function $[f^\star \triangleright (-g)^\star] \triangleright (-h)^\star$. In Theorem 4 below, we shall see that the brackets can be omitted in the latter expressions. Prior to this, we enunciate a theorem which gives an expression for the recession function of $g \triangleright f$.

**Theorem 3** *Let $f$ be a closed proper convex function on $\mathbb{R}^n$ such that $f(\mathbf{0}) \leq 0$. Let $g$ be a closed proper concave function on $\mathbb{R}^m$ such that $g(\operatorname{dom} g) \subset [0, \infty)$ and $\mathbf{0} \in \operatorname{dom} g$. Then $g0^+(\operatorname{dom} g0^+) \subset [0, \infty)$ and*

$$(g \triangleright f)0^+ = g0^+ \triangleright f.$$

**Theorem 4 (Associativity)** *Let $f$ be a closed proper convex function on $\mathbb{R}^n$ such that $f(\mathbf{0}) \leq 0$. Let $g$ and $h$ be closed proper concave functions on $\mathbb{R}^m$ and $\mathbb{R}^l$, respectively, such that $g(\operatorname{dom} g) \subset [0, \infty)$, $\mathbf{0} \in \operatorname{dom} g$ and $h(\operatorname{dom} h) \subset [0, \infty)$. Then*

(i) *$(g \triangleright f)(\mathbf{0}, \mathbf{0}) \leq 0$ and the function $h \triangleright g$ which is concave by Corollary 2(2') is non-negative on its domain;*

(ii) *$h \triangleright (g \triangleright f)$ is closed proper and convex, and we have*

$$h \triangleright (g \triangleright f) = (h \triangleright g) \triangleright f.$$

The proof of Theorem 4 relies in part on the previous theorem [8]. By duality and symmetry, similar statements can made in the case where $g \triangleright f$ is as in Corollaries 1(2), 2(1') or 2(2'). We also obtain, by induction, the following result:

**Corollary 3** *Let $f$ be a closed proper convex function on $\mathbb{R}^d$ and let $g_1, \ldots, g_n$ be closed proper concave functions on $\mathbb{R}^{m_1}, \ldots, \mathbb{R}^{m_n}$, respectively, such that $g_k(\operatorname{dom} g_k) \subset [0, \infty)$, for all $k = 1, \ldots, n$ and that $\mathbf{0} \in \operatorname{dom} g_k$ for all $k = 1, \ldots, n-1$. Then*

$$g_n \triangleright \cdots \triangleright g_1 \triangleright f \quad \text{and} \quad f^\star \triangleright (-g_1)^\star \triangleright \cdots \triangleright (-g_n)^\star$$

*are unambiguously defined closed proper convex functions which are conjugate to each other.*

When the functions involved in Corollary 3 are positively homogeneous of some degree, we obtain conjugation (and convexity) results for products and quotients. Suppose for example that $\alpha_1, \ldots, \alpha_n$ are positive numbers, that

$$\gamma > 1 + \sum_{k=1}^{n} \alpha_k,$$

and let $g_k$ be a closed proper concave function on $\mathbb{R}^{m_k}$ that is positively homogeneous of degree

$$\beta_k := \frac{\alpha_k}{\gamma - 1 - \sum_{j=1}^{k-1} \alpha_j}$$

for all $k = 1, \ldots, n$. For simplicity, assume that $g_k$ is positive on its domain, except at the origin. Clearly, $0 < \beta_k < 1$, and the domain of $g_k$ is a convex cone containing the origin in $\mathbb{R}^{m_k}$. Let $f$ be a closed proper convex function on $\mathbb{R}^d$ that is positively homogeneous of degree $\gamma$, and define

$$G = g_n \rhd \cdots \rhd g_1 \quad \text{and} \quad K = G \rhd f.$$

It can be easily shown, by induction, that

$$G(\mathbf{x}_1, \ldots, \mathbf{x}_n) = \begin{cases} \displaystyle\prod_{k=1}^{n} g_k^{1 - \frac{1}{\gamma - 1} \sum_{j=1}^{k-1} \alpha_j}(\mathbf{x}_k) & \text{if } \mathbf{x}_k \in \operatorname{dom} g_k \setminus \{\mathbf{0}\},\ \forall k, \\ 0 & \text{if } \mathbf{x}_k = \mathbf{0} \text{ for some } k, \\ -\infty & \text{if } \mathbf{x}_k \notin \operatorname{dom} g_k \text{ for some } k. \end{cases}$$

Each function in the above product is positively homogeneous of degree $(\gamma - 1)^{-1}\alpha_k$, so that $G$ is positively homogeneous of degree $(\gamma - 1)^{-1} \sum \alpha_j$. It is worth noticing that in this example, the functional operation $\rhd$ is commutative. This is not true, of course, in general. The function $G \rhd f$ is given by

$$(G \rhd f)(\mathbf{z}, \mathbf{x}) = \begin{cases} \dfrac{f(\mathbf{z})}{\prod_{k=1}^{n} g_k^{\frac{\alpha_k}{\beta_k}}(\mathbf{x}_k)} & \text{if } G(\mathbf{x}) > 0, \\ f0^+(\mathbf{z}) & \text{if } G(\mathbf{x}) = 0 \text{ (i.e. } \mathbf{x} = \mathbf{0}), \\ +\infty & \text{if } G(\mathbf{x}) = -\infty, \end{cases}$$

in which we have denoted by $\mathbf{x}$ the vector formed with the $\mathbf{x}_k$'s. By Corollary 3, $K$ is a closed proper convex function on $\mathbb{R}^{d+\sum m_j}$. It is easy to see that $K$ is positively homogeneous of degree $\gamma - 1 - \sum \alpha_j$ and that its domain is the convex cone

$$\operatorname{dom} K = \operatorname{dom} f \times \operatorname{dom} g_1 \times \ldots \times \operatorname{dom} g_n.$$

The function $K$ is a generalized version of the function $k$ defined in the introduction. If the $g_k$'s and $f$ are the functions defined on $\mathbb{R}$ by

$$g_k(x) := \begin{cases} x^{\beta_n} & \text{if } x \geq 0, \\ -\infty & \text{if } x < 0, \end{cases} \qquad \text{and} \quad f(z) := \frac{1}{\gamma}|z|^\gamma,$$

then the resulting function $K$ is given by

$$K(z, \mathbf{x}) = \begin{cases} \dfrac{1}{\gamma}\dfrac{|z|^\gamma}{\mathbf{x}^{\boldsymbol{\alpha}}} & \text{if } \mathbf{x} > \mathbf{0}, \\ \delta\,(z \mid 0) & \text{if } \mathbf{x} \geq \mathbf{0} \text{ but } \mathbf{x} \not> \mathbf{0}, \\ +\infty & \text{if } \mathbf{x} \not\geq \mathbf{0}. \end{cases}$$

The conjugation formula given in Corollary 3 provides an easy way to compute the conjugate of $K$:

$$K^\star(\zeta, \boldsymbol{\xi}) := \begin{cases} \left(\dfrac{\mathbf{a}}{c}\right)^{\mathbf{a}} \dfrac{1}{c}\dfrac{|\zeta|^c}{(-\boldsymbol{\xi})^{\mathbf{a}}} & \text{if } \zeta \neq 0 \text{ and } \boldsymbol{\xi} < \mathbf{0}, \\ +\infty & \text{if } \zeta \neq 0 \text{ and } \boldsymbol{\xi} \not< \mathbf{0}, \\ \delta\,(\boldsymbol{\xi} \mid (-\infty, 0]^n) & \text{if } \zeta = 0. \end{cases}$$

where $(c, \mathbf{a})$ is the multi-exponent *conjugate* to $(\gamma, \boldsymbol{\alpha})$:

$$\mathbf{a} := \frac{\boldsymbol{\alpha}}{\gamma - (1 + |\boldsymbol{\alpha}|)}, \qquad c := \frac{\gamma}{\gamma - (1 + |\boldsymbol{\alpha}|)}.$$

Note in passing that $c > 1 + |\mathbf{a}|$, that

$$\boldsymbol{\alpha} = \frac{\mathbf{a}}{c - (1 + |\mathbf{a}|)}, \qquad \gamma = \frac{c}{c - (1 + |\mathbf{a}|)},$$

and that $\gamma - |\boldsymbol{\alpha}|$ and $c - |\mathbf{a}|$ are conjugate exponents in the usual sense. In [7], more conjugation formulas can be found, and strict convexity of $k$ in the case where $\gamma > 1 + \sum \alpha_j$, which can be established by studying the Hessian, is shown via differentiability of the conjugate.

Another example is provided by the function $F$ defined on $\mathbb{R}^n$ by

$$F(\mathbf{x}) := \begin{cases} \dfrac{1}{1+\sum_{j=1}^{n} \dfrac{1}{x_j}} & \text{if } \mathbf{x} > \mathbf{0}, \\ 0 & \text{if } \mathbf{x} \geq \mathbf{0} \text{ but } \mathbf{x} \not> \mathbf{0}, \\ -\infty & \text{if } \mathbf{x} \not\geq \mathbf{0}. \end{cases}$$

The function $F$ can be built from the function $f$ defined on $\mathbb{R}$ by

$$f(x) := \frac{x}{1+x}, \quad x \geq 0.$$

Clearly, $f$ is closed proper concave, non-negative on its domain, and $0 \in \operatorname{dom} f$. It is readily seen that

$$F(\mathbf{x}) = (f \rhd \cdots \rhd f)(x_1, \ldots, x_n).$$

Applying Corollary 2(2') recursively then shows that $F$ is a closed proper concave function.

## 3 FURTHER PROPERTIES

It is rather immediate from Definition 1 that the functional operation $\rhd$ is *right distributive with respect to addition.* Since addition and infimal convolution are dual operations, the conjugation formula of Theorem 1 implies that, in essence, $\rhd$ is also *left distributive with respect to infimal convolution.* More precisely, the following holds [8]:

**Theorem 5** *Let $f_1, f_2$ be closed proper convex functions on $\mathbb{R}^n$ such that $f_i(\mathbf{0}) \leq 0$, $i = 1, 2$ and let $g$ be a closed proper concave function on $\mathbb{R}^m$ such that $g(\operatorname{dom} g) \subset [0, \infty)$. Then $g \rhd (f_1 + f_2)$ is a closed proper convex function, and we have*

$$g \rhd (f_1 + f_2) = (g \rhd f_1) + (g \rhd f_2). \tag{21.1}$$

In the case where $g$ is convex, as in Corollary 1(2), the same conclusion holds if we make sure that $f_1 + f_2$ is closed and proper, so that the recession function of the sum equals the sum of the recession functions. (This is garanteed, of course, by the assumptions of the previous theorem.)

**Theorem 6** *Let $f_1, f_2$ be closed proper convex functions on $\mathbb{R}^n$ such that $f_i \geq f_i0^+$, $i = 1, 2$ and $\operatorname{dom} f_1$ meets $\operatorname{dom} f_2$, and let $g$ be a closed proper convex non-negative function on $\mathbb{R}^m$. Then $g \rhd (f_1 + f_2)$ is a closed proper convex function, and Eq. (21.1) holds.*

Recall that the infimal convolution of two convex functions $\varphi_1$ and $\varphi_2$ on $\mathbb{R}^d$ is the convex function defined by

$$(\varphi_1 \,\square\, \varphi_2)(\mathbf{x}) := \inf\{\varphi_1(\mathbf{x}_1) + \varphi_2(\mathbf{x}_2) \mid \mathbf{x}_1 + \mathbf{x}_2 = \mathbf{x}\}.$$

We define likewise the supremal convolution of two concave functions by

$$(\varphi_1 \,\square\, \varphi_2)(\mathbf{x}) := \sup\{\varphi_1(\mathbf{x}_1) + \varphi_2(\mathbf{x}_2) \mid \mathbf{x}_1 + \mathbf{x}_2 = \mathbf{x}\}.$$

The next two theorems establish that $\triangleright$ is left distributive with respect to $\square$ in situations corresponding to Corollary 1.

**Theorem 7** *Let $f$ be a closed proper convex function such that $f(\mathbf{0}) \leq 0$, and let $g_1, g_2$ be closed proper concave functions such that $g_i(\operatorname{dom} g_i) \subset [0, \infty)$, $i = 1, 2$ and $\operatorname{ri}\operatorname{dom} g_{1\star}$ meets $\operatorname{ri}\operatorname{dom} g_{2\star}$. Then $(g_1 \,\square\, g_2) \triangleright f$ is a closed proper convex function and we have*

$$(g_1 \,\square\, g_2) \triangleright f = (g_1 \triangleright f) \,\square\, (g_2 \triangleright f). \tag{21.2}$$

**Theorem 8** *Let $f$ be a closed proper convex function such that $f \geq f0^+$, and let $g_1, g_2$ be closed proper convex non-negative functions. Then $(g_1 \,\square\, g_2) \triangleright f$ is a closed proper convex function and Eq. (21.2) holds in either of the following cases:*

(i) *$f$ is positively homogeneous (so that $f = f0^+$) and $\operatorname{ri}\operatorname{dom} g_1^\star 0^+$ meets $\operatorname{ri}\operatorname{dom} g_2^\star 0^+$;*

(ii) *$f$ is not positively homogeneous, and $\operatorname{ri}\operatorname{dom} g_1^\star$ meets $\operatorname{ri}\operatorname{dom} g_2^\star$.*

By symmetry, similar results can be obtained in the case where the generated functions are concave. Note also that we have restricted ourselves to cases where the generated functions are closed and proper. Distributivity may still hold in other cases.

## References


[1] M. Avriel, W. E. Diewert, S. Schaible and I. Zang, *Generalized Convexity*, (Plenum Press, New York and London, 1988).

[2] J.M. Borwein, "A generalization of Young's $l^p$ inequality," *Mathematical Inequalities and Applications* 1, (1998) 131–136.

[3] J.P. Crouzeix, J.A. Ferland and S. Schaible, "Generalized convexity on affine subspaces with an application to potential functions," Mathematical Programming 56, (1992) 223–232.

[4] I. Imai, "On the convexity of the multiplicative version of Karmarkar's potential function," *Mathematical Programming* 40, (1988) 29–32.

[5] M. Iri and I. Imai, "A multiplicative barrier function method for linear programming," *Algorithmica* 1, (1986) 455–482.

[6] J.-B. Hiriart-Urruty and C. Lemaréchal, *Convex Analysis and Minimization Algorithms, I and II*, (Springer-Verlag, Berlin, 1993).

[7] P. Maréchal, "On the convexity of the multiplicative potential and penalty functions and related topics," to appear in *Mathematical Programming*.

[8] P. Maréchal, "On a functional operation that generates new convex functions from old ones," submitted to *Transactions of the American Mathematical Society*.

[9] R.T. Rockafellar, *Convex Analysis*, (Princeton University Press, Princeton, N.J., 1970).

[10] S. Schaible, "Quasi-convex optimization in general real linear spaces," *Zeitschrift fuer Operations Research* 16, (1972) 205–213.

Chapter 22

# THE METHOD OF MOMENTS FOR NONCONVEX VARIATIONAL PROBLEMS

René Meziat[1,2], Juan José Egozcue[1] and Pablo Pedregal[2]

[1] *Facultad de Matemáticas y Estadística*
*Departamento de Matemática Aplicada III*
*Universidad Politécnica de Cataluña*
*Calle Gran Capitán s/n, 08034, Barcelona, Spain*

[2] *Departamento de Matemáticas*
*Escuela de Ingeniería de Telecomunicaciones de Cuenca*
*Universidad de Castilla La Mancha*
*Edificio Melchor Cano, 16071, Cuenca, Spain*
rmeziat@arq-cu.uclm.es

**Abstract** We present a method for solving the generalized formulation of one dimensional nonconvex variational problems. This method is based on Young measure theory, it works particularly well when the Lagrangian function has a precise polynomial structure since it uses the moments of the parametrized measures as variables. The most remarkable feature of this method is its ability to transform nonlinear nonconvex problems into convex ones. In addition, it solves a particular type of global optimization problems by estimating convex hulls of positive polynomials. We also use at some point the famous Carathéodory's theorem.

**Keywords:** Convex envelopes, Moments theory, Variational problems, Nonlinear elasticity, Young measures.



*N. Hadjisavvas and P.M. Pardalos (eds.), Advances in Convex Analysis and Global Optimization,* 371–382.

# 1 INTRODUCTION

We deal with one dimensional variational problems in the form

$$\min_u I(u) = \int_0^1 \psi(x, u(x), u'(x))\, dx \tag{22.1}$$

where the Lagrangian function $\psi(x, u, \lambda)$ is a polynomial in $\lambda$ given by the general expression

$$\psi(x, u, \lambda) = \sum_{k=0}^{2n} a_k(x, u)\, \lambda^k \tag{22.2}$$

where the coefficients $a_k$ are assumed to have a convex dependence on $x$ and $u$. When $\psi$ fails to be convex in $\lambda$, problem (22.1) may lack a classical solution. Consider the famous *Bolza's example,* in which the Lagrangian function takes the nonconvex form

$$\psi(x, u, \lambda) = \left(\lambda^2 - 1\right)^2 + u^2 \tag{22.3}$$

and the boundary conditions are $u(0) = u(1) = 0$. We refer the reader to [1, 2, 3] for the implications of Convexity on the Calculus of Variations.

It is easy to see that all minimizing sequences for Bolza's example exhibit an oscillatory behavior that is present in all spatial scales. This limit behavior has important implications on the physical realm. For example, in elasticity models it is linked with the kind of microstructure developed on materials [4, 5]. *Young Measure theory* has been used to analyze this limit behavior for nonconvex variational problems [6, 7]. In this work, we propose a systematic method to describe the limit behavior in minimizing sequences, when the Lagrangian function has the general polynomial form given in (22.2).

We start considering the *generalized* formulation of problem (22.1) given by

$$\min_v \widetilde{I}(v) = \int_0^1 \int_R \psi(x, u(x), \lambda)\, dv_x(\lambda)\, dx \tag{22.4}$$

where $v_x$ is a parametrized probability measure and $u(x)$ is given by

$$u(x) = \int_0^x \int_R \lambda\, dv_s(\lambda)\, ds\,.$$

The admissible objects for the generalized problem (22.4) are families of parametrized probability measures $v = \{v_x\}_{0 \le x \le 1}$ which are called *Young measures.* The minimizer $v^*$ for problem (22.4) provides information about the limit behavior of minimizing sequences for problem

(22.1). That is, for every minimizing sequence $\{u_n\}_{n=1}^{\infty}$ we have

$$\int_E dv_x^* = PR\left(\lim_{n\to\infty} u_n'(x) \in E\right) \tag{22.5}$$

where $PR$ means *probability*. These points are clarified in [7].

Since the Lagrangian function has the polynomial structure (22.2), the generalized formulation (22.4) is easily transformed into the optimization problem

$$\min_m J(m) = \int_0^1 \sum_{k=0}^{2n} a_k(x,u)\, m_k(x)\, dx \tag{22.6}$$

where $m_k(x)$ is the $k$-th algebraic moment of the parametrized measure $v_x$ and $u(x)$ is given by

$$u(x) = \int_0^x m_1(s)\, ds\ .$$

We have studied the optimization problem (22.6) as a new formulation for the nonconvex problem (22.1) [8]. It is worth noting that this new formulation preserves linearity and shrinks the admissible objects from a vast family of Young measures to a convex family of vectorial functions. Simultaneously, we devise a method for treating global optimization problems involving positive polynomials. We start our contribution by describing how Caratheodory's theorem can help in our analysis. Next, we treat simple examples of global optimization problems by calculating convex envelopes of positive polynomials. Then, we show some examples of nonconvex variational problems. Finally, we give some remarks about theory.

## 2 CARATHEODORY'S THEOREM

Let $\phi : R \to R$ be a continuous function such that $\lim_{|x|\to\infty} \phi(x) = \infty$, then the convex envelope of $\phi$ is given by the optimization problem

$$\phi_c(t) = \min_\mu \int_R \phi(\lambda)\, d\mu(\lambda) \tag{22.7}$$

where $\mu$ denotes the family of all probability measures with mean $t$. This fact follows as a corollary from the famous Caratheodory's theorem [9]. On the other hand, the variational problem (22.1) admits the *relaxed* formulation

$$\min_u \bar{I}(u) = \int_0^1 \psi_c(x, u(x), u'(x))\, dx \tag{22.8}$$

whose minimizer $u_*$ provides the solution for problem (22.4). We mean that the minimizer $v^* = \{v_x^*\}_{0 \le x \le 1}$ for problem (22.4) can be obtained by solving problem (22.7) when $\psi(x, u_*(x), \cdot)$ replace $\phi(\cdot)$ and $t = u_*'(x)$; that is,

$$u_*'(x) = \int_R \lambda \, dv_x^*(\lambda) \tag{22.9}$$

$$\psi_c(x, u_*(x), u_*'(x)) = \int_R \psi(x, u_*(x), \lambda) \, dv_x^*(\lambda)$$

for every point $x$ on the domain $0 < x < 1$. So, we must study problem (22.7) when $\phi$ becomes a polynomial. In other words, we will use problem (22.7) as a small laboratory to analyze and test our methods. We refer to [7] for the connections between Young measure theory and nonconvex variational problems.

## 3 CONVEX ENVELOPES OF POLYNOMIALS

We may find the convex envelopes of positive polynomials

$$\phi(\lambda) = \sum_{k=0}^{2n} a_k \lambda^k \qquad a_{2n} > 0 \tag{22.10}$$

by solving the optimization problem in measures given in (22.7). However, each integral in (22.7) can be written as $\sum_{k=0}^{2n} a_k m_k$ where $m_k$ is the $k$-th order moment of measure $\mu$. Therefore, we can pose problem (22.7) as the programming problem

$$\min_{m_k} \sum_{k=0}^{2n} a_k m_k \tag{22.11}$$

whose variables $m_k$ must be the moments of some probability measure. To describe the feasible region it is enough to demand that the Hankel matrix $H = (m_{i+j})_{i,j=0}^{n}$ be positive semidefinite [10, 11]. Further, this fact may be characterized by nonlinear inequalities using $H$ principal minors [12].

**Theorem 1** If $m_0, \ldots, m_{2n}$ are the moments of a positive measure, then $H = (m_{i+j})_{i,j=0}^{n}$ is a positive semidefinite Hankel matrix.

**Theorem 2** Let $S$ be the set of vectors $(m_k)_{k=0}^{2n}$ which form a semidefinite positive Hankel matrix, and let $F$ be the set of vectors $(m_k)_{k=0}^{2n}$ whose entries are the moments of some positive measure, then $S$ is convex and $F$ is a dense subset of $S$.

Thus, we can calculate the convex envelope of every positive polynomial by solving the corresponding programming problem (22.11) subject to a set of nonlinear inequality constraints. It is easy to see that this programming problem has a linear objective function $\sum_{k=0}^{2n} a_k m_k$ and a convex feasible region $S$.

## 3.1 ANALYTICAL EXAMPLE

The Hankel matrix

$$H = \begin{pmatrix} 1 & t & m_2 \\ t & m_2 & m_3 \\ m_2 & m_3 & m_4 \end{pmatrix}$$

is positive semidefinite if and only if its entries satisfy the following inequalities

$$m_2 \geq t^2 \qquad m_4 \geq m_2^2 \qquad |H| \geq 0 \,. \tag{22.12}$$

Therefore, the convex envelope for every four degree positive polynomial in the form

$$\phi(\lambda) = \sum_{k=0}^{4} a_k \lambda^k \qquad a_4 = 1 \tag{22.13}$$

is given by

$$\phi_c(t) = a_0 + a_1 t + a_2 m_2^* + a_3 m_3^* + m_4^*$$

where $m_k^*$ is the solution for the programming problem with objective function

$$f = a_0 + a_1 t + a_2 m_2 + a_3 m_3 + m_4 \tag{22.14}$$

subject to the constraints in (22.12). Because the feasible region is convex, the Kuhn-Tucker equations

$$\vec{\nabla} f - \lambda \vec{\nabla} |H| = 0 \qquad |H| = 0$$

yield the solution

$$\begin{aligned} m_2^* &= -\tfrac{a_2}{2} + \tfrac{a_3^2}{8} - \tfrac{a_3}{2} t \\ m_3^* &= \tfrac{a_2 a_3}{4} - \tfrac{a_3^3}{16} + \left(\tfrac{3}{8} a_3^2 - \tfrac{a_2}{2}\right) t \\ m_4^* &= \tfrac{a_2^2}{4} - \tfrac{a_2 a_3^2}{4} + \tfrac{3}{64} a_3^4 + \left(\tfrac{a_2 a_3}{2} - \tfrac{a_3^3}{4}\right) t \,. \end{aligned} \tag{22.15}$$

To obtain (22.15) we have assumed $m_2^* > t^2$, which means that we are in the case where $\phi_c(t) < \phi(t)$ ; that is, the region where the curve fails to be convex and its convex envelope is a straight line. Otherwise, we obtain $m_2^* = t^2$, $m_3^* = t^3$, $m_4^* = t^4$ as the optimum moments, which means that $\phi_c(t) = \phi(t)$; that is, the region where the curve is convex. Thus, the general expression for the convex envelope of any four degree positive polynomial in (22.13) is given by

$$\phi_c(t) = \begin{cases} a_0 - \frac{1}{4}a_2^2 + \frac{1}{8}a_2a_3^2 - \frac{1}{64}a_3^4 + \left(a_1 - \frac{1}{2}a_2a_3 + \frac{1}{8}a_3^3\right)t \\ \qquad if \quad |4t + a_3| \leq \sqrt{3a_3^2 - 8a_2} \quad , \\ \phi(t) \qquad otherwise \end{cases}$$

This example is interesting because it illustrates some important features of more complex situations. In order to calculate convex envelopes of nonconvex polynomial curves, we use Caratheodory's Theorem to pose the optimization problem in probability measures (22.7). Next, we transform it into the programming problem (22.11) by using the fact that $\phi$ is a positive polynomial. Further, we realize that the new programming problem has a linear objective function and a convex feasible region, and therefore it can be solved by standard optimization techniques.

## 3.2 NUMERICAL EXAMPLE

Once we have realized that the calculus of the convex envelope of the polynomial $\phi$ can be reduced to the programming problem (22.11), we assign this task to a particular computer routine. For the nonconvex positive polynomial

$$\phi(\lambda) = 1 + 7\lambda - 5\lambda^2 - 3\lambda^3 + \lambda^4$$

we obtain the next values for its convex envelope

| $t$ | $\phi_c(t)$ | $t$ | $\phi_c(t)$ |
|---|---|---|---|
| −2.00 | 7.00 | 1.00 | −16.01 |
| −1.50 | −5.56 | 1.50 | −17.95 |
| −1.00 | −8.25 | 2.00 | −19.89 |
| −0.50 | −10.20 | 2.50 | −21.82 |
| 0.00 | −12.14 | 3.00 | −23.00 |
| 0.50 | −14.07 | 3.50 | −14.31 |

It is very important to note that this method is an automatic way to calculate the convex envelope of a polynomial. Therefore, it may be used to find numerically minimizers for the relaxed and generalized problems associated with nonconvex variational problems. In addition, this automatic method finds the global minimum of polynomial curves eluding local-minima difficulties: Work in this direction has also been carried out by Shor, Lasserre and Nesterov [13, 14, 15].

# 4 NONCONVEX VARIATIONAL PROBLEMS

To understand the limit behavior in minimizing sequences for nonconvex variational problems (22.1) with Lagrangian function (22.2), we solve the optimization problem (22.6) where the variables $m_k(x)$ must be the moments of some probability measure for every point $x$ on $0 < x < 1$. Therefore, we impose nonlinear inequality constraints using the principal minors of each member within the parametrized family of Hankel matrices

$$H_x = [m_{i+j}(x)]_{i,j=0}^n .$$

Next, we translate the problem into discrete form and send it to an appropriate computer optimization environment. In spite of the lack of convexity of the original variational problem, we are copping with a programming problem with convex objective function and convex feasible region. Therefore, the optimum parametrized moments $m_k^*(x)$ can be obtained by standard optimization routines.

Next, we use some classical results about moments [10] to recover the optimum parametrized measure $v_x^*$ from the optimum moments $m_k^*(x)$. Thus, we can construct the minimum Young measure $v^*$ for the generalized problem (22.4). This step is extremely simple because Caratheodory's theorem guarantees that the optimum parametrized measures are supported in two points at the most [9].

Using the optimum Young measure $v^*$ obtained in (22.4) and the meaning given by (22.5), we discover the limit behavior present in all minimizing sequences for problem (22.1). Thus, we can describe the oscillatory behavior in the minimizing sequences and we clarify the existence of classical solution for problem (22.1). Next, we show some examples of nonconvex variational problems tackled by the method of moments.

## 4.1 BOLZA'S EXAMPLE

In the discrete form $x_i = \frac{i}{N}$, Bolza's example (22.3) with boundary conditions $u(0) = 0$ and $u(1) = a$, transforms into the programming problem

$$\min_{m_k(x_i)} \sum_{i=1}^{N} \left\{ 1 - 2m_2(x_i) + m_4(x_i) + \left( \sum_{j=1}^{i} m_1(x_j)\,\Delta x_j \right)^2 \right\} \Delta x_i$$

subject to the nonlinear constraints

$$\begin{vmatrix} 1 & m_1(x_i) & m_2(x_i) \\ m_1(x_i) & m_2(x_i) & m_3(x_i) \\ m_2(x_i) & m_3(x_i) & m_4(x_i) \end{vmatrix} \geq 0 \qquad \begin{vmatrix} m_2(x_i) & m_3(x_i) \\ m_3(x_i) & m_4(x_i) \end{vmatrix} \geq 0$$

$$\begin{vmatrix} 1 & m_1(x_i) \\ m_1(x_i) & m_2(x_i) \end{vmatrix} \geq 0 \qquad \begin{vmatrix} 1 & m_2(x_i) \\ m_2(x_i) & m_4(x_i) \end{vmatrix} \geq 0$$

for each point $i = 1, \ldots, N$ and the linear equality constraint

$$\sum_{i=1}^{N} m_1(x_i)\,\Delta x_i = a \quad .$$

This problem may be solved by standard computer procedures because the objective function and the feasible set are convex. In the first table below, we list the optimum moments when $a = 0$ and $N = 10$. From these values we obtain the optimum parametrized measures listed in the second table where $r_i$, $s_i$ are the possible slopes and $\alpha_i$, $\beta_i$ are respectively, their probabilities.

| $i$ | $m_1^*(x_i)$ | $m_2^*(x_i)$ | $m_3^*(x_i)$ | $m_4^*(x_i)$ |
|---|---|---|---|---|
| 1 | 0.00 | 1.00 | 0.00 | 1.00 |
| 2 | 0.00 | 1.00 | 0.00 | 1.00 |
| 3 | 0.00 | 1.00 | 0.00 | 1.00 |
| 4 | 0.00 | 1.00 | 0.00 | 1.00 |
| 5 | 0.00 | 1.00 | 0.00 | 1.00 |
| 6 | 0.00 | 1.00 | 0.00 | 1.00 |
| 7 | 0.00 | 0.99 | 0.00 | 0.99 |
| 8 | 0.00 | 1.00 | 0.00 | 1.00 |
| 9 | 0.00 | 1.00 | 0.00 | 1.00 |
| 10 | 0.00 | 1.00 | 0.00 | 1.00 |

| $r_i$ | $\alpha_i$ | $s_i$ | $\beta_i$ |
|---|---|---|---|
| −0.99 | 0.49 | 1.00 | 0.50 |
| −0.99 | 0.50 | 1.00 | 0.49 |
| −0.99 | 0.50 | 1.00 | 0.49 |
| −0.99 | 0.49 | 1.00 | 0.50 |
| −0.99 | 0.49 | 1.00 | 0.50 |
| −1.00 | 0.49 | 1.00 | 0.50 |
| −0.99 | 0.50 | 1.00 | 0.49 |
| −0.99 | 0.50 | 1.00 | 0.49 |
| −0.99 | 0.50 | 1.00 | 0.49 |
| −0.99 | 0.49 | 1.00 | 0.50 |

These results mean that the oscillatory behavior is present along all the domain $0 < x < 1$ and that the terms $u_n$ oscillate between $\pm 1$ slopes with equal probability when $n \to \infty$ . They agree with the known analytical solution [7, 6]

$$v_x^* = \frac{1}{2}\delta_{-1} + \frac{1}{2}\delta_1 \qquad \forall x : 0 < x < 1 \quad .$$

For the boundary condition $a = 0.5$ we obtain the following results

| $m_1^*(x_i)$ | $m_2^*(x_i)$ | $m_3^*(x_i)$ | $m_4^*(x_i)$ |
|---|---|---|---|
| 0.00 | 1.00 | 0.00 | 1.00 |
| −0.03 | 0.99 | −0.03 | 0.99 |
| 0.03 | 1.00 | 0.03 | 1.00 |
| −0.11 | 0.99 | −0.10 | 0.99 |
| 0.19 | 1.00 | 0.19 | 1.00 |
| 0.85 | 1.00 | 0.85 | 1.00 |
| 1.00 | 1.01 | 1.02 | 1.03 |
| 1.00 | 1.01 | 1.02 | 1.03 |
| 1.01 | 1.02 | 1.04 | 1.05 |
| 1.02 | 1.04 | 1.06 | 1.09 |

| $r_i$ | $\alpha_i$ | $s_i$ | $\beta_i$ |
|---|---|---|---|
| −0.99 | 0.49 | 1.00 | 0.50 |
| −0.99 | 0.51 | 1.00 | 0.48 |
| −0.99 | 0.48 | 1.00 | 0.51 |
| −0.99 | 0.55 | 1.00 | 0.44 |
| −0.99 | 0.40 | 1.00 | 0.59 |
| −0.99 | 0.07 | 1.00 | 0.92 |
| 0.00 | 0.00 | 1.00 | 1.00 |
| 0.00 | 0.00 | 1.00 | 0.99 |
| 0.00 | 0.00 | 1.01 | 0.99 |
| 0.00 | 0.00 | 1.02 | 1.00 |

These results mean that the oscillatory behavior takes place on $0 < x < \frac{1}{2}$ and that the terms $u_n$ alternate between $\pm 1$ slopes with equal probability. However, there is no oscillatory behavior on $\frac{1}{2} < x < 1$ where the terms $u_n$ prefer the straight line with slope 1. Again, these results agree with the analytical solution

$$v_x^* = \begin{cases} \frac{1}{2}\delta_{-1} + \frac{1}{2}\delta_1 & for \quad 0 < x < \frac{1}{2} \\ \delta_1 & for \quad \frac{1}{2} < x < 1 \end{cases} .$$

Note the connection between $\pm 1$ slopes supporting the oscillatory behavior and the position of the two wells of the nonconvex potential $\phi(t) = (t^2 - 1)^2$. Also observe how the boundary condition $a$ affects the zone where the oscillatory behavior takes place. The purpose of the method of moments is to discover this kind of relationships in more complicated cases.

## 4.2 MORE EXAMPLES

For the nonconvex variational problem

$$\min_u \int_0^1 \left\{ \phi(u'(x)) + (u(x) - x)^2 \right\} dx$$

with boundary conditions $u(0) = -0.5$, $u(1) = 0$, the procedure described yields the optimum parametrized moments and the optimum parametrized measures listed below

| $m_1^*(x_i)$ | $m_2^*(x_i)$ | $m_3^*(x_i)$ | $m_4^*(x_i)$ |
|---|---|---|---|
| 1.05 | 1.11 | 1.17 | 1.23 |
| 1.04 | 1.09 | 1.14 | 1.19 |
| 1.03 | 1.07 | 1.11 | 1.16 |
| 1.03 | 1.08 | 1.12 | 1.16 |
| 1.04 | 1.08 | 1.10 | 1.17 |
| 1.01 | 1.02 | 1.03 | 1.04 |
| 0.82 | 0.99 | 0.82 | 0.99 |
| −0.02 | 0.99 | 0.01 | 0.99 |
| −1.00 | 1.01 | −0.99 | 1.03 |
| −1.02 | 1.05 | −1.08 | 1.11 |

| $r_i$ | $\alpha_i$ | $s_i$ | $\beta_i$ |
|---|---|---|---|
| 1.05 | 1.00 | 0.00 | 0.00 |
| 1.04 | 1.00 | 0.00 | 0.00 |
| 1.03 | 1.00 | 0.00 | 0.00 |
| 1.03 | 1.00 | 0.00 | 0.00 |
| 1.04 | 1.00 | 0.00 | 0.00 |
| 1.01 | 1.00 | 0.00 | 0.00 |
| −1.00 | 0.08 | 0.99 | 0.91 |
| −0.97 | 0.52 | 1.02 | 0.47 |
| 0.00 | 0.00 | −1.00 | 1.00 |
| 0.00 | 0.00 | −1.02 | 1.00 |

These results show that the oscillatory behavior appears about the point 0.75 on the domain, a straight line with slope 1 is plausible before this point, and a straight line with slope −1 is preferred after this point. For the nonconvex variational problem

$$\min_u \int_0^1 \left\{\phi\left(u'(x)\right) + u(x)^2\right\} dx$$

with six degree polynomial potential $\phi(t) = 2t^2 - 4t^4 + t^6$ and boundary conditions $u(0) = u(1) = 0$, our procedure yields the following results

| $m_1^*(x_i)$ | $m_2^*(x_i)$ | $m_3^*(x_i)$ | | $m_6^*(x_i)$ |
|---|---|---|---|---|
| 0.09 | 2.38 | 0.18 | | 13.51 |
| −0.01 | 2.38 | −0.03 | | 13.52 |
| 0.01 | 2.38 | 0.04 | | 13.50 |
| 0.01 | 2.34 | −0.00 | | 12.95 |
| 0.05 | 2.38 | 0.14 | ··· | 13.53 |
| −0.30 | 2.32 | −0.58 | | 12.66 |
| 0.03 | 2.38 | 0.06 | | 13.53 |
| 0.03 | 2.38 | 0.11 | | 13.52 |
| −0.02 | 2.38 | −0.01 | | 13.55 |
| 0.09 | 2.38 | 0.19 | | 13.54 |

| $r_i$ | $\alpha_i$ | $s_i$ | $\beta_i$ |
|---|---|---|---|
| −1.55 | 0.46 | 1.53 | 0.53 |
| −1.54 | 0.50 | 1.54 | 0.49 |
| −1.54 | 0.49 | 1.54 | 0.50 |
| −1.53 | 0.49 | 1.52 | 0.50 |
| −1.54 | 0.48 | 1.54 | 0.51 |
| −1.50 | 0.60 | 1.55 | 0.39 |
| −1.55 | 0.48 | 1.53 | 0.51 |
| −1.53 | 0.49 | 1.54 | 0.50 |
| −1.53 | 0.50 | 1.55 | 0.49 |
| −1.54 | 0.46 | 1.54 | 0.53 |

Because Caratheodory's theorem guarantees the optimum measure is supported in two points at most, only the first three optimum moments are used to recover the optimum parametrized measures. We infer that the oscillatory behavior appears along all the domain and the terms $u_n$ prefer $\pm 1.53$ slopes with equal probability in the limit $n \to \infty$. Note that these slope values are connected with the two deepest wells in the nonconvex potential $\phi$.

# 5 THEORETICAL REMARKS

As we emphasize in the title of this contribution, we propose a new method for solving a particular family of variational problems. The method is strongly based on the theory of Young measures and its connections with nonconvex variational problems. These connections are analyzed in [7], from this work we use the essential fact that the optimum Young measure for the generalized problem (22.4) is given by the expressions (22.9) involving the minimizer of the relaxed problem (22.8) . Therefore, the central question is reduced to verify that the convex envelope of positive polynomials can be estimated by solving the programming problem (22.11) .

To clarify this point we study two optimization problems; namely, the optimization problem in measures (22.7) and the programming problem in moments (22.11) . We establish an equivalence between them using an old result on moments theory from E. Fisher [11]. A distinct alternative is to analyze the feasible sets for each problem. So, we have shown [8] that the moments of the admissible measures for (22.7) are a dense subset of the feasible region for the programming problem (22.11) .

This analysis implies that the minimizer for problem (22.11) provides the moments of the optimum measure for problem (22.7) . Therefore, we can find the optimum measure that defines the convex envelope of every positive polynomial by solving the programming problem (22.11) . This remark, Pedregal's work and Caratheodory's theorem are the theoretical basis of the method of moments.

## References

[1] G. Buttazzo, M. Giaquinta, *One-dimensional Variational Problems,* Clarendon Press, Oxford, 1998.

[2] L. Cesari, *Optimization Theory and Applications*, Springer-Verlag, 1983.

[3] B. Dacorogna, *Direct Methods in the Calculus of Variations,* Springer Verlag, 1989.

[4] J. Ball, R. James, *Fine phase mixtures as minimizers of energy,* Archives for Rational Mechanics and Analysis, vol 100, pages 13-52, 1987.

[5] R. James, *Co-existence phases in the one dimensional static theory of elastic bars,* Archives for Rational Mechanics and Analysis, vol. 72, pages 99-140, 1979.

[6] J. Muñoz, P. Pedregal, *Explicit solutions of nonconvex variational problems in dimension one,* AMO, 41, 129-140.

[7] P. Pedregal, *Parametrized Measures and Variational Principles,* Birkhauser, 1997.

[8] R. Meziat, J. Egozcue, P. Pedregal, *From a nonlinear, nonconvex variational problem to a linear, convex formulation,* submitted.

[9] R. Rockafellar, *Convex Analysis,* Princeton, 1970.

[10] N. Akhiezer, *The Classical Moment Problem,* Oliver and Boyd, 1965.

[11] J. Shohat, J. Tamarkin, *The Problem of Moments,* mathematical surveys 1, American Mathematical Society, 1943.

[12] F. Gantmacher, *The Theory of Matrices,* Chelsea, 1989.

[13] J. Lasserre, *New LMI relaxations for the general non convex quadratic problem,* International Conference on Convex Analysis and Global Optimization, Samos, Greece, June, 2000.

[14] N. Shor, *Nondifferentiable Optimization and Polynomial Problems,* Kluwer, 2000.

[15] Y. Nesterov, *Squared Functional Systems and Optimization Problems,* Higher Performance Optimization, Kluwer, 2000.

Chapter 23

# A PIVOTING-BASED HEURISTIC FOR THE MAXIMUM CLIQUE PROBLEM

Alessio Massaro and Marcello Pelillo
*Dipartimento di Informatica*
*Università Ca' Foscari di Venezia*
*Via Torino 155*
*30172 Venezia-Mestre, Italy*
{amassaro,pelillo}@dsi.unive.it

**Abstract** We tackle the maximum clique problem by reformulating it in terms of a standard quadratic program, which derives from a theorem of Motzkin and Straus. A heuristic is then obtained by solving with Lemke's method the linear complementarity problem that arises from the KKT conditions of the QP. We have added a pivoting rule that exploits degeneracy to reach good suboptimal solutions. Very positive results have been obtained on the DIMACS benchmark graphs.

**Keywords:** Maximum clique problem, Linear complementarity problem, Lemke's algorithm.

## 1 INTRODUCTION

Let $G = (V, E)$ be an undirected graph, where $V = \{1, \ldots, n\}$ is the vertex set and $E \subseteq V \times V$ is the edge set. The complement graph of $G$, denoted by $\bar{G}$, is a graph having the same vertex set as $G$ and $\bar{E} = \{(i, j) \in V \times V : i \neq j \text{ and } (i, j) \notin E\}$ as the edge set. Two distinct vertices $i, j \in V$ are said to be *adjacent* if they are connected by an edge, i.e. if $(i, j) \in E$. The *neighborhood* of a vertex $v$ will be indicated with $N(v) = \{i \in V : (i, v) \in E\}$ and its degree will be $\deg(v) = |N(v)|$.

A *clique* is a subset of $V$ in which all vertices are pair-wise adjacent. A clique $S$ is called *maximal* if no strict superset of $S$ is a clique. A maximum cardinality clique (or, simply, a *maximum clique*) is a clique whose cardinality is the largest possible. The maximum clique problem



*N. Hadjisavvas and P.M. Pardalos (eds.), Advances in Convex Analysis and Global Optimization,* 383–394.

(MAX-CLIQUE) is the problem of finding a maximum clique $S$ in $G$ (see [2] for a review). The maximum size of a clique in $G$ is called the *clique number* (of $G$) and is typically denoted by $\omega(G)$.

In 1965, Motzkin and Straus [6] established a remarkable connection between MAX-CLIQUE and a certain quadratic problem. Let $G = (V, E)$ be an undirected graph, and let $\Delta$ denote the standard simplex in the $n$-dimensional Euclidean space $\mathbb{R}^n$:

$$\Delta = \left\{x \in \mathbb{R}^n : x_i \geq 0 \text{ for all } i \in V,\ e^T x = 1\right\}$$

where $e$ is a vector of appropriate length, consisting of unit entries.

Now, consider the following quadratic function

$$g(x) = x^T A_G x$$

where $A_G = (a_{ij})_{i,j \in V}$ is the adjacency matrix of $G$, i.e. $a_{ij} = 1$ if $(i,j) \in E$, and $a_{ij} = 0$ if $(i,j) \notin E$, and let $x^*$ be a global maximizer of $g$ in $\Delta$. Motzkin and Straus proved that

$$\omega(G) = \frac{1}{1 - g(x^*)}$$

and that $S \subseteq V$ is a maximum clique of $G$ if and only if its *characteristic vector* $x^S$, which is the vector in $\Delta$ defined by $x_i^S = 1/|S|$ if $i \in S$ and $x_i^S = 0$ otherwise, is a global maximizer of $g$ in $\Delta$.

One drawback of the Motzkin-Straus formulation relates to the existence of spurious solutions, i.e. maximizers of $g$ that are not in the form of characteristic vectors. This was observed empirically by Pardalos and Phillips [7] and more recently formalized by Pelillo and Jagota [9]. In principle, spurious solutions represent a problem for we cannot extract the vertices comprising the clique directly from them; they just provide information about the clique number.

The spurious solution problem has recently been solved by Bomze [1]. He considers the following regularized version of function $g$:

$$\hat{g}(x) = x^T \hat{A}_G x \quad \text{where} \quad \hat{A}_G = A_G + \frac{1}{2} I,$$

$I$ being the $n \times n$ identity matrix. The following is the spurious-free counterpart of the original Motzkin-Straus theorem.

**Theorem 1** *Let $S$ be a subset of vertices of a graph $G = (V, E)$, and let $x^S$ be its characteristic vector. Then the following statements hold:*

- *$S$ is a maximal (maximum) clique of $G$ if and only if $x^S$ is a local (global) maximizer of $\hat{g}$ over the simplex $\Delta$*

- *All local maximizers $x$ of $\hat{g}$ over $\Delta$ are strict, and of the form $x = x^S$ for some $S \subseteq V$.*

Note that maximizing $\hat{g}$ over $\Delta$ is equivalent to minimizing $f(x) = x^T \left(\gamma ee^T - \hat{A}_G\right) x$, where $\gamma$ is an arbitrary constant. Let us thus take $\gamma = 1$ and define the matrix $\breve{A}_G = ee^T - \hat{A}_G = \hat{A}_{\bar{G}}$. In the sequel it will be more convenient to consider the following minimization program

$$\begin{array}{ll} \min & f(x) = x^T \breve{A}_G x \\ \text{s.t.} & x \in \Delta \end{array} \tag{23.1}$$

in order to facilitate the shift from quadratic programming to the theory of linear complementarity problems.

## 2 LINEAR COMPLEMENTARITY PROBLEMS

Computing the stationary points of (23.1) can be done by solving the corresponding linear complementarity problem (LCP, see [4]) $(q_G, M_G)$, which is the problem of finding a vector $x$ satisfying the system

$$y = q_G + M_G x \geq 0, \quad x \geq 0, \quad x^T y = 0$$

where

$$q_G = \begin{bmatrix} 0 \\ \vdots \\ 0 \\ -1 \\ 1 \end{bmatrix} \qquad M_G = \begin{bmatrix} \breve{A}_G & -e & e \\ e^T & 0 & 0 \\ -e^T & 0 & 0 \end{bmatrix}. \tag{23.2}$$

With the above definitions, it is well known that if $x$ is a complementary solution of $(q_G, M_G)$ with $x^T = \left(\alpha^T, \beta^T\right)$ and $\alpha \in \mathbb{R}^n$, then $\alpha$ is a stationary point of (23.1). Indeed, the matrix $\breve{A}_G$ is always strictly copositive, hence so is $M_G$ and that is enough to assure that $(q_G, M_G)$ always has a solution [4].

Of course, a stationary point of (23.1) is not necessarily a local minimum, but in practice this is not a problem since there are several techniques that can start from it to reach a nearby local optimum. An example is given by the replicator dynamics [8].

Technical literature supplies a large number of algorithms to solve an LCP [4]. The most popular is probably Lemke's method, largely for its ability to provide a solution for a large number of matrix classes. Lemke's "Scheme I" belongs to the family of pivoting algorithms. Given the generic LCP $(q, M)$, it deals with the augmented problem $(q, d, M)$ defined by

$$y = q + [M, d] \begin{bmatrix} x \\ \theta \end{bmatrix} \geq 0, \quad \theta \geq 0, \quad x \geq 0, \quad x^T y = 0. \qquad (23.3)$$

A solution of $(q, d, M)$ with $\theta = 0$ promptly yields a solution to $(q, M)$. Lemke's method intends to compute precisely such a solution.

In the sequel we will refer to column $n + 3$ of matrix $M$, meaning the auxiliary column that contains vector $d$ in (23.3). The latter vector $d$ is called *covering vector*, belongs to $\mathbb{R}^{n+2}$ and must satisfy $d_i > 0$ whenever $q_i < 0$. We will choose $d = e$.

As usually done for outlining pivoting algorithms, we will use an exponent for the problem data. In practice $q^\nu$ and $M^\nu$ will identify the state after $\nu$ pivots and $\breve{A}_G^\nu$ will indicate the $n \times n$ leading principal submatrix of $M_G^\nu$. Consistently, $y^\nu$ and $x^\nu$ will indicate the vectors of basic and non-basic variables, respectively, each made up of a combination of the original $x_i$ and $y_i$ variables. The notation $\left\langle x_i^\nu, y_j^\nu \right\rangle$ will be used to indicate pivoting transformations. The index-set of the basic variables that satisfy the min-ratio test at iteration $\nu$ will be denoted with $\Omega^\nu$, i.e.

$$\Omega^\nu = \arg\min_i \left\{ \frac{-q_i^\nu}{m_{is}^\nu} : m_{is}^\nu < 0 \right\}$$

where $s$ is the index of the driving column.

In general, assuming an LCP non-degenerate is a strategy commonly taken to prove finiteness of pivoting schemes. This assumption basically implies that $|\Omega^\nu| = 1$ for all $\nu$. This in turn excludes that a certain algorithm cycles by repeating a small sequence of basis. In particular, Lemke's method is guaranteed to process any non-degenerate problem $(q, M)$ where $M$ is strictly copositive, and to do so without terminating on a secondary ray [4].

Unfortunately $(q_G, M_G)$ is degenerate, but it is possible to give an equivalent formulation of (23.1) in order to obtain a non-degenerate LCP. To this end it is easy to see that program (23.1) is equivalent to the following QP:

$$\begin{array}{ll} \min & f'(x) = x^T \breve{A}'_G x + c^T x \\ \text{s.t.} & x \in \Delta. \end{array}$$

where $c \in \mathbb{R}^n$ and $\breve{A}'_G = \breve{A}_G - \left(ce^T + ec^T\right)$. If $c \leq 0$ then copositivity is maintained, and if all its entries are different, then the corresponding LCP is non-degenerate. Furthermore, if $c_i \leq -1$ for some $i$, it is straightforward to check that even the first pivot step of Lemke's method changes.

The above method for degeneracy removal has a characteristic in common with the lexicographic degeneracy resolution (LDR) method [4]. Namely, they both require the introduction of extra data: vector $c$ in the previous case; a nonsingular square matrix as big as $M$ with lexicographically positive rows in the case of LDR. Of course these objects have to be assigned values and there are many sensible methods to do it, each one having a different theoretical ground and/or performance impact on the final result. However, for the scope of this paper it shall be enough to say that all tests we performed using these methods with different assignment techniques led to unsatisfactory performances.

# 3 THE PIVOTING-BASED HEURISTIC

Rather than continuing to investigate the enormous variety of assignment techniques mentioned in the previous section, we carry on examining the original degenerate form of the LCP $(q_G, M_G)$. In doing so, we employ the least-index rule which amounts to blocking the driving variable with a basic one that has minimum index within a certain subset of $\Omega^\nu$, i.e. $r = \min \Phi^\nu$ for some $\Phi^\nu \subseteq \Omega^\nu$. The latter set is constructed in order to make the number of the degenerate variables decrease as slowly as possible, i.e.

$$\Phi^\nu = \arg\min_i \left\{ |\Omega^\nu| - \left|\Omega_i^{\nu+1}\right| > 0 : i \in \Omega^\nu \right\}$$

where $\Omega_i^{\nu+1}$ is the index-set of those variables that would satisfy the min-ratio test at iteration $\nu+1$ if the driving variable at iteration $\nu$ was blocked with $y^\nu_{i\in\Omega^\nu}$. The previous conditional implies that a pivot step is taken and then reset in a sort of "look-ahead" fashion, hence we will refer to this rule as the *look-ahead (pivot) rule*. Computing $\left|\Omega_i^{\nu+1}\right|$ is actually $O(n)$ since only the driving and $q$ columns are required.

Before actually proceeding to illustrate a version of Lemke's algorithm applied to MAX-CLIQUE, let us take a look at the tableaus that it generates. This will help us to identify regularities that are reflected in the behavior of the algorithm itself. The initial tableau follows.

| | $1$ | $x_1$ | $\cdots$ | $x_n$ | $x_{n+1}$ | $x_{n+2}$ | $\theta$ |
|---|---|---|---|---|---|---|---|
| $y_1$ | $0$ | | | | $-1$ | $1$ | $1$ |
| $\vdots$ | $\vdots$ | | $\breve{A}_G$ | | $\vdots$ | $\vdots$ | $\vdots$ |
| $y_n$ | $0$ | | | | $-1$ | $1$ | $1$ |
| $y_{n+1}$ | $-1$ | $1$ | $\cdots$ | $1$ | $0$ | $0$ | $1$ |
| $y_{n+2}$ | $1$ | $-1$ | $\cdots$ | $-1$ | $0$ | $0$ | $1$ |

As $q_{n+1}$ is the only negative entry for the column of $q$, the first pivot to occur during initialization is $\langle y_{n+1}, \theta\rangle$ thereby producing the following transformation.

| | $1$ | $x_1$ | $\cdots$ | $x_n$ | $x_{n+1}$ | $x_{n+2}$ | $y_{n+1}$ |
|---|---|---|---|---|---|---|---|
| $y_1$ | $1$ | | | | $-1$ | $1$ | $1$ |
| $\vdots$ | $\vdots$ | | $\breve{A}_G - ee^T$ | | $\vdots$ | $\vdots$ | $\vdots$ |
| $y_n$ | $1$ | | | | $-1$ | $1$ | $1$ |
| $\theta$ | $1$ | $-1$ | $\cdots$ | $-1$ | $0$ | $0$ | $1$ |
| $y_{n+2}$ | $2$ | $-2$ | $\cdots$ | $-2$ | $0$ | $0$ | $1$ |

The driving variable for the second pivot is $x_{n+1}$. Since $m^1_{i,n+1} = -1$ for all $i = 1, \ldots, n$ it is immediate to see that the relative blocking variable can be any one of $y_1, \ldots, y_n$. In this case we apply no degeneracy resolution criterion, but rather allow for user intervention by catering for the possibility of deciding the second driving variable a priori. Let thus $y_p$ be the (arbitrary) variable that shall block $x_{n+1}$. After performing $\langle y_p, x_{n+1}\rangle$, we have the following tableau.

| | $1$ | $x_1$ | $\cdots$ | $x_n$ | $y_p$ | $x_{n+2}$ | $y_{n+1}$ |
|---|---|---|---|---|---|---|---|
| $y_1$ | $0$ | | | | $1$ | $0$ | $0$ |
| $\vdots$ | $\vdots$ | | | | $\vdots$ | $\vdots$ | $\vdots$ |
| $y_{p-1}$ | $0$ | | | | $1$ | $0$ | $0$ |
| $x_{n+1}$ | $1$ | | $\breve{A}_{G,p}$ | | $-1$ | $1$ | $1$ |
| $y_{p+1}$ | $0$ | | | | $1$ | $0$ | $0$ |
| $\vdots$ | $\vdots$ | | | | $\vdots$ | $\vdots$ | $\vdots$ |
| $y_n$ | $0$ | | | | $1$ | $0$ | $0$ |
| $\theta$ | $1$ | $-1$ | $\cdots$ | $-1$ | $0$ | $0$ | $1$ |
| $y_{n+2}$ | $2$ | $-2$ | $\cdots$ | $-2$ | $0$ | $0$ | $1$ |

where $\breve{A}_{G,p}$ denotes the matrix whose rows are defined as

$$\left(\breve{A}_{G,p}\right)_i = \begin{cases} \left(\breve{A}_G\right)_p - e^T & \text{if } i = p \\ \left(\breve{A}_G\right)_i - \left(\breve{A}_G\right)_p & \text{otherwise.} \end{cases}$$

Algorithm 1 formalizes the above statements.

Empirical evidence indicated $p$ as a key parameter for the quality of the final result of Algorithm 1. Unfortunately we could not identify any effective means to restrict the choice of values in $V$ that can guarantee a good sub-optimal solution. We thus had to consider iterating for most, if not all, vertices of $V$ as outlined in Algorithm 2. There we employ a

**Algorithm 1** Lemke's "Scheme I" with the "look-ahead" rule applied to MAX-CLIQUE.

**Input**: A graph $G = (V, E)$ and $p \in V$.
Let $(q_G, e, M_G)$ be the augmented LCP
where $q_G$ and $M_G$ are defined in (23.2).
$\nu \leftarrow 0$, $\langle y_{n+1}, \theta \rangle$, $\nu \leftarrow \nu + 1$, $\langle x_p, x_{n+1} \rangle$
The driving variable is $x_p$.
Infinite loop
  $\nu \leftarrow \nu + 1$
  Let $x_s^\nu$ denote the driving variable.
  $\Omega^\nu = \arg\min_i \left\{ \frac{-q_i^\nu}{m_{is}^\nu} : m_{is}^\nu < 0 \right\}$
  If $|\Omega^\nu| = 1$ then $r = \min \Omega^\nu$
  else $\Phi^\nu = \arg\min_i \left\{ |\Omega^\nu| - \left| \Omega_i^{\nu+1} \right| > 0 : i \in \Omega^\nu \right\}$, $r = \min \Phi^\nu$.
  $\langle y_{r \in \Omega^\nu}^\nu, x_s^\nu \rangle$
  If $y_r^\nu \equiv \theta$ then
    Let $\bar{x}$ denote the complementary solution of $(q_G, M_G)$.
    The result is $\mathrm{supp}\,(\bar{x}) \cap V$
  The new driving variable is the complementary of $y_r^\nu$.

very simple criterion to avoid considering those nodes that cannot drive to larger cliques than the one we already have because they have a too small degree. It is easy to comment that such criterion is effective only for very sparse graphs.

We also observed that the schema is sensitive to the ordering of nodes and found that the best figures were obtained by reordering $G$ by decreasing degree of the nodes. This feature too is formalized in Algorithm 2. We will refer to that schema by the name Pivoting Based Heuristic (PBH).

Another remarkable empirical finding was that Algorithm 1 never failed to return a maximal clique. We then never needed to perturb the final KKT point in order to reach a nearby local minimizer. We tried to find exceptions by running PBH on random graphs with non-clique regular subgraphs. The latter subgraphs correspond in fact to stationary points of program (23.1) as shown in [1]. Hundreds of experiments were conducted on random instances with different degrees of noise, but PBH never failed to return a maximal clique. At the moment we cannot give a formal proof of this fact.

**Algorithm 2** The *pivoting-based heuristic* (PBH) for MAX-CLIQUE.

**Input**: A graph $G = (V, E)$.
Let $G' = (V', E')$ be a permutation of $G$
with $\deg(u') \geq \deg(v')$ for all $u', v' \in V'$ with $u' < v'$.
$K^\star \leftarrow \emptyset$
For $v' = 1, \ldots, n : \deg(v') > |K^\star|$ do
    Run Algorithm 1 with $G'$ and $v'$ as input.
    Let $K$ be the obtained result.
    If $|K| > |K^\star|$, then $K^\star \leftarrow K$.
The result is the mapping of $K^\star$ in $G$.

## 4 EXPERIMENTAL RESULTS

In this section we will discuss the results obtained and confront them with those relative to two other methods based on the Motzkin-Straus theorem: Annealed Replication (AR) [3] and the Continuous Based Heuristic (CBH) [5].

Tables 23.1 and 23.2 list the performance figures obtained by running PBH (column *PBH*) over a selection of the DIMACS benchmark graphs. These graphs have been gathered for the second *DIMACS Challenge* (`http://dimacs.rutgers.edu`) with the aim to constitute a group of hard problems to be solved. Their order, density and clique number are reported in columns *Order*, *Density* and $\omega$, respectively. The results that pertain to AR and CBH are reproduced in columns *AR* and *CBH*. Entries that correspond to the best result for a given graph are boldfaced. Computing times (column *Time*) are in seconds and refer to a C++ implementation for a Linux machine with a 550Mhz Pentium III CPU. Some of them are missing (marked with "> 1800") because the Unix `clock` system call could not time periods longer than approximately 30 minutes on our test machine.

Keller graphs (prefixed with *keller*) are searched for cliques in a part of a proof of Keller's conjecture. We could consider only two instances because bigger ones riquired too much memory for our test machine. PBH computed the biggest cliques in both cases. One clique was maximum.

The *p_hat* graphs are generalized random graphs with a wider node degree spread. In 10 out of 15 cases PBH has produced the best results and 7 of them are maximum cliques. The largest known clique was actually reached in 9 cases.

| Graph | Order | Density | $\omega$ | AR | CBH | PBH | Time (sec) |
|---|---|---|---|---|---|---|---|
| keller4 | 171 | 64.9% | 11 | 8 | 10 | **11** | 3.62 |
| keller5 | 776 | 75.2% | 27 | 16 | 21 | **26** | 1093.50 |
| p_hat300-1 | 300 | 24.4% | 8 | **8** | **8** | **8** | 13.96 |
| p_hat300-2 | 300 | 48.9% | 25 | **25** | **25** | **25** | 34.85 |
| p_hat300-3 | 300 | 74.4% | 36 | 35 | **36** | 35 | 61.01 |
| p_hat500-1 | 500 | 25.3% | 9 | **9** | **9** | **9** | 83.54 |
| p_hat500-2 | 500 | 50.5% | 36 | **36** | 35 | **36** | 282.51 |
| p_hat500-3 | 500 | 75.2% | $\geq 49$ | 47 | **49** | 48 | 485.65 |
| p_hat700-1 | 700 | 24.9% | 11 | 9 | **11** | 10 | 249.38 |
| p_hat700-2 | 700 | 49.8% | 44 | 41 | **44** | **44** | 1022.3 |
| p_hat700-3 | 700 | 74.8% | $\geq 62$ | 59 | 60 | **62** | 1804.00 |
| p_hat1000-1 | 1000 | 24.5% | 10 | **10** | **10** | **10** | 798.03 |
| p_hat1000-2 | 1000 | 49.0% | $\geq 46$ | 44 | **46** | **46** | $> 1800$ |
| p_hat1000-3 | 1000 | 74.4% | $\geq 65$ | 62 | **65** | 64 | $> 1800$ |
| p_hat1500-1 | 1500 | 25.3% | 12 | 10 | 11 | **12** | $> 1800$ |
| p_hat1500-2 | 1500 | 50.6% | $\geq 65$ | **64** | 63 | **64** | $> 1800$ |
| p_hat1500-3 | 1500 | 75.4% | $\geq 94$ | 91 | **94** | 91 | $> 1800$ |
| MANN_a9 | 45 | 92.8% | 16 | **16** | **16** | **16** | 0.12 |
| MANN_a27 | 378 | 99.0% | 126 | 117 | 121 | **125** | 699.69 |
| MANN_a45 | 1035 | 99.6% | 345 | — | — | 342 | $> 1800$ |
| johnson8-2-4 | 28 | 55.6% | 4 | — | — | **4** | 0.01 |
| johnson8-4-4 | 70 | 76.8% | 14 | — | — | **14** | 0.32 |
| johnson16-2-4 | 120 | 76.5% | 8 | — | — | **8** | 1.05 |
| johnson32-2-4 | 496 | 87.9% | 16 | — | — | **16** | 184.78 |

*Table 23.1* Performance of PBH on the DIMACS graphs (part I).

Graphs prefixed with *MANN* are a reduction to MAX-CLIQUE of MINSETCOVER. Our schema computed the largest cliques in all cases and for one of them the clique was maximum.

*Johnson* graphs are borrowed from coding theory. We could not find results for these graphs pertaining AR and CBH, but PBH managed to find a maximum clique for all instances.

These *brock* graphs use a form of degree equalization to hide a large clique in a moltitude of smaller ones. Also for this category PBH reached the largest cliques except for one instance, the same and only one for

| Graph | Order | Density | $\omega$ | AR | CBH | PBH | Time (sec) |
|---|---|---|---|---|---|---|---|
| brock200_1 | 200 | 74.5% | 21 | 19 | **20** | **20** | 9.65 |
| brock200_2 | 200 | 49.6% | 12 | 10 | **12** | 11 | 5.08 |
| brock200_3 | 200 | 60.5% | 15 | 13 | **14** | **14** | 6.36 |
| brock200_4 | 200 | 65.8% | 17 | 14 | **16** | **16** | 7.29 |
| brock400_1 | 400 | 74.8% | 27 | 20 | 23 | **24** | 111.61 |
| brock400_2 | 400 | 74.9% | 29 | 23 | **24** | **24** | 113.32 |
| brock400_3 | 400 | 74.8% | 31 | 23 | 23 | **24** | 111.16 |
| brock400_4 | 400 | 74.9% | 33 | 23 | **24** | **24** | 112.69 |
| brock800_1 | 800 | 64.9% | 23 | 18 | 20 | **21** | 858.59 |
| brock800_2 | 800 | 65.1% | 24 | 18 | 19 | **20** | 866.38 |
| brock800_3 | 800 | 64.9% | 25 | 19 | **20** | **20** | 864.48 |
| brock800_4 | 800 | 65.0% | 26 | 19 | 19 | **20** | 862.36 |
| san200_0.7_1 | 200 | 70.0% | 30 | 15 | 15 | **30** | 9.93 |
| san200_0.7_2 | 200 | 70.0% | 18 | 12 | 12 | **17** | 8.23 |
| san200_0.9_1 | 200 | 90.0% | 70 | 45 | 46 | **70** | 28.76 |
| san200_0.9_2 | 200 | 90.0% | 60 | 39 | 36 | **60** | 22.79 |
| san200_0.9_3 | 200 | 90.0% | 44 | 31 | 30 | **44** | 19.01 |
| san400_0.5_1 | 400 | 50.0% | 13 | 7 | 8 | **13** | 52.26 |
| san400_0.7_1 | 400 | 70.0% | 40 | 20 | 20 | **40** | 142.02 |
| san400_0.7_2 | 400 | 70.0% | 30 | 15 | 15 | **30** | 110.69 |
| san400_0.7_3 | 400 | 70.0% | 22 | 12 | 14 | **17** | 93.84 |
| san400_0.9_1 | 400 | 90.0% | 100 | 50 | 50 | **100** | 397.82 |
| sanr200_0.7 | 200 | 69.7% | 18 | 16 | **18** | **18** | 8.15 |
| sanr200_0.9 | 200 | 89.8% | $\geq 42$ | **41** | **41** | **41** | 21.44 |
| sanr400_0.5 | 400 | 50.1% | 13 | **13** | 12 | **13** | 59.49 |
| sanr400_0.7 | 400 | 70.0% | $\geq 21$ | **21** | 20 | 20 | 101.9 |
| san1000 | 1000 | 50.2% | 15 | 8 | 8 | **15** | 1185.00 |

*Table 23.2* Performance of PBH on the DIMACS graphs (part II).

which a maximum clique was computed by CBH. All other computed cliques are not maximum.

The Sanchis graphs (*san*) are those for which we can notice the biggest difference between PBH and the other methods. In only one case did we obtain a clique smaller than that of Annealed Replication. For all other instances PBH has computed the biggest cliques and for 11 out of 15 of them, the clique was maximum. It is significant to notice that AR and

CBH have obtained cliques that are, on average, half the size of those obtained by PBH.

The DIMACS collection also includes the *c-fat* and *hamming* graphs, but these have been omitted because a simple run of Lemke's algorithm with least-index degeneracy resolution was sufficient to compute a maximum clique for each instance. This fact induced us to consider them not experimentally significant.

# 5 CONCLUSIONS

We have presented an effective heuristic for MAX-CLIQUE by means of the novel idea of employing a pivoting based schema on its Motzkin-Straus-derived formulation. The remarked effectiveness of our approach and the empirical immunity of Lemke's method to saddle points seem to indicate pivoting-based methods as a new promising way to tackle these combinatorial problems.

W.r.t. to the wide selection of combinatorial algorithms that have been proposed in the past [2], our method has the valuable property of not requiring the tuning of any parameter.

Indeed, further subjects of investigations may try a formal proof of convergence to local minimizers of problem (23.1) and tackle the problem of reducing and optimizing the still high $O(n^4)$ time complexity of our method as well as reducing the $O(n^2)$ space complexity.

## Acknowledgments

We wish to thank Prof. Immanuel M. Bomze (Univ. of Vienna) and Prof. Richard W. Cottle (Stanford Univ.) for sharing with us their precious expertise.

## References

[1] Bomze, I. M. (1997). Evolution towards the maximum clique. *J. Global Optim.*, 10:143–164.

[2] Bomze, I. M., Budinich, M., Pardalos, P. M., and Pelillo, M. (1999). The maximum clique problem. In Du, D.-Z. and Pardalos, P. M., editors, *Handbook of Combinatorial Optimization - suppl. vol. A*, pages 1–74. Kluwer Academic Publishers.

[3] Bomze, I. M., Budinich, M., Pelillo, M., and Rossi, C. (2000). A new "annealed" heuristic for the maximum clique problem. In Pardalos, P. M., editor, *Approximation and Complexity in Numerical Optimization.* Kluwer Academic Publishers.

[4] Cottle, R. W., Pang, J., and Stone, R. E. (1992). *The Linear Complementarity Problem.* Academic Press, Boston, MA.

[5] Gibbons, L. E., Hearn, D. W., and Pardalos, P. M. (1996). A continuous-based heuristic for the maximum clique problem. In Johnson, D. and Trick, M. A., editors, *Cliques, colouring and satisfiability*, volume 26 of *DIMACS Series*, pages 103–124. AMS.

[6] Motzkin, T. S. and Straus, E. G. (1965). Maxima for graphs and a new proof of a theorem of Turán. *Canad. J. Math.*, 17(4):533–540.

[7] Pardalos, P. M. and Phillips, A. T. (1990). A global optimization approach for solving the maximum clique problem. *Int. J. of Comput. Math.*, 33:209–216.

[8] Pelillo, M. (2000). Replicator dynamics in combinatorial optimization. In Pardalos, P. M. and Floudas, C. A., editors, *Encyclopedia of Optimization*. Kluwer Academic Publishers, Boston, MA.

[9] Pelillo, M. and Jagota, A. (1995). Feasible and infeasible maxima in a quadratic program for the maximum clique problem. *J. Artif. Neural Networks*, 2(4):411–420.

Chapter 24

# AN ANALYTIC CENTER SELF-CONCORDANT CUT METHOD FOR THE CONVEX FEASIBILITY PROBLEM

Faranak Sharifi Mokhtarian*
*Department of Mathematics and Statistics*
*Concordia University*
*1400 de Maisonneuve Blvd. West, LB Building*
*Montreal, Quebec, Canada H3G 1M8*
faranak@alcor.concordia.ca

Jean-Louis Goffin*
*Faculty of Management*
*McGill University*
*1001 Sherbrooke West*
*Montreal, Quebec, Canada H3A 1G5*
goffin@management.mcgill.ca

**Abstract** We consider a case of the convex feasibility problem where the set is defined by an infinite number of certain strongly convex self-concordant inequalities. At each iteration, the algorithm adds a self-concordant cut through an approximate analytic center of the current set of localization until a feasible point is found. We show that the algorithm is a fully polynomial approximation scheme.

**Keywords:** Convex feasibility problem, strongly convex function, interior-point methods, analytic center, self-concordant function

*This research is supported by the Natural Sciences and Engineering Research Council of Canada, grant number OPG0004152 and by the FCAR of Quebec.



*N. Hadjisavvas and P.M. Pardalos (eds.), Advances in Convex Analysis and Global Optimization*, 395–407.

# 1 INTRODUCTION

The problem studied is that of finding a point in a convex set $\Gamma$ defined as follows:

$$\Gamma = \{y \in \mathbb{R}^m \ : \ f_j(y) \leq 0, \ \ j = 1, 2, \ldots\}.$$

We make the following assumptions:

1. The set $\Gamma$ has a nonempty interior and is contained in the unit ball
$$\Omega^0 = \{y \in \mathbb{R}^m \ : \ \|y\| \leq 1\}.$$
2. For each $j = 1, 2, \ldots$, $f_j$ is $\bar{\mu}_j$-self-concordant, i.e.,
$$\{f_j'''(y)[h, h, h]\}^2 \leq 4\bar{\mu}_j \langle h, f_j''(y)h\rangle^3$$
for all $h \in \mathbb{R}^m$, for all $y \in \Omega^0$ and for some $\bar{\mu}_j > 0$.
3. For each $j = 1, 2, \ldots$, $-\ln(z_j - f_j(y))$ is $\hat{\mu}_j$-self-concordant for any $z_j \geq f_j(y)$ and for some $\hat{\mu}_j > 0$.
4. Let $\mu_j = \max\{\bar{\mu}_j, \hat{\mu}_j\}$, then $\mu_j \leq \mu$ for all $j = 1, 2, \ldots$.
5. Without loss of generality we assume that $\mu \geq 1$.
6. For each $j = 1, 2, \ldots$, $f_j''(y) - \alpha I$ is positive semidefinite for some constant $\alpha > 0$ and for all $y \in \Omega^0$.
7. For each $j = 1, 2, \ldots$
$$\max\{\|f_j'\|, \|f_j''\|\} = 1$$
where $\|f_j'\| = \max_{y \in \Omega^0} \|f_j'(y)\|$ and $\|f_j''\|$ is defined similarly (Note that this assumption implies that $\alpha \leq 1$.).
8. The set $\Gamma$ contains a point $\hat{y}$ such that for each $j = 1, 2, \ldots$ and for some $\epsilon > 0$
$$f_j(\hat{y}) \leq -\epsilon, \qquad \|\hat{y}\| \leq 1 - \epsilon.$$

We consider a self-concordant cut method for solving the above problem. At each iteration, the algorithm finds an approximate analytic center of an outer approximation of $\Gamma$ defined by the intersection of self-concordant inequalities generated in the previous iterations. If this approximate center is a solution, then the algorithm stops; otherwise, an inequality, violated at the current approximate center, is selected and is placed through the approximate center.

Our method is an extension of our algorithm for the convex quadratic feasibility problem with infinitely many constraints in [8]. The difference is that the quadratic cuts are replaced by self-concordant cuts. As in [8], we show that the total number of (damped) Newton steps to update from one approximate analytic center to another is at most $O(1)$ and that the algorithm is a fully polynomial approximation scheme.

Our algorithm can also be viewed as an extension of the algorithm of [4]. They consider the case where the total number of constraints is finite and use "shallow" cuts; whereas, we consider the case of infinite cuts and use "central" cuts. The case of convex quadratic feasibility problem is studied in [3, 8] and the case of variational inequalities is studied by [1, 2].

We will adopt the following standard notations in our analysis. For any generic vector $u$, $u^T$ will denote the vector transpose. The interior of a convex set $\Omega$ will be denoted by $\Omega_{int}$. For two symmetric $m \times m$ matrices $A$ and $B$, we write $A \succeq B$ if the matrix $A - B$ is positive semi-definite. We will also use the following notation:

$$\begin{aligned} \|u\|_H &= \langle u, Hu\rangle^{1/2}, \\ \|u\|_H^* &= \langle u, H^{-1}u\rangle^{1/2} \end{aligned}$$

where $H$ is a given symmetric positive definite matrix.

Let $\Omega$ be a bounded set in $\mathbb{R}^m$ defined by $n$ self-concordant inequalities as follows:

$$\Omega = \{y \in \mathbb{R}^m \ : \ f_j(y) \leq 0, \ \ j = 1, \ldots, n\}. \tag{24.1}$$

For $j = 1, \ldots, n$, the potential function of $\Omega$ at $y$ is

$$\phi(\Omega, y) = -\sum_{j=1}^{n} \ln(-f_j(y))$$

and the min potential of $\Omega$ is

$$P(\Omega) = \min_{y \in \Omega_{int}} \phi(\Omega, y).$$

The analytic center of $\Omega$ is the unique point at which the above minimum is attained. The gradient and Hessian of $\phi(\Omega, y)$ are given by

$$g(\Omega, y) = \phi'(\Omega, y) = \sum_{j=1}^{n} \frac{f_j'(y)}{-f_j(y)}$$

and

$$H(\Omega, y) = \phi''(\Omega, y) = \sum_{j=1}^{n} \left( \frac{f_j'(y)(f_j'(y))^T}{(f_j(y))^2} + \frac{f_j''(y)}{-f_j(y)} \right).$$

The point $y$ is called a $\theta$-approximate analytic center, or $\theta$-center of $\Omega$ for short, if

$$\|g(\Omega, y)\|^*_{H(\Omega,y)} \leq \theta.$$

## 2 POTENTIAL CHANGES

In this section, we study the changes in the min-potential when a self-concordant cut is added directly through a current approximate center. We then use these results to find an upper bound on the total number of (damped) Newton steps required to move from one approximate center to another. Throughout this section, we assume that $\Omega$ is defined by (24.1) and $\Omega$ includes the constraints of $\Omega^0$. Suppose that $\Omega$ is changed to $\Omega^+$ as follows:

$$\Omega^+ = \{y \in \mathsf{R}^m \ : \ f_j(y) \leq 0, \ \ j = 1, \ldots, n, \ f_{n+1}(y) \leq f_{n+1}(y^a)\}$$

where $y^a$ is a $\theta$-center of $\Omega$ with $0 < \theta < 1$. Let

$$y^u = y^a - \frac{\beta}{\bar{r}}\bar{H}^{-1}f'_{n+1}(y^a) \tag{24.2}$$

where $0 < \beta < 1/\sqrt{\mu}$ and

$$\bar{H} \ = \ H(\Omega, y^a) + \frac{f''_{n+1}(y^a)}{\bar{r}} \tag{24.3}$$

$$\varphi(\bar{r}) \ = \ \langle f'_{n+1}(y^a), \bar{H}^{-1}f'_{n+1}(y^a)\rangle^{\frac{1}{2}} \tag{24.4}$$

and $0 < \bar{r} < 1$ satisfies

$$\sqrt{\frac{1}{2}}\bar{r} \leq \varphi(\bar{r}) \leq \bar{r}. \tag{24.5}$$

We assume that the constraint defining $\Omega^0$ in which $\Gamma$ is contained is repeated $\rho$ times, where $\rho \geq 1$ is an integer and will be determined later. This means that

$$\phi(\Omega^0, y) = -\rho\ln(1 - \|y\|^2)$$

which implies that

$$\begin{aligned} H(\Omega^0, y) \ &= \ \frac{2\rho}{1 - \|y\|^2}I + \frac{4\rho yy^T}{(1 - \|y\|^2)^2} \\ &\succeq \ 2\rho I. \end{aligned}$$

This leads to the following Lemma.

**Lemma 2.1** *For any* $n \geq 0$

$$\varphi(\bar{r}) \leq \frac{1}{\sqrt{2\rho}}.$$

**Proof.** Using Assumption 7, we have

$$\begin{aligned}\varphi(\bar{r}) &\leq \langle f'_{n+1}(y^a), [H(\Omega^0, y^a)]^{-1} f'_{n+1}(y^a)\rangle^{1/2} \\ &\leq \frac{\|f'_{n+1}(y^a)\|}{\sqrt{2\rho}} \\ &\leq \frac{1}{\sqrt{2\rho}}.\end{aligned}$$

**Theorem 2.1** *If $y^u$ is chosen as in (24.2), then $y^u \in \Omega^+_{int}$ and*

$$P(\Omega^+) \leq \phi(\Omega^+, y^u) \leq \phi(\Omega, y^a) - \ln \bar{r} + \alpha_1(\beta, \theta, \mu). \tag{24.6}$$

*Here*

$$\alpha_1(\beta, \theta, \mu) = -\ln(1-\beta) - \ln\left[\frac{\beta}{2}\left(1 - \frac{\beta}{2} - \frac{\eta\beta^2}{3}\right)\right] - (1-\theta)\beta \tag{24.7}$$

*with*

$$\eta = \frac{2\sqrt{\mu\rho}}{[\sqrt[4]{\rho} - \beta\sqrt{\mu}]^3}$$

*and $\rho \geq 1$ chosen such that $\eta < \frac{3}{2}$.*

**Proof.** Note that

$$\begin{aligned}(y^a - y^u)^T H(\Omega, y^a)(y^a - y^u) &= \frac{\beta^2}{\bar{r}^2}(f'_{n+1}(y^a))^T \bar{H}^{-1} H(\Omega, y^a) \bar{H}^{-1} f'_{n+1}(y^a) \\ &\leq \frac{\beta^2}{\bar{r}^2}(\varphi(\bar{r}))^2 \quad \text{as } \bar{H} \succeq H(\Omega, y^a) \\ &\leq \beta^2 < 1.\end{aligned}$$

It follows from [6, Theorem 2.1.1] that $y^u \in \Omega_{int}$. Furthermore,

$$\begin{aligned}f_{n+1}(y^a) - f_{n+1}(y^u) &= \langle f'_{n+1}(y^a), y^a - y^u\rangle - \frac{1}{2}\langle y^u - y^a, f''_{n+1}(y^a)(y^u - y^a)\rangle \\ &\quad - \frac{1}{3} f'''_{n+1}(\tilde{y})[y^u - y^a, y^u - y^a, y^u - y^a]\end{aligned}$$

where $\tilde{y} = \lambda y^u + (1-\lambda) y^a$ for some $0 < \lambda < 1$. Due to self-concordancy of $f_{n+1}$

$$|f'''_{n+1}(\tilde{y})[h, h, h]| \leq 2\sqrt{\mu}\langle h, f''_{n+1}(\tilde{y})h\rangle^{3/2}$$

where $h = y^u - y^a$. Furthermore

$$\begin{aligned}\|h\|^2_{f''_{n+1}(y^a)} &= \langle y^u - y^a, f''_{n+1}(y^a)(y^a - y^u)\rangle \\ &= \frac{\beta^2}{\bar{r}^2}(f'_{n+1}(y^a))^T \bar{H}^{-1} f''_{n+1}(y^a) \bar{H}^{-1} f'_{n+1}(y^a)\end{aligned}$$

$$\begin{aligned} &\le \frac{\beta^2(\varphi(\bar{r}))^2}{\bar{r}} \quad \text{as } \bar{r}\bar{H} \succeq f''_{n+1}(y^a) \qquad (24.8)\\ &\le \beta^2 \bar{r} \\ &\le \beta^2 . \end{aligned}$$

So $\|h\|_{f''_{n+1}(y^a)} \le \frac{1}{\sqrt{\mu}}$ which implies that $\|\tilde{y} - y^a\|_{f''_{n+1}(y^a)} \le \frac{1}{\sqrt{\mu}}$; it then follows from self-concordancy of $f_{n+1}$ that (see [6, Theorem 2.1.1])

$$\langle h, f''_{n+1}(\tilde{y})h\rangle \le \frac{\langle h, f''_{n+1}(y^a)h\rangle}{[1 - \sqrt{\mu}\langle h, f''_{n+1}(y^a)h\rangle^{1/2}]^2} .$$

Therefore,

$$\begin{aligned} |f'''_{n+1}(\tilde{y})[h,h,h]| &\le \frac{2\sqrt{\mu}\|h\|^3_{f''_{n+1}(y^a)}}{[1 - \sqrt{\mu}\|h\|_{f''_{n+1}(y^a)}]^3} \\ &\le \frac{2\sqrt{\mu}\left(\frac{\beta\varphi(\bar{r})}{\bar{r}^{1/2}}\right)^3}{[1 - \frac{\sqrt{\mu}\beta\varphi(\bar{r})}{\bar{r}^{1/2}}]^3} \quad \text{by (24.8)} \\ &\le \left[\frac{2\sqrt{\mu}\varphi(\bar{r})}{\bar{r}^{1/2}[1 - \frac{\sqrt{\mu}\beta\varphi(\bar{r})}{\bar{r}^{1/2}}]^3}\right]\left[\frac{(\varphi(\bar{r}))^2\beta^3}{\bar{r}}\right] \\ &\le \left[\frac{2\sqrt{\mu}\bar{r}^{1/2}}{[1 - \sqrt{\mu}\beta\bar{r}^{1/2}]^3}\right]\left[\frac{(\varphi(\bar{r}))^2\beta^3}{\bar{r}}\right] \quad \text{since } \varphi(\bar{r}) \le \bar{r} \\ &\le \left[\frac{2\sqrt{\mu}(\frac{1}{\rho})^{1/4}}{[1 - \sqrt{\mu}\beta(\frac{1}{\rho})^{1/4}]^3}\right]\left[\frac{(\varphi(\bar{r}))^2\beta^3}{\bar{r}}\right] \\ &= \frac{2\sqrt{\mu\rho}}{[\sqrt[4]{\rho} - \beta\sqrt{\mu}]^3}\left[\frac{(\varphi(\bar{r}))^2\beta^3}{\bar{r}}\right] \end{aligned}$$

where the last inequality follows from Lemma 2.1 and the fact that

$$\bar{r} \le \sqrt{2}\varphi(\bar{r}) \le \frac{1}{\sqrt{\rho}} .$$

Let

$$\eta = \eta(\mu, \beta, \rho) = \frac{2\sqrt{\mu\rho}}{[\sqrt[4]{\rho} - \beta\sqrt{\mu}]^3} . \qquad (24.9)$$

Note also that

$$\langle f'_{n+1}(y^a), y^a - y^u\rangle = \frac{\beta(\varphi(\bar{r}))^2}{\bar{r}} .$$

Therefore,

$$f_{n+1}(y^a) - f_{n+1}(y^u) \geq \frac{\varphi(\bar{r})^2}{\bar{r}} \left[\beta - \frac{\beta^2}{2} - \frac{\eta\beta^3}{3}\right] > 0$$

provided that $\eta(\mu, \beta, \rho) < \frac{3}{2}$. So if we choose $\rho$ large enough so that it satisfies the following inequality

$$2\sqrt{\mu\rho} < 3[\sqrt[4]{\rho} - \beta\sqrt{\mu}]^3$$

then $y^u \in \Omega^+_{int}$. Finally,

$$\begin{aligned} P(\Omega^+) &\leq \phi(\Omega^+, y^u) \\ &= \phi(\Omega, y^u) - \ln(f_{n+1}(y^a) - f_{n+1}(y^u)) \\ &\leq \phi(\Omega, y^u) - \ln\left[\left(\beta - \frac{\beta^2}{2} - \frac{\eta\beta^3}{3}\right)\frac{(\varphi(\bar{r}))^2}{\bar{r}}\right] \\ &\leq \phi(\Omega, y^u) - \ln\left[\left(\beta - \frac{\beta^2}{2} - \frac{\eta\beta^3}{3}\right)\frac{\bar{r}}{2}\right]. \end{aligned}$$

But by Theorem 2.3(3) of [5], we have the following:

$$\begin{aligned} \phi(\Omega, y^u) \leq \; & \phi(\Omega, y^a) + \langle g(\Omega, y^a), y^u - y^a\rangle \\ & - \ln(1 - \|y^u - y^a\|_{H(\Omega,y^a)}) - \|y^u - y^a\|_{H(\Omega,y^a)}. \end{aligned}$$

This, together with the fact that $y^a$ is a $\theta$-center of $\Omega$, implies that

$$\begin{aligned} \phi(\Omega, y^u) &\leq \phi(\Omega, y^a) + \|g(\Omega, y^a)\|^*_{H(\Omega,y^a)}\|y^u - y^a\|_{H(\Omega,y^a)} - \ln(1-\beta) - \beta \\ &\leq \phi(\Omega, y^a) + \theta\beta - \ln(1-\beta) - \beta. \end{aligned}$$

Therefore,

$$\begin{aligned} P(\Omega^+) \leq \phi(\Omega^+, y^u) \leq \; & \phi(\Omega, y^a) - \ln\bar{r} - \ln(1-\beta) - \ln\left[\tfrac{\beta}{2}\left(1 - \tfrac{\beta}{2} - \tfrac{\eta\beta^2}{3}\right)\right] \\ & -(1-\theta)\beta. \end{aligned}$$

*Q.E.D.*

We will also have the following lower bound on the new min-potential.

**Theorem 2.2** *Suppose that $y^u$ is as defined by (24.2). Then*

$$P(\Omega^+) \geq \phi(\Omega, y^a) - \ln\bar{r} + \alpha_2(\theta, \alpha), \tag{24.10}$$

*where*

$$\alpha_2(\theta, \alpha) = \min_{v>0}\{(1-\theta)v - \ln(1+v) - \ln(1+\sqrt{1+\alpha^2 v^2}) - \ln\alpha\}. \tag{24.11}$$

**Proof.** Note that, it follows from convexity of $f_{n+1}$ that

$$f_{n+1}(y^a) - f_{n+1}(y) \leq \langle f'_{n+1}(y^a), y^a - y\rangle. \tag{24.12}$$

Besides (see [7, page 21]), for all $d$ such that $y^a + d \in \Omega_{int}$

$$\phi(\Omega, y^a + d) \geq \phi(\Omega, y^a) + \langle g(\Omega, y^a), d\rangle + \bar{\rho}\left(-\|d\|_{H(\Omega,y^a)}\right)$$

where

$$\bar{\rho}(u) = -\ln(1-u) - u.$$

Using this result, for all $d$ such that $y^a + d \in \Omega_{int}$ we get the following

$$\begin{aligned}
\phi(\Omega^+, y^a + d) &= \phi(\Omega, y^a + d) - \ln(f_{n+1}(y^a) - f_{n+1}(y^a + d)) \\
&\geq \phi(\Omega, y^a + d) - \ln\langle f'_{n+1}(y^a), -d\rangle \quad \text{by (24.12)} \\
&\geq \phi(\Omega, y^a) + \langle g(\Omega, y^a), d\rangle \\
&\quad + \bar{\rho}\left(-\|d\|_{H(\Omega,y^a)}\right) - \ln\langle f'_{n+1}(y^a), -d\rangle \\
&\geq \phi(\Omega, y^a) - \|g(\Omega, y^a)\|^*_{H(\Omega,y^a)}\|d\|_{H(\Omega,y^a)} \\
&\quad - \ln(1 + \|d\|_{H(\Omega,y^a)}) + \|d\|_{H(\Omega,y^a)} \\
&\quad - \ln\left(\|f'_{n+1}(y^a)\|^*_{\bar{H}}\right) - \ln\|d\|_{\bar{H}} \\
&\geq \phi(\Omega, y^a) - \theta\|d\|_{H(\Omega,y^a)} - \ln(1 + \|d\|_{H(\Omega,y^a)}) \\
&\quad + \|d\|_{H(\Omega,y^a)} - \ln(\varphi(\bar{r})) - \ln(\|d\|_{\bar{H}}) \\
&= \phi(\Omega, y^a) - \ln(\varphi(\bar{r})) + [(1-\theta)v - \ln\|d\|_{\bar{H}} - \ln(1+v)] \\
&\quad (\text{where } v = \|d\|_{H(\Omega,y^a)}) \\
&\geq \phi(\Omega, y^a) - \ln\bar{r} + [(1-\theta)v - \ln\|d\|_{\bar{H}} - \ln(1+v)].
\end{aligned}$$

Now for all $d$ such that $y^a + d \in \Omega^+_{int}$ we have

$$f_{n+1}(y^a + d) \leq f_{n+1}(y^a).$$

On the other hand, it follows from strong convexity of $f_{n+1}$ that

$$\begin{aligned}
0 &\geq f_{n+1}(y^a + d) - f_{n+1}(y^a) \\
&\geq \langle f'_{n+1}(y^a), d\rangle + \frac{\alpha}{2}\|d\|^2 \\
&\geq \langle f'_{n+1}(y^a), d\rangle + \frac{\alpha}{2}\|d\|^2_{f''_{n+1}(y^a)}
\end{aligned}$$

where the last inequality follows from Assumption 7 and the fact that

$$\|d\|^2_{f''_{n+1}(y^a)} \leq \|d\|^2\|f''_{n+1}(y^a)\| \leq \|d\|^2.$$

Hence

$$\bar{r}\|d\|_{\bar{H}} \geq \varphi(\bar{r})\|d\|_{\bar{H}} = \|f'_{n+1}(y^a)\|^*_{\bar{H}}\|d\|_{\bar{H}} \geq \frac{\alpha}{2}\|d\|^2_{f''_{n+1}(y^a)}.$$

Dividing both sides by $\bar{r}$ and adding to both sides $\frac{\alpha}{2}\|d\|_H^2$ (Recall that $v = \|d\|_{H(\Omega,y^a)}$.), we get

$$\|d\|_{\bar{H}} + \frac{\alpha}{2}\|d\|_H^2 \geq \frac{\alpha}{2}\|d\|_{\bar{H}}^2$$

which implies that

$$\|d\|_{\bar{H}} \leq \frac{1+\sqrt{1+\alpha^2 v^2}}{\alpha}.$$

Therefore,

$$\begin{aligned}
&\min_{\{d\,:\, y^a+d\in\Omega^+_{int}\}} \phi(\Omega^+, y^a + d) \\
&\geq \phi(\Omega, y^a) - \ln \bar{r} \\
&\quad + \min_{v>0}\{(1-\theta)v - \ln(1+\sqrt{1+\alpha^2 v^2} - \ln\alpha - \ln(1+v)\}.
\end{aligned}$$

Defining $\alpha_2(\theta,\alpha) = \min_{v>0}\{(1-\theta)v - \ln(1+\sqrt{1+\alpha^2v^2}) - \ln\alpha - \ln(1+v)\}$ yields the result.

## 3 THE ALGORITHM AND ITS CONVERGENCE

The self-concordant cut algorithm is as follows:

### Algorithm I

**Step 0.** (Initialization) Let $0 < \theta < 1/3$, $0 < \beta < 1$ and $\Omega^0$ be defined by $\rho$ inequalities of the form $\|y\|^2 \leq 1$ where $\rho$ must satisfy the following inequality

$$2\sqrt{\mu\rho} < 3[\sqrt[4]{\rho} - \beta\sqrt{\mu}]^3.$$

Set $y^0 = 0$.

**Step 1.** ($k$-th iteration) Calculate $y^k$, a $\theta$-center of $\Omega^k$, the current outer approximation of $\Gamma$. Check if $y^k \in \Gamma$. If so stop, else there is $i$ such that the $i$-th inequality of $\Gamma$ is violated. Add this inequality to $\Omega^k$ through $y^k$.

**Step 2.** Set $k = k + 1$ and return to Step 1.

Using Theorems 2.1, 2.2 we can give a bound on the total the number of recentering steps as follows.

**Theorem 3.1** *In Algorithm I, the total number of (damped) Newton steps to update to a new $\theta$-center after addition of a new cut, is at most $O(1)$, with $O(1)$ depending on $\theta$, $\beta$, $\alpha$ and $\mu$.*

**Proof.** See Theorem 5.1 of [8].

**Lemma 3.1** *At the kth iteration of Algorithm I, let*

$$(\varphi(\bar{r}_k))^2 = (f'_{k+1}(y^k))^T(\bar{H}^k)^{-1}f'_{k+1}(y^k)$$

*where* $\bar{H}^k = H^k + \frac{f''_{k+1}(y^k)}{\bar{r}_k}$ *with* $H^k = H^k(\Omega^k, y^k)$, *and* $0 < \bar{r}_k < 1$ *satisfies*

$$\sqrt{\frac{1}{2}}\bar{r}_k \leq \varphi(\bar{r}_k) \leq \bar{r}_k\,. \tag{24.13}$$

*Then*

$$\bar{r}_k^2 \leq \frac{4}{\alpha(k+1)}\,. \tag{24.14}$$

**Proof.** Note that

$$\begin{aligned}
\bar{H}^k &= H^k + \frac{f''_{k+1}(y^k)}{\bar{r}_k} \\
&\succeq H^k \\
&= H^0(\Omega^0, y^k) + \sum_{j=1}^{k}\left[\frac{f'_j(y^k)(f'_j(y^k))^T}{s_j^2} + \frac{f''_j(y^k)}{s_j}\right] \\
&\succeq 2\rho I + \sum_{j=1}^{k}\frac{\alpha I}{s_j} \qquad (\text{since } H^0 \succeq 2\rho I,\ \ f''_j(y^k) \succeq \alpha I)
\end{aligned}$$

with

$$\begin{aligned}
s_j &= f_j(y^{j-1}) - f_j(y^k) \\
&\leq \langle f'_j(y^{j-1}), y^{j-1} - y^k\rangle \\
&\leq \|f'_j(y^{j-1})\|\|y^{j-1} - y^k\| \\
&\leq 2\,.
\end{aligned}$$

Thus, since $\alpha \leq 1 < 4$ (see Assumption 7) and $\rho \geq 1$,

$$\bar{H}^k \succeq H^k \succeq 2I + \frac{k\alpha}{2}I \succeq \frac{\alpha(k+1)}{2}I\,.$$

This together with $\|y^k\| \leq 1$ and $\|f'_{k+1}(y^k)\| \leq 1$ imply that

$$\begin{aligned}
(\varphi(\bar{r}_k))^2 &= (f'_{k+1}(y^k))^T(\bar{H}^k)^{-1}f'_{k+1}(y^k) \\
&\leq \frac{2\|f'_{k+1}(y^k)\|^2}{\alpha(k+1)} \\
&\leq \frac{2}{\alpha(k+1)}\,.
\end{aligned}$$

On the other hand, it follows from (24.13) that

$$\bar{r}_k^2 \leq 2(\varphi(\bar{r}_k))^2$$

yielding the result.

**Lemma 3.2** *Suppose $\epsilon$ and $\hat{y}$ are as defined by Assumption 8. Then for all $k \geq 0$*

$$P(\Omega^k) \leq (k+\rho)\ln\frac{1}{\epsilon}.$$

**Proof.**

$$\begin{aligned} P(\Omega^k) &= \min_{y\in\Omega^k_{int}} -\sum_{j=1}^{k+\rho} \ln(-f_j(y)) \\ &\leq -\sum_{j=1}^{k+\rho} \ln(-f_j(\hat{y})) \\ &\leq (k+\rho)\ln\frac{1}{\epsilon}. \end{aligned}$$

The complexity of Algorithm I can now be given.

**Theorem 3.2** *Algorithm I finds a solution as soon as $k$ satisfies*

$$\epsilon^{2\left(\frac{k+1+\rho}{k+1}\right)} \geq \left[\frac{4(1+\ln(k+1))}{\alpha(k+1)}\right] e^{-2\alpha_2}$$

*where $\alpha_2 = \alpha_2(\theta,\alpha)$ is as defined by (24.11) and $\epsilon > 0$ is as defined by Assumption 8.*

**Proof.** It follows from Theorem 2.2 that

$$P(\Omega^+) \geq P(\Omega) - \ln\bar{r} + \alpha_2.$$

Applying this result to the $k$-th iteration, we get

$$P(\Omega^{k+1}) \geq P(\Omega^k) - \ln\bar{r}_k + \alpha_2.$$

Furthermore by Lemma 3.2 we have

$$P(\Omega^{k+1}) \leq (k+1+\rho)\ln\frac{1}{\epsilon}.$$

Hence

$$(k+1+\rho)\ln\frac{1}{\epsilon} \geq P(\Omega^0) - \frac{1}{2}\sum_{j=0}^{k}\ln(\bar{r}_j)^2 + (k+1)\alpha_2$$

or

$$
\begin{aligned}
\ln \epsilon + \frac{k+1}{k+1+\rho}\alpha_2 &\leq \frac{1}{2(k+1+\rho)} \sum_{j=0}^{k} \ln(\bar{r}_j)^2 \\
&= \frac{k+1}{2(k+1+\rho)} \left[ \frac{\sum_{j=0}^{k} \ln(\bar{r}_j)^2}{k+1} \right] \\
&\leq \frac{k+1}{2(k+1+\rho)} \ln \left[ \frac{\sum_{j=0}^{k} (\bar{r}_j)^2}{k+1} \right] \qquad \text{by concavity of ln} \\
&\leq \frac{k+1}{2(k+1+\rho)} \ln \left[ \frac{\sum_{j=0}^{k} \frac{4}{\alpha(j+1)}}{k+1} \right] \qquad \text{(by Lemma 3.1)} \\
&\leq \frac{k+1}{2(k+1+\rho)} \ln \left[ \frac{\sum_{j=1}^{k+1} \frac{4}{\alpha j}}{k+1} \right] \\
&\leq \frac{k+1}{2(k+1+\rho)} \ln \left[ \frac{4(1+\ln(k+1))}{\alpha(k+1)} \right]
\end{aligned}
$$

where the last inequality follows from the fact that

$$\sum_{j=1}^{k+1} \frac{1}{j} < 1 + \ln(k+1)\,.$$

Hence

$$\ln \epsilon^{2\left(\frac{k+1+\rho}{k+1}\right)} + 2\alpha_2 \leq \ln \left[ \frac{4(1+\ln(k+1))}{\alpha(k+1)} \right]$$

or,

$$\epsilon^{2\left(\frac{k+1+\rho}{k+1}\right)} e^{2\alpha_2} \leq \frac{4(1+\ln(k+1))}{\alpha(k+1)}$$

yielding the result.

Q.E.D.

Finally, an upper bound on the amount of time required to find a suitable $\bar{r}$ in each iteration of the algorithm is as follows.

**Theorem 3.3** *At the kth iteration of the algorithm, the number of arithmetic operations required to find a suitable $\bar{r}$ is at most*

$$O\left(\log_2\left(\log_2\left(\frac{4(3k+1-3\theta)}{\epsilon(1-3\theta)}\right)\right)\right).$$

**Proof.** See [8, Section 6].

## References

[1] H.J. Lüthi and B Büeler (2000). The Analytic Center Quadratic Cut Method (ACQCM) for Strongly Monotone Variational Inequality Problems, *SIAM Journal on Optimization*, Vol. 10, No. 2, 415–426

[2] H.J. Lüthi and B Büeler (1998). The Analytic Center Quadratic Cut Method (ACQCM) for Strongly Monotone Variational Inequality Problems with Approximate Centers, Institute for Operations Research ETH, Zurich.

[3] Z.Q. Luo and J. Sun (1998). An Analytic Center Based Column Generation Algorithm for Convex Quadratic Feasibility Problems. *SIAM Journal on Optimization*, Vol. 9, No. 1, 217–235.

[4] Z.Q. Luo and J. Sun (2000). A Polynomial Cutting Surfaces Algorithm for the Convex Feasibility Problem Defined by Self-Concordant Inequalities. *Computational Optimization and Applications*, Vol. 15, Issue 2, 167–191.

[5] Yu. Nesterov (1995). Complexity Estimates of some Cutting Plane Methods Based on the Analytical Barrier. *Mathematical Programming*, 69, 149–176.

[6] Yu. Nesterov and A.S. Nemirovskii (1994). Interior-Point Polynomial Algorithms in Convex Programming (SIAM, Philadelphia).

[7] A.S. Nemirovskii, (1994). Interior-Point Polynomial Methods in Convex Programming, Lecture notes, Faculty of Industrial Engineering and Management, Israel Institute of Technology, Israel.

[8] F. S. Mokhtarian and J.L. Goffin (April 2000). An Analytic Center Quadratic Cut Method for a the Convex Quadratic Feasibility Problem, Technical Report G-2000-18, Groupe d'Études et de Recherche en Analyse des Décision, Université de Montréal, URL:ftp://ftp.crt.umontreal.ca/pub/users/jlg/fquadcut.ps.gz

Chapter 25

# STRENGTHENED SEMIDEFINITE PROGRAMMING RELAXATIONS FOR THE MAX-CUT PROBLEM

Miguel F. Anjos and Henry Wolkowicz
*Combinatorics and Optimization*
*University of Waterloo*
*Ontario, Canada N2L 3G1*
anjos@stanfordalumni.org
hwolkowi@orion.math.uwaterloo.ca

**Abstract** In this paper we summarize recent results on finding tight semidefinite programming relaxations for the Max-Cut problem and hence tight upper bounds on its optimal value. Our results hold for every instance of Max-Cut and in particular we make no assumptions on the edge weights. We present two strengthenings of the well-known semidefinite programming relaxation of Max-Cut studied by Goemans and Williamson. Preliminary numerical results comparing the relaxations on several interesting instances of Max-Cut are also presented.

**Keywords:** Semidefinite programming, Discrete optimization, Lagrangian relaxation, Max-Cut problem, Semidefinite relaxations, Cut polytope, Metric polytope.

## 1 INTRODUCTION

The Max-Cut problem (MC) is a combinatorial optimization problem on undirected graphs with weights on the edges. Given such a graph, the problem consists in finding a partition of the set of vertices into two parts so as to maximize the sum of the weights on the edges that have one end in each part of the partition. It is well known that MC is an NP-hard problem [17], although some special cases can be solved efficiently, for example those where the graph is not contractible to $K_5$, the complete graph on five vertices [8]. We consider the general case where the graph



*N. Hadjisavvas and P.M. Pardalos (eds.), Advances in Convex Analysis and Global Optimization,* 409–420.

is complete and we require no restriction on the type of edge weights. So, in particular, negative or zero edge weights are permitted.

Following [19], we can formulate the MC problem as follows. Let the given graph $G$ have vertex set $\{1,\ldots,n\}$ and let it be described by its weighted adjacency matrix $A(G)$. We tacitly assume that the graph in question is complete (if not, missing edges can be given weight 0 to complete the graph). Let $L$ denote the *Laplacian matrix* associated with the graph; hence $L := \operatorname{Diag}(A(G)\cdot e) - A(G)$, where the linear operator Diag returns a diagonal matrix with diagonal formed from the vector given as its argument, and $e$ denotes the vector of all ones. Let the vector $v \in \{\pm 1\}^n$ represent any cut in the graph via the interpretation that the sets $\{i : v_i = +1\}$ and $\{i : v_i = -1\}$ form a partition of the vertex set of the graph. Then we can formulate MC as:

$$\text{(MC1)}\qquad \begin{array}{rll} \mu^* = & \max & \frac{1}{4}v^T L v \\ & \text{s.t.} & v_i^2 = 1, i = 1,\ldots,n. \end{array}$$

Define $Q := \frac{1}{4}L$ and consider the change of variable $X := vv^T, v \in \{\pm 1\}^n$. Then $v^T L v = \operatorname{trace} LX$ and in fact an equivalent formulation for MC is:

$$\text{(MC2)}\qquad \begin{array}{rll} \mu^* = & \max & \operatorname{trace} QX \\ & \text{s.t.} & \operatorname{diag}(X) = e \\ & & \operatorname{rank}(X) = 1 \\ & & X \succeq 0, X \in \mathcal{S}^n \end{array}$$

where diag denotes the linear operator that returns a vector with the diagonal elements of its matrix argument, $X \succeq 0$ denotes that $X$ is positive semidefinite and $\mathcal{S}^n$ denotes the space of symmetric $n \times n$ matrices. Note that $\mu^*$ denotes the optimal value of MC. Our interest is in finding tight (upper) bounds on $\mu^*$ using semidefinite programming.

The best previously known bound was proved by Goemans and Williamson [11]. They considered the relaxation of MC obtained by dropping the constraint $\operatorname{rank}(X) = 1$ from MC2. The resulting problem is a semidefinite programming problem (SDP) and its solution can be approximated in polynomial-time to within any given (fixed) precision using interior-point methods. This follows from the seminal work of Nesterov and Nemirovskii much of which is summarized in [21]. They also implemented the first interior-point method for SDP in [20]. Independently, Alizadeh extended interior-point polynomial-time algorithms from linear programming to SDP and studied applications to discrete optimization [1, 2]. Non-smooth optimization methods for solving semidefinite programming problems have also been proposed (see e.g. [13]).

Let us denote the feasible set of MC2, less the rank constraint, by $\mathcal{E}_n$, so

$$\mathcal{E}_n := \{X \in \mathcal{S}^n : \operatorname{diag}(X) = e, X \succeq 0\}.$$

With this notation, Goemans and Williamson consider the semidefinite programming problem SDP1:

$$\text{(SDP1)} \quad \begin{array}{rcll} \nu_1^* & = & \max & \operatorname{trace} QX \\ & & \text{s.t.} & X \in \mathcal{E}_n, \end{array}$$

and give an algorithm that uses the optimal solution of SDP1 to find a cut whose weight is guaranteed to be within 14% of the weight of the maximum cut. Since SDP1 is a relaxation of MC, $\nu_1^*$ is a bound on $\mu^*$ and the analysis of Goemans and Williamson implies that it is at most 14% above the value of $\mu^*$. More precisely:

$$\mu^* \geq \alpha\, \nu_1^*,$$

where $\alpha := \min_{0 \leq \theta \leq \pi} \frac{2}{\pi} \frac{\theta}{1-\cos\theta} \approx 0.87856$, and since $\frac{1}{\alpha} \leq 1.14$, we have that $\nu_1^* \leq 1.14\, \mu^*$. It is important to note that this result requires the assumption that there are no negative edge weights.

Observe that $\mathcal{E}_n$ is a convex relaxation of the feasible set of problem MC2 since we dropped the rank constraint. Other convex relaxations of the feasible set of MC2 have been proposed in the literature. The smallest convex set containing all the matrices $X$ which are feasible for MC2 is their convex hull. This relaxation is known as the *cut polytope*:

$$C_n := \operatorname{Conv}\{X : X = vv^T, v \in \{\pm 1\}^n\}.$$

Optimizing $\operatorname{trace} Q\,X$ over $C_n$ would yield exactly $\mu^*$, but an efficient description of the cut polytope is not known. Therefore one approach is to seek good relaxations of $C_n$, that is, to approximate the cut polytope by a larger convex $T$ such that

- $C_n \subseteq T$;
- $T \backslash C_n$ is as "small" as possible; and
- we can optimize $\operatorname{trace} QX$ over $T$ efficiently.

Note that $\mathcal{E}_n$ is such a set.

A well-known relaxation of the cut polytope is the *metric polytope* $M_n$, defined as the set of all matrices satisfying the triangle inequalities:

$$\begin{aligned} M_n := \{X \in \mathcal{S}^n :\ & \operatorname{diag}(X) = e, \text{ and} \\ & X_{ij} + X_{ik} + X_{jk} \geq -1, X_{ij} - X_{ik} - X_{jk} \geq -1, \\ & -X_{ij} + X_{ik} - X_{jk} \geq -1, -X_{ij} - X_{ik} + X_{jk} \geq -1, \\ & \forall 1 \leq i < j < k \leq n\}. \end{aligned}$$

The triangle inequalities model the fact that for any three mutually connected vertices in the graph, it is only possible to cut either zero or two of the edges joining them. In fact, the triangle inequalities are sufficient to describe the cut polytope for graphs with less than five vertices, i.e. $C_n = M_n$ for $n \leq 4$; however, $C_n \subsetneq M_n$ for $n \geq 5$, see for example [10]. Nonetheless, if the graph $G$ is not contractible to $K_5$, the complete graph on 5 vertices, then Barahona [9] proved that the linear programming problem

$$\begin{array}{ll} \max & \operatorname{trace} Q X \\ \text{s.t.} & X \in M_n \end{array}$$

has optimal value equal to $\mu^*$.

A natural idea is to consider optimizing over the set $\mathcal{E}_n \cap M_n$. This was proposed in [22]. Adding triangle inequalities to SDP1 is considered in e.g. [12, 14, 15]. However, it is not the case that adding a certain subset of the triangle inequalities will improve every instance of MC. Furthermore, Karloff [16] has shown that it is impossible to improve the performance guarantee of the approximation algorithm of Goemans and Williamson by adding valid linear inequalities to SDP1.

In this paper we summarize recent results on finding tight SDP relaxations of the cut polytope and hence better bounds on the optimal value of MC. Let us note that these results hold for every instance of MC; in particular, we make no assumptions on the edge weights. In the next section we present two strengthenings of SDP1. The first one, denoted SDP2, is proved to give a tighter bound than SDP1 whenever the optimal value of SDP1 is not equal to the optimal value of MC [4]. The second one, denoted SDP3, is a further tightening of SDP2 and its feasible set corresponds, in a way that is described later, to a convex set $F_n \subseteq S^n$ that is larger than $C_n$ but nonetheless is strictly contained in $\mathcal{E}_n \cap M_n$ [5]. In the last section we report some preliminary numerical results comparing these relaxations on several interesting graphs.

## 2 THE STRENGTHENED SDP RELAXATIONS

The derivation of SDP2 begins by adding to MC2 the *redundant* quadratic constraints

- $X^2 - nX = 0$ and
- $X \circ X = E$, where $\circ$ denotes the Hadamard (elementwise) product of matrices and $E$ denotes the matrix of all ones.

One motivation for adding these redundant constraints to the formulation MC2 is that we will be using Lagrangian duality to obtain new

SDP relaxations. To obtain tighter bounds we therefore want the duality gap incurred in the process to be as small as possible. The interest in these redundant quadratic constraints comes from the results in [6, 7] where the addition of redundant constraints of this type was shown to guarantee strong duality for certain problems where duality gaps can exist.

The validity of the constraint $X^2 - nX = 0$ follows from the observation that $X^2 = vv^Tvv^T$ and $v^Tv = n$ for all $v \in \{\pm 1\}^n$. Since we can simultaneously diagonalize $X$ and $X^2$, the eigenvalues of $X$ must satisfy $\lambda^2 - n\lambda = 0$, which implies that the only eigenvalues of $X$ are 0 and $n$. This shows that the constraint $X \succeq 0$ becomes redundant and may be removed. Moreover, since the diagonal constraint implies that the trace of $X$ is $n$, we conclude that $X$ must be rank-one and the rank constraint can also be removed. Finally, if $v \in \{\pm 1\}^n$ then all the entries of $X = vv^T$ are $\pm 1$ and clearly $X \circ X = E$ holds.

The resulting problem MC3 is thus another formulation of MC:

$$
\text{(MC3)} \qquad \begin{array}{rll} \mu^* = & \max & \operatorname{trace} QX \\ & \text{s.t.} & \operatorname{diag}(X) = e \\ & & X \circ X = E \\ & & X^2 - nX = 0. \end{array}
$$

Taking the Lagrangian dual of MC3 and then the dual of the dual yields the relaxation SDP2. To express this relaxation let us define $t(i) := \frac{i(i+1)}{2}$ and

$$
T(i,j) := \begin{cases} t(j-1) + i, & \text{if } i \le j \\ t(i-1) + j, & \text{otherwise.} \end{cases}
$$

Further define sMat to be the linear operator that, given a vector $x \in \Re^{t(n)}$, returns a matrix $X \in \mathcal{S}^n$ obtained by filling in columnwise the upper triangular part of X with the $t(n)$ components of $x$ and completing the strictly lower triangle by symmetry. Then SDP2 is:

$$
\text{(SDP2)} \qquad \begin{array}{rll} \nu_2^* = & \max & \operatorname{trace} Q \operatorname{sMat}(x) \\ & \text{s.t.} & \operatorname{diag}(Y) = e \\ & & Y_{0,t(i)} = 1, i = 1, \ldots, n \\ & & Y_{0,T(i,j)} = \frac{1}{n} \sum_{k=1}^{n} Y_{T(i,k),T(k,j)}, \\ & & \qquad \forall i,j \text{ s.t. } 1 \le i < j \le n \\ & & Y \succeq 0, \\ & & Y = \begin{pmatrix} 1 & x^T \\ x & \bar{Y} \end{pmatrix} \in \mathcal{S}^{t(n)+1}, \end{array}
$$

where $Y$ is partitioned so that $x$ is its first column less the top left element and $\bar{Y} \in \mathcal{S}^{t(n)}$. Note that $T(i,i) = t(i)$ and that we index the rows and columns of $Y$ by $0, 1, \ldots, t(n)$.

The following results for SDP2 are proved in [4].

**Lemma 1** *[4, Lemma 3.2] Suppose that* $Y = \begin{pmatrix} 1 & x^T \\ x & \bar{Y} \end{pmatrix}$ *is feasible for SDP2. Then*

$$\mathrm{sMat}\,(x) \succeq 0$$

*and so* $\mathrm{sMat}\,(x)$ *is feasible for SDP1.* ■

**Lemma 2** *[4, Theorem 3.1] The optimal values satisfy*

$$\nu_2^* \leq \nu_1^*.$$

■

It turns out that SDP2 can be further strengthened by the following observation. Consider again the rank-one matrices $X = vv^T, v \in \{\pm 1\}^n$. We know that these matrices $X$ have all their entries equal to $\pm 1$. Hence if $x$ satisfies $X = \mathrm{sMat}\,(x)$, then the matrix

$$Y = \begin{pmatrix} 1 \\ x \end{pmatrix} \begin{pmatrix} 1 \\ x \end{pmatrix}^T = \begin{pmatrix} 1 & x^T \\ x & xx^T \end{pmatrix}$$

is feasible for SDP2 and has all its entries equal to $\pm 1$.

From SDP2, consider the constraints

$$Y_{0,T(i,j)} = \frac{1}{n} \sum_{k=1}^{n} Y_{T(i,k),T(k,j)}, \quad \forall 1 \leq i < j \leq n.$$

For all the above rank-one matrices $Y$, the entry $Y_{0,T(i,j)}$ is equal to 1 in magnitude. The corresponding constraint says that it must be equal to the average of $n$ entries in the block $\bar{Y}$. But each of these $n$ entries has magnitude at most 1, so for equality to hold, they must all have magnitude equal to 1, and in fact they must all equal $Y_{0,T(i,j)}$.

We can state this observation in a different way. If $Y$ is rank-one, then the block $\bar{Y} = xx^T$, and therefore $Y_{T(i,k),T(k,j)} = x_{T(i,k)}x_{T(k,j)} = v_i v_k \cdot v_k v_j$. But if $v_k^2 = 1$, then $Y_{T(i,k),T(k,j)} = v_i v_j = X_{ij} = Y_{0,T(i,j)}$.

This discussion leads us to define the relaxation SDP3:

$$
\text{(SDP3)} \qquad
\begin{array}{rll}
\nu_3^* = & \max & \operatorname{trace} Q \operatorname{sMat}(Z_{1:t(n),0}) \\
& \text{s.t.} & \operatorname{diag}(Z) = e \\
& & Z_{0,t(i)} = 1, i = 1, \ldots, n \\
& & Z_{0,T(i,j)} = Z_{T(i,k),T(k,j)}, \\
& & \quad k = 1, \ldots, n, \ \forall 1 \leq i < j \leq n \\
& & Z \succeq 0, Z \in \mathcal{S}^{t(n)+1}.
\end{array}
$$

The linear operator sMat allows us to map the feasible set of SDP3 into $\mathcal{S}^n$ by defining

$$F_n := \{X \in \mathcal{S}^n : X = \operatorname{sMat}(Z_{1:t(n),0}), Z \text{ feasible for SDP3}\}.$$

Since the feasible set of SDP3 is convex and compact, and since $F_n$ is the image of that feasible set under a linear transformation, it follows that $F_n$ is also convex and compact. Furthermore

**Lemma 3** *[5, Lemma 4.1]* $C_n \subseteq F_n$. ■

Clearly, every $Z$ feasible for SDP3 is feasible for SDP2. Therefore, by Lemma 1 above, we have the inclusion

**Corollary 1** *[5, Corollary 4.1]* $F_n \subseteq \mathcal{E}_n$. ■

It turns out that the matrices in $F_n$ also satisfy the triangle inequalities.

**Theorem 1** *[5, Theorem 4.1]* $F_n \subseteq M_n$. ■

By the result of Barahona in [9, Theorem 6.4], this guarantees that the optimal value of SDP3 is exactly equal to the optimal value of MC if the graph $G$ is not contractible to $K_5$.

In fact, appropriate examples are given in [5] to prove the following stronger result:

**Theorem 2** *[5, Corollary 6.1]* $C_n \subsetneq F_n \subsetneq \mathcal{E}_n \cap M_n$ *for* $n = 5$, *and hence for* $n \geq 5$. ■

The examples are:

**Example 1** *Consider the matrix*

$$X = \begin{pmatrix} 1 & -\frac{1}{4} & -\frac{1}{4} & -\frac{1}{4} & -\frac{1}{4} \\ -\frac{1}{4} & 1 & -\frac{1}{4} & -\frac{1}{4} & -\frac{1}{4} \\ -\frac{1}{4} & -\frac{1}{4} & 1 & -\frac{1}{4} & -\frac{1}{4} \\ -\frac{1}{4} & -\frac{1}{4} & -\frac{1}{4} & 1 & -\frac{1}{4} \\ -\frac{1}{4} & -\frac{1}{4} & -\frac{1}{4} & -\frac{1}{4} & 1 \end{pmatrix}.$$

*It is known that* $X \notin C_5$ *[18]. However, there exists a* $16 \times 16$ *matrix* $P^*$ *which is feasible for SDP3 and such that* $\text{sMat}\,(P^*_{1:15,0}) = X$. *Hence* $X \in F_5$.

**Example 2** *Consider the matrix*

$$X = \begin{pmatrix} 1 & -0.65 & -0.65 & -0.65 & 0.93 \\ -0.65 & 1 & 0.3 & 0.3 & -0.65 \\ -0.65 & 0.3 & 1 & 0.3 & -0.65 \\ -0.65 & 0.3 & 0.3 & 1 & -0.65 \\ 0.93 & -0.65 & -0.65 & -0.65 & 1 \end{pmatrix}.$$

*It is easy to check that* $X \in \mathcal{E}_5 \cap M_5$. *However, there is no matrix* $P$ *feasible for SDP3 such that* $\text{sMat}\,(P_{1:15,0}) = X$. *Hence* $X \notin F_5$.

## 3 NUMERICAL RESULTS

The relaxations SDP1, SDP2 and SDP3 were compared for several interesting problems using the software package SDPPACK (version 0.9 Beta) [3]. It turns out that the dimension of the variables $Y$ and $Z$ of SDP2 and SDP3 respectively can be slightly reduced by the fact that their feasible sets as described above have no strictly feasible point and thus can be projected onto an appropriate face of the positive semidefinite cone. More details can be found in [4]. For completeness we also solved the linear relaxation over the metric polytope:

$$\begin{array}{ll} \max & \text{trace}\, Q\, X \\ \text{s.t.} & X \in M_n. \end{array}$$

This relaxation is easily formulated as an LP and we solved it using the Matlab solver LINPROG. The results are summarized in the following table. The value $\rho$ equals the value of the optimal cut divided by the bound, and R.E. denotes the relative error with respect to the optimal cut.

We report numerical results for the following problems:

1. The first line of results corresponds to solving the three SDP relaxations for a 5-cycle with unit edge-weights.

*Table 25.1* Numerical comparison of relaxations for selected test problems

| *Graph* | $\mu^*$ | *SDP1* *bound* | *SDP2* *bound* | $M_n$ *bound* | $\mathcal{E}_n \cap M_n$ *bound* | *SDP3* *bound* |
|---|---|---|---|---|---|---|
| $C_5$ | 4 | 4.5225<br>$\rho = 0.8845$<br>R.E.: 13.06% | 4.2889<br>$\rho = 0.9326$<br>R.E.: 7.22% | 4.0000<br>$\rho = 1.0000$<br>R.E.: 0% | 4.0000<br>$\rho = 1.0000$<br>R.E.: 0% | 4.0000<br>$\rho = 1.0000$<br>R.E.: 0% |
| $K_5\backslash e$ | 6 | 6.2500<br>$\rho = 0.9600$<br>R.E.: 4.17% | 6.1160<br>$\rho = 0.9810$<br>R.E.: 1.93% | 6.0000<br>$\rho = 1.0000$<br>R.E.: 0% | 6.0000<br>$\rho = 1.0000$<br>R.E.: 0% | 6.0000<br>$\rho = 1.0000$<br>R.E.: 0% |
| $K_5$ | 6 | 6.2500<br>$\rho = 0.9600$<br>R.E.: 4.17% | 6.2500<br>$\rho = 0.9600$<br>R.E.: 4.17% | 6.6667<br>$\rho = 0.9000$<br>R.E.: 11.11% | 6.2500<br>$\rho = 0.9600$<br>R.E.: 4.17% | 6.2500<br>$\rho = 0.9600$<br>R.E.: 4.17% |
| Given by $A(G)$ | 9.28 | 9.6040<br>$\rho = 0.9663$<br>R.E.: 3.49% | 9.4056<br>$\rho = 0.9866$<br>R.E.: 1.35% | 9.3867<br>$\rho = 0.9886$<br>R.E.: 1.15% | 9.2961<br>$\rho = 0.9983$<br>R.E.: 0.17% | 9.2800<br>$\rho = 1.0000$<br>R.E.: 0% |
| $\mathrm{AW}_9^2$ | 12 | 13.5<br>$\rho = 0.8889$<br>R.E.: 12.50% | 12.9827<br>$\rho = 0.9243$<br>R.E.: 8.19% | 12.8571<br>$\rho = 0.9333$<br>R.E.: 7.14% | 12.6114<br>$\rho = 0.9515$<br>R.E.: 5.10% | 12.4967<br>$\rho = 0.9603$<br>R.E.: 4.14% |
| Pet. | 12 | 12.5<br>$\rho = 0.9600$<br>R.E.: 4.17% | 12.3781<br>$\rho = 0.9695$<br>R.E.: 3.15% | 12.0000<br>$\rho = 1.0000$<br>R.E.: 0% | 12.0000<br>$\rho = 1.0000$<br>R.E.: 0% | 12.0000<br>$\rho = 1.0000$<br>R.E.: 0% |
| Rand. gen. | 88 | 90.3919<br>$\rho = 0.9735$<br>R.E.: 2.72% | 89.5733<br>$\rho = 0.9824$<br>R.E.: 1.79% | 89.3333<br>$\rho = 0.9851$<br>R.E.: 1.52% | 88.0029<br>$\rho = 1.0000$<br>R.E.: $3.3E-5$ | 88.0000<br>$\rho = 1.0000$<br>R.E.: $9.9E-7$ |

2. The second line corresponds to the complete graph on 5 vertices with unit edge-weights on all edges except one, which is assigned weight zero.

3. The third line corresponds to the complete graph on 5 vertices with unit edge-weights. In this example, none of the four SDP relaxations attains the MC optimal value, and in fact they are not distinguishable numerically. Only the polyhedral relaxation $M_n$ gives a noticeably weaker bound.

4. The fourth line corresponds to the graph defined by the weighted adjacency matrix

$$A(G) = \begin{pmatrix} 0 & 1.52 & 1.52 & 1.52 & 0.16 \\ 1.52 & 0 & 1.60 & 1.60 & 1.52 \\ 1.52 & 1.60 & 0 & 1.60 & 1.52 \\ 1.52 & 1.60 & 1.60 & 0 & 1.52 \\ 0.16 & 1.52 & 1.52 & 1.52 & 0 \end{pmatrix}.$$

This problem is interesting because it shows a significant difference between SDP3 and all the other relaxations; in this case, SDP3 is the only relaxation that attains the MC optimal value.

5. The fifth line corresponds to the graph in the figure with unit edge weights. This graph is the antiweb $AW_9^2$ and it is the hardest example[1] that the authors know for the relaxation $\mathcal{E}_n \cap M_n$. It is interesting that SDP3 performs better on this example than on the $K_5$ with unit edge weights.

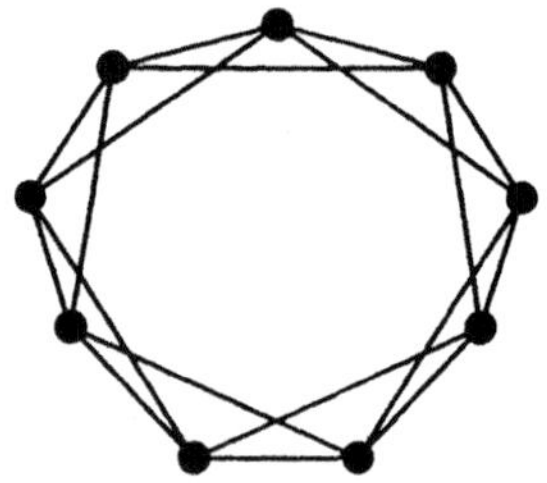

*Figure 25.1* Antiweb $AW_9^2$

6. The last two lines correspond to slightly larger graphs. The graph on 10 vertices is the Petersen graph with unit edge-weights. The graph on 12 vertices is a randomly generated graph that gives slightly different results for each relaxation (the exact description of the graph can be found in [5]).

In all cases a relative error equal to zero means that the relative error was below $10^{-11}$, the value of the smallest default stopping criteria used by SDPPACK.

Numerical tests have so far been done only for small problems. Because SDP2 has $O(n^2)$ equality constraints and SDP3 has $O(n^3)$ equality constraints, solving them using an interior-point method becomes very time-consuming and requires large amounts of memory, even for moderate values of $n$. However, the constraints of both problems, and especially those of SDP3, are sparse and have a very special structure so it should be possible to develop a specialized algorithm that exploits the structure of the problem.

## References

[1]Thanks to Franz Rendl for suggesting this interesting example.

[1] F. ALIZADEH. *Combinatorial optimization with interior point methods and semidefinite matrices.* PhD thesis, University of Minnesota, 1991.

[2] F. ALIZADEH. Interior point methods in semidefinite programming with applications to combinatorial optimization. *SIAM J. Optim.*, 5:13–51, 1995.

[3] F. ALIZADEH, J.-P. HAEBERLY, M.V. NAYAKKANKUPPAM, M.L. OVERTON, and S. SCHMIETA. SDPpack user's guide – version 0.9 Beta. Technical Report TR1997–737, Courant Institute of Mathematical Sciences, NYU, New York, NY, June 1997.

[4] M.F. ANJOS and H. WOLKOWICZ. A strengthened SDP relaxation via a second lifting for the Max-Cut problem. Research Report CORR 99-55, University of Waterloo, Waterloo, Ontario, 1999. 28 pages.

[5] M.F. ANJOS and H. WOLKOWICZ. A tight semidefinite relaxation of the cut polytope. Research Report CORR 2000-19, University of Waterloo, Waterloo, Ontario, 2000. 24 pages.

[6] K.M. ANSTREICHER, X. CHEN, H. WOLKOWICZ, and Y. YUAN. Strong duality for a trust-region type relaxation of the quadratic assignment problem. *Linear Algebra Appl.*, 301(1-3):121–136, 1999.

[7] K.M. ANSTREICHER and H. WOLKOWICZ. On Lagrangian relaxation of quadratic matrix constraints. *SIAM J. Matrix Anal. Appl.*, 22(1):41–55, 2000.

[8] F. BARAHONA. The max-cut problem on graphs not contractible to $K_5$. *Oper. Res. Lett.*, 2(3):107–111, 1983.

[9] F. BARAHONA. On cuts and matchings in planar graphs. *Math. Programming*, 60(1, Ser. A):53–68, 1993.

[10] M.M. DEZA and M. LAURENT. *Geometry of cuts and metrics.* Springer-Verlag, Berlin, 1997.

[11] M.X. GOEMANS and D.P. WILLIAMSON. .878-approximation algorithms for MAX CUT and MAX 2SAT. In *ACM Symposium on Theory of Computing (STOC)*, 1994.

[12] C. HELMBERG. *An interior point method for semidefinite programming and max-cut bounds.* PhD thesis, Graz University of Technology, Austria, 1994.

[13] C. HELMBERG and F. OUSTRY. Bundle methods to minimize the maximum eigenvalue function. In H. Wolkowicz, R. Saigal, and L. Vandenberghe, editors, *HANDBOOK OF SEMIDEFINITE*

*PROGRAMMING: Theory, Algorithms, and Applications.* Kluwer Academic Publishers, Boston, MA, 2000.

[14] C. HELMBERG and F. RENDL. A spectral bundle method for semidefinite programming. *SIAM J. Optim.*, 10(3):673 – 696, 2000.

[15] C. HELMBERG, F. RENDL, R. J. VANDERBEI, and H. WOLKOWICZ. An interior-point method for semidefinite programming. *SIAM J. Optim.*, 6(2):342–361, 1996.

[16] H. KARLOFF. How good is the Goemans-Williamson MAX CUT algorithm? *SIAM J. on Computing*, 29(1):336–350, 1999.

[17] R. M. KARP. Reducibility among combinatorial problems. In R. E. Miller and J.W. Thatcher, editors, *Complexity of Computer Computation*, pages 85–103. Plenum Press, New York, 1972.

[18] M. LAURENT, S. POLJAK, and F. RENDL. Connections between semidefinite relaxations of the max-cut and stable set problems. *Math. Programming*, 77:225–246, 1997.

[19] B. MOHAR and S. POLJAK. Eigenvalues in combinatorial optimization. In *Combinatorial Graph-Theoretical Problems in Linear Algebra*, IMA Vol. 50. Springer-Verlag, 1993.

[20] Y.E. NESTEROV and A.S. NEMIROVSKII. *Optimization over positive semidefinite matrices: Mathematical background and user's manual.* USSR Acad. Sci. Centr. Econ. & Math. Inst., 32 Krasikova St., Moscow 117418 USSR, 1990.

[21] Y.E. NESTEROV and A.S. NEMIROVSKII. *Interior Point Polynomial Algorithms in Convex Programming.* SIAM Publications. SIAM, Philadelphia, USA, 1994.

[22] S. POLJAK and F. RENDL. Nonpolyhedral relaxations of graph-bisection problems. *SIAM J. Optim.*, 5(3), 1995. 467-487.

# Chapter 26

# SUPERVISED TRAINING USING GLOBAL SEARCH METHODS

V.P. Plagianakos
*Department of Mathematics, University of Patras, GR-26110, Patras, Greece.*
vpp@math.upatras.gr

G.D. Magoulas
*Department of Information Systems and Computing, Brunel University, Uxbridge, UB8 3PH, United Kingdom.*
George.Magoulas@brunel.ac.uk

M.N. Vrahatis
*Department of Mathematics, University of Patras, GR-26110, Patras, Greece.*
vrahatis@math.upatras.gr

**Abstract** Supervised learning in neural networks based on the popular backpropagation method can be often trapped in a local minimum of the error function. The class of backpropagation-type training algorithms includes local minimization methods that have no mechanism that allows them to escape the influence of a local minimum. The existence of local minima is due to the fact that the error function is the superposition of nonlinear activation functions that may have minima at different points, which sometimes results in a nonconvex error function. This work investigates the use of global search methods for batch-mode training of feedforward multilayer perceptrons. Global search methods are expected to lead to "optimal" or "near-optimal" weight configurations by allowing the neural network to escape local minima during training and, in that sense, they improve the efficiency of the learning process. The paper reviews the fundamentals of simulated annealing, genetic and evolutionary algorithms as well as some recently proposed deflection procedures. Simulations and comparisons are presented.

**Keywords:** Local minima, simulated annealing, genetic algorithms, evolutionary algorithms, deflection procedure, feedforward neural networks, supervised training.



*N. Hadjisavvas and P.M. Pardalos (eds.), Advances in Convex Analysis and Global Optimization,* 421–432.

# 1 INTRODUCTION

In neural network training the objective is usually to minimize a cost function defined as the multi-variable error function of the network. More specifically, supervised training of a feed-forward neural network (FNN) can be viewed as the minimization of an error function that depends on the weights of the network [9]. This perspective gives some advantage to the development of effective training algorithms, because the problem of minimizing a function is well known in the field of numerical analysis.

It is well known that the supervised training using a Backpropagation (BP) based learning rule can be trapped in a local minimum of the error function [17]. Intuitively, the existence of local minima is due to the fact that the error function is the superposition of nonlinear activation functions that may have minima at different points, which sometimes results in a nonconvex error function [4]. The insufficient number of hidden nodes as well as improper initial weight settings can cause convergence to a local minimum, which prevents the network from learning the entire training set and results in inferior network performance. Gradient-based backpropagation training algorithms are local minimization methods and have no mechanism that allows them to escape the influence of a local minimum.

Recently, several researchers have presented conditions on the network architecture, the training set and the initial weight vector that allow BP to reach the optimal solution [4, 8, 22]. However, conditions such as the linear separability of the patterns and the pyramidal structure of the FNN [4] as well as the need for a great number of hidden neurons (as many neurons as patterns to learn) make these interesting results not easily interpretable in practical situations even for simple problems.

This paper presents techniques that alleviate the problem of occasional convergence to local minima in supervised training. Global search methods are expected to lead to "optimal" or "near-optimal" weight configurations by allowing the network to escape local minima during training. The remaining of this paper is organized as follows. In section 2 the training problem is formulated. In section 3 a recently proposed deflection procedure [10] is briefly presented. The fundamentals of simulated annealing [2, 3, 7] are presented in section 4, while genetic and evolutionary algorithms [12] are reviewed in section 5 and 6 respectively. Section 7 summarizes and discusses our results as well as presents simulations and comparisons with the standard backpropagation algorithm.

## 2 PROBLEM FORMULATION

Let us consider an FNN whose $l$-th layer contains $N_l$ neurons, for $l = 1, \ldots, L$. The network operation is based on the following equations:

$$net_j^l = \sum_{i=1}^{N_{l-1}} w_{ij}^{l-1,l} y_i^{l-1}, \qquad y_j^l = f(net_j^l),$$

where $net_j^l$ is for the $j$-th neuron in the $l$-th layer ($j = 1, \ldots, N_l$), the sum of its weighted inputs. The weights from the $i$-th neuron at the $(l-1)$ layer to the $j$-th neuron at the $l$-th layer are denoted by $w_{ij}^{l-1,l}$, $y_j^l$ is the output of the $j$-th neuron that belongs to the $l$-th layer, and $f(net_j^l)$ is the $j$-th's neuron activation function.

If there is a fixed, finite set of input–output pairs, the square error over the training set, which contains $P$ representative cases, is:

$$E(w) = \sum_{p=1}^{P} \sum_{j=1}^{N_L} \left(y_{j,p}^L - t_{j,p}\right)^2 = \sum_{p=1}^{P} \sum_{j=1}^{N_L} \left[\sigma^L\left(net_j^L + \theta_j^L\right) - t_{j,p}\right]^2.$$

This equation formulates the error function to be minimized, in which $t_{j,p}$ specifies the desired response at the $j$–th neuron of the output layer at the input pattern $p$ and $y_{j,p}^L$ is the output at the $j$–th neuron of the output layer $L$ that depends on the weights of the network and $\sigma$ is a nonlinear activation function, such as the well known sigmoid $\sigma(x) = (1 + e^{-x})^{-1}$. The weights in the network can be expressed using vector notation as:

$$w = \left(\ldots, w_{ij}^{l-1,l}, w_{i+1\,j}^{l-1,l}, \ldots, w_{N_{l-1}\,j}^{l-1,l}, \theta_j^l, w_{i\,j+1}^{l-1,l}, w_{i+1\,j+1}^{l-1,l}, \ldots\right)^\top,$$

where $\theta_j^l$ denotes the bias of the $j$–th neuron ($j = 1, \ldots, N_l$) at the $l$–th layer ( $l = 2, \ldots, L$). This formulation defines the weight vector as a point in the $N$–dimensional real Euclidean space $\mathbb{R}^N$, where $N$ denotes the total number of weights and biases in the network.

Minimization of $E(w)$ is attempted by updating the weights using a training algorithm. The weight update vector describes a direction in which the weight vector will move in order to reduce the network training error. The commonly used training methods are gradient–based algorithms such as the popular Backpropagation algorithm [17]. The BP procedure minimizes the error function $E(w)$ using the steepest descent with constant stepsize $\mu$:

$$w^{k+1} = w^k - \mu \nabla E(w^k), \quad k = 0, 1, \ldots .$$

The optimal value of the stepsize $\mu$ depends on the shape of the $N$–dimensional error function. The gradient, $\nabla E$, is computed by applying the chain rule on the layers of the FNN (see [17]). It is well known that the BP algorithm leads to slow training and often yields suboptimal solutions [4].

Attempts to speed up back–propagation training have been made by dynamically adapting the stepsize $\mu$ during training [9, 20], or by using second derivative related information [11, 13, 19]. However, these BP–like training algorithms occasionally converge to local minima which affect the efficiency of the learning process.

# 3 THE DEFLECTION PROCEDURE

It is well known that in order to minimize the error function $E$ we require a sequence of weight vectors $\{w^k\}_0^\infty$, where $k$ indicates iterations, that converges to a minimizer of $E$. Following the *deflection procedure* proposed in [10], when this sequence converges to a local minimum $r \in \mathbb{R}^N$ the following function is formulated:

$$F(w) = S(w; r, \lambda)^{-1} E(w),$$

where $S(w; r, \lambda)$ is a function depending on a weight vector $w$ and on the local minimizer $r$ of $E$; $\lambda$ is a relaxation parameter. In case there exist $m$ local minima $r_1, \ldots, r_m \in \mathbb{R}^N$, the above relation is reformulated as:

$$F(w) = S(w; r_1, \lambda_1)^{-1} \cdots S(w; r_m, \lambda_m)^{-1} E(w).$$

The deflection procedure suggests to find a "proper" $S(\cdot)$ such that $F(w)$ will not have a minimum at $r_i, i = 1, \ldots, m$, while keeping all other minima of $E$ locally "unchanged". In other words, we have to construct functions $S$ that provide $F$ with the property that any sequence of weights converging to $r_i$ (a local minimizer of $E$) will not produce a minimum of $F$ at $w = r_i$. In addition, this function $F$ will retain all other minima of $E$. This is *the deflection property* [10], and, as it will be explained below, the function:

$$S(w; r_i, \lambda_i) = \tanh\left(\lambda_i \|w - r_i\|\right),$$

provides $F$ with this property.

Let us assume that a local minimum $r_i$ has been determined, then

$$\lim_{w \to r_i} \frac{E(w)}{\tanh\left(\lambda \|w - r_i\|\right)} = +\infty,$$

which means that $r_i$ is no longer a local minimizer of $F$. Moreover, it is easily verified that for $\|w - r_i\| \geq \varepsilon$, where $\varepsilon$ is a small positive constant, it holds that:

$$\lim_{\lambda \to +\infty} F(w) = \lim_{\lambda \to +\infty} \frac{E(w)}{\tanh(\lambda \|w - r_i\|)} = E(w), \tag{26.1}$$

since the denominator tends to unity. This means that the error function remains unchanged in the whole weight space.

It is worth noticing that the effect of the deflection procedure is problem–dependent and is related to the value of $\lambda$. For an arbitrary value of $\lambda$ there is a small neighborhood $\mathcal{R}(r, \rho)$ with center $r$ and radius $\rho$, with $\rho \propto \lambda^{-1}$, that for any $x \in \mathcal{R}(r, \rho)$ it holds that $F(x) > E(x)$. To be more specific, when the value of $\lambda$ is small (say $\lambda < 1$) the denominator in the above relation becomes one for $w$ "far" from $r$. Thus, the deflection procedure affects a large neighborhood around $r$ in the weight space. On the other hand, when the value of $\lambda$ is large, new local minima is possible to be created near the computed minimum $r$, like a Mexican hat. These minima have function values greater than $F(r)$ and can be easily avoided by taking a proper stepsize or by changing the value of $\lambda$.

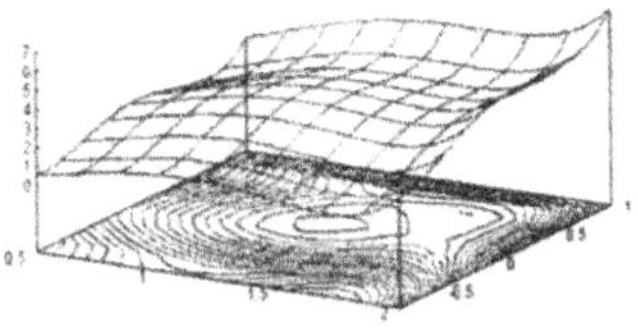

*Figure 26.1* The original Six Hump Camel Back test function

To better visualize the effect of the deflection procedure, we provide an application example on the well–known Six Hump Camel Back test function:

$$f(x_1, x_2) = 4x_1^2 - 2.1x_1^4 + \frac{1}{3}x_1^6 + x_1x_2 - 4x_2^2 + 4x_2^4$$

The plot of the original function is shown in Fig. 26.1. The application of the deflection procedure for $\lambda = 1.5$ is illustrated in Fig. 26.2 (left) and for $\lambda = 10$ in Fig. 26.2 (right); note the Mexican hat that is generated.

Notice that the deflection procedure can be incorporated in any training algorithm to help escaping the influence of local minima. In the experiments reported below, the classical BP method has been equipped with the deflection procedure. The resulting scheme is named BP with deflection (BPD).

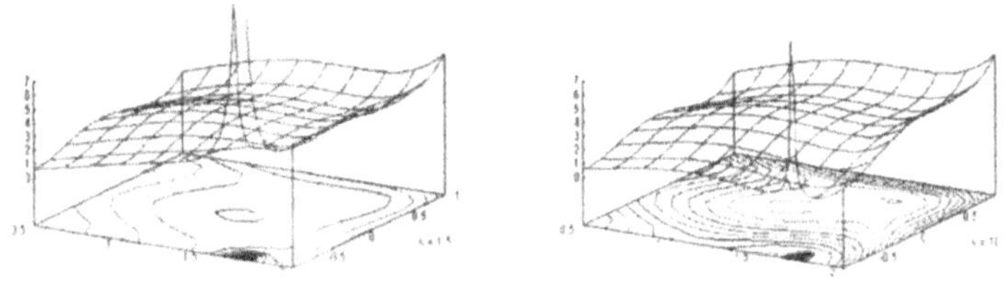

*Figure 26.2* Applying the deflection procedure to the Six Hump Camel Back test function for $\lambda = 1.5$ (left) and $\lambda = 10$ (right)

# 4 THE SIMULATED ANNEALING

*Simulated Annealing* (SA) [7] refers to the process in which random noise in a system is systematically decreased at a constant rate so as to enhance the response of the system. In the numerical optimization framework, SA is a procedure that has the capability to move out of regions near local minima. SA is based on random evaluations of the cost function, in such a way that transitions out of a local minimum are possible. It does not guarantee, of course, to find the global minimum, but if the function has many good near–optimal solutions, it should find one. In particular, the method is able to discriminate between "gross behavior" of the function and finer "wrinkles". First, it reaches an area in the function domain where a global minimum should be present, following the gross behavior irrespectively of small local minima found on the way. It then develops finer details, finding a good, near–optimal local minimum, if not the global minimum itself.

The performance of the SA [3], as observed on typical neural network training problems, is not the appropriate one. SA is characterized by the need for a number of function evaluations greater than that commonly required for a single run of common training algorithms and by the absence of any derivative related information. In addition, the problem with minimizing the neural network error function is not the well defined local minima but the broad regions that are nearly flat. In this case. the so–called Metropolis move is not strong enough to move the algorithm out of these regions [21].

In [2], it has been suggested to incorporate SA in the weight update rule:

$$w^{k+1} = w^k - \mu \nabla E(w^k) + nc2^{-dk},$$

where $n$ is a constant controlling the initial intensity of the noise, $c \in (-0.5, +0.5)$ is a random number and $d$ is the noise decay constant. In our experiments we have applied this technique for updating the weights from the beginning of the training as proposed by Burton *et al.* [2]. Alternatively, we update the weights using BP until convergence to a global or local minimum is achieved. In the latter case, we switch to SA. This combined BP with SA is named BPSA.

# 5 GENETIC ALGORITHMS

*Genetic Algorithms* (GA) are simple and robust search algorithms based on the mechanics of natural selection and natural genetics. The mathematical framework of GAs was developed in the 1960s and is presented in Holland's pioneering book [5]. GAs have been used primarily in optimization and machine learning problems.

Briefly, a simple GA processes a finite population of fixed length binary strings called *genes.* GAs have two basic operators, namely: *crossover* of genes and *mutation* for random change of genes. Another operator associated with each of these operators is the *selection* operator, which produces survival of the fittest in the GA. The crossover operator explores different structures by exchanging genes between two strings at a crossover position and the mutation operator is primarily used to escape the local minima in the weight space, by altering a bit position of the selected string; thus introducing diversity in the population. The combined action of crossover and mutation is responsible for much of the effectiveness of GA's search. The parallel noise–tolerant nature of GAs, as well as their hill–climbing capability, make GAs eminently suitable for training neural networks, as they seem to search the weight space efficiently. The "Genetic Algorithm for Optimization Toolbox (GAOT)" [6] has been used for the experiments reported in this paper. GAOT's default crossover and mutation schemes, and a real–valued encoding of the FNN's weights have been employed.

# 6 EVOLUTIONARY ALGORITHMS

*Evolutionary algorithms* (EA) are stochastic search methods which mimic the metaphor of natural biological evolution. Evolutionary algorithms operate on a population of potential solutions applying the principle of survival of the fittest to produce better and better approximations to a solution. At each generation, a new set of approximations is created by the process of selecting individuals according to their level of fitness in the problem domain and breeding them together using operators borrowed from natural genetics. This process leads to the evolution of populations of individuals that are better suited to their environment than their progenitors, just as in natural adaptation.

To demonstrate the efficiency of the Evolutionary Algorithms (EA) in alleviating the local minima problem, we have used the *Differential Evolution* (DE) strategies [18]. This is because DEs can efficiently handle non differentiable, nonlinear and multimodal objective functions, and require few, easily chosen control parameters. Experimental results have shown that DEs have good convergence properties and outperform other evolutionary algorithms [14, 15, 16]. To apply DEs to neural network training we start with a specific number (*NP*) of $N$–dimensional weight vectors, as an initial weight population, and evolve them over time. *NP* is fixed throughout the training process. The weight vectors population is initialized randomly following a uniform probability distribution.

As in GAs, at each iteration, called *generation*, new weight vectors are generated by the combination of weight vectors randomly chosen from the population. This operation is called *mutation*. The outcoming weight vectors are then mixed with another predetermined weight vector, the *target* weight vector, and this operation is called *crossover*. This operation yields the so–called *trial* weight vector. The trial vector is accepted for the next generation if and only if it reduces the value of the error function $E$. This last operation is called *selection*.

We now briefly review the two basic DE operators used for FNN training. The first DE operator, we consider, is mutation. Specifically, for each weight vector $w_g^i$, $i = 1,\ldots,NP$, where $g$ denotes the current generation, a new vector $v_{g+1}^i$ (mutant vector) is generated according to one of the following relations:

$$v_{g+1}^i = w_g^{r_1} + \xi\left(w_g^{r_1} - w_g^{r_2}\right) \tag{26.2}$$
$$v_{g+1}^i = w_g^{\text{best}} + \xi\left(w_g^{r_1} - w_g^{r_2}\right) \tag{26.3}$$
$$v_{g+1}^i = w_g^{r_1} + \xi\left(w_g^{r_2} - w_g^{r_3}\right) \tag{26.4}$$
$$v_{g+1}^i = w_g^i + \xi\left(w_g^{\text{best}} - w_g^i\right) + \xi\left(w_g^{r_1} - w_g^{r_2}\right) \tag{26.5}$$
$$v_{g+1}^i = w_g^{\text{best}} + \xi\left(w_g^{r_1} - w_g^{r_2}\right) + \xi\left(w_g^{r_3} - w_g^{r_4}\right) \tag{26.6}$$
$$v_{g+1}^i = w_g^{r_1} + \xi\left(w_g^{r_2} - w_g^{r_3}\right) + \xi\left(w_g^{r_4} - w_g^{r_5}\right) \tag{26.7}$$

where $w_g^{\text{best}}$ is the best member of the previous generation, $\xi > 0$ is a real parameter, called mutation constant, which controls the amplification of the difference between two weight vectors, and

$$r_1, r_2, r_3, r_4, r_5 \in \{1, 2, \ldots, i-1, i+1, \ldots, NP\}$$

are random integers mutually different and different from the running index $i$.

Relation (26.2) has been introduced as crossover operator for genetic algorithms [12] and is similar to relations (26.3) and (26.4). The remaining relations are modifications which can be obtained by the combination of (26.2), (26.3) and (26.4). It is clear that many more relations of this type can be generated using the above ones as building blocks. In recent works [14, 15, 16], we have shown that the above relations can efficiently be used to train FNNs with arbitrary integer weights as well.

The second DE operator, i.e. the crossover, is applied to increase the diversity of the mutant weight vector. Specifically, for each component $j$ ($j = 1, 2, \ldots, N$) of the mutant weight vector $v^i_{g+1}$, we randomly choose a real number $r$ in the interval $[0, 1]$. Then, this number is compared with the crossover constant $\rho$; if $r \leq \rho$ we replace the $j$-th component of the trial vector $u^i_{g+1}$ with the $j$-th component of the mutant vector $v^i_{g+1}$; otherwise, we pick the $j$-th component of the target vector $w^i_g$.

# 7 EXPERIMENTS AND DISCUSSION

Several experiments have been performed to evaluate the training methods mentioned in the previous sections and compare their performance. Below, we exhibit preliminary results on two notorious for their local minima problems. The algorithms have been tested using initial weights chosen from the uniform distribution in the interval $(-1, +1)$. Note that BPSA and BPD update the weights using BP until convergence to a global or local minimum is obtained: the weight vector $w^k$ is considered as a global minimizer when $E(w^k) \leq 0.04$. Convergence to a local minimizer is related to the magnitude of the gradient vector, i.e. when the stopping condition $|\nabla E(w^k)| \leq 10^{-3}$ is met, $w^k$ is taken as a local minimizer $r_i$ of the error function $E$.

We call DE1 the algorithm that uses relation (26.2) as mutation operator, DE2 the algorithm that uses relation (26.3), and so on. We note here that a key feature of the GAs and the DE algorithms is that *only* error function values are needed. No gradient information is required, so there is no need of backward passes. We made no effort to tune the mutation and crossover parameters, $\xi$ and $\rho$ respectively. We have used the fixed values $\xi = 0.5$ and $\rho = 0.7$, instead. The weight population size *NP* has been chosen to be twice the dimension of the problem, i.e. *NP*$= 2N$, for all the simulations considered. Some experimental results have shown that a good choice for *NP* is $2N \leq NP \leq 4N$. It is obvious that the exploitation of the weight space is more effective for large values of *NP*, but sometimes more error function evaluations are required. On the other hand, small values of *NP* make the algorithm inefficient and more generations are required in order to converge to the minimum.

*1) The XOR classification problem:* classification of the four XOR patterns in one of two classes, $\{0,1\}$, using a 2–2–1 FNN is a classical test problem [17, 19]. The XOR problem is sensitive to initial weights and presents a multitude of local minima [1]. The stepsize is taken equal to 1.5 and the heuristics for SA and BPSA are tuned to $n = 0.3$ and $d = 0.002$. In all instances, 100 simulations have been run and the results are summarized in Table 26.1.

*2) The three bit parity problem* [17]: a 3–3–1 FNN receives eight, 3-dimensional binary input patterns and must output an "1" if the inputs have an odd number of ones and "0" if the inputs have an even number of ones. This is a very difficult problem for an FNN because the network must determine the proper parity (the value at the output) for input patterns which differ only by Hamming distance 1. It is well known that the network's weight space contains "bad" local minima. The stepsize has been taken equal to 0.5 and the heuristics for SA and BPSA have been tuned to $n = 0.1$ and $d = 0.00025$. In all instances, 100 simulations have run and the results are summarized in Table 26.1.

| Training | *XOR Problem* | | | *Parity Problem* | | |
|---|---|---|---|---|---|---|
| Method | Succ. | Mean | s.d. | Succ. | Mean | s.d. |
| BP | 42% | 144.1 | 112.6 | 91% | 932.0 | 1320.8 |
| SA | 43% | 424.2 | 420.8 | 22% | 805.4 | 2103.1 |
| BPSA | 65% | 1661.9 | 2775.7 | 66% | 2634.0 | 6866.8 |
| GA | 95% | 422.3 | 397.5 | 73% | 1091.5 | 766.2 |
| DE1 | 75% | 192.9 | 124.7 | 91% | 622.6 | 522.1 |
| DE2 | 80% | 284.9 | 216.2 | 61% | 1994.1 | 657.6 |
| DE3 | 97% | 583.9 | 256.3 | 99% | 896.3 | 450.6 |
| DE4 | 98% | 706.1 | 343.7 | 98% | 1060.2 | 716.6 |
| DE5 | 85% | 300.5 | 250.2 | 26% | 2112.0 | 644.9 |
| DE6 | 93% | 482.9 | 264.9 | 44% | 2062.5 | 794.8 |
| BPD | 100% | 575.1 | 387.3 | 100% | 760.0 | 696.4 |

*Table 26.1* Comparative results

The results of Table 26.1 suggest that combination of local and global search methods like BPSA and BPD provide a better probability of success than the BP. Note that the performance of SA is not the appropriate one although derivative related information has been used. On the other hand, BPD escapes local minima and converges to the global minimum in all cases. A consideration that is worth mentioning is that the number of function evaluations in BPSA and BPD contains the additional evaluations required for BP to satisfy the local minima stopping condition. The results also indicate that the GA and the DE are promising and effective, even when compared with other methods that require the gradient of the error function, in addition to the error function values. For example, GAs as well as DE3 and DE4 have exhibited very good performance for the test problems considered. On the other hand, there have been cases where a discrepancy has been found in DE's behavior; see for example DE5 and DE6. For a discussion on the generalization capabilities of the networks generated by the DE algorithms see [14, 15].

In conclusion, global search methods provide techniques that alleviate the problem of occasional convergence to local minima in feedforward neural network training. Escaping from local minima is not always possible, however these methods exhibit a better chance in locating appropriate solutions and, in that sense, they improve the efficiency of the learning process. Preliminary results on two notorious for their local minima problems are promising.

## References


[1] E.K. Blum, "Approximation of Boolean functions by sigmoidal networks: Part I: XOR and other two variable functions", Neural Computation, vol.1, 1989, 532–540.

[2] M. Burton Jr. and G.J. Mpitsos, "Event dependent control of noise enhances learning in neural networks", Neural Networks, vol.5, 1992, 627–637.

[3] A. Corana, M. Marchesi, C. Martini, and S. Ridella, "Minimizing multimodal functions of continuous variables with the Simulated Annealing algorithm", ACM Trans. Math. Soft., vol.13, 1987, 262–280.

[4] M. Gori and A. Tesi, "On the problem of local minima in backpropagation", IEEE Trans. Pattern Analysis and Machine Intelligence, vol.14, 1992, 76–85.

[5] J.H. Holland, "Adaptation in Neural and Artificial Systems", University of Michigan Press, 1975.

[6] C. Houck, J. Joines, and M. Kay, "A Genetic Algorithm for Function Optimization: A Matlab Implementation", NCSU–IE TR, 1995, 95–09.

[7] S. Kirkpatrick, C.D. Gelatt Jr., and M.P. Vecchi, "Optimization by simulated annealing", Science, vol.220, 1983, 671–680.

[8] Y. Lee, S.H. Oh, and M. Kim, "An analysis of premature saturation in backpropagation learning", Neural Networks, vol.6, 9993, 719–728.

[9] G.D. Magoulas, M.N. Vrahatis, and G.S. Androulakis, "Xffective backpropagation training with variable stepsize", Neural Networks, vol.10, No.1, 6997, 63–12.

[10] G.D. Magoulas, M.N. Vrahatis, and G.S. Androulakis, "On the alleviation of local minimi in backpropagation", Nonlinear Analysis, Theory, Methods and Applications, vol.30, 1997, 4545–4550.

[11] G.D. Magoulas, M.N. Vrahatis, T.N. Grapsa, and G.S. Androulakis, "Neural network supervised training based on a dimension reducing method", Mathematics of Neural Networks, Models, Algorithms and Applications, S.W. Ellacott, J.C. Mason, I.J. Anderson Eds., Kluwer Academic Publishers, Boston, 1997, 245–249.

[12] Z. Michalewicz, "Genetic algorithms + data structures = evolution programs", Springer, 1996.

[13] M.F. Möller, "A scaled conjugate gradient algorithm for fast supervised learning", Neural Networks, vol.6, 1993, 525–533.

[14] V.P. Plagianakos and M.N. Vrahatis, "Training neural networks with 3–bit integer weights", Proceedings of Genetic and Evolutionary Computation Conference (GECCO'99), 1999, 910–915.

[15] V.P. Plagianakos and M.N. Vrahatis, "Neural network training with constrained integer weights", Proceedings of Congress on Evolutionary Computation (CEC'99), 1999, 2007-2013.

[16] V.P. Plagianakos and M.N. Vrahatis, "Training Neural Networks with Threshold Activation Functions and Constrained Integer Weights", Proceedings of the IEEE International Joint Conference on Neural Networks (IJCNN'2000), (2000).

[17] D.E. Rumelhart, G.E. Hinton, and R.J. Williams, "Learning internal representations by error propagation", Parallel Distributed Processing: Explorations in the Microstructure of Cognition 1, D.E. Rumelhart, J.L. McClelland Eds., MIT Press, 1986, 318–362.

[18] R. Storn and K. Price, "Differential Evolution – A Simple and Efficient Heuristic for Global Optimization over Continuous Spaces", Journal of Global Optimization, vol.11, 1997, 341–359.

[19] P.P. Van der Smagt, "Minimisation methods for training feedforward neural networks", Neural Networks, vol.7, 1994, 1–11.

[20] T.P. Vogl, J.K. Mangis, A.K. Rigler, W.T. Zink, and D.L. Alkon, "Accelerating the convergence of the back–propagation method", Biological Cybernetics, vol.59, 1988, 257–263.

[21] S.T. Weslstead, "Neural network and fuzzy logic applications in C/C++", Wiley, 1994.

[22] X.-H. Yu, G.-A. Chen, "On the local minima free condition of backpropagation learning", IEEE Trans. Neural Networks, vol.6, 1995, 1300–1303.

Chapter 27

# LEARNING RATE ADAPTATION IN STOCHASTIC GRADIENT DESCENT

V. P. Plagianakos
*Department of Mathematics, University of Patras,*
*GR–26110, Patras, Greece.*
vpp@math.upatras.gr

G. D. Magoulas
*Department of Information Systems and Computing, Brunel University,*
*Uxbridge, UB8 3PH, United Kingdom.*
George.Magoulas@brunel.ac.uk

M. N. Vrahatis
*Department of Mathematics, University of Patras,*
*GR–26110, Patras, Greece.*
vrahatis@math.upatras.gr

**Abstract** The efficient supervised training of artificial neural networks is commonly viewed as the minimization of an error function that depends on the weights of the network. This perspective gives some advantage to the development of effective training algorithms, because the problem of minimizing a function is well known in the field of numerical analysis. Typically, deterministic minimization methods are employed, however, in several cases, significant training speed and alleviation of the local minima problem can be achieved when stochastic minimization methods are used. In this paper a method for adapting the learning rate in stochastic gradient descent is presented. The main feature of the proposed learning rate adaptation scheme is that it exploits gradient–related information from the current as well as the two previous pattern presentations. This seems to provide some kind of stabilization in the value of the learning rate and helps the stochastic gradient descent to exhibit fast convergence and a high rate of success. Tests in various



*N. Hadjisavvas and P.M. Pardalos (eds.), Advances in Convex Analysis and Global Optimization,* 433–444.

problems validate the above mentioned characteristics of the new algorithm.

**Keywords:** Backpropagation neural networks, Batch training, On–line training, Learning rate adaptation, Stochastic gradient descent.

# 1 INTRODUCTION

In the neural network research field, BackPropagation Neural Networks (BPNNs) are the most popular models. The efficient supervised training of BPNNs is a subject of considerable ongoing research and numerous algorithms have been proposed to this end. A common training approach is to minimize the network learning error, which is a measure of its performance, and is usually based on the difference between the actual output vector of the network and the desired output vector (supervised learning). The rapid computation of a set of weights that minimizes this error is a rather difficult task since, in general, the number of network weights is high and the error function generates a complicated surface in the weight space, possessing multitudes of local minima and having broad flat regions adjoined to narrow steep ones that need to be searched to locate an "optimal" weight set.

Applications of supervised learning can be divided in two categories: stochastic (also called on–line) and batch (also called off–line) learning. *Batch* supervised learning is the classical approach in machine learning: a set of examples is obtained and used in order to learn a good approximating function (i.e. train the network), before the network is used in the application. On the other hand, in *on–line* learning, data gathered during the normal operation of the system are used to continuously adapt the learned function.

Batch training is consistent with the theory of unconstrained optimization, since the information from all the training set is used. It can be viewed as the minimization of the error function $E$; that is to find a minimizer $w^* = (w_1^*, w_2^*, \dots, w_n^*) \in \mathbb{R}^n$, such that:

$$w^* = \min_{w \in \mathbb{R}^n} E(w),$$

where $E$ is the batch error measure defined as the sum–of–squared–differences error function over the entire training set:

$$E(w) = \frac{1}{2} \sum_{p=1}^{P} \sum_{j=1}^{N_L} \left(y_{j,p}^L - t_{j,p}\right)^2 = \sum_{p=1}^{P} E_p. \qquad (27.1)$$

Note that $p$ is an index over input–output patterns; $P$ denotes the total number of patterns in the training set; $y_{j,p}^L$ is the output of the $j$-th neuron that belongs to the $L$-th (output) layer; $N_L$ is the number of neurons of the output layer; $t_{j,p}$ is the desired response at the $j$-th neuron of the output layer at the input pattern $p$, and $E_p$ is the pattern–based error of the network. This minimization corresponds to updating the weights by epoch and to be successful it requires a sequence of weight iterates $\{w^k\}_{k=0}^{\infty}$, where $k$ indicates epochs, which converges to a minimizer $w^*$.

In on–line training, the network weights are updated after the presentation of each training pattern, which may be sampled with or without repetition. This corresponds to the minimization of the instantaneous error of the network $E_p$ (see relation (27.1)).

On–line training may be chosen for a learning task either because of the very large (or even redundant) training set or because we want to model a slowly time–varying system. Although batch training seems faster for small training sets and networks, on–line training is probably faster for large training sets and BPNNs, it helps escaping local minima and provides a more natural approach for learning non–stationary tasks. Given the inherent efficiency of stochastic gradient descent, various schemes have been proposed recently [1, 19, 20, 21, 23]. Unfortunately, on–line training suffers from several drawbacks such as sensitivity to learning parameters [19]. Another disadvantage is that most advanced optimization methods, such as conjugate gradient, variable metric, simulated annealing etc., rely on a fixed error surface, and thus there are difficult to use in on–line training [19].

However, on–line learning has several advantages over batch learning. On–line methods seem more robust as errors, omissions or redundant data in the training set can be corrected or ejected during the training phase. Additionally, training data can often be generated easily and in great quantities when the system is in operation, whereas it is usually scarce and precious before. Finally, on–line training is necessary in order to learn and track time varying functions and to continuously adapt in a changing environment. In a broad sense, on–line learning is essential if we want to obtain *learning* systems as opposed to merely *learned* ones, as pointed out in [24].

The paper is focused on learning rate adaptation schemes for stochastic gradient methods. The next section of the paper briefly reviews the deterministic learning rate adaptation strategies. A new on–line training algorithm with adaptive learning rate is presented in Section 3. Experimental results are reported in Section 4 to evaluate the performance of the proposed algorithm and compare it with several on–line and batch training algorithms. Finally, in Section 5, conclusions are presented.

## 2 DETERMINISTIC LEARNING RATE ADAPTATION

BPNN training research usually focuses on deterministic gradient–based algorithms with adaptive learning rate that aim to accelerate the learning process. The following strategies are usually suggested: (i) start with a small learning rate and increase it exponentially, if successive epochs reduce the error, or rapidly decrease it, if a significant error increase occurs [3, 25], (ii) start with a small learning rate and increase it, if successive epochs keep gradient direction fairly constant, or rapidly decrease it, if the direction of the gradient varies greatly at each epoch [6], (iii) for each weight, an individual learning rate is given, which increases if the successive changes in the weights are in the same direction and decreases otherwise [10, 15, 17, 22], and (iv) use a closed formula to calculate a common learning rate for all the weights at each iteration [9, 12, 16] or a different learning rate for each weight [7, 13]. Note that all the above–mentioned strategies employ heuristic parameters in an attempt to enforce the decrease of the learning error at each iteration and to secure the converge of the training algorithm.

A different approach is based on Goldstein's and Armijo's work on steepest–descent and gradient methods. The method of Goldstein [8] requires the assumption that $E$ is twice continuously differentiable on $\mathcal{S}(w^0)$, where $\mathcal{S}(w^0) = \{w \colon E(w) \leq E(w^0)\}$ is bounded, for some initial vector $w^0$. It also requires that $\eta$ is chosen to satisfy the relation $\sup \|H(w)\| \leq \eta^{-1} < \infty$, where $H(w)$ denotes the Hessian of $E$ at $w$, in some bounded region, where the relation $E(w) \leq E(w^0)$ holds.

The $k$th iteration of an algorithm model that follows this approach consists of the following steps:

1. **Choose** $\eta_0$ to satisfy $\sup \|H(w)\| \leq \eta_0^{-1} < \infty$ and $\delta$ to satisfy $0 < \delta \leq \eta_0$ .
2. **Set** $\eta^k = \eta$, where $\eta$ is such that $\delta \leq \eta \leq 2\eta_0 - \delta$ and **go to** the next step.
3. **Update** the weights $w^{k+1} = w^k - \eta^k \nabla E(w^k)$.

However, the manipulation of the full Hessian is too expensive in computation and storage for BPNNs with several hundred weights [5]. In [11], a technique based on appropriate perturbations of the weights has been proposed for the on–line estimation of the extreme eigenvalues and eigenvectors of the Hessian without calculating the full matrix $H$. According to experiments reported in [11], the largest eigenvalue of the Hessian is mainly determined by the BPNN's architecture, the initial weights and by short–term low–order statistics of the training data. This technique can be used to determine $\eta$ requiring additional presentations of the training set in the early training.

Armijo's method, [2], suggests that the value of the learning rate $\eta$ is related to the value of the Lipschitz constant $L$, which depends on the morphology of the error surface. In this case, the weights are updated using the formula:

$$w^{k+1} = w^k - \frac{1}{2L}\nabla E(w^k), \tag{27.2}$$

and convergence to the point $w^*$ which minimizes $E$ is obtained (see [2] for conditions under which convergence occurs and a convergence proof). In [12] a local estimation of the Lipschitz constant has been proposed in a learning rate adaptation strategy, which provides increased rate of convergence, and guarantees the stability of the learning process.

# 3 STOCHASTIC LEARNING RATE ADAPTATION

Despite the abundance of methods for learning from examples, there are only few that can be used effectively for on–line learning. For example, the classic batch training algorithms cannot straightforwardly handle nonstationary data. Even when some of them are used in on–line training there exists the problem of "catastrophic interference", in which training on new examples interferes excessively with previously learned examples, leading to saturation and slow convergence [24].

Methods suited to on–line learning are those that can handle nonstationary (time–varying) data, while at the same time, require relatively little additional memory and computation, in order to process one additional example. Examples of such methods are the variants of the stochastic gradient descent proposed in [1]. The first method, named $\text{ALAP}_1$, uses at each iteration a common learning rate for all the weights:

$$\eta_i^k = \eta_i^{k-1} + \gamma\Big\langle \nabla E_{p-1}(w^{k-1}), \nabla E_p(w^k)\Big\rangle, \tag{27.3}$$

where $\eta_i^0 = c$ for all network weights ($c$ is a small positive constant), and $\langle\cdot,\cdot\rangle$ stands for the usual inner product in $\mathbb{R}^n$.

The other two methods introduced in [1] (named $\text{ALAP}_2$ and $\text{ALAP}_3$), use a different learning rate for each weight. This feature makes these on–line training algorithms able to realize variants of the gradient procedure, which move along a direction that does not necessarily coincide with the gradient direction, to accelerate the minimization process. The $\text{ALAP}_2$ uses the learning rate update formula:

$$\eta_i^k = \eta_i^{k-1}\Big[1 + \gamma\, \partial_i E_{p-1}(w^{k-1})\; \partial_i E_p(w^k)\Big]. \tag{27.4}$$

The ALAP3 is a normalized version of ALAP2 update rule, given by:

$$\eta_i^k = \eta_i^{k-1}\left[1+\gamma\,\frac{\partial_i E_{p-1}(w^{k-1})\;\partial_i E_p(w^k)}{u_i^k}\right], \tag{27.5}$$

where $u_i^k$ is an exponential average of the square of $\partial_i E_p(w^k)$, obtained through:

$$u_i^k = \mu u_i^{k-1} + (1-\mu)\left[\partial_i E_p(w^k)\right]^2,$$

where $\mu$ and $\gamma$ are positive constants ($\mu = 0.9$ and $\gamma = 0.01$ seem appropriate for most cases [1]).

We propose a new relation for adapting a common learning rate for all weights in the context of stochastic gradient descent:

$$\begin{aligned}\eta^{k+1} = \eta^k &+ \gamma_1\left\langle \nabla E_{p-1}(w^{k-1}), \nabla E_p(w^k)\right\rangle \\ &+ \gamma_2\left\langle \nabla E_{p-2}(w^{k-2}), \nabla E_{p-1}(w^{k-1})\right\rangle.\end{aligned} \tag{27.6}$$

The main feature of the proposed learning rate adaptation scheme is that it exploits gradient–related information from the current as well as the two previous pattern presentations. This seems to provide some kind of stabilization in the values of the learning rate and helps the stochastic gradient descent to exhibit fast convergence and a high rate of success.

A high level description of the proposed algorithm is given in Algorithm 1. In this algorithm model, $\eta$ is the learning rate, $\gamma_1$ and $\gamma_2$ are

STOCHASTIC DESCENT WITH ADAPTIVE LEARNING RATE

0: Initialize the weights $w^0$, $\eta^0$, $\gamma_1$, and $\gamma_2$.
1: **Repeat**
2: Set $k = k+1$
3: Randomly choose a pattern $p$ from the training set.
4: Using this pattern, calculate $E_p(w^k)$ and then $\nabla E_p(w^k)$.
5: Calculate the new weights using:
$w^{k+1} = w^k - \eta^k \nabla E_p(w^k)$
6: Calculate the new learning rate using:
$\eta^{k+1} = \eta^k + \gamma_1\langle \nabla E_{p-1}(w^{k-1}), \nabla E_p(w^k)\rangle$
$+\gamma_2\langle \nabla E_{p-2}(w^{k-2}), \nabla E_{p-1}(w^{k-1})\rangle$
7: **Until** the *termination condition* is met.
8: **Return** the final weights $w^{k+1}$.

Algorithm 1: The Proposed Algorithm in Pseudocode.

the meta–learning rates. As the termination condition the classification error or an upper limit to the error function evaluations can be used.

# 4 EXPERIMENTAL TESTS

Simulations have been conducted to study the performance of the proposed training method. To this end, Algorithm–1 has been evaluated and compared with stochastic as well as batch training methods. More specifically, in the simulations reported below, we have compared Algorithm–1 with three stochastic learning rate adaptation methods proposed by Almeida *et al.* in [1] ($ALAP_1$, $ALAP_2$ and $ALAP_3$). Moreover, for the comparisons we have also tested the on–line Back Propagation (On–line BP) [18], the batch Back Propagation (Batch BP) [18], and the batch adaptive BP with adaptive momentum (Batch ABP) [25].

The algorithms were tested using the same initial weights, initialized by the Nguyen–Widrow method [14], and received the same sequence of input patterns. For each test problem described below, we present a table summarizing the performance of the algorithms for simulations that reached solution. The training phase was considered successful when the network exhibited zero misclassifications on the training set. The reported parameters are: $Min$ the minimum number of pattern presentations, $Mean$ the mean value of pattern presentations, $Max$ the maximum number of pattern presentations, and *Succ.* the number of simulations succeeded out of 100 runs. If an algorithm fails to converge within a predetermined error function evaluation limit, it is considered that it fails to train the BPNN, and its pattern presentations are not included in the statistical analysis of its results.

The values of the meta–learning parameters $\gamma_1$ and $\gamma_2$ were chosen as $\gamma_1 \ll \gamma_2 = 1$. It seems that the choice of $\gamma_1$ and $\gamma_2$ is not critical for successful training. However, one may achieve faster convergence, if the meta–learning rates are fine–tuned, which is not the case in our experiments. On the other hand, much effort has been made to properly tune the heuristic learning parameters of the Batch BP and Batch ABP, but there is no guarantee that our final choice is optimal. However, our experience with simulations indicates that the behavior of the algorithms described in the examples to follow is characteristic.

## 4.1 THE XOR PROBLEM

The first test case is the eXclusive–OR (XOR) Boolean function problem, which has historically been considered as a good test of a neural network model and training algorithm. The XOR function maps two binary inputs to a single binary output. This simple Boolean function is not linearly separable (i.e. it cannot be solved by a simple mapping directly from the inputs to the output), and, thus, the neural network requires the use of extra hidden units to learn the task. Moreover, XOR learning is sensitive to initial weights as well as to learning rate variations and presents a multitude of local minima with certain weight vectors.

A 2–2–1 BPNN (six weights, three biases) has been used to learn the XOR mapping. The network is based on neurons with logistic activations. The error function evaluation limit was 4000, i.e. only 4000 pattern presentations were allowed. Comparative results are shown in Table 27.1. From those results it is evident that the proposed algorithm clearly outperforms the $ALAP_1$, $ALAP_2$ and $ALAP_3$ algorithms, the on–line and the batch versions of the BP algorithm, but the ABP method exhibits a slightly higher rate of success. This was expected since, in general, the batch algorithms are very good with problems that have small training sets and/or small network topologies, but are slower than on–line methods.

| Algorithm | Min | Mean | Max | Succ. |
|---|---|---|---|---|
| Batch BP | 176 | 1693.9 | 3840 | 17% |
| Batch ABP | 144 | 1430.4 | 3708 | 49% |
| On–line BP | 72 | 724.2 | 2972 | 43% |
| $ALAP_1$ | 56 | 736.1 | 3900 | 38% |
| $ALAP_2$ | 40 | 816.9 | 3960 | 37% |
| $ALAP_3$ | 52 | 1000.5 | 3636 | 43% |
| Algorithm–1 | 44 | 680.2 | 3388 | 48% |

*Table 27.1* Results for the XOR problem

## 4.2 THE NUMERIC FONT LEARNING PROBLEM

In the second experiment, a network with 64 input, 6 hidden and 10 output nodes (444 weights, 16 biases) is trained to recognize $8 \times 8$ pixel machine printed numerals from 0 to 9 in helvetica italic [12]. The network is based on neurons of the logistic activation model. The termination condition for all algorithms tested is to exhibit zero misclassifications on the training set, within 1000 error function evaluations. Detailed results regarding the training performance of the algorithms are presented in Table 27.2.

| Algorithm | Min | Mean | Max | Succ. |
|---|---|---|---|---|
| Batch BP | 210 | 500.8 | 980 | 90% |
| Batch ABP | 420 | 789.2 | 990 | 51% |
| On–line BP | 230 | 507.7 | 950 | 99% |
| $ALAP_1$ | 130 | 475.5 | 990 | 90% |
| $ALAP_2$ | 190 | 433.6 | 860 | 90% |
| $ALAP_3$ | 210 | 486.3 | 990 | 96% |
| Algorithm–1 | 170 | 436.3 | 870 | 99% |

*Table 27.2* **Results for the numeric font learning problem**

The on–line BP method exhibited very high success rate, but the $ALAP_1$, $ALAP_2$ and $ALAP_3$ methods were faster. On the other hand, the proposed method and the On–line BP algorithms had almost perfect success rate (99%). Moreover, Algorithm–1 exhibited fast convergence since it needed on average only 436 pattern presentations in order to train the network.

## 4.3 THE ALPHABETIC FONT LEARNING PROBLEM

For this problem, 26 matrices with the capital letters of the English alphabet are presented to a 35–30–26 BPNN (1830 weights, 56 biases). Each letter has been defined in terms of binary values on a grid of size $5 \times 7$. The BPNN was based on neurons with logistic activations. The results are exhibited in Table 27.3. Once again, the proposed method exhibited a very high success rate (96%) and was faster than all the other methods considered. On average it needed only 749 pattern presentations in order to complete the task.

# 5 CONCLUSIONS

In this paper, a new on–line learning algorithm for neural network training has been proposed. The algorithm is a variant of the stochastic gradient descent that uses a common adaptive learning rate for all weights. Such algorithms are able to train large networks capable to adapt to data unknown at the time of the first training, and are better suited for tasks with large, redundant or slowly time varying training sets.

| Algorithm | Min | Mean | Max | Succ. |
|---|---|---|---|---|
| Batch BP | 4498 | 21375.9 | 41860 | 79% |
| Batch ABP | 3588 | 3815.7 | 4212 | 98% |
| On–line BP | 1404 | 1861.1 | 2418 | 87% |
| $ALAP_1$ | 494 | 1519.4 | 2548 | 72% |
| $ALAP_2$ | 338 | 756.6 | 1846 | 94% |
| $ALAP_3$ | 338 | 754.5 | 2418 | 79% |
| Algorithm–1 | 364 | 749.6 | 1872 | 96% |

*Table 27.3* Results for the alphabetic font learning problem

The simulation results suggest that the proposed algorithm provides fast and stable learning, when compared with other on–line as well as batch training methods, and, therefore, a greater possibility of good performance. Further work must be done to optimize the algorithm's performance and extensive testing on bigger and more complex real–life learning tasks is necessary to fully evaluate its performance.

## References

[1] L.B. Almeida, T. Langlois, J.D. Amaral, A. Plankhov (1998). Parameter adaptation in Stochastic Optimization, In: *On–line Learning in Neural Networks*, D. Saad, ed., 111–134, Cambridge University Press.

[2] L. Armijo (1966). Minimization of functions having Lipschitz–continuous first partial derivatives. *Pacific Journal of Mathematics*, **16**, 1–3.

[3] R. Battiti (1989). Accelerated backpropagation learning: two optimization methods. *Complex Systems*, **3**, 331–342.

[4] R. Battiti (1992). First- and second-order methods for learning: between steepest descent and Newton's method. *Neural Computation*, **4**, 141–166.

[5] S. Becker and Y. Le Cun, (1988). Improving the convergence of the back–propagation learning with second order methods. In: D.S. Touretzky, G.E. Hinton and T.J. Sejnowski (Eds.), *Proceedings of the 1988 Connectionist Models Summer School*, 29–37. San Mateo: Morgan Koufmann.

[6] L.W. Chan and F. Fallside (1987). An adaptive training algorithm for back–propagation networks. *Computers Speech and Language*, **2**, 205–218.

[7] S.E. Fahlman (1988). Faster–learning variations on back–propagation: an empirical study. In: D.S. Touretzky, G.E. Hinton and T.J. Sejnowski (Eds.), *Proceedings of the 1988 Connectionist Models Summer School*, 38–51. San Mateo: Morgan Koufmann.

[8] A.A. Goldstein (1962). Cauchy's method of minimization. *Numerische Mathematik*, **4**, 146–150.

[9] H.C. Hsin, C.C. Li, M. Sun, and R.J. Sclabassi (1995). An adaptive training algorithm for back–propagation neural networks. *IEEE Transactions on System, Man and Cybernetics*, **25**, 512–514.

[10] R.A. Jacobs (1988). Increased rates of convergence through learning rate adaptation. *Neural Networks*, **1**, 295–307.

[11] Y. Le Cun, P.Y. Simard, and B.A. Pearlmutter (1993). Automatic learning rate maximization by on–line estimation of the Hessian's eigenvectors. In: S.J. Hanson, J.D. Cowan, and C.L. Giles (Eds.), *Advances in Neural Information Processing Systems 5*, 156–163. San Mateo: Morgan Koufmann.

[12] G.D. Magoulas, M.N. Vrahatis and G.S. Androulakis (1997). Effective back–propagation with variable stepsize, *Neural Networks*, **10**, 69–82.

[13] G.D. Magoulas, M.N. Vrahatis, and G.S. Androulakis (1999). Improving the convergence of the back–propagation algorithm using learning rate adaptation methods. *Neural Computation*, **11**, 1769–1796.

[14] D. Nguyen and B. Widrow (1990). Improving the learning speed of 2–layer neural network by choosing initial values of the adaptive weights, In: *IEEE First International Joint Conference on Neural Networks*, 21–26.

[15] M. Pfister and R. Rojas (1993). Speeding–up backpropagation – A comparison of orthogonal techniques. In *Proceedings of the Joint Conference on Neural Networks*. 517–523. Nagoya, Japan.

[16] V.P. Plagianakos, M.N. Vrahatis, and G.D. Magoulas (1999). Nonmonotone Methods for Backpropagation Training with Adaptive Learning Rate. In: *Proceedings of the IEEE International Joint Conference on Neural Networks (IJCNN'99)*. Washington D.C.

[17] M. Riedmiller and H. Braun (1993). A direct adaptive method for faster backpropagation learning: the Rprop algorithm. In: *Proceed-*

*ings of the IEEE International Conference on Neural Networks*, 586–591. San Francisco.

[18] D.E. Rumelhart and J.L. McClelland (1986). Parallel Distributed Processing: Explorations in the Microstructure of Cognition, Vol. 1: Foundations, MIT Press.

[19] D. Saad (1998). *On–line Learning in Neural Networks*, Cambridge University Press.

[20] N.N. Schraudolph (1998). Online Local Gain Adaptation for Multi–layer perceptrons, Technical Report, IDSIA–09–98, IDSIA, Lugano, Switzerland.

[21] N.N. Schraudolph (1999). Local Gain Adaptation in Stochastic Gradient Descend, Technical Report, IDSIA–09–99, IDSIA, Lugano, Switzerland.

[22] F. Silva and L. Almeida (1990). Acceleration techniques for the back–propagation algorithm. *Lecture Notes in Computer Science*, **412**, 110–119. Berlin: Springer–Verlag.

[23] R.S. Sutton (1992). Adapting Bias by Gradient Descent: an Incremental Version of Delta–Bar–Delta, In: *Proc. of the Tenth National Conference on Artificial Intelligence*, MIT Press, 171–176.

[24] R.S. Sutton and S.D. Whitehead (1993). Online Learning with Random Representations, In: *Proc. of the Tenth International Conference on Machine Learning*, Morgan Kaufmann, 314–321.

[25] T.P. Vogl, J.K. Mangis, J.K. Rigler, W.T. Zink and D.L. Alkon (1988). Accelerating the Convergence of the Back–propagation Method, Biological Cybernetics, **59**, 257–263.

Chapter 28

# IMPROVING THE PARTICLE SWARM OPTIMIZER BY FUNCTION "STRETCHING"


K. E. Parsopoulos
*Department of Mathematics, University of Patras, GR–26110, Patras, Greece.*
kostasp@math.upatras.gr

V. P. Plagianakos
*Department of Mathematics, University of Patras, GR–26110, Patras, Greece.*
vpp@math.upatras.gr

G. D. Magoulas
*Department of Information Systems and Computing, Brunel University, Uxbridge, UB8 3PH, United Kingdom.*
George.Magoulas@brunel.ac.uk

M. N. Vrahatis
*Department of Mathematics, University of Patras, GR–26110, Patras, Greece.*
vrahatis@math.upatras.gr



**Abstract** In this paper a new technique, named Function "Stretching", for the alleviation of the local minima problem is proposed. The main feature of this technique is the usage of a two–stage transformation of the objective function to eliminate local minima, while preserving the global ones. Experiments indicate that combined with the Particle Swarm Optimizer method, the new algorithm is capable of escaping from local minima and effectively locate the global ones. Our experience is that the modified algorithm behaves predictably and reliably and the results were quite





*N. Hadjisavvas and P.M. Pardalos (eds.), Advances in Convex Analysis and Global Optimization,* 445–457.

satisfactory. The function "Stretching" technique provides stable convergence and thus a better probability of success to the method with which it is combined.

**Keywords:** Global optimization, Derivative free methods, Evolutionary algorithms, Local minima, Particle swarm optimizer, Swarm intelligence.

# 1 INTRODUCTION

The global optimization problem is usually stated as finding the *global optimum* $x^*$ of a real valued objective function $f : \mathcal{E} \to \mathbb{R}$. In many practical optimization problems the search is usually focused on locating the global minimizer, i.e. finding a point $x^* \in \mathcal{E}$ such that

$$f(x^*) \leq f(x), \quad \forall x \in \mathcal{E}, \tag{28.1}$$

where the compact set $\mathcal{E} \subset \mathbb{R}^D$ is a $D$–dimensional parallelepiped.

There are many Global Optimization (GO) methods developed so far to deal with this problem, which can be classified in two main categories: deterministic methods and probabilistic methods. Most of the deterministic methods apply heuristics, such as modifying the trajectory in trajectory methods or adding penalties in penalty–based methods, to help escape from local minima. On the other hand, probabilistic methods rely on probability to indicate whether the search should move away from the neighborhood of a local minimum, or not. Some of the most important GO strategies, see [11] for details, are listed below:

- *Adaptive partition and search strategies*, e.g., branch–and–bound algorithms, interval arithmetic based methods and Bayesian approaches [4, 7, 8, 12, 13, 14, 19].
- *Enumerative strategies* which are used for solving combinatorial problems, or certain "structured", e.g. concave, optimization problems [3, 7, 8].
- *"Globalized" local search methods* that apply a grid search or random search type global phase, and a local search algorithm [7, 13].
- *Heuristic strategies*, such as deflation, tunneling, filled function methods, approximate convex global underestimation, tabu search, etc. [5, 6, 7, 13].
- *Homotopy (parameter continuation) methods* and analogous approaches. These among others include pivoting algorithms and fixed point methods [7].
- *Passive (simultaneous) strategies*, such as uniform grid search, pure random search [7, 13, 19].

- *Successive approximation (relaxation) methods*, such as cutting plane, more general cuts, minorant construction approaches, certain nested optimization and decomposition strategies [3, 7, 13].
- *Trajectory methods* which are differential equation model–based, or path–following search strategies [7].
- *Adaptive stochastic search algorithms.* These include simulated annealing, random search, evolution and genetic algorithms [7, 10, 13, 17, 19].

Differently from other adaptive stochastic search algorithms, evolutionary computation techniques work on a set of potential solutions, which is called *population*, and find the optimal problem solution through cooperation and competition among the potential solutions. These techniques can often find optima in complicated optimization problems more quickly than traditional optimization methods. The most commonly used population–based evolutionary computation techniques, such as Genetic Algorithms and Artificial Life methods, are motivated from the evolution of nature and the social behavior.

It is worth noting that, in general, GO strategies possess strong theoretical convergence properties, and, at least in principle, are straightforward to implement and apply. Issues related to their numerical efficiency are considered by equipping GO algorithms with a "traditional" local optimization phase. Global convergence, however, needs to be guaranteed by the global–scope algorithm component which, theoretically, should be used in a complete, "exhaustive" fashion. These remarks indicate the inherent computational demand of the GO algorithms, which increases non–polynomially, as a function of problem–size, even in the simplest cases.

In practical applications, most of the above–mentioned methods can detect just *sub–optimal solutions* of the function $f$. In many cases these sub–optimal solutions are acceptable but there are applications where the optimal solution is not only desirable but also indispensable. Therefore, the development of robust and efficient GO methods is a subject of considerable ongoing research.

Recently, Eberhart and Kennedy [15, 16] proposed the *Particle Swarm Optimization* (PSO) algorithm: a new, simple evolutionary algorithm, which differs from other evolution–motivated evolutionary computation techniques in that it is motivated from the simulation of social behavior. Although, in general, PSO results good solutions, in high–dimensional spaces it stumbles on local minima.

In this paper we propose a new technique, named *Function "Stretching"*, and we show through simulation experiments that this strategy provides a way of escape from the local minima when PSO's convergence stalls.

The paper is organized as follows: the background of the PSO is presented in Section 2. The proposed technique of function "Stretching" for escaping local minima is derived in Section 3. In Section 4, the results of the improved PSO algorithm that incorporates the function "Stretching" strategy are presented and discussed, and finally conclusions are drawn in Section 5.

## 2 THE PARTICLE SWARM OPTIMIZATION METHOD

As already mentioned, PSO is different from other evolutionary algorithms. Indeed, in PSO the population dynamics simulates a bird flock's behavior where social sharing of information takes place and individuals can profit from the discoveries and previous experience of all other companions during the search for food. Thus, each companion, called *particle*, in the population, which is now called *swarm*, is assumed to "fly" over the search space in order to find promising regions of the landscape. For example, in the minimization case, such regions possess lower functional values than other visited previously. In this context, each particle is treated as a point in a $D$–dimensional space which adjusts its own "flying" according to its flying experience as well as the flying experience of other particles (companions).

There are many variants of the PSO proposed so far, after Eberhart and Kennedy introduced this technique [2, 9]. In our experiments we used a new version of this algorithm, which is derived by adding a new inertia weight to the original PSO dynamics [1]. This version is described in the following paragraphs.

First let us define the notation adopted in this paper: the $i$-th particle of the swarm is represented by the $D$–dimensional vector $X_i = (x_{i1}, x_{i2}, \ldots, x_{iD})$ and the best particle in the swarm, i.e. the particle with the smallest function value, is denoted by the index $g$. The best previous position (the position giving the best function value) of the $i$-th particle is recorded and represented as $P_i = (p_{i1}, p_{i2}, \ldots, p_{iD})$, and the position change (velocity) of the $i$-th particle is $V_i = (v_{i1}, v_{i2}, \ldots, v_{iD})$.

The particles are manipulated according to the equations

$$v_{id} = w * v_{id} + c_1 * r_1 * (p_{id} - x_{id}) + c_2 * r_2 * (p_{gd} - x_{id}), \quad (28.2)$$

$$x_{id} = x_{id} + v_{id}, \quad (28.3)$$

where $d = 1, 2, \ldots, D$; $i = 1, 2, \ldots, N$ and $N$ is the size of population; $w$ is the inertia weight; $c_1$ and $c_2$ are two positive constants; $r_1$ and $r_2$ are two random values in the range $[0, 1]$.

The first equation is used to calculate $i$-th particle's new velocity by taking into consideration three terms: the particle's previous velocity, the distance between the particle's best previous and current position, and, finally, the distance between swarm's best experience (the position of the best particle in the swarm) and $i$-th particle's current position. Then, following the second equation, the $i$-th particle flies toward a new position. In general, the performance of each particle is measured according to a predefined fitness function, which is problem–dependent.

The role of the inertia weight $w$ is considered very important in PSO convergence behavior. The inertia weight is employed to control the impact of the previous history of velocities on the current velocity. In this way, the parameter $w$ regulates the trade–off between the global (wide–ranging) and local (nearby) exploration abilities of the swarm. A large inertia weight facilitates global exploration (searching new areas), while a small one tends to facilitate local exploration, i.e. fine–tuning the current search area. A suitable value for the inertia weight $w$ usually provides balance between global and local exploration abilities and consequently a reduction on the number of iterations required to locate the optimum solution. A general rule of thumb suggests that it is better to initially set the inertia to a large value, in order to make better global exploration of the search space, and gradually decrease it to get more refined solutions, thus a time decreasing inertia weight value is used.

From the above discussion it is obvious that PSO, to some extent, resembles evolutionary programming. However, in PSO, instead of using genetic operators, each individual (particle) updates its own position based on its own search experience and other individuals (companions) experience and discoveries. Adding the velocity term to the current position, in order to generate the next position, resembles the mutation operation in evolutionary programming. Note that in PSO, however, the "mutation" operator is guided by particle's own "flying" experience and benefits by the swarm's "flying" experience. In another words, PSO is considered as performing mutation with a "conscience", as pointed out by Eberhart and Shi [1].

## 3 THE FUNCTION "STRETCHING" TECHNIQUE

The local minima problem which is considered in this paper can be stated as follows. Let a point $\bar{x}$ such that there exists a neighborhood $\mathcal{B}$

of $\bar{x}$ with

$$f(\bar{x}) \leq f(x), \quad \forall x \in \mathcal{B}. \tag{28.4}$$

This point is a local minimizer of the objective function and many GO methods get stuck in such points. In order to alleviate this problem the following two–stage transformation in the form of the original function $f(x)$ can be applied soon after a local minimum $\bar{x}$ of the function $f$ has been detected:

$$G(x) = f(x) + \frac{\gamma_1}{2} \|x - \bar{x}\| (\text{sign}(f(x) - f(\bar{x})) + 1), \tag{28.5}$$

$$H(x) = G(x) + \frac{\gamma_2 (\text{sign}(f(x) - f(\bar{x})) + 1)}{2 \tanh(\mu(G(x) - G(\bar{x})))}, \tag{28.6}$$

where $\gamma_1, \gamma_2$ and $\mu$ are arbitrary chosen positive constants, and sign$(\cdot)$ defines the well known three valued sign function.

The first transformation stage elevates the function $f(x)$ and makes disappear all the local minima which are located above $\bar{x}$. The second stage stretches the neighborhood of $\bar{x}$ upwards, since it assigns higher function values to those points. Both stages do not alter the local minima located below $\bar{x}$; thus, the location of the global minimum is left unchanged.

It is worth noting that the sign function, which appears in the above transformation, can be approximated by the well known logistic function:

$$\text{sign}(x) \approx \text{logsig}(x) = \frac{2}{1 + \exp(-\lambda_1 x)} - 1 \simeq \tanh(\lambda_2 x),$$

for large values of $\lambda_1$ and $\lambda_2$. This sigmoid function is continuously differentiable and is widely used as a transfer function in artificial neurons.

At this point it is useful to provide an application example of the proposed technique in order to illustrate its effect. The problem considered is a notorious two dimensional test function, called the *Levy No. 5*:

$$\begin{aligned} f(x) &= \sum_{i=1}^{5} i \cos[(i+1)x_1 + i] \times \sum_{j=1}^{5} j \cos[(j+1)x_2 + j] + \\ &\quad + (x_1 + 1.42513)^2 + (x_2 + 0.80032)^2, \end{aligned} \tag{28.7}$$

where $-10 \leq x_i \leq 10, i = 1, 2$. There are about 760 local minima and one global minimum with function value $f^* = -176.1375$ located at $x^* = (-1.3068, -1.4248)$. The large number of local optimizers makes extremely difficult for any method to locate the global minimizer. In Fig. 28.1, the original plot of the *Levy No. 5* into the cube $[-2, 2]^2$ is shown.

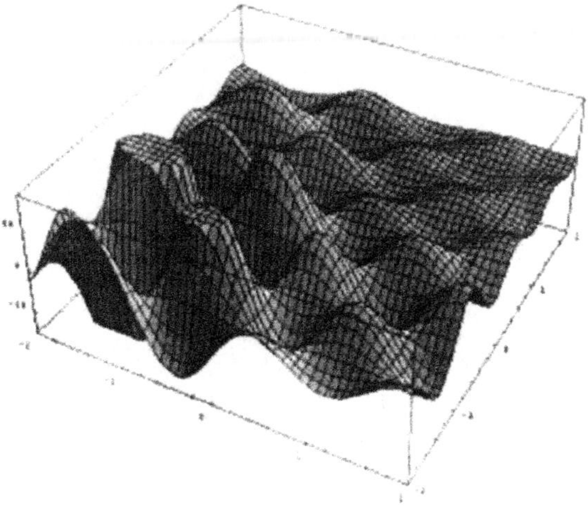

*Figure 28.1* The original plot of the function *Levy No. 5*.

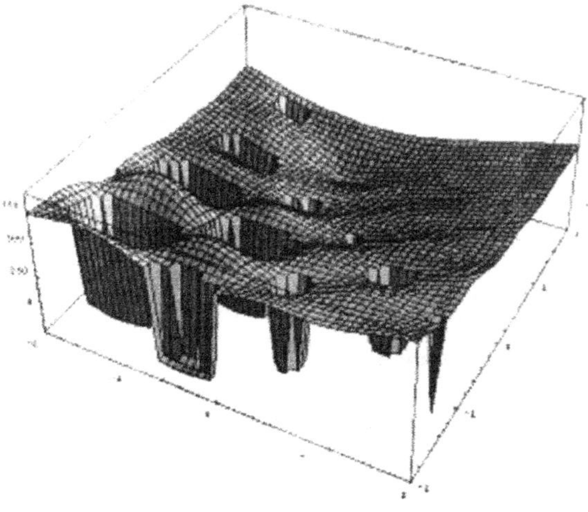

*Figure 28.2* Plot of the *Levy No. 5* after the first stage of the "Stretching" technique.

After applying the transformation of Eq. 28.5 (first stage of function "Stretching") to the *Levy No. 5*, the new form of the function is shown in Fig. 28.2. As one can see, local minima with higher functional values than the "stretched" local minimum (which looks as if a pin is positioned over it and the rest of the function is stretched around it) disappeared, while lower minima as well as the global one have been left unaffected.

In Fig. 28.3, the final landscape, derived after applying the second transformation stage to the *Levy No. 5*, is presented. It is clearly shown

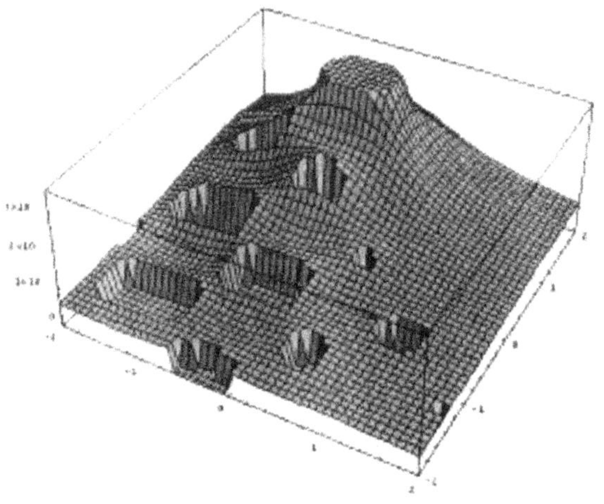

*Figure 28.3* Plot of the *Levy No. 5* after the second stage of the "Stretching" technique.

how the whole neighborhood of the local minimum has been elevated; thus, the former local minimum has now turned to be a local maximum of the function. Details on the performance of the function "Stretching" technique on some well known test problems, as well as suggestions for selecting parameter values are presented in the next section.

# 4 EXPERIMENTAL RESULTS

In this section, we present results from testing a combination of the function "Stretching" technique and the PSO algorithm. This optimization strategy is named "Stretched" PSO (SPSO) and is initialized by applying the PSO algorithm for the minimization of the fitness function. In case the PSO converges to a local minimum, the function "Stretching" technique is applied to the original function and the PSO is re–initialized.

The performance of the SPSO has been evaluated by means of simulation runs in hard–optimization problems, like the minimization of the functions *Levy No. 5* and *Corana*, and in training Artificial Neural Networks (ANNs) for pattern classification problems, like the classification of the classical eXclusive–OR (XOR) patterns.

In all the simulations reported, the values of $\gamma_1, \gamma_2$ and $\mu$ were fixed: $\gamma_1 = 10000, \gamma_2 = 1$ and $\mu = 10^{-10}$. Default values for the parameters $c_1$ and $c_2$ have been used: $c_1 = c_2 = 0.5$. Although the choice of the parameter values seems to be not critical for the success of the method, faster convergence can be obtained by proper fine–tuning. The balance between the global and local exploration abilities of the SPSO is mainly controlled by the inertia weights, since the particles' positions are updated according to the classical PSO strategy. A time decreasing inertia weight value, i.e. start from 1 and gradually decrease towards 0.4, has been found to work better than using a constant value. This is because large inertia weights help to find good seeds at the beginning of the search, while, later, small inertia weights facilitate a finer search.

The first test refers to the minimization of the *Levy No. 5* function that has been described in the previous section. The results of Table 28.1 have been obtained after 100 runs using the SPSO, with a swarm of size 20, initialized into the cube $[-2,2]^2$. The average performance is exhibited in terms of the mean value and standard deviation of the number of function evaluations, and the percentage of SPSO success. Information concerning the performance of the PSO algorithm with and without the function "Stretching" technique is also illustrated. As can be seen from

| | "Stretched" | No "Stretched" | Overall |
|---|---|---|---|
| Mean Value | 3854.2 | 1049.4 | 1245.8 |
| Standard Deviation | 1630.1 | 235.1 | 854.2 |
| Success | 7/7 | 93/93 | 100% |

*Table 28.1* Analysis of the results for the minimization of the Levy No.5 function.

Table 28.1, in 93 out of 100 cases PSO found the global minimum without any help, while in 7 cases it got stuck in a local minimum. In these cases function "Stretching" has been applied and the global minimum had finally been detected. Thus the success rate of PSO increased by 7%.

The second experiment concerns the minimization of the *Corana* function:

$$f(x) = \sum_{j=1}^{4} \begin{cases} 0.15 \times \left(z_j - 0.05 \times \mathrm{sgn}(z_j)\right)^2 \times d_j, & \text{if } |x_j - z_j| < 0.05, \\ d_j \times x_j^2, & \text{otherwise,} \end{cases}$$

where $x_j \in [-1000, 1000]$, $d_j = 1, 1000, 10, 100$ and

$$z_j = \left\lfloor \left| \frac{x_j}{0.2} \right| + 0.49999 \right\rfloor \times \mathrm{sgn}(x_j) \times 0.2.$$

SPSO has been tested in 100 simulation runs with a swarm consisted of 40 particles, initialized and constrained inside the hypercube $[-1,1]^4$. The success rate of SPSO for the minimization of this function is 100% (see Table 28.2), but the success percentage of the plain PSO is just 74%.

| | "Stretched" | No "Stretched" | Overall |
|---|---|---|---|
| Mean Value | 13704.6 | 2563.2 | 5460.0 |
| Standard Deviation | 7433.5 | 677.5 | 6183.8 |
| Success | 26/26 | 74/74 | 100% |

*Table 28.2* Analysis of the results for the minimization of the Corana function with constrained population.

In this case, the "Stetching" technique increased the success percentage by 26%, which is a significant improvement on the performance of the PSO.

Results for the unconstrained–swarm case are exhibited in Table 28.3. If free movement of population in the search space is allowed, the plain PSO has a success of 96%. Regarding the rest of the cases, in 2 of them the "Stetching" technique has been successfully applied leading the SPSO to reach the global minimum within the predefined limit of 2000 iterations. However, in 2 cases the SPSO didn't reach the desirable accuracy within the 2000 limit. This may suggest that in high–dimensional spaces the constrained movement of the swarm helps the algorithm to exhibit better performance; further investigation is necessary to extract useful conclusions.

| | "Stretched" | No "Stretched" | Overall |
|---|---|---|---|
| Mean Value | 72200.0 | 3219.1 | 8737.6 |
| Standard Deviation | 12269.1 | 590.1 | 19154.5 |
| Success | 2/4 | 96/96 | 98% |

*Table 28.3* Analysis of the results for the minimization of the Corana function without constraint.

In the third experiment an ANN has been trained using the SPSO to learn the XOR Boolean classification problem. The XOR function maps two binary inputs to a single binary output and the ANN that was trained to solve the problem had 2 linear input nodes, two hidden nodes with logistic activations and one linear output node. This task corresponds to the minimization of the following objective function [18]:

$$
\begin{aligned}
f(x) = & \left[1+\exp\left(-\frac{x_7}{1+\exp(-x_1-x_2-x_5)}-\frac{x_8}{1+\exp(-x_3-x_4-x_6)}-x_9\right)\right]^{-2} \\
& +\left[1+\exp\left(-\frac{x_7}{1+\exp(-x_5)}-\frac{x_8}{1+\exp(-x_6)}-x_9\right)\right]^{-2} \\
& +\left[1-\left\{1+\exp\left(-\frac{x_7}{1+\exp(-x_1-x_5)}-\frac{x_8}{1+\exp(-x_3-x_6)}-x_9\right)\right\}^{-1}\right]^{2} \\
& +\left[1-\left\{1+\exp\left(-\frac{x_7}{1+\exp(-x_2-x_5)}-\frac{x_8}{1+\exp(-x_4-x_6)}-x_9\right)\right\}^{-1}\right]^{2}.
\end{aligned}
$$

In the context of ANNs, the parameters $x_1, x_2, \ldots, x_9$ are called weights and are usually initialized in the interval $[-1, 1]$. It is well known from the neural networks literature that successful training in this case, i.e. reaching a global minimizer, strongly depends on the initial weight values and that the above–mentioned function presents a multitude of local minima. It is obvious from the results reported in Table 28.4 that the function "Stretching" technique helped to increase significantly the success percentage of the PSO, i.e. the success rate has been increased from 77% to 100%.

| | "Stretched" | No "Stretched" | Overall |
|---|---|---|---|
| Mean Value | 29328.6 | 1459.7 | 7869.6 |
| Standard Deviation | 15504.2 | 1143.1 | 13905.4 |
| Success | 23/23 | 77/77 | 100% |

*Table 28.4* **Analysis of the results for the XOR problem.**

# 5 CONCLUSIONS

A new technique, named function "Stretching", for the alleviation of the local minima problem was introduced. The main feature of this technique is that it uses a two–stage transformation of the fitness function to eliminate local minima, while preserving the global ones.

Experiments indicate that the PSO method when equipped with the proposed technique is capable to escape from local minima and locate the global one effectively. The function "Stretching" technique provides stable convergence and thus a better probability of success for the PSO.

Further work is focused on optimizing the performance of the proposed modification of the PSO algorithm. In addition, extensive testing on high–dimensional and more complex real–life optimization tasks is necessary to fully investigate the properties and evaluate the performance of the function "Stretching" technique.

## References

[1] R.C. Eberhart and Y.H. Shi (1998). Evolving Artificial Neural Networks. *Proc. International Conference on Neural Networks and Brain*, Beijing, P.R. China.

[2] R.C. Eberhart, P.K. Simpson and R.W. Dobbins (1996). *Computational Intelligence PC Tools*, Academic Press Professional, Boston.

[3] F. Forgó (1988). Nonconvex Programming. Akadémiai Kiadó, Budapest.

[4] E.R. Hansen (1992). Global Optimization Using Interval Analysis. Marcel Dekker, New York.

[5] F. Glover (1989). Tabu Search – part I, *ORSA Journal on Computing*, 1(3), 190–206.

[6] F. Glover (1990). Tabu Search – part II, *ORSA Journal on Computing*, 2(1), 4–32.

[7] R. Horst and P.M. Pardalos (1995). Handbook of Global Optimization. Kluwer Academic Publishers, London.

[8] R. Horst and H. Tuy (1996). Global Optimization – Deterministic Approaches. Springer, New York.

[9] J. Kennedy and R.C. Eberhart (1995). Particle Swarm Optimization. *Proc. IEEE International Conference on Neural Networks*, Piscataway, NJ, IV:1942–1948.

[10] Z. Michalewicz (1996). Genetic Algorithms + Data Structures = Evolution Programs. Springer, New York.

[11] A. Neumaier (2000). On the server of the Computational Mathematics group at the University of Vienna, Austria, http://solon.cma.univie.ac.at/~neum/glopt.html, accessed 26/06/2000.

[12] A. Neumaier (1990). Interval Methods for Systems of Equations. Cambridge University Press, Cambridge.

[13] J.D. Pintér (1996). Global Optimization in Action. Kluwer Academic Publishers, Dordrecht, Boston, London.

[14] H. Ratschek and J.G. Rokne (1988). New Computer Methods for Global Optimization. Ellis Horwood, Chichester.

[15] Y.H. Shi and R.C. Eberhart (1998). Parameter Selection in Particle Swarm Optimization. *Proc. Annual Conference on Evolutionary Programming*, San Diego.

[16] Y.H. Shi and R.C. Eberhart (1998). A Modified Particle Swarm Optimizer, *Proc. IEEE International Conference on Evolutionary Computation*, Anchorage, Alaska.

[17] P.J.M. Van Laarhoven and E.H.L. Aarts (1987). Simulated Annealing: Theory and Applications. Kluwer Academic Publishers, London.

[18] M.N. Vrahatis, G.S. Androulakis, J.N. Lambrinos and G.D. Magoulas (2000), A class of gradient unconstrained minimization algorithms with adaptive stepsize. *Journal of Computational and Applied Mathematics*, 114, 367–386.

[19] A.A. Zhigljavsky (1991). Theory of Global Random Search. Kluwer Academic Publishers, London.

Chapter 29

# SOME CONVERGENCE PROPERTIES OF THE STEEPEST DESCENT ALGORITHM REVEALED BY RENORMALISATION

Luc Pronzato
*Laboratoire I3S, CNRS/Université de Nice-Sophia Antipolis, France*
pronzato@i3s.unice.fr

Henry P. Wynn
*Dept. of Statistics, University of Warwick, UK*
hpw@stats.warwick.ac.uk

Anatoly A. Zhigljavsky
*School of Mathematics, Cardiff University, UK*
ZhigljavskyAA@Cardiff.ac.uk

**Abstract** The behaviour of the standard steepest-descent algorithm for a quadratic function in $\mathbb{R}^d$ is investigated. By rescaling the iterates to remain always on the unit sphere one can reveal special features of this behaviour: the renormalized algorithm converges to a two-point cycle on the unit circle; the cycle depends on the starting point in a complicated manner (the set of points converging to the same cycle is fractal), but all cycles belong to a particular plane, given by certain eigenvectors of the Hessian matrix of the objective function. The stability of the attractor is analysed. The rate of convergence of the algorithm is investigated. It is shown that the worst value of this rate is obtained only for some particular starting points. The introduction of a relaxation coefficient in the steepest-descent algorithm completely changes its behaviour, which may become chaotic. Different attractors are presented. We show that relaxation allows a significantly improved rate of convergence.

**Keywords:** Asymptotic behaviour, attractors, relaxation, renormalisation, steepest descent



*N. Hadjisavvas and P.M. Pardalos (eds.), Advances in Convex Analysis and Global Optimization,* 459–471.

# 1 INTRODUCTION

For a general smooth function $f(\cdot)$ the steepest-descent algorithm is

$$x^{(k+1)} = x^{(k)} - \alpha_k \nabla f(x^{(k)}), \tag{29.1}$$

where $x^{(k)} = \left(x_1^{(k)}, \ldots, x_d^{(k)}\right)^T$ is the $k$-th iterate of the algorithm,

$$\nabla f(x) = \left(\frac{\partial f}{\partial x_1}, \ldots, \frac{\partial f}{\partial x_d}\right)^T$$

is the gradient of $f(\cdot)$, and $\alpha_k = \arg\min_\alpha f(x^{(k)} - \alpha \nabla f(x^{(k)}))$. In what follows we study the asymptotic behaviour of this algorithm for a quadratic function in $\mathbb{R}^d$: $f(x) = (1/2)(x - x^*)^T A(x - x^*)$, where $x, x^* \in \mathbb{R}^d$ and $A$ is a positive definite $d \times d$ matrix. This seemingly simple problem has some very complex aspects which are uncovered by the following renormalization idea. Define

$$v^{(k)} = \frac{x^{(k)} - x^*}{||x^{(k)} - x^*||}, \tag{29.2}$$

which has the effect of rescaling the iterates to remain always on the unit sphere: $||v|| = 1$. We are then interested in the behaviour of the process $\{v^{(k)}\}$ as $k \to \infty$. Except for pathological cases, in the limit the process is found to belong to the two dimensional space spanned by the eigenvectors $(u_1, u_d)$ corresponding to the minimal and maximum eigenvalues $\lambda_1$ and $\lambda_d$ of the matrix $A$ under the assumption $0 < \lambda_1 < \lambda_2 \leq \ldots \leq \lambda_{d-1} < \lambda_d$, where $\lambda_i$ is the $i$-th ordered eigenvalue of $A$. Ordering the $v_i^{(k)}$ compatibly with the eigenvalues, the property states that $v_i^{(k)} \to 0$ for $i = 2, \ldots, d-1$ as $k \to \infty$, and that the algorithm, in its renormalised form, converges to a two-point set on the circle $\{||v_1||^2 + ||v_d||^2 = 1\}$ in the plane spanned by $u_1$ and $u_d$, see Section 2 Numerical simulations show that the same convergence property holds for a convex differentiable function locally quadratic around its minimum $x^*$. It has been known for many years that the asymptotic convergence rate of the algorithm depends on the ratio $\rho = \lambda_d/\lambda_1$ (for instance, see [4], p. 152). Thus for any $x^{(k)} \in \mathbb{R}^d$

$$f(x^{(k+1)}) \leq \left(\frac{\rho - 1}{\rho + 1}\right)^2 f(x^{(k)}) \tag{29.3}$$

(see also Section 4). However, it is the worst-case rate, given by (29.3), not the actual rate, which simply depends on $\lambda_1$ and $\lambda_d$. We shall return to the discussion of the asymptotic rate of convergence in Section 4 The behaviour of the steepest-descent algorithm with relaxation is considered in Section 5

## 2 ATTRACTION THEOREM

Consider the quadratic function

$$f(x) = \frac{1}{2}(x - x^*)^T A(x - x^*), \tag{29.4}$$

where $A$ is a positive definite matrix with ordered eigenvalues

$$0 < \lambda_1 < \lambda_2 \le \ldots \le \lambda_{d-1} < \lambda_d. \tag{29.5}$$

The gradient of $f(\cdot)$ at $x^{(k)}$ equals $g^{(k)} = \nabla f(x^{(k)}) = A(x^{(k)} - x^*)$, and

$$\alpha_k = \frac{(g^{(k)})^T g^{(k)}}{(g^{(k)})^T A g^{(k)}}.$$

Therefore, the algorithm (29.1) can be rewritten as

$$x^{(k+1)} = x^{(k)} - \frac{(g^{(k)})^T g^{(k)}}{(g^{(k)})^T A g^{(k)}} g^{(k)}. \tag{29.6}$$

Without loss of generality we can assume that $x^* = 0$ and the matrix $A$ is diagonal: $A = \mathrm{diag}(\lambda_1, \ldots \lambda_d)$ where $0 < \lambda_1 < \lambda_2 \le \ldots \le \lambda_{d-1} < \lambda_d$. With this assumption, the iteration of the algorithm (29.6) can be written as

$$x_i^{(k+1)} = x_i^{(k)} - \frac{\sum_{j=1}^d (g_j^{(k)})^2}{\sum_{j=1}^d \lambda_j (g_j^{(k)})^2} g_i^{(k)} \quad \text{for } i = 1, \ldots, d. \tag{29.7}$$

Multiplying both sides of $i$-th equation of (29.7) by $\lambda_i$ we obtain

$$g_i^{(k+1)} = g_i^{(k)} - \frac{\sum_{j=1}^d (g_j^{(k)})^2}{\sum_{j=1}^d \lambda_j (g_j^{(k)})^2} \lambda_i g_i^{(k)} \quad \text{for } i = 1, \ldots, d. \tag{29.8}$$

Introduce now the variables $y_i^{(k)} = (g_i^{(k)})^2$, and their renormalised versions

$$z_i^{(k)} = \frac{y_i^{(k)}}{\sum_{j=1}^d y_j^{(k)}}. \tag{29.9}$$

Note that $z_i^{(k)} \ge 0$ for $i = 1, \ldots, d$ and $\sum_{j=1}^d z_j^{(k)} = 1$. The equations (29.8) then imply

$$y_i^{(k+1)} = \left(1 - \frac{\lambda_i \sum_{j=1}^d y_j^{(k)}}{\sum_{j=1}^d \lambda_j y_j^{(k)}}\right)^2 y_i^{(k)} \quad \text{for } i = 1, \ldots, d, \tag{29.10}$$

and

$$z_i^{(k+1)} = z_i^{(k)} \frac{\left(\sum_{j=1}^d \lambda_j z_j^{(k)} - \lambda_i\right)^2}{\sum_{l=1}^d \left(\sum_{j=1}^d \lambda_j z_j^{(k)} - \lambda_l\right)^2 z_l^{(k)}} \quad \text{for } i = 1, \ldots, d. \quad (29.11)$$

We can consider $z_i^{(k)}$ as a weight on the eigenvalue $\lambda_i$, so that (29.11) defines a transformation on discrete probability measures supported on $\{\lambda_1, \ldots, \lambda_d\}$. Note that to have the convergence to the two-dimensional plane $V$ for the sequence $\{x^{(k)}/||x^{(k)}||\}$, it is enough to have the same property for the sequence $\{z^{(k)}\}$, $k\to\infty$. The asymptotic behaviour of the sequence $\{z^{(k)}\}$ generated by (29.11) is studied in ([1, 2]) and chapter 7 of ([5]). The main result is that, assuming $0 < \lambda_1 < \lambda_2 \leq \cdots \leq \lambda_{d-1} < \lambda_d$, the sequence converges to a two-dimensional plane, spanned by the eigenvectors $u_1$, $u_d$ associated with $\lambda_1$ and $\lambda_d$. More precisely, the attraction property can be stated as follows.

**Theorem 1** *Let the objective function $f$ be given by* (29.4), *with the eigenvalues of $A$ satisfying* (29.5). *Let $V = \text{span}(u_1, u_d)$ be the two-dimensional plane generated by the (distinct orthonormal) eigenvectors $u_1$, $u_d$ corresponding to $\lambda_1$ and $\lambda_d$, respectively. Then for any starting vector $x^{(1)}$ for which*

$$u_1{}^T v^{(1)} > 0 \text{ and } u_d{}^T v^{(1)} > 0, \quad (29.12)$$

*with $v^{(k)}$ defined by* (29.2), *the algorithm attracts to the plane $V$ in the following sense:*

$$w^T v^{(k)} \to 0, \quad k\to\infty,$$

*for any non-zero vector $w \in V^\perp$. Moreover, the sequence $\{z^{(k)}\}$ of renormalised variables defined by* (29.9) *converges to a two-point cycle:*

$$z^{(2k)} \to p\, u_1 + (1-p)\, u_d, \quad z^{(2k+1)} \to (1-p)\, u_1 + p\, u_d \quad \text{when } k\to\infty, \quad (29.13)$$

*where $p$ is some number in* $(0, 1)$ *depending on $z^{(1)}$ (see Section 2 for the range of possible values for $p$).*

This property has important consequences on the asymptotic rate of convergence of the steepest descent algorithm, see Section 4 The proofs used in ([1, 2]) cannot be extended easily to the case where the space $\mathcal{H}$ for $x$ is infinite dimensional. An extension of the proof given in Chapter 7 of ([5]) to this infinite dimensional case will be presented in a forthcoming paper. Note that the updating formula (29.11) is exactly the same for the weights $z_i^{(k)}$ and $z_j^{(k)}$ corresponding to equal $\lambda_i = \lambda_j$, and therefore these weights can be summed. It follows that a proof based on the assumption that all $\lambda_i$ are different can easily be extended to the case of ties among $\lambda_2, \ldots, \lambda_{d-1}$. In the case when at least one of the eigenvalues $\lambda_1$ or $\lambda_d$ is not simple, we still have convergence to a cycle (29.13), where $u_1$ and $u_d$ are certain eigenvectors associated with $\lambda_1$ and $\lambda_d$. These eigenvectors are determined uniquely only in the case when the eigenvalues $\lambda_1$ and $\lambda_d$ are simple and (29.12) holds. Also, Theorem 1 obviously generalizes to the case where (29.12) is not satisfied (at least one of the scalar products is equal to zero). The algorithm then attracts to a two-dimensional plane $V'$ and $\{v^{(k)}\}$ converges to a two-point cycle. $V'$ is defined by the eigenvectors $u_i$, $u_j$ associated with the smallest and largest eigenvalues such that ${u_i}^T v^{(1)} > 0$, ${u_j}^T v^{(1)} > 0$. For the sake of simplicity of notations, we shall assume in what follows that (29.12) is satisfied, and therefore $u_i = u_1$, $u_j = u_d$.

Although the limiting behaviour of the algorithm is simple, its behaviour en route to the attractor is fairly complicated and presents a fractal structure, see [5].

# 3 STABILITY

From Theorem 1, the algorithm attracts to two conjugate points on the circle $\{||v_1||^2 + ||v_d||^2 = 1\}$, characterized by $p$. However, some values of $p$ correspond to unstable points. We shall use the following definition of stability, see [3], p. 444.

**Definition 1** *A fixed point $\pi$ for a mapping $T(\cdot)$ is called stable if $\forall \epsilon > 0$, $\exists \alpha > 0$ such that $\forall \pi^0$ for which $\|\pi^0 - \pi\| < \alpha$, $\|T^n(\pi^0) - \pi\| < \epsilon$ for all $n > 0$. A fixed point $\pi$ is unstable if it is not stable.*

Consider a two-step iteration for $z_i^{(k)}$, $1 \le i \le d$:

$$
\begin{aligned}
z_i^{(k+2)} &= z_i^{(k+1)} \frac{(\sum_{j=1}^d \lambda_j z_j^{(k+1)} - \lambda_i)^2}{D_{k+1}} \\
&= z_i^{(k)} \frac{(\mu_1 - \lambda_i)^2}{D_k} \frac{1}{D_{k+1}} \left( \frac{\mu_1^3 - 2\mu_1\mu_2 + \mu_3}{D_k} - \lambda_i \right)^2 \quad (29.14)
\end{aligned}
$$

where

$$
\begin{aligned}
D_k &= \sum_{l=1}^d \left( \sum_{j=1}^d \lambda_j z_j^{(k)} - \lambda_l \right)^2 z_l^{(k)} = \mu_2 - \mu_1^2 , \\
D_{k+1} &= \sum_{l=1}^d \left( \sum_{j=1}^d \lambda_j z_j^{(k+1)} - \lambda_l \right)^2 z_l^{(k+1)}
\end{aligned}
$$

$$= \frac{1}{D_k}(\mu_1^2\mu_2 - 2\mu_1\mu_3 + \mu_4) - \frac{1}{D_k^2}(\mu_1^3 - 2\mu_1\mu_2 + \mu_3)^2,$$

with $\mu_m$ the $m$th moment of the discrete measure

$$\pi_k = \left\{ \begin{array}{ccc} \lambda_1 & \dots & \lambda_d \\ z_1^{(k)} & \dots & z_d^{(k)} \end{array} \right\}$$

which places the weights $z_i^{(k)}$ at the points $\lambda_i$, $i = 1, \dots, d$, that is,

$$\mu_m = \mu_m(\pi_k) = \int \lambda^m d\pi_k(\lambda) = \sum_{j=1}^{d} \lambda_j^m z_j^{(k)}, \quad m = 0, \pm 1, \pm 2, \dots \quad (29.15)$$

Also consider the corresponding transformation $\psi^2(\cdot)$ defined by $\pi_{k+2} = \psi^2(\pi_k)$. Since $\sum_{i=1}^{d} z_i^{(k)} = 1$ for all $k$, we substitute $1 - \sum_{i=1}^{d-1} z_i^{(k)}$ for $z_d^{(k)}$ in (29.14). This defines an operator $\phi(\cdot) : \mathcal{S}_{d-1} \mapsto \mathcal{S}_{d-1}$ on the $(d-1)$-dimensional canonical simplex

$$\mathcal{S}_{d-1} = \{z = (z_1, \dots, z_{d-1}) \mid z_i \geq 0, \sum_{i=1}^{d-1} z_i \leq 1\},$$

which maps $(z_1^{(k)}, \dots, z_{d-1}^{(k)})$ to $(z_1^{(k+2)}, \dots, z_{d-1}^{(k+2)})$. Studying the properties of $\psi^2(\cdot)$ is equivalent to studying the properties of $\phi(\cdot)$.

**Theorem 2**
**(i)** Stability. *All fixed points $z(p) = (p, 0, \dots, 0) \in \mathcal{S}_{d-1}$ with $p$ in the interval $\mathcal{I}_S$ defined by*

$$\mathcal{I}_S = ]\frac{1}{2} - s(\lambda_{i^*}),\ \frac{1}{2} + s(\lambda_{i^*})[,$$

*where*

$$s(\lambda) = \frac{\sqrt{(\lambda_d - \lambda)^2 + (\lambda_1 - \lambda)^2}}{2(\lambda_d - \lambda_1)}, \qquad (29.16)$$

*and $i^*$ is such that $|\lambda_{i^*} - \frac{\lambda_1 + \lambda_d}{2}|$ is minimum over all $\lambda_i$'s, $i = 2, \dots, d-1$, are stable for $\phi$.*
**(ii)** Instability. *All fixed points $z(p)$ with $p$ in the set $\mathcal{I}_U$ defined by*

$$\mathcal{I}_U = [0,\ \frac{1}{2} - s(\lambda_{i^*})[ \cup ]\frac{1}{2} + s(\lambda_{i^*}),\ 1]$$

*are unstable for $\phi$.*

*Proof.*
(i) Stability. Take $z^{(k)} = (z_1^{(k)}, \ldots, z_{d-1}^{(k)})$ in $\mathcal{S}_{d-1}$ such that $z_1^{(k)} \in \mathcal{I}_S$ and $|z_i^{(k)}| < \alpha$, $i = 2, \ldots, d-1$. The two-step iteration (29.14) can be written as

$$z_i^{(k+2)} = z_i^{(k)} F(\lambda_i, \mu_1, \mu_2, \mu_3, \mu_4), \tag{29.17}$$

with $\mu_m = \mu_m(\pi_k)$, $m = 1, \ldots, 4$, see (29.15). Define $\mathbf{z}^{(k)} = (z_1^{(k)}, 0, \ldots, 0) \in \mathcal{S}_{d-1}$ and $\bar{\mu}_m = \lambda_1^m z_1^{(k)} + \lambda_d^m (1 - z_1^{(k)})$, $m = 1, \ldots, 4$. Then, for $i = 2, \ldots, d-1$,

$$z_i^{(k+2)} = z_i^{(k)} f(\lambda_i, z_1^{(k)})[1 + O(\alpha)], \ \alpha \to 0,$$

with $f(\lambda, z_1^{(k)}) = F(\lambda, \bar{\mu}_1, \bar{\mu}_2, \bar{\mu}_3, \bar{\mu}_4)$. Straightforward calculations show that $f[\lambda, 1/2 \pm s(\lambda)] = 1$, with $s(\lambda)$ given by (29.16), and $z_1^{(k)} \in \mathcal{I}_S$ implies $f(\lambda_i, z_1^{(k)}) < 1$, $i = 2, \ldots, d-1$. Therefore, for $\alpha$ smaller than some $\alpha_0$, there exist $L^{(k)} < 1$ such that

$$z_i^{(k+2)} < L^{(k)} z_i^{(k)}, \ i = 2, \ldots, d-1.$$

Similarly, (29.17) gives

$$z_1^{(k+2)} = z_1^{(k)} F(\lambda_1, \bar{\mu}_1, \bar{\mu}_2, \bar{\mu}_3, \bar{\mu}_4) + \sum_{j=2}^{d-1} z_j^{(k)} \frac{\partial F(\lambda_1, \mu_1', \mu_2', \mu_3', \mu_4')}{\partial z_j}_{|\zeta^{(k)}},$$

with $\zeta^{(k)} = (1-\gamma)\mathbf{z}^{(k)} + \gamma z^{(k)}$ for some $\gamma \in [0,1]$, and $\mu_m' = \sum_{i=1}^{d-1} \lambda_i^m z_i + \lambda_d^m (1 - \sum_{i=1}^{d-1} z_i)$, $m = 1, \ldots, 4$. Now, $F(\lambda_1, \bar{\mu}_1, \bar{\mu}_2, \bar{\mu}_3, \bar{\mu}_4) = 1$ and direct calculation shows that

$$\frac{\partial F(\lambda_1, \mu_1', \mu_2', \mu_3', \mu_4')}{\partial z_j}_{|\mathbf{z}^{(k)}} = \frac{(\lambda_j - \lambda_1)(\lambda_d - \lambda_j)^2 [2\lambda_d (1 - z_1^{(k)}) + 2\lambda_1 z_1^{(k)} - \lambda_1 - \lambda_j]}{(z_1^{(k)})^2 (1 - z_1^{(k)})^2 (\lambda_d - \lambda_1)^4}$$

is positive for $z_1^{(k)} = 1/2 - s(\lambda_j)$, negative for $z_1^{(k)} = 1/2 + s(\lambda_j)$ and its absolute value reaches its maximum at one of these two points. By continuity arguments, the same is true for $\partial F(\lambda_1, \mu_1', \mu_2', \mu_3', \mu_4')/\partial z_{j|\zeta^{(k)}}$. By choosing $\alpha$ small enough, we thus guarantee that $z_1^{(k+2)} \in \mathcal{I}_S$, with

$$\min\{z_1^{(k)}, 1/2 - s(\lambda_{i^*}) + \delta\} \le z_1^{(k+2)} \le \max\{z_1^{(k)}, 1/2 + s(\lambda_{i^*}) - \delta\}$$

for some $\delta > 0$. This ensures that exist $L < 1$ such that $L^{(k+2)} < L^* = \max(L^{(k)}, L) < 1$. Repeating the same arguments, we thus obtain

$$z_i^{(k+2m)} < (L^*)^m z_i^{(k)}\,,\ i = 2, \ldots, d-1\,.$$

Moreover, $|z_1^{(k+2)} - z_1^{(k)}| < B \sum_{i=2}^{d-1} z_i^{(k)}$ for some constant $B$, and therefore

$$|z_1^{(k+2m)} - z_1^{(k)}| < B \sum_{i=2}^{d-1} \sum_{n=0}^{m-1} z_i^{(k+2n)}$$

so that $|z_1^{(k+2m)} - z_1^{(k)}| < B(d-2)\alpha L^*/(1-L^*)$ for any $m$.

**(ii) Instability.** The Jacobian matrix $J_\phi$ of $\phi(\cdot)$ at points $z(p) = (p, 0, \ldots, 0)$ can be computed analytically, and is given by:

$$\begin{aligned}(J_\phi)_{ij} &= \frac{\partial \phi_i(z)}{\partial z_j}_{|z=z(p)} \\ &= \begin{cases} 1 & \text{if } i = j = 1 \\ \frac{(\lambda_j-\lambda_1)(\lambda_d-\lambda_j)^2[2\lambda_d(1-p)+2\lambda_1 p-\lambda_1-\lambda_j]}{p(1-p)^2(\lambda_d-\lambda_1)^4} & \text{if } j > 1 \text{ and } i = 1 \\ \frac{[\lambda_d(1-p)+\lambda_1 p-\lambda_j]^2[\lambda_d p+\lambda_1(1-p)-\lambda_j]^2}{p^2(1-p)^2(\lambda_d-\lambda_1)^4} & \text{if } i = j \\ 0 & \text{otherwise}\,. \end{cases}\end{aligned}$$

One can easilly check that for $p \in \mathcal{I}_U$, $(J_\phi)_{ii} > 1$ for at least one $i$ in $\{2, \ldots, d-1\}$. Since the $(J_\phi)_{ii}$'s are eigenvalues of $J_\phi$, Theorem 15.16 in ([3]) indicates that $z(p)$ is unstable. ■

Note that $\forall \lambda_1, \ldots, \lambda_d$ the stability interval $\mathcal{I}_S$ contains the interval

$$\left]\frac{1}{2} - \frac{1}{2\sqrt{2}},\ \frac{1}{2} + \frac{1}{2\sqrt{2}}\right[ \approx\ ]0.14645,\ 0.85355[\,.$$

When $d = 3$, numerical simulations show that for any initial density of $x^{(1)}$ in $\mathbb{R}^d$ associated with a density of $z^{(1)}$ on $\mathcal{S}_{d-1}$ reasonably spread, the density of attractors $z(p)$ characterized by $p$ can be approximated by the density

$$\varphi(p) = C \log \min\{1, (J_\phi)_{22}\} = \begin{cases} C \log (J_\phi)_{22} & \text{if } p \in \mathcal{I}_S \\ 0 & \text{otherwise}\,, \end{cases}$$

where $C$ is a normalisation constant. Figure 29.1 shows the empirical density of attractors (full line) together with $\varphi(p)$ (dashed line) in the case $\lambda_1 = 1$, $\lambda_2 = 4$, $\lambda_3 = 10$. The support of this density coincides with the stability interval $\mathcal{I}_S$.

When $d > 3$, the density of attractors depends on the initial density of $x^{(1)}$.

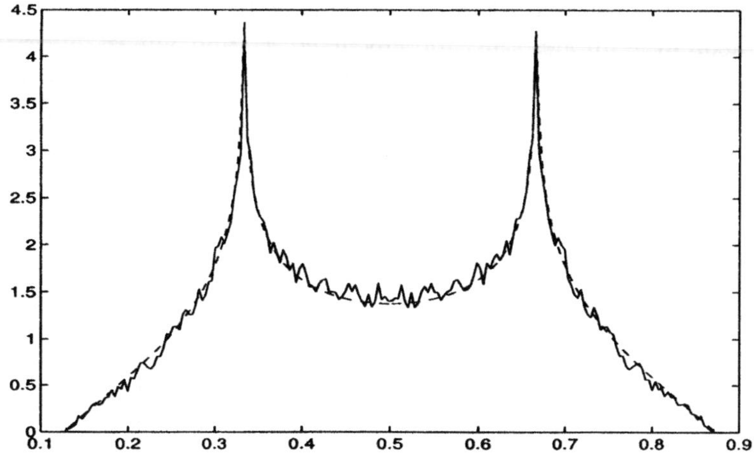


*Figure 29.1* Empirical density of attractors (full line) and $\varphi(p)$ (dashed line) ($d = 3, \lambda_1 = 1, \lambda_2 = 4, \lambda_3 = 10$)

# 4 RATE OF CONVERGENCE

Define the convergence rate at iteration $k$ by

$$r_k = \frac{f(x^{(k+1)})}{f(x^{(k)})}, \tag{29.18}$$

where $f(x)$ is given by (29.4). Without any loss of generality, one can take $f(x) = \sum_{i=1}^d \lambda_i x_i^2$. First, we rewrite $r_k$ as

$$r_k = \frac{\sum_{i=1}^d y_i^{(k+1)}/\lambda_i}{\sum_{i=1}^d y_i^{(k)}/\lambda_i}.$$

Using (29.10), we get

$$r_k = 1 - \frac{\left(\sum_{i=1}^d y_i^{(k)}\right)^2}{\left(\sum_{i=1}^d \lambda_i y_i^{(k)}\right)\left(\sum_{i=1}^d y_i^{(k)}/\lambda_i\right)},$$

so that, using the definition (29.9) of $z^{(k)}$, $r_k = 1 - 1/(\mu_1\mu_{-1})$, where $\mu_m$ is defined by (29.15). Define the asymptotic rate as

$$R = R(x^{(1)}, x^*) = \lim_{k\to\infty} \left(\prod_{j=1}^k r_j\right)^{1/k}. \tag{29.19}$$

Generally, $R$ depends on the initial point $x^{(1)}$ and the optimal point $x^*$. Theorem 1 implies that for any fixed $x^*$ and almost all $x^{(1)}$ the asymptotic rate depends only on the value of $p$ and is given by $R(p) = [f(x^{(k+2)})/f(x^{(k)})]^{1/2}$, where $x^{(k)}$ is associated with $p\,e_1 + (1-p)\,e_d$ or $(1-p)\,e_1 + p\,e_d$. This gives

$$R(p) = \frac{p(1-p)(\rho-1)^2}{[p+\rho(1-p)][(1-p)+\rho p]},$$

with $\rho = \lambda_d/\lambda_1$ the condition number of the matrix $A$. The function $R(p)$ is symmetric with respect to $1/2$ and monotonously increasing from 0 to $1/2$. The worst asymptotic rate is thus obtained at $p = 1/2$:

$$R_{\max} = \left(\frac{\rho-1}{\rho+1}\right)^2,$$

see (29.3). Note that from the Kantorovich inequality, see [4], p. 151, $\mu_1\mu_{-1} \leq (1+\rho)^2/(4\rho)$, and therefore $\forall x^{(k)}$, $r_k \leq R_{\max}$. The worst rate is thus achieved only when $x_1^{(k)} = \pm\rho x_d^{(k)}$, $x_2^{(k)} = \cdots = x_{d-1}^{(k)} = 0$.

Consider now another convergence rate, defined by

$$R' = \lim_{k\to\infty}\left(\prod_{j=1}^{k} r'_j\right)^{1/k}.$$

where $r'_k = \|x^{(k+1)}\|^2/\|x^{(k)}\|^2$. In a way similar to the derivation of $r_k$, by rewriting $r'_k$ in terms of $z^{(k)}$, one gets

$$r'_k = 1 - \frac{2\mu_{-1}}{\mu_1\mu_{-2}} + \frac{1}{\mu_1^2\mu_{-2}}.$$

One can easilly check that for almost all $x^{(1)}$ the asymptotic rate $R'$ is equal to $R(p)$, where $p$ defines the attractor. Similarly, we can define $R''$ from $r''_k$, with $r''_k = \|g_{k+1}\|/\|g_k\|$, which gives $r''_k = \mu_2/\mu_1^2 - 1$ and again $R'' = R(p)$ for almost all $x^{(1)}$.

# 5 STEEPEST DESCENT WITH RELAXATION

The introduction of a relaxation coefficient $\gamma$, with $0 < \gamma < 1$, in the steepest-descent algorithm totally changes its behaviour. The algorithm (29.6) then becomes

$$x^{(k+1)} = x^{(k)} - \gamma\frac{(g^{(k)})^T g^{(k)}}{(g^{(k)})^T A g^{(k)}} g^{(k)}.$$

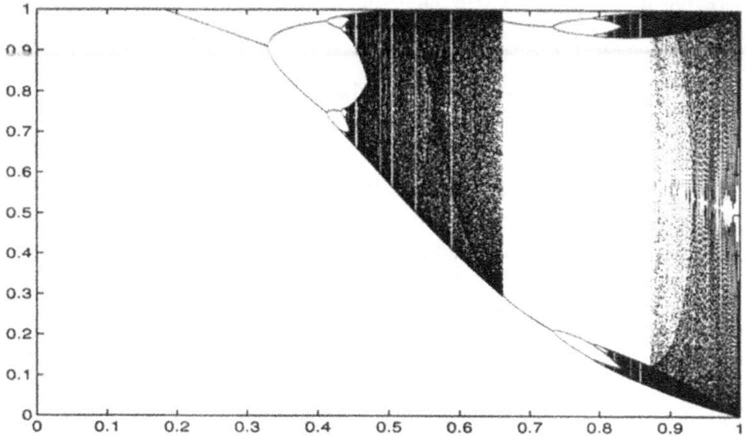


*Figure 29.2* Attractors for $z_1$ as a function of $\gamma$ ($d = 2, \lambda_1 = 1, \lambda_2 = 10$)

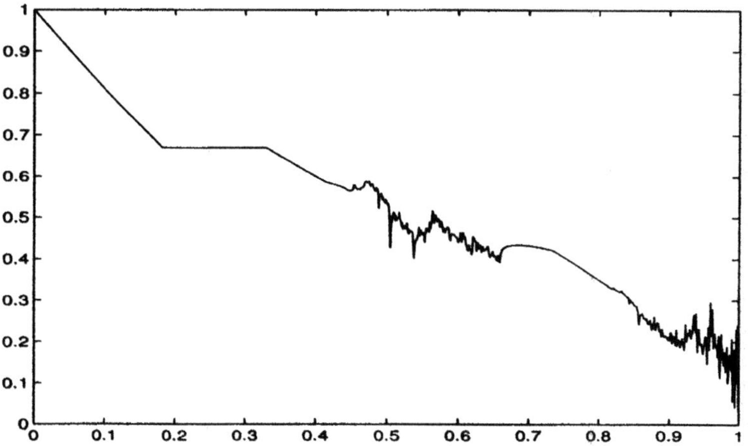


*Figure 29.3* Asymptotic rate (29.19) as a function of $\gamma$ ($d = 2, \lambda_1 = 1, \lambda_2 = 10$)

For fixed $A$, depending on the value of $\gamma$, the renormalized process either attracts to periodic orbits (the same for almost all starting points) or exhibits a chaotic behaviour. Figures 29.2 presents the classical period-doubling phenomenon in the case $d = 2$ when $\lambda_1 = 1$ and $\lambda_2 = 10$. Figures 29.3 gives the asymptotic rate (29.19) as a function of $\gamma$ in the same situation.

We get now instead of (29.18): $r_k(\gamma) = 1 - \gamma(2-\gamma)/(\mu_1\mu_{-1})$. Note that from the Kantorovich inequality the worst value of the rate is

$$1 - \gamma(2-\gamma)\frac{4\rho}{(1+\rho)^2} > R_{\max},$$

if $\gamma < 1$. However, numerical results show that for $\gamma$ large enough the asymptotic rate is significantly better than $R_{\max}$. A detailed analysis of the 2-dimensional case gives the following results.

(i) If $0 < \gamma \leq \frac{2}{\rho+1}$, the process attracts to the fixed point $p = 1$ and $R = R(\gamma) = 1 - \frac{2\rho\gamma}{\rho+1}$.

(ii) If $\frac{2}{\rho+1} < \gamma \leq \frac{4\rho}{(\rho+1)^2}$, the process attracts to the fixed point $p = \frac{2\rho-\gamma(\rho+1)}{2(\rho-1)}$, and $R(\gamma) = R_{\max}$.

(iii) If $\leq \frac{4\rho}{(\rho+1)^2} < \gamma \leq \frac{2(\sqrt{2}+1)\rho}{(\rho+1)^2}$, the process attracts to the 2-point cycle $(p_1, p_2)$, with

$$p_{1,2} = \frac{2\rho - \gamma(\rho+1) \pm \sqrt{\gamma(\gamma\rho^2 - 4\rho + 2\gamma\rho + \gamma)}}{2(\rho-1)},$$

and $R(\gamma) = 1 - \gamma$.

(iv) For larger values of $\gamma$ one observes a classical period-doubling phenomenon, see Figure 29.2.

(v) If $\rho > 3 + 2\sqrt{2} \approx 5.828427$, the process attracts again to a 2-point cycle for values of $\gamma$ larger than $\gamma_\rho = \frac{8\rho}{(\rho+1)^2}$, see Figure 29.2. For the limiting case $\gamma = \gamma_\rho$, the cycle is given by $(p_1', p_2')$, with

$$p_{1,2}' = \frac{\rho\left(\rho^2 - 2\rho + 5 \pm 2\sqrt{(\rho^2 - 2\rho + 5)(5\rho^2 - 2\rho + 1)}\right)}{(\rho-1)(\rho+1)^3},$$

and the associated asymptotic rate is $R(\gamma_\rho) = (\rho^2 - 6\rho + 1)/(\rho^2 - 1)$.

In higher dimensions, repeated numerical trials show that the process typically no longer attracts to the 2-dimensional plane spanned by $(u_1, u_d)$.

## References

[1] Akaike, H. (1959). On a successive transformation of probability distribution and its application to the analysis of the optimum gradient method. *Ann. Inst. Statist. Math. Tokyo*, 11:1–16.

[2] Forsythe, G. (1968). On the asymptotic directions of the $s$-dimensional optimum gradient method. *Numerische Mathematik*, 11:57–76.

[3] Hale, J. and Koçak, H. (1991). *Dynamics and Bifurcations.* Springer-Verlag, Heidelberg.

[4] Luenberger, D. (1973). *Introduction to Linear and Nonlinear Programming.* Addison-Wesley, Reading, Massachusetts.

[5] Pronzato, L., Wynn, H., and Zhigljavsky, A. (2000). *Dynamical Search.* Chapman & Hall/CRC, London et al.

# Chapter 30

# INTERIOR–POINT ALGORITHMS FOR DANTZIG AND WOLFE DECOMPOSITION PRINCIPLE


Marcos Augusto dos Santos
*Departamento de Ciência da Computação–ICEx*
*Universidade Federal de Minas Gerais*
*Belo Horizonte, Brazil*
marcos@dcc.ufmg.br

Paulo Roberto Oliveira
*Coordenação dos Programas de Pós–graduação em Engenharia–COPPE*
*Universidade Federal do Rio de Janeiro*
*Rio de Janeiro, Brazil*
poliveir@cos.ufrj.br



**Abstract** We present algorithms to solve the problems originated from the application of the Dantzig and Wolfe decomposition principle to linear programming problems. One algorithm requires the sub–problems (slaves) completely solved at each iteration and the other uses sub–optimal solutions. All algorithms partially solve the master problem at each iteration. Also, we present a two–cut approach to save the excessive computation time in the restarting phase.



**Keywords:** Decomposition, Interior–point, Analytic center, Cutting plane


## 1 INTRODUCTION

The Dantzig and Wolfe decomposition principle [3] gave rise to a family of algorithms through which efficient solutions of large scale problems were expected to be obtained. Although the methods have interesting ideas of distribution and decentralization, they have poor convergence in some problems; most of the effort gained in the solution of easy sub–




*N. Hadjisavvas and P.M. Pardalos (eds.), Advances in Convex Analysis and Global Optimization,* 473–485.

problems is wasted due to a high number of oracle calls required to achieve a point close to the solution.

One major class of interior-point algorithms is the analytic center method (**ACM**) [5], [13], [19], [23]. Some authors gave new life to the decomposition methods using analytic centers; see, for example, the numerical experiments in [8]. In our algorithms, we partially apply the **ACM** to problems originated from the application of Dantzig–Wolfe decomposition principle to linear programming problems.

This paper has four sections. In section 2 we review the analytic center concept, exploiting its use in a cutting plane environment. The algorithms are described in section 3, followed by the conclusions and the references.

## 2 THE CUTTING PLANE IDEA AND THE ANALYTIC CENTER

In the cutting plane method [14], a concave function $D(u)$ approximation is improved by an expanding set of linear forms given by:

$$D(u) \approx D^{\Delta}(u) := min_{u \in S}\{D(u_i) + s_i^T(u - u_i), i = 1, 2, \ldots, k\}, \quad (30.1)$$

where $S$ is a compact convex set with nonempty interior, denoted by $int(S)$, that contains $u^* = argMaxD(u)$. At each iteration, $u_{k+1}$ is computed by

$$u_{k+1} := argMax[min_{u \in S}\{D(u_i) + s_i^T(u - u_i), i = 1, 2, \ldots, k\}],$$

which is equivalent to

$$(u_{k+1}, r^{k+1}) := argMax_{u \in S}\{r : r \leq D(u_i) + s_i^T(u - u_i), i = 1, 2, \ldots, k\}.$$

If $u_{k+1}$ does not approximate to $u^*$, a supergradient $s_i$ with the value of the function will supply one more hyperplane that will enrich the model function (30.1).

The fundamental aspect of the cutting plane method in an interior–point environment lies in the choice of the test point $u_{k+1}$. As suggested in [23] and used in [1], [2],[9], [15], [17], [18], [22] and others, the test point is the analytic center. It represents an idea of center which is a convenient place to locate the iterations in case of lack of information about the optimal solution. It is a balance among all forms that define the model. This is particularly important when the current iteration is far from the optimal solution.

Given a strictly lower bound $\rho^k$ of the optimal value of $D$ ($\rho^k < D(u^*)$), we define an auxiliary function $\phi(u, r, \rho^k)$ as

$$\phi(u, r, \rho^k) := -q \ln (r - \rho^k) - \sum_{i=1}^{k} \ln (s_i^T u - r - d_i) - \phi_S(u),$$

where $q$ is a positive integer number, $q > m_1 + 1$, $d_i = s_i^T x - D(u_i)$ and $\phi_S(u)$ is the logarithmic function of the slacks of the constraints that describe $S$. The function $\phi(u, r, \rho^k)$ is defined in the interior of $S_{\rho^k}$,

$$S_{\rho^k} := S \cap \{(u, r) \in R^{m_1+1} : s_i^T u - r \geq d_i, i = 1, 2, \ldots, k\} \cap$$

$$\cap \{(u, r) \in R^{m_1+1} : \underbrace{r \geq \rho^k, \ldots, r \geq \rho^k}_{q \quad times}\}.$$

The analytic center is the unique minimizer of $\phi(u, r, \rho^k)$ in the interior of $S_{\rho^k}$.

Consider a large scale linear programming problem given by:

$$\text{Minimize} \quad c^T x$$
$$\text{subject to:}$$

$$Ax = b$$
$$Dx \geq d$$
$$x \geq 0$$

where $c, x \in R^n, b \in R^{m_1}, d \in R^{m_2}, A \in R^{m_1 Xn}, B \in R^{m_2 Xn}$. We define

$$X := \{x \in R^n : Dx \geq d, x \geq 0\}$$

and we suppose it is a compact set with nonempty interior. For $u \in R^{m_1}$, the dual function of the optimization problem is

$$D(u) := u^T b + min\{(c^T - u^T A)x : x \in X\}.$$

It is well known that $D(u)$ is nonsmooth, concave and $s = b - Ax^*(u)$, where $x^*(u) := argMin_{x \in X}\{(c^T - u^T A)x\}$ is a supergradient for $D(u)$ at $u$.

Our approach is to reach $\overline{u}$ and $D^{\Delta}(u)$ such that $D^{\Delta}(\overline{u}) - D(u^*) \leq \delta$ for a threshold value $\delta > 0$.

# 3 ALGORITHMS

This section presents two algorithms in which the master problem of Dantzig–Wolfe is not solved to optimality. At iteration $k$, they generate one oracle call in a point close to the analytic center of the region $S_{\rho^k}$. The polytope that contains $(u^*, r^*)$ is updated and it is kept unchanged, in the sense that the algorithms do not make new oracle calls, during a number $\tau$, $(\tau \geq 1)$ of steps of the ACM. The steps increase one lower bound of the optimal value of $D$ from $\rho^k$ to $\rho^{k+1}$, which is close to the best known value of the function $D$.

The contrast between the two algorithms lies in the solution of the sub–problems; the first requires the slaves completely solved at each iteration and the second can use sub–optimal solutions.

## 3.1 A CUTTING PLANE ANALYTIC CENTER METHOD FOR DECOMPOSITION OF DANTZIG AND WOLFE

We present the algorithm (see figure 30.1) that demands the optimal solutions of the sub–problems. The formulation is rather conceptual.

At iteration $k$, we know a point $(u_k, r^k)$ near the analytic center of $S_{\rho^k}$. An oracle call returns an arbitrary supergradient $s_{k+1} \in \partial D(u_k)$ and $D(u_k)$ which produces a cut

$$s_{k+1}^T u - r \geq d_{k+1} = s_{k+1} k^T u_k - D(u_k). \tag{30.2}$$

Next, we restart the problem, that is, we compute a point close to the analytic center after (30.2) is added to the bundle of information.

Once proximity to the central path is recovered, we go on increasing $\rho^k$ with iterations of the ACM applied to the perturbed problem. Note that if we solve it entirely, that is, if we achieve

$$(u_{k+1}^*, r_{k+1}^*) = argMax_{u \in S_{\rho^k}} D^{\Delta}(u),$$

where

$$S_{\rho^k} := S_{\rho^k} \cap \{(u, r) \in R^{m_1+1} : s_{k+1}^T u - r \geq d_{k+1}\},$$

we will follow the cutting plane method of Kelley [14]. Appart from this, we will go through a number $\tau$, $(\tau \geq 1)$ of steps such that we reach $\rho^{k+1}$, close to

$$D_{max}^k := max\{D(u_i), i = 1, 2, \ldots, k\}.$$

We set the target value $\rho^{k+1}$ as

$$\rho^{k+1} := \rho^k + \theta(D_{max}^k - \rho^k),$$

where $\theta \in (0,1)$. The optimality test is carried out checking

$$(\rho^k - r^k) < \epsilon,$$

where $\epsilon$ is a positive tolerance constant.

*algorithm DWACM*
Starting:
Set $D_{max}^0 = D(u^0)$. Choose $\rho^0 < D_{max}^0$, $\theta \in (0,1)$, and $\epsilon > 0$
Compute $\left((u^0)^T, r^0\right)^T$ near the center of $S_{\rho^0}$
$k = 0$

*repeat*
*if* $r^k - \rho^k \leq \epsilon$
*then stop*
*end if*
New cut: compute $s_{k+1} \in \partial D(u_k)$ and $d^{k+1} = \langle s_{k+1}, u_k \rangle - D(u_k)$
Restarting procedure:
Update the polytope: $S_{\rho^k} = S_{\rho^k} \bigcap$

$$\bigcap \left\{ \left((u)^T, r\right)^T \in R^{n+1} : \langle s_{k+1}, u \rangle - r \geq d^{k+1} \right\}$$

Update the upper bound: $\rho^{k+1} = \rho^k + \theta\left(D_{max}^k - \rho^k\right)$
Apply ACM to $S_{\rho^k}$, until $\rho^k$ is increased to $\rho^{k+1}$,
achieving $\left((u_{k+1})^T, r_{k+1}\right)$, close to the center of $S_{\rho^k}$

Compute $D(u_{k+1})$
Compute $D_{max}^{k+1} = \max\left\{D_{max}^k, D(u_{k+1})\right\}$
$k = k + 1$
*end repeat*
*end algorithm.*

*Figure 30.1* Algorithm that uses optimal solutions of the slaves sub–problems

## 3.2 USING APPROXIMATED SOLUTIONS OF THE SUB–PROBLEMS

The algorithm in figure 30.1 is modified to address two important issues concerning decomposition of large scale problems:

- how exactly the sub–problem has to be solved and

- how the master problem has to be structured.

Our answer to the first question is to use sub–optimal solutions of the slaves with a criterion to account for their effectiveness. One sub–optimal solution will be accepted if it produces one cut that separates the current iteration; that is $(u_k, r^k)$ will not be verified by the new form to be added to the bundle.

We have not a precise answer to the second question. It is related to the criteria to be used in the decomposition. Usually, the decomposition is driven by the structure of the problem; the embedded easy sub–problems are likely to be the slaves. If the algorithm that uses sub-optimal solutions, proves to be effective in practice, in the sense that the number of sub–optimal solutions outnumbers the required optimal ones, it will weaken such use of the structure as a unique criterion for modularization, since it will enable us to work with complicated slaves.

So, consider sub–optimal solutions $x_k(u_k)$ of

$$Minimize_{x \in X}\{(c^T - u_k^T A)x\}. \tag{30.3}$$

We set a bound to be achieved by (30.3); for $x_k(u_k)$ such that

$$(c^T - u_k^T A)x_k(u_k) < r^k - u_k^T b, \tag{30.4}$$

the vector $s_k = b - Ax_k(u_k)$ is an $\epsilon$-supergradient for $D$ at $u_k$. Obviously, if (30.3) is entirely solved without achieving (30.4), we will be able to compute one supergradient and the algorithm will run as the former one.

It is impossible to compute $D(u_k)$ from sub-optimal solutions of (30.3). We compute $D_l(u_k) \leq D(u_k)$ and we employ it to update $D^k_{max}$:

$$D^{k+1}_{max} := max\{D^k_{max}, D_l(u_k)\}.$$

Every algorithm that computes the analytic center in a primal space, also minimizes a potential function assorted with the dual polytope, as it is in [19] and [9]. Let $z_k$ be this point. Suppose $x_k(u_k)$ close to the analytic center of a region associated with the ACM applied to solve (30.3) and consider $\zeta_k$ the value of the corresponding dual objective function at point $z_k$. From the weak duality theorem,

$$(c^T - u_k^T)x^* \geq \zeta_k,$$

or,

$$u_k^T b + (c^T - u_k^T)x^* \geq u_k^T b + \zeta_k$$

that implies

$$D(u^k) \geq u_k^T b + \zeta_k.$$

We adopt $D_l(u_k) = u_k^T b + \zeta_k$.

The new algorithm, not stated here, is obtained with the substitution of the step related to the oracle call (that furnishes the ingredients to make the cut) and the step that computes $D_{max}^{k+1}$.

## 3.3 A TWO–CUT APPROACH

A crucial aspect in these methods lies in the restarting procedure, where the (approximate) center must be recalculated after the adding of a new constraint. Several authors have contributed to the build-up problem with solutions, in a direct form, or as a column generation technique, see, e. g. [4],[5], [6], [7], [10], [11], [12], [16], [24], [25] and [26].

From another point of view, the algorithms in sections 3.1 and 3.2, at each iteration, reach a point close to the analytic center of a region that is perturbed by two cuts; one cut is provided by one oracle call and the other is originated from the increasing of the upper bound $\rho^k$. In spite of applying the same restarting procedure twice, we present an approach that solves the restarting problem in a unique sequence of centralizations.

We change the notation to simplify the exposition. Let $x = (u^T, r)^T$, $a_i = (s_i^T, -1)^T$ and $S = \{x \in R^n : g_i(x) \leq 0, i = 1, 2, \ldots, m\}$ be a convex compact region with nonempty interior. The analytic center of $S$ is the unique point $x_c$ belonging to $int(S)$ which maximizes the product of the slacks of the constraints that define $S$; that is,

$$x_c = argMax_{x \in int(S)} \prod_{i=1}^{m} [-g_i(x)].$$

It is easily computed applying Newton steps to the auxiliary problem

$$Minimize_{x \in int(S)} - P(x), \tag{30.5}$$

where $P(x)$ is the logarithmic barrier function given by

$$P(x) = \sum_{i=1}^{m} \ln[-g_i(x)]. \tag{30.6}$$

As it is impossible to exactly determine $x_c$, a precise criterion is used to establish a neighborhood of this point, sufficient to guarantee the convergence of interior-point algorithms that use it as a test point. Consider $H = H(x)$ and $g = g(x)$ the first and second order derivatives of (30.6). The hessian $H$ is defined positive for $x \in int(S)$ and it induces a norm $||x||_H = \sqrt{x^T H x}$, for all $x \in R^n$. We say that $\overline{x}_c \in int(S)$ is close to the analytic center o $S$ if the direction of Newton in the solution of (30.5) will be such that

$$\delta = \delta(x) = ||\sqrt{H^{-1}g}||_H < \overline{\delta},$$

where $\overline{\delta} \in (0,1)$.

Now, assume that $\overline{x}_c$ is close to the analytic center of $S$ and two cuts;

$$C_i = \{x \in R^n : a_i^T x \leq d_i\}, i = 1, 2;$$

is being incorporated to the ones that define $S$, resulting in the perturbed region $(S \cap C_1 \cap C_2)$. Our restarting problem concerns the reaching of an interior point $x$, such that $x$ is close to the analytic center of $(S \cap C_1 \cap C_2)$.

Now, we show how to expand the perturbed region keeping the iterate close to the analytic center. Given $\overline{x}_c$, Theorem 1 establishes the relaxation parameter $\gamma$ such that the perturbation of the region $S$ by $C_\gamma = \{x \in R^n : a^T x \leq \gamma\}$ maintains $\overline{x}_c$ close to the analytic center of $S_\gamma$. Although our demonstration presented in [20] is equivalent to [7], our result has a fundamental property: it lets the choice of the proximity of $\overline{x}_c$ to the analytic center of $S_\gamma$ free.

With Lemma 1, we clarify a relationship between two problems associated with the computation of analytic center.

Let $H$, $g$, $S$ and $\overline{x}_c$ as before. A new form $a^T x \leq d$ defines a semispace that perturbes the region $S$, producing

$$S_d = S \cap \{x \in R^n : a^T x \leq d\}$$

We assume $int(S_d) \neq \emptyset$.

**Theorem 1 [7].**

Let

$$\theta = a^T H^{-1} a, \quad \alpha = a^T H^{-1} g, \quad \epsilon^2 < 1 - \delta^2,$$

and

$$\beta = \sqrt{p^2\alpha^2 - \epsilon^2 p[\alpha^2 - \theta(p - \epsilon^2)]},$$

where $\delta < 1$ is the proximity of x to the analytic center of the perturbed region. Then for

$$\gamma \geq a^T x + \frac{1}{\epsilon^2}(p\alpha + \beta),$$

$$\text{if} \quad \alpha^2 - \theta(p - \epsilon^2) < 0;$$

$$\gamma \geq a^T x + \frac{1}{\epsilon^2}(p\alpha + \beta) \quad \text{or} \quad 0 \leq \gamma \leq a^T x + \frac{1}{\epsilon^2}(p\alpha - \beta),$$

$$if \quad \alpha^2 - \theta(p - \epsilon^2) > 0 \quad \text{and} \quad \alpha > 0;$$

$$\gamma \geq a^T x,$$

$$if \quad \alpha^2 - \theta(p - \epsilon^2) \geq 0 \quad \text{and} \quad \alpha < 0$$

and

$$\gamma \geq a^T x + \frac{2p\alpha}{\epsilon^2},$$

$$\text{if} \quad \alpha^2 - \theta(p - \epsilon^2) = 0 \quad \text{and} \quad \alpha > 0$$

the proximity of $x$ to the analytic center of the perturbed region is $\epsilon$.

**Proof.**

See [20].

Suppose Theorem 1 is applied to perturb $S$ by adding two cuts

$$C_{\gamma_i} = \{x \in R^n : a_i^T x \leq \gamma_i\}, i = 1, 2;$$

$\overline{x}_c$ is still close to the analytic center of

$$S_\gamma = S \cap_{i=1}^2 C_{\gamma_i}$$

Now, consider the problem

$$Minimize_{x \in int(S)} \{g_0(x) = -\prod_{i=1}^{2} (\gamma_i - a_i^T x)\} \tag{30.7}$$

Given $\rho$, an upper bound of (30.7), the analytic center assorted with the region

$$\hat{S}_\rho = S \cap \{x \in R^n : \underbrace{g_0(x) \leq \rho \ldots, g_0(x) \leq \rho}_{p \quad copies}\},$$

where $p > m + 2$, is

$$\hat{x}_c = argMin_{x \in S}\{-p \ln [\rho - g_0(x)] - P(x)\}.$$

The Lemma in the following shows that if $\hat{x}_c$ is close to the analytic center of $S_\gamma$ it will still be close to the analytic center of $\hat{S}_\rho$ which is associated with problem (30.7).

**Lemma 1.**

If $\overline{x}_c \in R^n$ is close to the analytic center of $S_\gamma$, then $\overline{x}_c$ will be close to the analytic center of $\hat{S}_\rho$, for

$$\rho = 2\prod_{i=1}^{2}(\gamma_i - a_i^T\overline{x}_c)$$

**Proof.**

See [20].

Now, we are able to state the main steps of a new restarting procedure to be used in the algorithms of the sections 3.1 and 3.2. We will partially apply the ACM to the problem

$$Minimize_{x\in S_\gamma}\{-\prod_{l=1,2}(\gamma_l - a_l^T x)\}.$$

We suppose it is given an upper bound $\gamma^0$ and a point $x^0$, close to the analytic center of the region $S_\gamma$ that is computed using Theorem 1 and Lemma 1. The iteration $k+1$ is given by

$$\overline{x}_c^{k+1} = argMin_{x\in S_\gamma}\{\psi(x,\rho^k)\}$$

where

$$\psi(x,\rho^k) = -qLn[\rho^k + \prod_{l=1,2}(\gamma_l - a_l^T x)] -$$

$$-P(x) - \sum_{l=1,2} Ln(\gamma_l - a_l^T x).$$

We compute

$$\gamma_l^{k+1} := \gamma_l^k + \theta(a_l^T\overline{x}_c^k - \gamma_l^k),$$

for $l = 1,2$ and $\theta \in (0,1)$ and

$$\rho^{k+1} = \prod_{l=1,2}\gamma_l^{k+1},$$

where $\gamma_l^0 = \gamma_l$ (see Theorem 1) and $\rho^0 = \rho$ (see Lemma 1), to perform the solution improvement. The stopping criterion is carried out by checking the feasibility of the point $\overline{x}_c^{k+1}$ and the values of $\gamma_l^{k+1}$.

# 4 CONCLUSIONS

The convergence results of the first algorithm are found in [18] with some complexity issues. The convergence of the algorithm in section 3.2 is in a forthcoming paper [21]. We would like to point out that the restarting procedure presented in section 3.3 is easily generalized to solve the linear algebraic feasibility problem [20].

We have been planning numerical experiments with the algorithms; particularly we intend to apply them to nonlinear multicommodity cost flow problems. Up to now, we can only report tests with the algorithm that uses supergradients. It was run in a set of nonsmooth convex academic examples, where its performance indicates its effectiveness; see [18].

## Acknowledgments

The authors wish to thank CNPq–INRIA, MEC and FAPEMIG for their support. We are indebted to Leda Maria de Moraes for her help and comments on this paper.

## References

[1] Atkinson, D. S. and Vaidya, P. (1995). A cutting plane algorithm for convex programming that uses analytic centers. *Mathematical Programming*, (69):1–4.

[2] Bahn, O., Merle O. du, Goffin, J. L., and Vial, J. P. (1995). A cutting plane method from analytic centers for stochastic programming. *Mathematical Programming*, (69):45–73.

[3] Dantzig, G. B. and Wolfe, P. (1960). The decomposition algorithm for linear programs. *Operations Research*, (8):101–111.

[4] Dantzig, G. B. and Ye, Y. (1990). A build-up interior method for linear programming: affine scaling form. Technical Report Sol 90-4, Dept. of Operations Research, Stanford, USA.

[5] Den Hertog, D. (1992). *Interior point approach to linear, quadratic and convex programming: algorithms and complexity.* PhD thesis, University of Technology, Delft, The Netherlands.

[6] Den Hertog, D., Roos, C., and Terlaky, T. (1992). A build-up variant of the path following method for lp. *Operations Research Letters*, (12):181–186.

[7] Feijoo, B., Sanchez, A., and Gonzaga, C. C. (1997). Maintaining closedness to the analytic center of a polytope by perturbing added hyperplanes. *Applied Mathematics and Optimization*, 2(35):139–144.

[8] Goffin, J. L., , Gondzio, J., R., S., and Vial, J. P. (1994). Solving nonlinear multicommodity flow problems by the analytic center cutting plane method. Technical Report LOGILAB 1994.21, GERARD, Faculty of management, McGill University.

[9] Goffin, J. L., Haurie, A., and Vial, J. P. (1992). Decomposition and nondifferentiable optimization with the projective algorithm. *Management Science*, (38):284–302.

[10] Goffin, J. L. and Vial, J. P. (1990). Cutting planes and column generation techniques with projective algorithm. *Journal of Optimization Theory and Applications*, (65):409–429.

[11] Goffin, J. L. and Vial, J. P. (1997). A two-cut approach in the analytic center cutting plane method. Technical Report LOGILAB 97-20, Department of Management Studies, University of Geneva.

[12] Goffin, J. L. and Vial, J. P. (1998). Multiple cuts in the analytic center method. Working Paper Logilab 98.10, Department of Management Studies, University of Geneva.

[13] Huard, P. (1967). Resolution of mathematical programming problems with nonlinear constraints by the method of centres. In Abadie, J., editor, *NonLinear Programming*, pages 207–21. North Holland Publishing Co. and Amsterdan.

[14] Kelley, J. E. (1960). The cutting plane method for solving convex programming. *J. Soc. Indus. Appl. Math.*, VIII(4):703–712.

[15] Kiwel, K. C. (1996). Complexity of some cutting plane methods that use analytic centers. *Mathematical Programming*, (74):47–54.

[16] Mitchell, J. E. (1988). *Karmakar"s algorithm and combinatorial optimization problems.* PhD thesis, Cornell University.

[17] Nesterov, Y. (1995). Complexity estimates of some cutting plane methods based on analytic barrier. *Mathematical Programming*, (69):149–176.

[18] Oliveira, P. R. and Santos, M. A. d. (2000). Using analytic center and cutting planes methods for nonsmooth convex programming. In Nguyen, V., Strodiot, J., and Tossings, P., editors, *Lecture Notes in Economics and Mathematical Systems–Optimization*, number 481, pages 339–356. Springer-Verlag Heidelberg.

[19] Roos, C. and Vial, J.-P. (1988). Analytic centers in linear programming. Report 88-68, Delft University of Technology, Delft, Netherlands.

[20] Santos, M. A. d., Lemos, A., and Oliveira, P. R. (2000). A multicut approach to the linear algebraic feasibility problem. Working paper, DCC-ICEx-UFMG, Belo Horizonte/MG, Brazil.

[21] Santos, M. A. d. and Oliveira, P. R. (2000). Interior-point algorithms in decomposition schemes. Working paper, DCC-ICEx-UFMG, Belo Horizonte/MG,Brazil.

[22] Sapatnekar, S. S., Rao, V. B., and Vaidya, P. M. (1991). A convex optimization approach to transistor sizing for cmos circuits. In *Proceedings of the IEEE International Conference on Computer Aided Design*, pages 482–485, Los Alamitos, CA. IEEE Computer Society Press.

[23] Sonnevend, G. (1985). An analytical centre for polyhedrons and new classes of global algorithms for linear (smooth, convex) programming. *Lecture Notes in Control and Information Sciences*, (84):866–876.

[24] Tone, K. (1991). An active-set strategy in interior point methods for linear programming. Working paper, Graduate school, Saita University, Urawa, Japan.

[25] Ye, Y. (1992). A potential reduction algorithm allowing column generation. *SIAM Journal of Optimization*, (2):7–29.

[26] Ye, Y. (1997). Complexity analysis of the analytic center cutting plane method that uses multiple cuts. *Mathematical Programming*, (78):85–104.

Chapter 31

# STOCHASTIC PERTURBATION METHODS FOR AFFINE RESTRICTIONS

M. Bouhadi, R. Ellaia

*Ecole Mohammadia d'ingénieurs, Universit Mohammed V*
*L.E.R.M.A.*
*BP 765, Agdal, Rabat,*
*Morocco*
ellaia@emi.ac.ma

J. E. Souza de Cursi

*Institut National des Sciences Appliquées*
*Avenue de l'Université BP 8*
*76800 Saint Etienne du Rouvray, France*
souza@insa-rouen.fr

**Abstract** This work concerns the global optimization of a nonconvex objective function under linear restrictions. A suitable random perturbation of the projected gradient descent method leads to global convergence. Numerical examples are given.

**Keywords:** Global optimization, Random perturbation, Projected gradient.

## 1 INTRODUCTION

This work considers the following global optimization problem:

$$x^* = Arg \min_{Ax \leq b} J \ ,$$

where $J$ is a regular but a non convex function. A wide range of deterministic methods for the numerical resolution of this problem has been developed and can be found in the literature. Fundamental difficulty arises when $J$ *is not* assumed to be convex. In order to prevent from



*N. Hadjisavvas and P.M. Pardalos (eds.), Advances in Convex Analysis and Global Optimization,* 487–499.

convergence to local minima, variations of the basic methods have been introduced in the literature. For instance, modifications of the projected gradient method have been considered by [14] or [16]; the sequential convex programming in [17], [15]; sequential quadratic programming in [7]. Moreover, stochastic or evolutionary search can be introduced, but these methods are usually considered as expensive and not accurate: on the one hand, the flexibility introduced by randomization implies often a large number of evaluations of the objective function and, on the other hand, the pure random search generates many points which do not improve the value of the objective function. We introduce in this paper a different approach, inspired from the method of random perturbations introduced by [11] for unconstrained minimization of continuously differentiable functions. After the introduction of the notations and the formulation of the basic problem (section 2), we consider the projected gradient method (section 3). We shall present numerical results in section 4

## 2 THE BASIC PROBLEM, NOTATIONS AND ASSUMPTIONS

We denote by $\mathbb{R}$ the set of the real numbers $]-\infty,+\infty[$, $\mathbb{R}^+$ the set of positive real numbers $]0,+\infty[$, $E=\mathbb{R}^n$, the $n$-dimensional real euclidean space. For $x=(x_1,\cdots,x_n)\in E$, we denote by $\|x\|=(x_1^2+\cdots+x_n^2)^{\frac{1}{2}}$ the euclidean norm of $x$. We denote also by $\|\bullet\|$ the matricial norm induced by $\|\bullet\|$: if $T=(T_{ij})$, $(1\le i\le p,\ 1\le j\le n)$ is a $p\times n$ matrix formed by real numbers, we have $\|Tx\|\le\|T\|\,\|x\|$ and

$$\|T\|=\sup\{|Tx\|:\|x\|=1\}.$$

The feasible set is

$$C=\left\{x\in E\mid\sum_{j=1}^{n}a_{ij}x_j-b_i\le 0,\ i=1,2,...,m\right\}.$$

Let us introduce the vector $b=(b_1,b_2,....,b_m)^T\in\mathbb{R}^m$ and the $m\times n$ real matrix $A=(a_{ij})$ $(1\le i\le m,\ 1\le j\le n)$. We have

$$A\equiv[a_1\ \ a_2\ \dots\ a_m]^T\quad;\quad a_i=(a_{i1},a_{i2},....,a_{in})^T\in\mathbb{R}^n,\quad i=1,....,m\quad.$$

and

$$C=\{x\in E\mid Ax\le b\}\ . \tag{31.1}$$

We observe that $C$ is closed and convex . Without loss of generality we assume that

$$\|a_i\| = 1,, \qquad i = 1, ...., m \quad . \tag{31.2}$$

For $x \in C$, the constraint $i$ is *active* at the point $x$ if and only if $\eta_i(x) = a_i^T x - b_i = 0$. The set of the active constraints $I(x)$ and the number of active constraints $p(x)$ at the point $x$ are, respectively

$$I(x) = \{i : 1 \le i \le m \text{ and } \eta_i(x) = 0\} \quad ; \quad p(x) = \text{card}(I_a(x)) \quad .$$

If $I(x) = \emptyset$, we set $A_T = 0$ $(n \times n)$. If $I(x) \neq \emptyset$, we denote by $A_N$ the $p \times n$ submatrix of $A$ formed by the rows corresponding to the active constraints at the point $x$. Without loss of generality we assume that $\text{rank}(A_N) = p$ (otherwise, we extract from $A_N$ a maximal rank submatrix and the associated vectors). From Eq. (31.1), the tangent cone to $C$ at the point $x$ is $T(x) = \{\delta x \in E \mid A_N \delta x = 0\}$ and the normal cone to $C$ at the point $x$ is the linear subspace spanned by rows corresponding to the active constraints $N(x) = \{0\} \cup [\{a_i \colon i \in I(x)\}]$. In order to simplify the notations, we drop the argument $x$ in the sequel. We have $T = N^\perp$.

The numerical algorithm considered in section 3 involves projections onto $T$. We will denote by $\Pi_T$ both this projection operator and its associated matrix. Since $E = T \oplus N$, the natural way to construct $\Pi_T$ consists in introducing orthogonal projections onto $T$ and $N$. For instance, the standard projected gradient method uses

$$\Pi_T = Id - \Pi_N \quad ; \quad \Pi_N = A_N^T \left(A_N A_N^T\right)^{-1} A_N \quad . \tag{31.3}$$

The objective function is $J : E \to \mathbb{R}$. Its lower bound on $C$ is denoted by $\ell^*$:

$$\ell^* = \inf_C J \quad .$$

Let us introduce

$$C_\alpha = S_\alpha \cap C \quad ; \quad S_\alpha = \{x \in E \mid J(x) \le \alpha\} \quad .$$

We assume that

$$J \text{ is twice continuously differentiable on } E \, . \tag{31.4}$$

$$\forall \ \alpha > \ell^* \ : \quad C_\alpha \ \text{ is not empty, closed and bounded .} \tag{31.5}$$

$$\forall \ \alpha > \ell^* \ : \quad \text{meas}\,(C_\alpha) > 0. \tag{31.6}$$

Since E is a finite-dimensional space, the assumption (31.5) is verified when $C$ is bounded *or* $J$ is coercive, *i.e.*,

$$J(x) \longrightarrow +\infty, \text{ if } \ \|x\| \longrightarrow +\infty \quad .$$

Assumption (31.6) is verified when $C$ contains a sequence of neighborhoods of a point of optimum $x^*$ having strictly positive measure, i. e., when $x^*$can be approximated by a sequence of points of the interior of $C$.

We observe that the assumptions (31.4)-(31.5) yield to

$$C = \cup_{\alpha > \ell^*} C_\alpha \ , \quad i.e., \quad \forall \ x \in C : \ \exists \ \alpha > \ell^* \ \text{ such that } \ x \in C_\alpha \ .$$

The gradient of $J$ at the point $x$ is denoted by $\nabla J(x)$. From (31.4)-(31.5):

$$\gamma_1(\alpha) = \sup\left\{\|\nabla J\,(x))\| : x \in C_\alpha\right\} \ < +\infty \ .$$

We consider the problem of determining

$$x^* \in C \ \ , \ \ x^* = Arg \min_C \ J \ . \tag{31.7}$$

The set of solutions of this problem is denoted by $C^*$:

$$C^* = \{x^* \in C \mid J(x^*) = \ell^*\} \ .$$

# 3 RANDOM PERTURBATIONS OF THE PROJECTED GRADIENT

## 3.1 PROJECTED GRADIENT

*We do not develop here the theory concerning the projected gradient method: the reader interested in these aspects is invited to refer to the given references.*

The basic iterations of usual descent methods for unconstrained problems may lead to infeasible points, even if the initial point belongs to $C$. A simple idea in order to maintain feasibility consists in the use of projections onto the boundary $C$. For instance, the descent direction can be projected onto the boundary of $C$, in order to ensure the feasibility of each new generated point . [12] has proposed such an algorithm where the descent direction is projected onto the intersection of the active constraints, by using the projection matrix $\Pi_T$.

Let us recall briefly the essential points of the projected gradient method: an initial feasible guess $x_o \in C$ is given and a sequence $\{x_k\}_{k\geq 0} \subset C$ is generated by using iterations of the general form

$$\forall\, k \geq 0 \qquad x_{k+1} = Q_k\left(x_k\right) = x_k + \eta_k d_k \ . \tag{31.8}$$

The choice of the descent direction must simultaneously maintains feasibility and improves the value of the objective function $J$: $d_k$ must be a *feasible direction*, i.e., there exists $\delta > 0$ such that

$$\forall\, \eta \in [0, \delta] : \ x_k + \eta d_k \in C \quad . \tag{31.9}$$

The optimal choice for $\eta_k$ is

$$\eta_k = \operatorname{argmin}\{j_k(\eta) : 0 \leq \eta \leq \delta\} \quad ; \quad j_k(\eta) = J(x_k + \eta d_k) \quad . \tag{31.10}$$

We have that $J(x_k + \eta_k d_k) \leq J(x_k)$ ,

$$U_k = J(X_k) \ , \tag{31.11}$$

defines a decreasing sequence, bounded from below by $\ell^*$:

$$\forall \ \ k \geq 0 \quad : \quad \ell^* \leq U_{k+1} \leq U_k \ . \tag{31.12}$$

The feasible direction $d_k$ is $-\nabla J\left(x_k\right)$, if $x_k$ belongs to the interior of $C$ and is given by the theorem 3.1 below for the boundary points (see, for instance, [2], [5]):

**Theorem 3.1** *Let $x_k$ be a feasible point such that* $\operatorname{rank}(A_N) = q$, $1 \leq q \leq n$; *and* $r = \left(A_N A_N^T\right)^{-1} A_N \nabla J\left(x_k\right)$. *Then, there are real constants $\mu_1 > 0$ and $\mu_2$ such that*

$$\mu_1 \geq \left\| \left(A_N A_N^T\right)^{-1} \right\| \quad ; \quad \mu_2 = \max\left\{r_i : 1 \leq i \leq q\right\} = r_j \quad .$$

*If*

$$\left\|\Pi_T \nabla J\left(x_k\right)\right\| > \max\left(0, \mu_2 / 2\sqrt{\mu_1}\right)$$

*then* $d_k = -\Pi_T \nabla J(x_k)$ *is a feasible direction. If*

$$0 < \|\Pi_T \nabla J(x_k)\| \leq \mu_2 / 2\sqrt{\mu_1}$$

*then* $d_k = -\Pi_\Lambda \nabla J(x_k)$ *is a feasible direction, where* $\Pi_\Lambda$ *is given by Eq. (31.3) but using* $A_\Lambda$, *the matrix obtained by eliminating the row number* $j$ *from* $A_N$.

Usually, the iterations are stopped according to the following result:

**Theorem 3.2** *Let* $x$ *be a feasible point such that* $\text{rank}(A_N) = q$, $1 \leq q \leq n$; *and* $r = \left(A_N A_N^T\right)^{-1} A_N \nabla J(x)$. *If* $\Pi_T \nabla J(x) = 0$ *and* $\mu_2 = \max\{r_i : 1 \leq i \leq q\} \leq 0$ *then* $x$ *is a Kuhn-Tucker point.*

The main difficulty is the lack of convexity: if $J$ is not convex, the Kuhn-Tucker points may do not correspond to global minima. In the sequel, we shall improve this point by using convenient random perturbations.

## 3.2 RANDOM PERTURBATIONS

In order to have global convergence, the sequence of real numbers $\{x_k\}_{k\geq 0}$ is replaced by a sequence of random variables $\{X_k\}_{k\geq 0}$ involving a random perturbation $P_k$ of the deterministic iterations (31.8). The basic random perturbation algorithms involve the following three operations at each step $k \geq 0$:

- *unperturbed descent:* generation of a point $Y_k$ by using the basic descent method $Q_k$. We assume that $Y_k$ is feasible, i. e., $Y_k \in C$.

  This assumption is satisfied by the projected gradient method;

- *perturbation:* generation of feasible trial points by using a random variable $T_k \in C$, related to the random perturbation $P_k$ and $Y_k$.

  We assume that $T_k$ spans $Y_k$, i. e., $Y_k$ is a value from $T_k$. This assumption is satisfied by all perturbations introduced below;

- *dynamics*: the definition of $X_{k+1}$, which is usually a function of the actual feasible values $\mathcal{P}_k = \{T_k, X_k\}$. We denote this function by $S_k$. Then, $X_{k+1} = S_k(\mathcal{P}_k)$.

From the numerical standpoint, random perturbation algorithms start at the initial value $X_0 = x_0 \in C$. At step $k \geq 0$, $X_k$ is known and $X_{k+1}$ is determined by performing the following three operations:

- *unperturbed descent:* generate a point $Y_k = Q_k(X_k)$;

- *perturbation:* Since $Y_k$ is a feasible point and $T_k$ spans $Y_k$, we set $T_k^0 = Y_k$. In addition, we generate a $NT$ supplementary values from $T_k$, denoted by $T_k^i : 1 \leq i \leq NT$. Then, $\mathcal{T}_k = \left(T_k^0, T_k^1, ..., T_k^{NT}\right)$ is a sample of the random variable $T_k$. This sample is used in order to approach $T_k$. The case $NT = 0$ corresponds to the unperturbed descent method;

- *dynamics*: the available feasible points form the set $\mathcal{P}_k = \{X_k\} \cup \mathcal{T}_k$. $X_{k+1}$ is obtained by applying a rule $S_k$, *i. e.*, $X_{k+1} = S_k(\mathcal{P}_k)$.

A simple way for the generation of the perturbation consists in the use of $\Pi_T$ :

$$X_0 = x_0 \in C \;\; ; \;\; \forall\, k \geq 0 \qquad T_k = X_k - \eta_k \Pi_T \left(\nabla J\left(X_k\right) + P_k\right) \; . \tag{31.13}$$

The modified method can be viewed as a perturbation of the descent direction. Here, $P_k$ can be generated by $P_k = \xi_k \mathcal{Z}_k$, , where $\{\mathcal{Z}_k\}_{k \geq 0} \subset \mathbb{R}^n$ is a sequence of random vectors and $\{\xi_k\}_{k \geq 0}$ is a suitable *decreasing* sequence of *strictly positive* real numbers *converging to* 0 and such that $\xi_0 \leq 1$. In the numerical experiments, we consider $\mathcal{Z}_k$ an unit normal random vector. Its density is

$$g_k\left(\|z\|\right) = \frac{1}{(2\pi)^{n/2}} \exp(-\|z\|^2/2) \quad , \forall\, z \in \mathbb{R}^n.$$

We shall consider an *elitist dynamics*, which deterministically rejects $T_k$ that increases the value of the objective function. Thus,

$$X_{k+1} = T_k, \quad \text{if} \quad J(T_k) \leq J(X_k) \quad ; \quad X_{k+1} = X_k, \quad \text{otherwise.}$$

From the numerical standpoint, only a sample $\mathcal{T}_k$ from $T_k$ is disponible. We consider the best trial point, i.e., these dynamics are applied to

$$T_k = \arg\min \left\{J(X) : X \in \mathcal{P}_k - \{X_k\}\right\}.$$

$Q_k \equiv 0$ corresponds to the stochastic descent procedure. The algorithm generates a sequence $U_k = J(X_k)$. By construction, this sequence is decreasing and bounded below by $\ell^*$. Thus, there exists $U \geq \ell^*$ such that $U_k \longrightarrow U$ for $k \longrightarrow +\infty$. The global convergence of the elitist dynamics is ensured by the following result, which follows from the Borel-Cantelli's lemma (see, for instance, [11], [10]) :

**Theorem 3.3** *Let $\gamma = J(x_0)$. Assume that $x_o \in C_{\ell^*+\mu}$ and*

$$\sum_{k=0}^{+\infty} g_k\left(\frac{\beta(\gamma,\varepsilon)}{\xi_k}\right) = +\infty \quad . \tag{31.14}$$

*Then $U = \ell^*$ almost surely.*

**Corollary 3.1** *Let $a > 0$, $k_0 \geq 2$ and*

$$\xi_k = \sqrt{\frac{c}{\log(k + k_o)}} \tag{31.15}$$

*If $x_o \in C$ then $U = \ell^*$ almost surely.*

**Remark 3.1** *Another convenient choice is $\xi_k = \xi_0 =$ constant $> 0$.*

# 4 NUMERICAL EXPERIMENTS

The method has been tested for examples suggested by [4], [8], [9], [1] and [6]. By reasons of limitation of the room, we do not present the complete set of performed experiments.

We shall introduce a maximum iteration number $KMAX$ : the algorithm stops when $k = KMAX$. Thus, even if $x_k$ is a Kuhn-Tucker point ( $k \leq KMAX$), algorithm does not stop. For each descent direction, the optimal step $\eta_k$ is calculated by unidimensional exhaustive search with a fixed step, which may imply a large number of calls of the objective function. The global number of function evaluations is $n_{eval}$ and includes the evaluations for unidimensional search. The optimal value obtained is $f_{opt}$. The perturbation is normally distributed and samples are generated by using the log-trigonometric generator $s_{n+1} = (as_n + b) mod\, m$, with $s_0 = 0$, $a = 69069$, $b = 1$, $m = 2147483647$.

## 4.1 EXPERIMENT 1

This example has been considered in [4]. The dimension is $n = 20$. The function is $J(x) = -10\,\|x\|^2$. The restrictions are $x \geq 0$, and $\widetilde{A}x \leq \widetilde{b}$ where

$$\widetilde{A}^T = \begin{bmatrix} -3 & 7 & 0 & -5 & 1 & 1 & 0 & 2 & -1 & 1 \\ 7 & 0 & -5 & 1 & 1 & 0 & 2 & -1 & -1 & 1 \\ 0 & -5 & 1 & 1 & 0 & 2 & -1 & -1 & -9 & 1 \\ -5 & 1 & 1 & 0 & 2 & -1 & -1 & -9 & 3 & 1 \\ 1 & 1 & 0 & 2 & -1 & -1 & -9 & 3 & 5 & 1 \\ 1 & 0 & 2 & -1 & -1 & -9 & 3 & 5 & 0 & 1 \\ 0 & 2 & -1 & -1 & -9 & 3 & 5 & 0 & 0 & 1 \\ 2 & -1 & -1 & -9 & 3 & 5 & 0 & 0 & 1 & 1 \\ -1 & -1 & -9 & 3 & 5 & 0 & 0 & 1 & 7 & 1 \\ -1 & -9 & 3 & 5 & 0 & 0 & 1 & 7 & -7 & 1 \\ -9 & 3 & 5 & 0 & 0 & 1 & 7 & -7 & -4 & 1 \\ 3 & 5 & 0 & 0 & 1 & 7 & -7 & -4 & -6 & 1 \\ 5 & 0 & 0 & 1 & 7 & -7 & -4 & -6 & -3 & 1 \\ 0 & 0 & 1 & 7 & -7 & -4 & -6 & -3 & 7 & 1 \\ 0 & 1 & 7 & -7 & -4 & -6 & -3 & 7 & 0 & 1 \\ 1 & 7 & -7 & -4 & -6 & -3 & 7 & 0 & -5 & 1 \\ 7 & -7 & -4 & -6 & -3 & 7 & 0 & -5 & 1 & 1 \\ -7 & -4 & -6 & -3 & 7 & 0 & -5 & 1 & 1 & 1 \\ -4 & -6 & -3 & 7 & 0 & -5 & 1 & 1 & 0 & 1 \\ -6 & -3 & 7 & 0 & -5 & 1 & 1 & 0 & 2 & 1 \end{bmatrix} \quad \widetilde{b} = \begin{bmatrix} -5 \\ 2 \\ -1 \\ -3 \\ 5 \\ 4 \\ -1 \\ 0 \\ 0 \\ 40 \end{bmatrix}$$

The optimal value stated in the literature is $\ell^\star = -8695$, attained at the point $x^\star = (0, 0, 2.88, 0, 0, 4.18, 0, 0, 0, 0, 0, 0, 0, 0, 0.62, 4.09, 0, 2.31, 0)$.

The method runs with $(x_0)_{2i+1} = 2$, $i = 0, \ldots, 9$, $(x_0)_{2i} = 1$, $i = 1, \ldots, 10$ ; $k_0 = 2$. The tables 4.1 and 31.1 show the values of $n_{eval}$, $f_{opt}$ and $dist = \left\| x_{k_{stop}} - x^\star \right\|$.

| $c$ | 0.1 | 1 | 5 | 10 | 100 | 1000 |
|---|---|---|---|---|---|---|
| $n_{eval}$ | 27051 | 27201 | 27448 | 27251 | 27009 | 26691 |
| $f_{opt}$ | $-8695$ | $-8695$ | $-8695$ | $-8694$ | $-8695$ | $-8695$ |
| $dist$ | $3E-4$ | $1E-7$ | $1E-7$ | $4E-4$ | $1E-7$ | $1E-7$ |

*Table 31.1* Influence of $c$ in Experiment 1 ($NT = 50$, $KMAX = 50$)

Tests with other values $NT$ have led to analogous results. The results are stable when the random sequence is modified by other choices of $s_0$.

## 4.2 EXPERIMENT 2

This example is given in [8]. The dimension is $n = 5$. The objective function is

| $NT$ | 0 | 1 | 10 | 25 | 50 | 100 |
|---|---|---|---|---|---|---|
| $n_{eval}$ | 50 | 563 | 5491 | 13351 | 27251 | 54551 |
| $f_{opt}$ | $-500$ | $-1600$ | $-5995$ | $-4418$ | $-8695$ | $-8695$ |
| $dist$ | 6.3 | 6.3 | 8.4 | 6.7 | $1E-7$ | $1E-7$ |

*Table 31.2* Influence of $NT$ in Experiment 1 ($c = 1$, $KMAX = 50$)

| $KMAX$ | 10 | 20 | 30 | 40 | 50 | 100 |
|---|---|---|---|---|---|---|
| $n_{eval}$ | 6061 | 11571 | 17081 | 22191 | 27201 | 52251 |
| $f_{opt}$ | $-1600$ | $-2964$ | $-8548$ | $-8695$ | $-8695$ | $-8695$ |
| $dist$ | 6.3 | 5.7 | 0.2 | $1E-7$ | $1E-7$ | $1E-7$ |

*Table 31.3* Influence of $KMAX$ in Experiment 1 ($c = 1$, $NT = 50$)

$$J(x) = -\left(x_1 + 0.5x_2 + 0.667x_3 + 0.75x_4 + 0.8x_5\right)^{1.5}.$$

The restrictions are $x \geq 0$ and $\widetilde{A}x \leq \widetilde{b}$, where

$$\widetilde{A} = \begin{bmatrix} 0.795137 & 0.225733 & 0.371307 & 0.225064 & 0.878756 \\ -0.905037 & -0.638848 & -0.134430 & -0.921211 & 0.150370 \\ 0.905037 & 0.248231 & 0.278197 & 0.376265 & -0.597468 \\ 0.762043 & -0.304755 & -0.012345 & -0.394312 & -0.792129 \\ 0.564347 & 0.746523 & -0.822105 & -0.892331 & -0.922916 \\ -0.954276 & -0.196016 & 0.242000 & 0.797313 & -0.147119 \\ 0.747682 & 0.912055 & -0.529338 & 0.243496 & 0.279402 \\ -0.109599 & 0.727219 & -0.741781 & -0.058455 & 0.749470 \\ 0.209106 & -0.074202 & -0.022484 & -0.144214 & -0.735169 \end{bmatrix}; \ \widetilde{b} = \begin{bmatrix} 4.242372 \\ -1.785220 \\ 3.213560 \\ 1.205676 \\ -0.891062 \\ -0.066698 \\ 2.286079 \\ 0.521564 \\ -0.730516 \end{bmatrix}$$

| $NT$ | 0 | 1 | 10 | 25 | 50 | 100 |
|---|---|---|---|---|---|---|
| $n_{eval}$ | 10 | 5515 | 55041 | 137601 | 275356 | 550716 |
| $f_{opt}$ | $-7.16$ | $-19.14$ | $-21.12$ | $-21.12$ | $-21.12$ | $-21.12$ |
| $dist$ | 3.1 | 0.5 | 0.01 | $1E-7$ | $1E-7$ | $1E-7$ |

*Table 31.4* Results for Experiment 2 ($c = 1$, $KMAX = 10$)

The optimal value stated in the literature is $\ell^\star = -21.12$, attained at the point $x^\star = (0.41, 5.6, 6.1, 0, 0.42)$. The method runs with $(x_0)_i = 1$, $i = 1, \ldots, 5$; $KMAX = 10$, $c = 1$. The results are summarized in Table 4.2.

# 5 CONCLUDING REMARKS

The numerical examples show that the adjonction of the stochastic perturbations improve the results furnished by the projected gradient, but involve a larger number of evaluations of the objective function. The resulting method has shown to be less sensitive to variations of the parameters than purely stochastic ones, but the problem of the optimal choice of the parameters related to the stochastic perturbation is still open.

The method has been tested for examples suggested by [4], [8], [9], [1] and [6], involving a number of unknowns $n$ going from 4 to 20 and number of restrictions going from 10 to 30, with results analogous to those presented above.

Improvements leading to less function evaluations may be obtained by using, for instance,

- a dynamics analogous to the *Metropolis dynamics* instead of the elitist one (see, for instance [3]). The first results teobtained by using this approach are encouraging and show a significant gain in the number of function evaluations. This point will be matter of a futur work.

- As previously observed, the optimal step $\eta_k$ has been calculated by unidimensionnal exhaustive search with a fixed step, what may imply a large number of calls of the objective function. A more efficient unidimensionnal search may improve this point.

- an evolutionary version of the algorithm. For unconstrained problems, the use of population based methods reduces significantly the number of function evaluations. This will be matter of futur work.

Tests involving other distributions, such as, for instance, Cauchy, uniform or Tsallis have led to analogous results.

Extension to nonlinear constraints using the ideas of [13] will be matter of futur work.

## References

[1] Bazaraa, M. S. and Shetty, C. M. (1979). *Nonlinear Programming : Theory and Algorithms.* John Wiley, Atlanta (USA).

[2] Bouhadi, M. (1997). *Contribution à l'Optimisation Globale avec Contraintes. Approche Stochastique.* PhD thesis, Université Mohammed V, Faculté des Sciences, Rabat (Morocco).

[3] de Cursi, J. E. S. (1992). Recuit simulé et application à l'approximation par des sommes d'exponentielles. Research report, Ecole Centrale de Nantes. can be downloaded at http://lmrserv.insa-rouen.fr/~souza.

[4] Floudas, C. and Pardalos, P. (1990). *A Collection of Test Problems for Constrained Global Optimization Algorithms*. Springer, Berlin. Lecture Notes in Computer Science.

[5] Forsythe, G. F. (1995). Computing constrained minima with lagrange multipliers. *SIAM*, 3:173–198.

[6] Hamma, S. B. (1992). *Etude de méthodes numériques d'optimisation globale*. PhD thesis, Université Paul Sabatier, Toulouse.

[7] Heinkenschloss, M. (1996). Projected sequential quadratic programming methods. *SIAM Journal on Optimization*, 6:373–417.

[8] Horst, R. and Thoai (1989). Modification, implementation and comparison of three algorithms for globally solving linearly constrained concave minimization problems. *Computing*, 42:271–289.

[9] Levy, A. and Montalvo, A. (1977). The tunneling algorithm for the global minimization of functions. In *Conference on Numerical Analysis*, Dundee (Scotland).

[10] Pinter, J. (1996). *Global Optimization in Action*. Kluwer, Dordrecht.

[11] Pogu, M. and de Cursi, J. E. S. (1994). Global optimization by random perturbation of the gradient method with a fixed parameter. *Journal of Global Optimization*, 5:159–180.

[12] Rosen, J. B. (1960a). The gradient projection method for nonlinear programming. part i. linear constraints. *J. Soc. Indust. Appl. Math.*, 1(8).

[13] Rosen, J. B. (1960b). The gradient projection method for nonlinear programming. part ii. nonlinear constraints. *J. Soc. Indust. Appl. Math.*, 9(4).

[14] Rubinstein, R. (1996). *Simulation and the Monte Carlo Method*. John Wiley & Sons, USA.

[15] Schittkowski, K. and Zillober, C. (1992). Sequential convex programming methods. In Marti, K. and Kall, P., editors, *Stochastic Programming*, volume 423, Berlin. Springer. Lecture Notes in Economics and Mathematical Systems.

[16] Torn, A. and Zilinskas, A. E. (1989). *Global Optimisation*. Springer, Berlin.

[17] Walters, F. H., Jr., L. R. P., Morgan, S. L., and Deming, S. N. (1991). *Global Optimisation.* CRC Press LLC, Berlin.

Chapter 32

# DIRECTED DERIVATIVES OF CONVEX COMPACT-VALUED MAPPINGS

Robert Baier
*Chair of Applied Mathematics*
*University of Bayreuth*
*D-95440 Bayreuth, Germany*
robert.baier@uni-bayreuth.de

Elza M. Farkhi
*School of Mathematical Sciences*
*Sackler Faculty of Exact Sciences*
*Tel Aviv University*
*69978 Tel Aviv, Israel*
elza@post.tau.ac.il

**Abstract** Convex compact sets can be embedded into the Banach space of directed sets. Directed sets allow a visualization as possibly non-convex, compact sets in $\mathbb{R}^n$ and hence, this space could be used to visualize differences of embedded convex compact sets. The main application is the visualization as well as the theoretical and numerical calculation of set-valued derivatives. Known notions of affine, semi-affine and quasi-affine maps and their derivatives are studied.

**Keywords:** Directed sets, Set-valued derivatives, Differences of convex sets and their visualization, Affine, Semi-affine, Quasi-affine maps, Embedding of convex compact sets into a vector space, Directed intervals

## 1 INTRODUCTION

Directed sets in $\mathbb{R}^n$ presented in [1, 2, 3] form a Banach space $\mathcal{D}(\mathbb{R}^n)$ in which the convex cone of all compact, convex, nonempty sets can be embedded. A directed set $\overrightarrow{A}$ is parametrized by unit vectors $l \in$



*N. Hadjisavvas and P.M. Pardalos (eds.), Advances in Convex Analysis and Global Optimization,* 501–514.

$S_{n-1}$ and consists of two components, a describing continuous function $a_n(\cdot)$ and a $(n-1)$-dimensional directed set function $\overrightarrow{A_{n-1}(\cdot)}$ which is uniformly bounded. The hyperplane

$$H(l, a_n(\cdot)) = \{x \in \mathbb{R}^n \mid <l, x> = a_n(l)\} \tag{32.1}$$

is fixed by $a_n(\cdot)$ and contains the reprojection of the visualization of $\overrightarrow{A_{n-1}(l)}$. This image forms the boundary part of the visualized $\overrightarrow{A}$ in direction $l$.

For an embedded convex compact set $C$, the hyperplane is determined by the value of its support function in direction $l$ and $\overrightarrow{A_{n-1}(l)}$ is the embedded projection of its supporting face $C^l$ (seen as a $(n-1)$-dimensional set) into $\mathcal{D}(\mathbb{R}^{n-1})$. Generalized intervals in [6, 7, 9] are equivalent to one-dimensional directed sets and form the basis of the recursive definition of directed sets.

The proposed embedding is different to other approaches and is not minimal in the sense that Hörmander's embedding in [5] uses only the second component of our embedding. Nevertheless, we focus here on visualization of differences of convex, compact sets as eventually non-convex, compact sets in $\mathbb{R}^n$, not as real-valued functions with arguments in $S_{n-1}$ resp. $\mathbb{R}^n$. For further references on embeddings and related articles see [2].

The visualization of directed sets consists of three parts, namely the convex part, the concave part (both are convex sets, one of them being empty, if they don't equal the same point) and the (non-convex) mixed-type part. The main application studies the visualization of derivatives of simple convex compact-valued mappings such as affine, semi-affine and quasi-affine maps.

Further on, the following notations are used throughout this paper: The set of convex, compact, nonempty sets in $\mathbb{R}^n$ is denoted by $\mathcal{C}(\mathbb{R}^n)$. $S_{n-1}$ denotes the unit sphere of the $n$-dimensional (closed) Euclidean ball $B_1(0)$.

The *support function* of $C \in \mathcal{C}(\mathbb{R}^n)$ and the *supporting face* of $C$ in direction $l \in \mathbb{R}^n$ are denoted by

$$\delta^*(l, C) := \max_{c \in C} <l, c> \quad \text{resp.} \quad C^l := \{c \in C \mid <l, c> = \delta^*(l, C)\}$$

and the *Hausdorff metric* on $\mathcal{C}(\mathbb{R}^n)$ as

$$\mathrm{d_H}(C, D) := \max_{l \in S_{n-1}} |\delta^*(l, C) - \delta^*(l, D)| \quad (C, D \in \mathcal{C}(\mathbb{R}^n)).$$

The *scalar multiplication* for $\lambda \geq 0$, the *Minkowski addition*, the *pointwise negative* (multiplication by $-1$) and the *algebraic difference* are defined as

$$\lambda C := \{\lambda c \,|\, c \in C\}, \quad C + D := \{c + d \,|\, c \in C, d \in D\} \text{ for } C, D \in \mathcal{C}(\mathbb{R}^n),$$
$$\ominus C := \{-c \,|\, c \in C\}, \quad C \ominus D := \{c - d \,|\, c \in C,\ d \in D\}.$$

The operation $\ominus$, frequently used also in interval analysis, does not even fulfill $C \ominus C = \{0_{\mathbb{R}^n}\}$ in general.

The *geometric difference* and the *Demyanov difference*

$$C \stackrel{*}{-} D := \{x \in \mathbb{R}^n \,|\, \forall l \in S_{n-1} :\ <l, x> \leq \delta^*(l, C) - \delta^*(l, D)\},$$
$$C \dot{-} D := \overline{\text{co}}\{C^l - D^l \,|\, l \in S_{n-1},\ C^l \text{ and } D^l \text{ are singletons}\} \neq \emptyset$$

in [4, 10] resp. [12] are convex, compact sets. $C \stackrel{*}{-} D$ could be empty and is usually too small, whereas $C \dot{-} D$ is usually too big in the sense that

$$(C \stackrel{*}{-} D) + D \underset{\neq}{\subset} C \quad \text{resp.} \quad (C \dot{-} D) + D \underset{\neq}{\supset} C \tag{32.2}$$

is only satisfied in general. The last two differences have close connections to the visualization of differences of embedded convex compact sets (see (32.6)).

## 2 DIRECTED SETS

Most of the definitions presented here are recursive and recalled from [1, 2].

**Definition 2.1** *$\overrightarrow{A}$ is called* ***directed interval/one-dimensional directed set****, if*

$$\overrightarrow{A} = (a_1(l))_{l=\pm 1}$$

*with $a_1(l) \in \mathbb{R}$ for $l = \pm 1$. Set $\|\overrightarrow{A}\|_1 := \max\limits_{l=\pm 1} |a_1(l)|$.*
*The space of directed intervals is denoted by $\mathcal{D}(\mathbb{R})$.*

*For $n \geq 2$, $\overrightarrow{A}$ is called* ***directed set****, if there exists a uniformly bounded function $\overrightarrow{A_{n-1}} : S_{n-1} \to \mathcal{D}(\mathbb{R}^{n-1})$ with respect to $\|\cdot\|_{n-1}$ and a continuous function $a_n : S_{n-1} \to \mathbb{R}$. Set $\overrightarrow{A} = (\overrightarrow{A_{n-1}(l)}, a_n(l))_{l \in S_{n-1}}$ and*

$$\|\overrightarrow{A}\|_n := \max\{\sup_{l \in S_{n-1}} \|\overrightarrow{A_{n-1}(l)}\|_{n-1}, \max_{l \in S_{n-1}} |a_n(l)|\}$$

*and denote by $\mathcal{D}(\mathbb{R}^n)$ the space of directed sets in $\mathbb{R}^n$.*

The definition is motivated by the fact that the supporting face of each convex compact set $C \in \mathcal{C}(\mathbb{R}^n)$ in direction $l \in S_{n-1}$ lies on the hyperplane given by the support function in this direction. To enable a recursive approach, the support function is saved separately from the supporting face and the latter is seen as a $(n-1)$-dimensional set.

**Definition 2.2** *The* ***embedding*** $J_n : \mathcal{C}(\mathbb{R}^n) \to \mathcal{D}(\mathbb{R}^n)$ *is defined for* $n = 1$ *as*

$$J_1([\alpha_1, \alpha_2]) := (\delta^*(l, [\alpha_1, \alpha_2]))_{l=\pm 1} \quad \textit{for } [\alpha_1, \alpha_2] \in \mathcal{C}(\mathbb{R})$$

*and for* $n \geq 2$ *as*

$$J_n(C) := (J_{n-1}(\Pi_{n-1}^{l,\delta^*(\cdot,C)}(C^l)), \delta^*(l, C))_{l \in S_{n-1}} \quad \textit{for } C \in \mathcal{C}(\mathbb{R}^n)$$

*embeds all convex, compact, nonempty sets into the space of directed sets. Hereby,* $\Pi_{n-1}^{l,f(\cdot)}(X)$ *is the projection of a* $(n-1)$*-dimensional set* $X \subset \mathbb{R}^n$ *contained in the hyperplane* $H(l, f(\cdot))$ *in* (32.1), *which is determined by the function* $f : \mathbb{R}^n \to \mathbb{R}$, *to* $\mathbb{R}^{n-1}$.

The embedding $J_n$ is an isometric, injective map from $\mathcal{C}(\mathbb{R}^n)$ to $\mathcal{D}(\mathbb{R}^n)$ with respect to a new metric $d_V(\cdot,\cdot)$ on $\mathcal{C}(\mathbb{R}^n)$ introduced in [2] which is equivalent to Demyanov's metric.

The operations introduced on $\mathcal{D}(\mathbb{R}^n)$ are defined recursively and act separately on both components of the directed set.

**Definition 2.3** *Consider* $\overrightarrow{A} = (\overrightarrow{A_{n-1}(l)}, a_n(l))_{l \in S_{n-1}}$ *and* $\overrightarrow{B} = (\overrightarrow{B_{n-1}(l)}, b_n(l))_{l \in S_{n-1}} \in \mathcal{D}(\mathbb{R}^n)$. *Define the* ***operations*** *recursively (for* $n = 1$ *the first components of the directed sets do not exist) as*

$$\overrightarrow{A} + \overrightarrow{B} := (\overrightarrow{A_{n-1}(l)} + \overrightarrow{B_{n-1}(l)}, a_n(l) + b_n(l))_{l \in S_{n-1}}, \tag{32.3}$$

$$\lambda \cdot \overrightarrow{A} := (\lambda \cdot \overrightarrow{A_{n-1}(l)}, \lambda \cdot a_n(l))_{l \in S_{n-1}} \qquad (\lambda \in \mathbb{R}), \tag{32.4}$$

$$\overrightarrow{A} - \overrightarrow{B} := \overrightarrow{A} + ((-1) \cdot \overrightarrow{B}). \tag{32.5}$$

The space $\mathcal{D}(\mathbb{R}^n)$ has remarkable properties with these operations which were studied in [2]. E.g., $\|\cdot\|$ in Definition 2.1 is a norm on $\mathcal{D}(\mathbb{R}^n)$ for which $\mathcal{D}(\mathbb{R}^n)$ is a Banach space. Since we are interested in visualizations of differences of embedded elements of $\mathcal{C}(\mathbb{R}^n)$, we restrict our attention to the closed linear hull of $J_n(\mathcal{C}(\mathbb{R}^n))$ in this paper.

**Proposition 2.4 ([2])** *The above mentioned subspace*

$$\overrightarrow{\mathcal{C}}(\mathbb{R}^n) := \overline{\text{lin}(J_n(\mathcal{C}(\mathbb{R}^n)))} \subset \mathcal{D}(\mathbb{R}^n)$$

*is a Banach space as* $\mathcal{D}(\mathbb{R}^n)$ *itself.*

The following result proven in [2] shows that the well-known and often used operations on $\mathcal{C}(\mathbb{R}^n)$ commute with the embedding.

**Proposition 2.5** ([2]) *Consider $C, D \in \mathcal{C}(\mathbb{R}^n)$ and $\lambda \geq 0$. Then,*

$$J_n(C+D) = J_n(C) + J_n(D) \quad \textit{and} \quad J_n(\lambda C) = \lambda J_n(C).$$

The visualization for a directed set $\overrightarrow{A} \in \overrightarrow{\mathcal{C}}(\mathbb{R}^n)$ must incorporate the convex case, i.e. $\overrightarrow{A} \in J_n(\mathcal{C}(\mathbb{R}^n))$, the concave case, i.e. $\overrightarrow{A} \in -J_n(\mathcal{C}(\mathbb{R}^n))$, and the indefinite case, i.e. $\overrightarrow{A}$ is a difference of two embedded convex compact sets or the limit of such a sequence. Therefore, the visualization splits in three parts, the convex part, the concave part and the mixed-type part. In the definition of the mixed-type part, the boundary part is involved.

The projection involved in Definition 2.2 splits into an orthonormal, linear part $\mathcal{L}_{n-1}^l$ and a translation and allows an inverse reprojection on $H(l, a_n(\cdot))$, i.e.

$$\Pi_{n-1}^{l,a_n(\cdot)}(x) = \mathcal{L}_{n-1}^l(x - a_n(l)l), \quad Q_n^{l,a_n(\cdot)}(y) = (\mathcal{L}_{n-1}^l)^{-1}(y) + a_n(l)l$$

for $x \in H(l, a_n(\cdot))$, $y \in \mathbb{R}^{n-1}$. It fulfills the inverse property

$$Q_n^{l,a_n(\cdot)}(\Pi_{n-1}^{l,a_n(\cdot)}(x)) = x \quad (x \in H(l, a_n(\cdot))).$$

**Definition 2.6** *Let $\overrightarrow{A} = (\overrightarrow{A_{n-1}(l)}, a_n(l))_{l \in S_{n-1}} \in \mathcal{D}(\mathbb{R}^n)$. Define the* ***convex (positive) part*** *$P_n$, the* ***concave (negative) part*** *$N_n$ and the* ***visualization*** *$V_n$ as*

$$P_n(\overrightarrow{A}) := \bigcap_{l \in S_{n-1}} \{x \in \mathbb{R}^n \mid \langle l, x\rangle \leq a_n(l)\}, \quad N_n(\overrightarrow{A}) := \ominus P_n(-\overrightarrow{A}),$$

$$V_n(\overrightarrow{A}) := P_n(\overrightarrow{A}) \cup N_n(\overrightarrow{A}) \cup M_n(\overrightarrow{A}).$$

*The* ***mixed-type part*** *$M_n$ and the* ***boundary part*** *$B_n$ are defined recursively:*

$$\begin{aligned} &B_1(\overrightarrow{A}) := \partial P_1(\overrightarrow{A}) \cup \partial N_1(\overrightarrow{A}), \quad M_1(\overrightarrow{A}) := \emptyset \qquad \textit{for } n = 1, \\ &\left.\begin{aligned} B_n(\overrightarrow{A}) &:= \bigcup_{l \in S_{n-1}} Q_n^{l,a_n(\cdot)}(V_{n-1}(\overrightarrow{A_{n-1}(l)})), \\ M_n(\overrightarrow{A}) &:= B_n(\overrightarrow{A}) \setminus (\partial P_n(\overrightarrow{A}) \cup \partial N_n(\overrightarrow{A})) \end{aligned}\right\} \quad \textit{for } n \geq 2. \end{aligned}$$

There is no mixed-type part for directed intervals, this part could appear only for dimensions $n \geq 2$. The convex and the concave part are both convex, compact sets, but could be empty as in one example in [3].

The directions $l \in S_{n-1}$ for which a boundary point lies on the hyperplane $H(l, a_n(\cdot))$ are elements of the orientation bundle.

**Definition 2.7** *The* ***orientation bundle*** *of a directed set $\overrightarrow{A}$ depends on a boundary point and forms a nonempty subset of $S_{n-1}$.*
*Let $n = 1$, $\overrightarrow{A} = (a_1(l))_{l=\pm 1} \in \mathcal{D}(\mathbb{R})$. Then,*

$$\begin{aligned} \mathcal{O}_1(a_1(1), \overrightarrow{A}) &:= \{1\}, \quad && \text{if } a_1(1) \neq -a_1(-1), \\ \mathcal{O}_1(-a_1(-1), \overrightarrow{A}) &:= \{-1\}, \quad && \text{if } a_1(1) \neq -a_1(-1), \\ \mathcal{O}_1(a_1(1), \overrightarrow{A}) &:= \{-1, 1\}, \quad && \text{if } a_1(1) = -a_1(-1). \end{aligned}$$

*Let $n \geq 2$, $\overrightarrow{A} = (\overrightarrow{A_{n-1}(l)}, a_n(l))_{l \in S_{n-1}} \in \mathcal{D}(\mathbb{R}^n)$ and $x \in B_n(\overrightarrow{A})$. Then,*

$$\mathcal{O}_n(x, \overrightarrow{A}) := \{ l \in S_{n-1} \,|\, x \in Q_n^{l, a_n(\cdot)}(V_{n-1}(\overrightarrow{A_{n-1}(l)})) \}$$

Let us collect special properties of the visualization parts and the boundary part which are proven in [3]. Especially, the boundary part and the visualization are never empty and the three parts of the visualization in Definition 2.6 are disjoint, except in the special case that the convex and the concave part equal to the same point.

Theorem 2.8 states that the visualized difference of two embedded convex compacts lies between the geometric and the Demyanov difference (compare (32.2)).

**Theorem 2.8 (see [3])** *Let $\overrightarrow{A} = J_n(C) - J_n(D) \in \mathcal{D}(\mathbb{R}^n)$ with $C, D \in \mathcal{C}(\mathbb{R}^n)$. Then,*

$$\begin{aligned} &P_n(\overrightarrow{A}) = C \stackrel{*}{-} D, \quad N_n(\overrightarrow{A}) = \ominus(D \stackrel{*}{-} C), \\ &B_n(\overrightarrow{A}) = \partial(C \stackrel{*}{-} D) \cup (\ominus \partial(D \stackrel{*}{-} C)) \cup M_n(\overrightarrow{A}). \end{aligned}$$

*The formula for the mixed-type part is rather complicated and is stated in a recursive way in [3]. Furthermore,*

$$C \stackrel{*}{-} D \subset V_n(\overrightarrow{A}) \subset C \dot{-} D, \quad \ominus(D \stackrel{*}{-} C) \subset V_n(\overrightarrow{A}) \subset \ominus(D \dot{-} C) \qquad (32.6)$$

Theorem 2.8 shows that the visualization of an embedded convex compact set $C$ is the set itself, the boundary part is the boundary of $C$ (set $D = \{0_{\mathbb{R}^n}\}$). There is no mixed-type part in this situation. The theorem also informs on the visualization of the inverse of $J_n(D)$ (set $C = \{0_{\mathbb{R}^n}\}$). Although, the visualization is the pointwise negative set $\ominus D$, the operations (32.3)–(32.5) in $\mathcal{D}(\mathbb{R}^n)$ avoid this operation.

From Definition 2.7, $J_n(C)$ has outer normals and $-J_n(D)$ inner normals.

# 3 DERIVATIVES OF SET-VALUED MAPPINGS

In this section, notions of simple convex-valued mappings are recalled from [8, 13]. In the following, $F : [t_0, T] \Rightarrow \mathbb{R}^n$ is a set-valued mapping with images in $\mathcal{C}(\mathbb{R}^n)$. Derivatives of the following simple maps are studied in section 4. Please see the above citations for a rich list of references.

**Definition 3.1** *$F$ is an **affine map**, if there exist $F_0, F_1 \in \mathcal{C}(\mathbb{R}^n)$ with*

$$F(t) = F_0 + \frac{t - t_0}{T - t_0} F_1 \qquad (t \in I = [t_0, T]). \tag{32.7}$$

*$F$ is a **semi-affine map**, if there exist $F_0, F_1 \in \mathcal{C}(\mathbb{R}^n)$ with*

$$F(t) = (1 - \frac{t - t_0}{T - t_0}) F_0 + \frac{t - t_0}{T - t_0} F_1 \qquad (t \in [t_0, T]). \tag{32.8}$$

*$F$ is **quasi-affine**, if a positive homogeneous function $c : \mathbb{R}^n \to \mathbb{R}$ exists with*

$$\delta^*(l, F(t)) = \mathrm{co}(\underbrace{\delta^*(\cdot, F(t_0)) + (t - t_0)c(\cdot)}_{=:f(\cdot,t)})(l) \quad (l \in S_{n-1},\ t \in I), \tag{32.9}$$

*where $\mathrm{co}(g)$ denotes the convex hull of a function $g : \mathbb{R}^n \to \mathbb{R}$ (see [11]). In this case, $\mathrm{co}(f(\cdot, t))$ equals the pointwise supremum of all affine functions $h$ with $h \leq f(\cdot, t)$ (see [11]).*
*$F$ is called **convex**, if for all $t_1, t_2 \in [t_0, T]$ and $\lambda \in [0, 1]$*

$$\lambda F(t_1) + (1 - \lambda) F(t_2) \subset F(\lambda t_1 + (1 - \lambda) t_2). \tag{32.10}$$

Since both affine and semi-affine maps are quasi-affine ([13, Proposition 2.2]) and quasi-affine maps are convex and continuous ([13, Lemma 2.7]), quasi-affine maps are an important class of simple set-valued mappings.

**Definition 3.2** *Let $\overrightarrow{F} : [t_0, T] \to \mathcal{D}(\mathbb{R}^n)$ be a map with directed sets as images. $\overrightarrow{F}$ is a **finitely generated directed map**, if there exist $\overrightarrow{F_i} \in \mathcal{D}(\mathbb{R}^n)$, $i = 1, 2, ..., k$, and scalar functions $f_i : [t_0, T] \to \mathbb{R}$, $i = 1, 2, ..., k$, with*

$$\overrightarrow{F(t)} = \sum_{i=1}^{k} f_i(t) \overrightarrow{F_i} \quad (t \in [t_0, T]). \tag{32.11}$$

$\vec{F}$ *is called* **directed differentiable in** $\mathbf{t} \in I = [t_0, T]$, *if the limit*

$$\overrightarrow{DF(t)} := \lim_{\substack{h \to 0 \\ t+h \in I}} \underbrace{\frac{\overrightarrow{F(t+h)} - \overrightarrow{F(t)}}{h}}_{=:\Delta_h(\vec{F},t)}$$

*exists in* $\mathcal{D}(\mathbb{R}^n)$. *The k. derivative* $\overrightarrow{F^{(k)}(t)}$ *in* $t \in I$ *could be defined as the directed derivative of* $\overrightarrow{F^{(k-1)}(\cdot)}$ *for* $k \geq 2$ *with* $\overrightarrow{F^{(1)}(\cdot)} = \overrightarrow{DF(\cdot)}$.

$F$ *is called* **directed differentiable in** $\mathbf{t} \in [t_0, T]$, *if the function* $t \mapsto J_n(F(t))$ *is directed differentiable in* $t$.

$\vec{F}$ *is a* **linear directed map**, *if there exists* $\vec{F_0}, \vec{F_1} \in \mathcal{D}(\mathbb{R}^n)$ *with*

$$\overrightarrow{F(t)} = \vec{F_0} + t\vec{F_1} \qquad (t \in [t_0, T]). \tag{32.12}$$

**Proposition 3.3** *Let* $F$ *and* $\vec{F}$ *be given with* $\overrightarrow{F(t)} := J_n(F(t))$, $t \in [t_0, T]$.

*(i) If* $F(\cdot)$ *is affine, then* $\overrightarrow{F(\cdot)}$ *is linear.*

*(ii) If* $F(\cdot)$ *is semi-affine, then* $\overrightarrow{F(\cdot)}$ *is linear.*

**Proof:** (i) Let $F$ be affine in the form (32.7). Then, Proposition 2.5 and Proposition 2.4, guaranteeing properties of a vector space, yield:

$$\begin{aligned}\overrightarrow{F(t)} &= J_n(F_0 + \frac{t-t_0}{T-t_0}F_1) = J_n(F_0) + \frac{t-t_0}{T-t_0}J_n(F_1) \\ &= \underbrace{(J_n(F_0) - \frac{t_0}{T-t_0}J_n(F_1))}_{=:\vec{F_0}} + t\underbrace{(\frac{1}{T-t_0}J_n(F_1))}_{=:\vec{F_1}}\end{aligned}$$

(ii) is proven similarly. ■

**Proposition 3.4** *Let us consider a linear directed map* $\vec{F}$ *in the form* (32.12). *Let* $\vec{C} \in \mathcal{D}(\mathbb{R}^n)$ *and* $F_0 \in \mathcal{C}(\mathbb{R}^n)$ *be arbitrarily chosen. If*

$$\vec{F_0} = J_n(F_0) - t_0\vec{C}, \quad \vec{F_1} = \vec{C}$$

*and* $P_n(\overrightarrow{F(t)})$ *is nonempty for all* $t \in [t_0, T]$, *then the function of the positive part of* $\vec{F}$, $P_n(\overrightarrow{F(\cdot)})$, *is quasi-affine with images in* $\mathcal{C}(\mathbb{R}^n)$.

Thus, positive parts of special linear directed maps are quasi-affine maps.

**Proposition 3.5** *A finitely generated directed map $\overrightarrow{F}$ in the form (32.11), where $f_i(t)$, $i = 1, ..., k$, are differentiable scalar functions with derivative $f_i'(t)$, is directed differentiable for all $t \in I = [t_0, T]$ with*

$$\overrightarrow{DF(t)} = \sum_{i=1}^{k} f_i'(t)\overrightarrow{F_i},$$

**Proof:** $\mathcal{D}(\mathbb{R}^n)$ is a normed vector space by Proposition 2.4, hence

$$\begin{aligned}\overrightarrow{DF(t)} &= \lim_{\substack{h\to 0\\ t+h\in I}} \frac{1}{h}\sum_{i=1}^{k}(f_i(t+h)\overrightarrow{F_i} - f_i(t)\overrightarrow{F_i}) \\ &= \sum_{i=1}^{k}(\lim_{\substack{h\to 0\\ t+h\in I}} \frac{f_i(t+h)-f_i(t)}{h})\overrightarrow{F_i} = \sum_{i=1}^{k} f_i'(t)\overrightarrow{F_i}.\end{aligned}$$

■

Proposition 3.3 states that affine and semi-affine convex-valued maps lead to linear directed maps which are special cases of finitely generated directed maps so that the next proposition can be easily derived from Proposition 3.5.

**Proposition 3.6** *An affine $F$ in the form (32.7) is directed differentiable with*

$$\overrightarrow{DF(t)} = \frac{1}{T-t_0}J_n(F_1) = J_n(\frac{1}{T-t_0}F_1) \quad \textit{for all } t \in [t_0, T].$$

*A semi-affine map $F$ in the form (32.8) is directed differentiable with*

$$\overrightarrow{DF(t)} = \frac{1}{T-t_0}(J_n(F_1) - J_n(F_0)) \qquad \textit{for all } t \in [t_0, T].$$

Eclipsing maps introduced in [8] could be discontinuous (see [13]), so that directed differentiability for every eclipsing map could not be expected.

## 4 EXAMPLES OF DIRECTED DERIVATIVES

In all following examples, set $\overrightarrow{F(t)} := J_2(F(t))$. The numerical calculation of the derivative with difference quotients is neither very reliable nor very efficient and is chosen here only for simplicity of the presentation.

**Example 4.1** *Consider $F : [0,1] \Rightarrow \mathbb{R}^2$ with*

$$F(t) = B_1(0) + t[0,2]^2 \quad (t \in [0,1]).$$

*$F$ is affine, hence $\overrightarrow{F}$ is linear and directed differentiable in $t \in [0,1]$ with*

$$\overrightarrow{DF(t)} = \Delta_h(\overrightarrow{F}, t) = J_2([0,2]^2)$$

*(see Proposition 3.6). The derivative is the embedded set $C = [0,2]^2 \in \mathcal{C}(\mathbb{R}^2)$, so that the visualization is simply $C$ itself with outer normals (see Theorem 2.8 and Figure 32.1). Figure 32.1 shows the two values of $\overrightarrow{F}$ involved in $\Delta_1(\overrightarrow{F}, 0)$.*

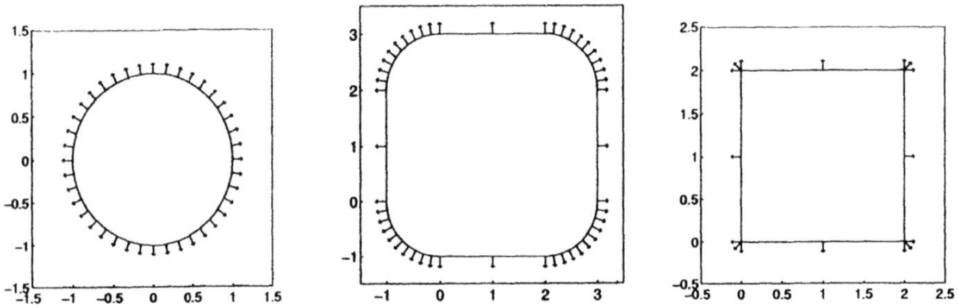


*Figure 32.1* $J_2(F(0))$, $J_2(F(1))$ and $\overrightarrow{DF(0)} = \Delta_1(\overrightarrow{F}, 0)$

**Example 4.2** *Consider $F : [0,1] \Rightarrow \mathbb{R}^2$ with*

$$F(t) = (1-t)[0,2]^2 \quad (t \in [0,1]).$$

*$F$ is semi-affine, hence $\overrightarrow{F}$ is linear and directed differentiable in $t \in [0,1]$ with*

$$\overrightarrow{DF(t)} = \Delta_h(\overrightarrow{F}, t) = -J_2([0,2]^2)$$

*(see Proposition 3.6). In this case, the derivative is the inverse of the embedded set $C = [0,2]^2 \in \mathcal{C}(\mathbb{R}^2)$ visualized in Figure 32.1, so that the visualization is simply $\ominus C$ with inner normals (see Theorem 2.8 and Figure 32.2).*

**Example 4.3** *Consider $F : [0,1] \Rightarrow \mathbb{R}^2$ with*

$$F(t) = (1-t)B_{0.5}(0) + t[-1,1]^2 \quad (t \in I = [0,1]).$$

*$F$ is also semi-affine, hence $\overrightarrow{F}$ is linear and directed differentiable with*

$$\overrightarrow{DF(t)} = \Delta_h(\overrightarrow{F}, t) = J_2([-1,1]^2) - J_2(B_{0.5}(0)) \quad (t \in I)$$

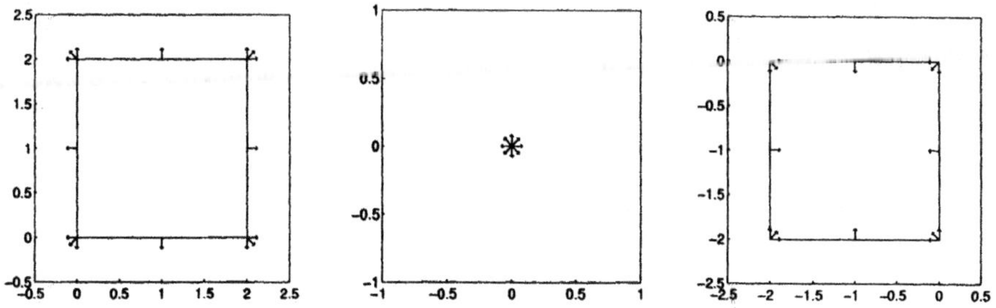

*Figure 32.2* $J_2(F(0))$, $J_2(F(1))$ and $\overrightarrow{DF(0)} = \Delta_1(\overrightarrow{F}, 0)$

*(see again Proposition 3.6). In this example, the derivative is the difference of the two embedded sets $C = [-1,1]^2$ and $D = B_{0.5}(0) \in \mathcal{C}(\mathbb{R}^2)$, so that the positive part is $C \stackrel{*}{-} D$ with outer normals and a nonempty, non-convex mixed-type part appears (see Theorem 2.8 and Figure 32.3).*

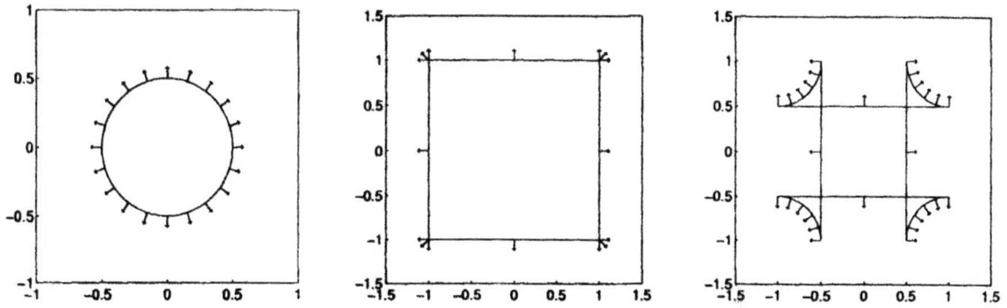

*Figure 32.3* $J_2(F(0))$, $J_2(F(1))$ and $\overrightarrow{DF(0)} = \Delta_1(\overrightarrow{F}, 0)$

**Example 4.4** *Consider the convex map $F : [-1,1] \Rightarrow \mathbb{R}^2$ (see (32.10)) with*

$$F(t) = (-t^2 + 1)B_1(0) \quad (t \in I = [-1,1]).$$

*($t \mapsto -t^2 + 1$ is concave and $B_1(0)$ is absorbing). Proposition 3.5 and direct calculations show that $\overrightarrow{F}$ is finitely generated and directed differentiable with*

$$\overrightarrow{DF(t)} = -2tJ_2(B_1(0)), \quad \Delta_h(\overrightarrow{F}, t) = -(2t+h)J_2(B_1(0)) \quad (t \in I).$$

*Here, the difference quotients $\Delta_h(\overrightarrow{F}, 0.5)$ approach the inverse of the embedded set $B_1(0)$ with inner normals (see Theorem 2.8 and Figures 32.4, 32.5).*

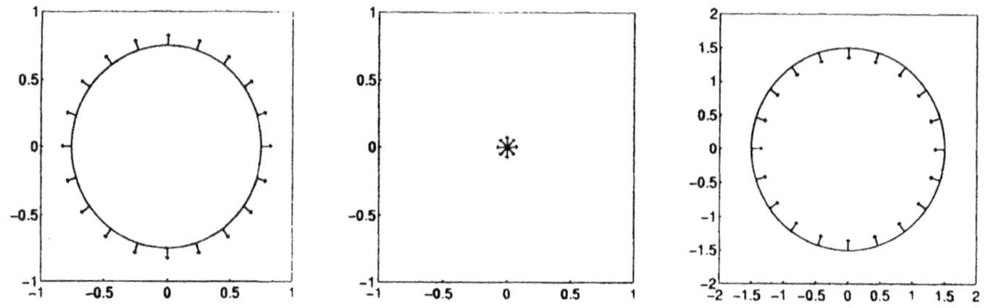

*Figure 32.4* $J_2(F(0.5))$, $J_2(F(1))$ and $\Delta_{0.5}(\overrightarrow{F}, 0.5)$

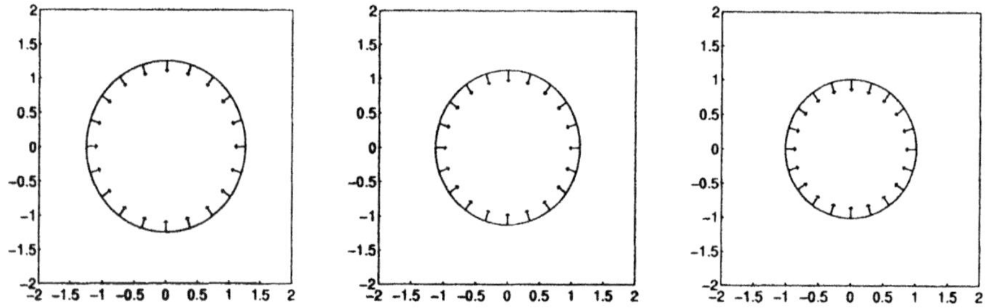

*Figure 32.5* $\Delta_{0.25}(\overrightarrow{F}, 0.5)$, $\Delta_{0.125}(\overrightarrow{F}, 0.5)$ and $\Delta_{0.01}(\overrightarrow{F}, 0.5)$

**Example 4.5** *Consider the quasi-affine map* $F : [0, 3] \Rightarrow \mathbb{R}^2$ *with*

$$F(t) = \{x \in \mathbb{R}^2 \,|\, \forall\, l \in S_1 : <l, x> \leq \delta^*(l, C) + t \cdot (\delta^*(l, D) - \delta^*(l, C))$$

*and* $C = \{0\} \times [-1, 1]$, $D = B_1(0)$ *(see (32.9)). F equals the positive part of the linear directed map* $\overrightarrow{G(t)} = J_2(C) + t \cdot (J_2(D) - J_2(C))$ *in Proposition 3.4. The approximated derivative for* $t = 2$ *is visualized in Figure 32.6.*

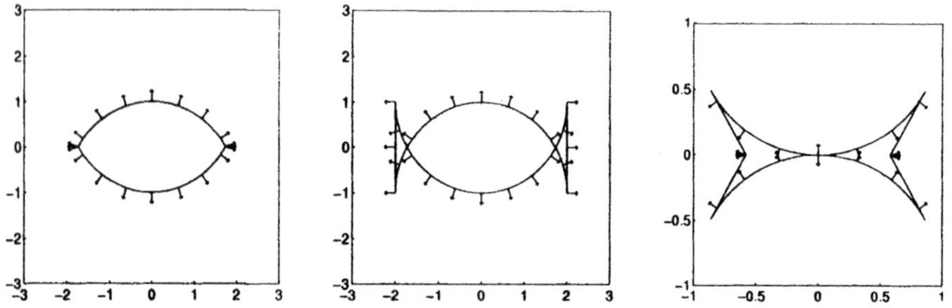

*Figure 32.6* $J_2(F(2))$, $\overrightarrow{G(2)}$ and $\overrightarrow{DF(2)} = \Delta_{0.01}(\overrightarrow{F}, 2)$

# 5 SUMMARY

Embedded affine and semi-affine maps are linear directed maps. The situation for quasi-affine maps is not so easy, but specially chosen linear directed maps have positive parts which, seen as a convex-valued maps, are quasi-affine.

The concept of linear directed maps tries to unify the different notions of simple mappings with convex, compact images in one hand and allows on the other hand to study their differences by visualizing their derivative. Directed sets offer a convenient tool for studying the possible directed differentiability of set-valued mappings with the help of the visualization of difference quotients.

## Acknowledgments

The authors would like to express their thanks to the organizers of this conference. This work was partially supported by the Hermann Minkowski Center for Geometry at Tel Aviv University and by the EU research project MINGLE, EU Contract No. HPRN-1999-00117.

## References

[1] R. Baier and E. Farkhi. Directed sets and differences of convex compact sets. In *Systems modelling and optimization, Proceedings of the* $18^{\text{th}}$ *IFIP TC7 Conference held in Detroit, Michigan, July 22-25, 1997*, pages 135–143, Boca Raton et al., 1999. Chapman and Hall/CRC.

[2] R. Baier and E. Farkhi. Differences of Convex Compact Sets in the Space of Directed Sets - Part I: The Space of Directed Sets. *Set-Valued Anal.*, to appear:1–31, 2000.

[3] R. Baier and E. Farkhi. Differences of Convex Compact Sets in the Space of Directed Sets - Part II: Visualization of Directed Sets. *Set-Valued Anal.*, to appear:1–28, 2000.

[4] H. Hadwiger. Minkowskische Addition und Subtraktion beliebiger Punktmengen und die Theoreme von Erhard Schmidt. *Math. Z.*, 53(3):210–218, 1950.

[5] P. L. Hörmander. Sur la fonction d'appui des ensembles convexes dans un espace localement convexe. *Ark. Mat.*, 3(12):181–186, 1954.

[6] E. Kaucher. Über Eigenschaften und Anwendungsmöglichkeiten der erweiterten Intervallrechnung und des hyperbolischen Fastkörpers über $\mathbb{R}$. *Comput. Suppl.*, 1:81–94, 1977.

[7] E. Kaucher. Interval Analysis in the Extended Interval Space $\mathbb{R}$. *Comput. Suppl.*, 2:33–49, 1980.

[8] C. Lemaréchal and J. Zowe. The Eclipsing Concept to Approximate a Multi-Valued Mapping. *Optimization*, 22(1):3–37, 1991.

[9] S. M. Markov. On the presentation of ranges of monotone functions using interval arithmetic. *Interval Comput.*, 4(6):19–31, 1992.

[10] L. S. Pontryagin. Linear differential games. II. *Sov. Math., Dokl.*, 8(4):910–912, 1967.

[11] R. T. Rockafellar. *Convex Analysis.* Princeton Mathematical Series 28. Princeton University Press, Princeton, New Yersey, 1970.

[12] A. M. Rubinov and I. S. Akhundov. Difference of compact sets in the sense of Demyanov and its application to non-smooth analysis. *Optimization*, 23(3):179–188, 1992.

[13] D. B. Silin. On Set-Valued Differentiation and Integration. *Set-Valued Anal.*, 5(2):107–146, 1997.

# Chapter 33

# A PERTURBED AUXILIARY PROBLEM METHOD FOR PARAMONOTONE MULTIVALUED MAPPINGS

G. Salmon, J.J. Strodiot and V.H. Nguyen
*Département de Mathématique*
*Facultés Universitaires Notre Dame de la Paix*
*Namur, Belgium*
genevieve.salmon@fundp.ac.be
Jean-Jacques.Strodiot@fundp.ac.be

**Abstract** We consider the auxiliary problem method to solve general variational inequalities with a multivalued mapping in a Hilbert space. Perturbed versions of this method have already been studied in the literature. They allow to consider for example barrier functions and interior approximations of the feasible domain. In a preceding paper, convergence is ensured assuming strong monotonicity of the operator defining the problem. In this paper, we present convergence results under weaker monotonicity assumptions. First, when the operator is paramonotone and satisfies a kind of continuity property, we obtain that at least one weak limit point of the sequence generated by the algorithm is a solution of the original problem. In a second part, we prove weak convergence under a condition satisfied for example if the operator is compact-valued and paramonotone, or is strongly monotone, or is the subdifferential of a lower semicontinuous proper convex function. Finally, we present a strong convergence result. In the particular case of non differentiable convex optimization, our scheme is a generalization of the projected subgradient procedure.

**Keywords:** General variational inequality, Multivalued mapping, Auxiliary problem principle, Variational convergence, Paramonotonicity, Projected subgradient method.



*N. Hadjisavvas and P.M. Pardalos (eds.), Advances in Convex Analysis and Global Optimization,* 515–529.

# 1 INTRODUCTION

Let $F$ be a monotone multivalued operator defined in a Hilbert space $H$, and let $\varphi : H \to I\!R \cup \{+\infty\}$ be a lower semicontinuous proper convex function whose effective domain is included in the domain of $F$. We consider the following general variational inequality problem:

$$(P)\begin{cases} \text{find } x^* \in H \text{ and } r(x^*) \in F(x^*) \text{ such that, for all } x \in H, \\ \langle r(x^*), x - x^* \rangle + \varphi(x) - \varphi(x^*) \geq 0. \end{cases}$$

Existence results for problem $(P)$ can be found, for example, in [2], ch.6, section 7, theorem 12. To be convinced of the importance to treat problems like $(P)$ with a multivalued mapping $F$ and a function $\varphi$ which is not necessarily the indicator function of a closed convex subset of $H$, we refer to the applications considered, for example, in [11], [12]. When $F$ is the subdifferential operator of a finite-valued convex continuous function $f$ defined on $H$, problem $(P)$ reduces to the nondifferentiable convex optimization problem:

$$(OP) \quad \min_{x \in H} \{f(x) + \varphi(x)\}.$$

For solving problem $(P)$, Cohen developed in [5] a general algorithmic framework based on the so-called auxiliary problem principle. Let $K : H \to I\!R$ be an auxiliary function continuously differentiable and strongly convex, and let $\{\mu_k\}_{k \in I\!N}$ be a sequence of positive numbers. The problem considered at iteration $k$ in the auxiliary problem scheme is the following:

$$(AP^k)\begin{cases} \text{choose } r(x^k) \in F(x^k) \text{ and find } x^{k+1} \in H \text{ such that, } \forall x \in H, \\ \langle r(x^k) + \mu_k^{-1}(\nabla K(x^{k+1}) - \nabla K(x^k)), x - x^{k+1} \rangle \\ +\varphi(x) - \varphi(x^{k+1}) \geq 0. \end{cases}$$

This problem can also be equivalently written under the following minimization form:

$$\begin{cases} x^{k+1} \in \operatorname{argmin}_{x \in H}\{\mu_k^{-1} K(x) + \varphi(x) + \langle r(x^k) - \mu_k^{-1} \nabla K(x^k), x - x^k \rangle\}, \\ \text{with } r(x^k) \in F(x^k). \end{cases}$$

Note that $r(x^k)$ can be given by a black box called at $x^k$.

In [9], this method is combined with a perturbation of the function $\varphi$ in the case where $F$ is a singlevalued operator. More precisely, the original function $\varphi$ is replaced at each iteration $k$ by an approximate function

$\varphi^k$ such that the sequence $\{\varphi^k\}_{k\in\mathbb{N}}$ converges in the Mosco sense to $\varphi$ (see [10]). In [16], an additional condition on the speed of convergence of the sequence $\{\varphi^k\}_{k\in\mathbb{N}}$ is introduced to allow weaker monotonicity conditions on the mapping $F$. In [15], this perturbed scheme is extended to the case where $F$ is a strongly monotone multivalued operator defined on a Hilbert space. However, in many applications, $F$ is monotone but not strongly monotone. Our purpose in this paper is to present convergence results in the perturbed setting under weaker monotonicity assumptions on $F$. Though, as observed in Remark 2.2 of [5], monotonicity of $F$ without any additional condition is not sufficient to ensure convergence. So, we need an adequate concept that lies strictly between monotonicity and strong monotonicity. The notion of paramonotonicity was introduced in [4], further studied in [8], and used, for example, in [3], [4] to ensure convergence of interior point methods with generalized distances when $\varphi$ is the indicator function of a closed convex set. A multivalued operator $F$ is said to be paramonotone on a convex subset $C$ of $H$ if it is monotone on $C$ and for all $x, y \in C, r(x) \in F(x), r(y) \in F(y)$, if $\langle r(x) - r(y), x - y\rangle = 0$, then $r(y) \in F(x)$ and $r(x) \in F(y)$. Observe that paramonotonicity is strictly weaker than strict monotonicity and thus than strong monotonicity. An important subclass of paramonotone mappings is that of subdifferentials of proper lower semicontinuous convex functions (see [8]). Paramonotonicity has been used very recently in [17] to obtain some convergence results for the scheme characterized by subproblems $(AP^k)$ when $\varphi$ is the indicator function of a closed convex set. In this paper, we extend these results to the case where $\varphi$ is any lower semicontinuous convex function and $\varphi$ is replaced in the subproblems $(AP^k)$ by an interior approximation $\varphi^k$ (i.e. $\varphi^k \geq \varphi$). The sequence $\{\varphi^k\}_{k\in\mathbb{N}}$ is assumed to converge sufficiently fast to $\varphi$ in the sense that, for each solution $x^*$ of problem $(P)$, there exists a sequence $\{w^k\}_{k\in\mathbb{N}}$ such that

$$\sum_{k=0}^{+\infty} \|w^k - x^*\| < +\infty, \quad \text{and} \quad \sum_{k=0}^{+\infty} |\varphi^k(w^k) - \varphi(x^*)| < +\infty. \tag{33.1}$$

For example, if $\varphi$ is the indicator function of a non empty closed convex set, we can choose for $\{\varphi^k\}_{k\in\mathbb{N}}$ interior approximations of the feasible set or barrier functions with barrier parameters converging to infinity fast enough. More details can be found in [16].

The paper is organized as follows. First, we obtain a convergence result for the perturbed scheme under the conditions that $F$ is paramonotone and satisfies a kind of continuity property. This result ensures that at least one weak limit point of the sequence generated by the algorithm is a solution of problem $(P)$. To go further and obtain that each weak limit point is a solution, we have to impose a condition on the mapping $F$ that will be satisfied for example if $F$ is compact-valued and paramonotone, or if $F$ is strongly monotone, or if $F$ is the subdifferential of a lower semicontinuous proper convex function. Afterwards, weak and strong convergence results are presented. In the case where $\varphi$ is the indicator function of a closed convex subset of $H$ and if we do not perturb $\varphi$, we recover the results of [17]. The application of our analysis to solve the optimization problem $(OP)$ will be straightforward since the conditions imposed on $F$ are satisfied by the subdifferential mapping of any finite valued convex continuous function $f$.

Throughout this paper, we denote by $\Gamma_0(H)$ the set of lower semicontinuous proper convex functions from $H$ into $\mathbb{R} \cup \{+\infty\}$. Any other undefined term or usage should be taken as in the books [2] and [14].

## 2 CONVERGENCE RESULTS

The perturbed auxiliary subproblem at iteration $k$ is the following:

$$(P^k) \begin{cases} \text{choose } r(x^k) \in F(x^k) \text{ and find } x^{k+1} \in H \text{ such that, } \forall x \in H, \\ \langle r(x^k) + \mu_k^{-1}(\nabla K(x^{k+1}) - \nabla K(x^k)), x - x^{k+1}\rangle \\ +\varphi^k(x) - \varphi^k(x^{k+1}) \geq 0. \end{cases}$$

In our context, the sequence $\{\mu_k\}_{k\in\mathbb{N}}$ will be chosen as, for example, in [1], i.e.

$$\begin{cases} \mu_k = \lambda_k/\eta_k, \forall k \in \mathbb{N}, \text{ with } \{\lambda_k\}_{k\in\mathbb{N}} \text{ a sequence of positive numbers,} \\ \text{and } \eta_k = \begin{cases} \max\{1, \|r(x^0)\|\}, & \text{if } k = 0; \\ \max\{\eta_{k-1}, \|r(x^k)\|\}, & \text{if } k \geq 1. \end{cases} \end{cases}$$

The introduction of the sequence $\{\eta_k\}_{k\in\mathbb{N}}$ allows us to prove that the sequence $\{x^k\}_{k\in\mathbb{N}}$ is bounded without any additional assumption on the mapping $F$. Moreover, as it is classically assumed in the multivalued case (see for example [5]), the positive sequence $\{\lambda_k\}_{k\in\mathbb{N}}$ will be such that $\sum_{k=0}^{+\infty} \lambda_k^2 < +\infty$ , and $\sum_{k=0}^{+\infty} \lambda_k = +\infty$. This rule is also considered in the literature for nonsmooth minimization problems (see, for example, [1], [6]).

In the convergence proofs, we consider the sequence of Lyapunov functions $\{\Gamma^k(x^*, \cdot)\}_{k \in \mathbb{N}}$ defined on H by

$$\begin{aligned}\Gamma^k(x^*, x) &= K(x^*) - K(x) - \langle \nabla K(x), x^* - x \rangle \\ &+ (\lambda_k/\eta_k)[\langle r(x^*), x - x^* \rangle + \varphi(x) - \varphi(x^*)],\end{aligned} \tag{33.2}$$

where $x^*$ denotes a solution of problem $(P)$ and $r(x^*)$ is the element in $F(x^*)$ such that $\langle r(x^*), x - x^* \rangle + \varphi(x) - \varphi(x^*) \geq 0$, for all $x$ in $H$. Lemma 2.1 below gives an upper bound on $\Gamma^{k+1}(x^*, x^{k+1}) - \Gamma^k(x^*, x^k)$ which will be often used in the sequel.

**Lemma 2.1** *Assume that $F$ is a monotone multivalued mapping defined on $H$, that problem $(P)$ admits at least one solution denoted by $x^*$, and that the following conditions are satisfied:*

**(i)** *$K : H \to \mathbb{R}$ is continuously differentiable and strongly convex with modulus $\beta > 0$ over dom $\varphi$;*

**(ii)** *$\nabla K$ is a Lipschitz continuous mapping with Lipschitz constant $\Lambda$ over dom $\varphi$;*

**(iii)** *$\{\lambda_k\}_{k \in \mathbb{N}}$ is a decreasing sequence of positive numbers;*

**(iv)** *$\{\varphi^k\}_{k \in \mathbb{N}}, \varphi \in \Gamma_0(H)$ are such that $\varphi \leq \varphi^k$ for all $k$, and there exists a sequence $\{w^k\}_{k \in \mathbb{N}}$ strongly converging to $x^*$ such that (33.1) holds.*

*Then, if $\{x^k\}_{k \in \mathbb{N}}$ denotes the sequence generated by solving subproblems $(P^k)$, we have for all $k \in \mathbb{N}$,*

$$\begin{aligned}\Gamma^{k+1}(x^*, x^{k+1}) - \Gamma^k(x^*, x^k) &\leq -c\|x^{k+1} - x^k\|^2 + T^k + \lambda_k^2 u \\ &+ (\lambda_k/\eta_k)[\langle r(x^k), x^* - x^k \rangle + \varphi(x^*) - \varphi(x^k)],\end{aligned} \tag{33.3}$$

*with $c, u > 0$, $T^k \geq 0$, and $\sum_{k=0}^{+\infty} T^k < +\infty$.*

**Proof.** The necessary and sufficient optimality conditions satisfied by $x^*$ and $x^{k+1}$ are respectively

$$\langle r(x^*), x - x^* \rangle + \varphi(x) - \varphi(x^*) \geq 0, \forall x \in H, \tag{33.4}$$

and

$$\begin{aligned}&\langle \eta_k^{-1} r(x^k) + \lambda_k^{-1}(\nabla K(x^{k+1}) - \nabla K(x^k)), x - x^{k+1} \rangle \\ &+ \eta_k^{-1}(\varphi^k(x) - \varphi^k(x^{k+1})) \geq 0, \forall x \in H, \text{ with } r(x^k) \in F(x^k).\end{aligned} \tag{33.5}$$

We consider the sequence of Lyapunov functions $\{\Gamma^k(x^*,\cdot)\}_{k\in I\!N}$ defined in (33.2). From the strong convexity of $K$ and inequality (33.4), we obtain that, for all $x \in \operatorname{dom}\varphi$, and all $k \in I\!N$,

$$\Gamma^k(x^*,x) \geq (\beta/2)\|x-x^*\|^2. \tag{33.6}$$

Using the definition of the Lyapunov function and the facts that $\lambda_{k+1} \leq \lambda_k$, and $\eta_{k+1} \geq \eta_k$ for all $k \in I\!N$, we can write

$$\Gamma^{k+1}(x^*,x^{k+1}) - \Gamma^k(x^*,x^k) \leq \Gamma^k(x^*,x^{k+1}) - \Gamma^k(x^*,x^k) = s_1 + s_2 + s_3, \tag{33.7}$$

$$\begin{aligned} \text{with} \quad s_1 &= K(x^k) - K(x^{k+1}) + \langle \nabla K(x^k), x^{k+1} - x^k \rangle, \\ s_2 &= \langle \nabla K(x^k) - \nabla K(x^{k+1}), x^* - x^{k+1} \rangle, \\ s_3 &= (\lambda_k/\eta_k)[\langle r(x^*), x^{k+1} - x^k \rangle + \varphi(x^{k+1}) - \varphi(x^k)]. \end{aligned}$$

For $s_1$, we derive easily from the strong convexity of $K$ that

$$s_1 \leq -(\beta/2)\|x^{k+1} - x^k\|^2. \tag{33.8}$$

Now, using the sequence $\{w^k\}_{k\in I\!N}$ given in assumption (iv), we can write $s_2$ as the sum of the two following terms:

$$\begin{aligned} s_{21} &= \langle \nabla K(x^k) - \nabla K(x^{k+1}), x^* - w^k \rangle, \\ s_{22} &= \langle \nabla K(x^k) - \nabla K(x^{k+1}), w^k - x^{k+1} \rangle. \end{aligned}$$

From the Lipschitz continuity of $\nabla K$, we deduce that

$$s_{21} \leq \Lambda\|x^{k+1}-x^k\|\,\|x^*-w^k\| \leq (\tau/2)\|x^{k+1}-x^k\|^2 + (\Lambda^2/(2\tau))\|x^*-w^k\|^2, \tag{33.9}$$

where the second inequality holds for any $\tau > 0$.
Using (33.5) with $x = w^k$, we obtain

$$\begin{aligned} s_{22} &\leq (\lambda_k/\eta_k)[\langle r(x^k), w^k - x^{k+1} \rangle + \varphi^k(w^k) - \varphi^k(x^{k+1})] \\ &= (\lambda_k/\eta_k)[\langle r(x^k), w^k - x^* \rangle \\ &\quad + \langle r(x^k), x^* - x^k \rangle + \varphi(x^*) - \varphi(x^k) \\ &\quad + \langle r(x^k), x^k - x^{k+1} \rangle \\ &\quad + \varphi^k(w^k) - \varphi(x^*) + \varphi(x^k) - \varphi^k(x^{k+1})]. \end{aligned} \tag{33.10}$$

From the definition of the sequence $\{\eta_k\}_{k\in I\!N}$ and the fact that $\lambda_k \leq \lambda_0$ for all $k$, we have

$$(\lambda_k/\eta_k)\langle r(x^k), w^k - x^* \rangle \leq \lambda_0\|w^k - x^*\|, \tag{33.11}$$

$$
\begin{aligned}
(\lambda_k/\eta_k)\langle r(x^k), x^k - x^{k+1}\rangle &\le \lambda_k\|x^{k+1} - x^k\| \\
&\le \lambda_k^2/(2\gamma) + (\gamma/2)\|x^{k+1} - x^k\|^2,
\end{aligned}
\tag{33.12}
$$

where the last inequality holds for any $\gamma > 0$. And we also have, for the first term in $s_3$, that

$$
\begin{aligned}
(\lambda_k/\eta_k)\langle r(x^*), x^{k+1} - x^k\rangle &\le \lambda_k\|r(x^*)\|\ \|x^{k+1} - x^k\| \\
&\le (\lambda_k^2/(2\mu))\|r(x^*)\|^2 + (\mu/2)\|x^{k+1} - x^k\|^2,
\end{aligned}
\tag{33.13}
$$

with $\mu$ any positive number.
Gathering the fact that $\varphi \le \varphi^k$ for all $k$ with inequalities (33.7)–(33.13) and rearranging the terms, we deduce that inequality (33.3) holds with

$$
\begin{aligned}
c &= (1/2)(\beta - \tau - \gamma - \mu), \\
T^k &= \lambda_0\|w^k - x^*\| + \lambda_0|\varphi^k(w^k) - \varphi(x^*)| + (\Lambda^2/(2\tau))\|w^k - x^*\|^2, \\
u &= (1/(2\gamma)) + (1/(2\mu))\|r(x^*)\|^2, \\
\tau,\ \ &\gamma,\ \ \mu > 0 \ \text{ such that } \tau + \gamma + \mu < \beta.
\end{aligned}
$$

Since the sequence $\{w^k\}_{k\in\mathbb{N}}$ has been chosen such that (33.1) holds, we have that $\sum_{k=0}^{+\infty} T^k < +\infty$ and the proof is complete. ∎

Theorem 2.1 below gives the conditions that ensure boundedness of the sequence $\{x^k\}_{k\in\mathbb{N}}$.

**Theorem 2.1** *Assume that all assumptions of Lemma 2.1 hold. If $\sum_{k=0}^{+\infty} \lambda_k^2 < +\infty$, then, provided that $x^0 \in \operatorname{dom}\varphi$, the sequence $\{x^k\}_{k\in\mathbb{N}}$ is bounded and $\lim_{k\to\infty} \|x^{k+1} - x^k\| = 0$.*

**Proof.** Since $r(x^k) \in F(x^k)$ for all $k$ and $F$ is monotone, we have that

$$(\lambda_k/\eta_k)[\langle r(x^k), x^* - x^k\rangle + \varphi(x^*) - \varphi(x^k)] \le 0.$$

So, we derive from (33.3) that

$$\Gamma^{k+1}(x^*, x^{k+1}) - \Gamma^k(x^*, x^k) \le -c\|x^{k+1} - x^k\|^2 + T^k + u\lambda_k^2. \tag{33.14}$$

Since the series $\sum_{k=0}^{+\infty} T^k$, $\sum_{k=0}^{+\infty} \lambda_k^2$ are convergent, it follows that the sequence $\{\Gamma^k(x^*, x^k)\}_{k\in\mathbb{N}}$ is a Cauchy sequence. This implies in turn that it is convergent in $H$. Using inequality (33.6), we conclude that the sequence $\{x^k\}_{k\in\mathbb{N}}$ is bounded. And by passing to the limit in (33.14), we also have that $\lim_{k\to\infty} \|x^{k+1} - x^k\| = 0$. ∎

To prove that at least one weak limit point of the sequence $\{x^k\}_{k\in\mathbb{N}}$ is a solution of problem $(P)$, we have to impose that $F$ is paramonotone and satisfies the following condition.

**Condition (A):**
*For any bounded sequences* $\{z^k\}_{k\in I\!N}$ *of dom* $\varphi$, *and* $\{r^k\}_{k\in I\!N}$ *with* $r^k \in F(z^k)$,

**(A.1)** *the sequence* $\{r^k\}_{k\in I\!N}$ *is bounded, and*
**(A.2)** *there exist subsequences* $\{z^{k'}\}_{k'\in K\subset I\!N}$ *and* $\{r^{k'}\}_{k'\in K\subset I\!N}$ *such that* $z^{k'} \rightharpoonup \bar{z}$, $r^{k'} \rightharpoonup \bar{r}, \bar{r} \in F(\bar{z}), \lim_{k'\to\infty}\langle r^{k'}, z^{k'} - \bar{z}\rangle + \varphi(z^{k'}) - \varphi(\bar{z}) \geq 0.$

Condition (A) is satisfied if $F$ is monotone, bounded on bounded subsets, and weakly closed (i.e. $z^k \rightharpoonup \bar{z}$, $r^k \rightharpoonup \bar{r}, r^k \in F(z^k)$ implies that $\bar{r} \in F(\bar{z})$). Observe also that, when $F$ is maximal monotone and dom $F$ is closed, $F$ is locally bounded in the interior of its domain (see [13]) and $F$ is demi–closed (i.e. $z^k \rightharpoonup \bar{z}$ (resp.$z^k \to \bar{z}$), $r^k \to \bar{r}$ (resp.$r^k \rightharpoonup \bar{r}$), $r^k \in F(z^k)$ implies that $\bar{r} \in F(\bar{z})$). In finite dimension, this implies that Condition (A) is satisfied for any maximal monotone operator with closed domain provided that dom $\varphi \subset$ int( dom $F$) . Note also that, if $\varphi$ is the indicator function of a closed convex set, then Condition (A) is just Property (A) of [17]. In order to prove our first convergence theorem, we need the following lemma which generalizes Proposition 2.3 of [8].

**Lemma 2.2** *Assume that $F$ is paramonotone on dom $\varphi$, and let $x^*$ be a solution of problem $(P)$. Then $\bar{x}$ solves problem $(P)$ if there exists $\bar{r} \in F(\bar{x})$ such that*

$$\langle \bar{r}, x^* - \bar{x}\rangle + \varphi(x^*) - \varphi(\bar{x}) \geq 0.$$

We can now state our first convergence result.

**Theorem 2.2** *Suppose that all assumptions of Lemma 2.1 are satisfied and that the sequence $\{x^k\}_{k\in I\!N}$ is bounded. If $\lim_{k\to\infty}\lambda_k = 0$, $\sum_{k=0}^{+\infty}\lambda_k = +\infty$, $F$ is paramonotone over dom $\varphi$ and satisfies Condition (A), then at least one weak limit point of the sequence $\{x^k\}_{k\in I\!N}$ is a solution of problem $(P)$.*

**Proof.** Since the sequence $\{x^k\}_{k\in I\!N}$ is bounded and Condition (A.1) holds, we derive that the sequence $\{r(x^k)\}_{k\in I\!N}$ is also bounded and there exists a constant $\bar{\eta} > 1$ such that $\|r(x^k)\| \leq \bar{\eta}$ for all $k$. Therefore, we have also that $\eta_k \leq \bar{\eta}$ for all $k$.
Now, we show that $\underline{\lim}_{k\to\infty}\langle r(x^k), x^k - x^*\rangle + \varphi(x^k) - \varphi(x^*) = 0$. Since $r(x^k) \in F(x^k)$ for all $k$ and $F$ is monotone, we have that $\underline{\lim}_{k\to\infty}\langle r(x^k), x^k - x^*\rangle + \varphi(x^k) - \varphi(x^*) \geq 0$. To prove the equality, assume for contradiction that the inferior limit is strictly positive, i.e. there exist $\delta > 0$ and

$k_0 \in I\!N$ such that

$$\langle r(x^k), x^k - x^* \rangle + \varphi(x^k) - \varphi(x^*) > \delta, \ \forall k \geq k_0.$$

Since the sequence $\{\lambda_k\}_{k \in I\!N}$ converges to zero, there also exists $k_1 \in I\!N$ such that

$$\lambda_k u < \delta/(2\bar{\eta}), \ \forall k \geq k_1.$$

So, we derive from these two last inequalities and from (33.3) that

$$\Gamma^{k+1}(x^*, x^{k+1}) - \Gamma^k(x^*, x^k) \leq T^k - \lambda_k \delta/(2\bar{\eta}), \ \forall k \geq \bar{k} = \max(k_0, k_1).$$

Summing up, this gives for all $N \geq \bar{k}$:

$$0 \leq \Gamma^{N+1}(x^*, x^{N+1}) \leq \Gamma^{\bar{k}}(x^*, x^{\bar{k}}) + \sum_{k=\bar{k}}^{N} T^k - (\delta/(2\bar{\eta})) \sum_{k=\bar{k}}^{N} \lambda_k.$$

If we take the limit on $N$, we obtain a contradiction because the series $\sum_{k=\bar{k}}^{+\infty} T^k$ is convergent and the series $\sum_{k=0}^{+\infty} \lambda_k$ is divergent. Hence, we deduce that there exists a subsequence $\{x^{k'}\}_{k' \in K \subset I\!N}$ such that

$$\lim_{k' \to \infty} \langle r^{k'}(x^{k'}), x^{k'} - x^* \rangle + \varphi(x^{k'}) - \varphi(x^*) = 0.$$

In addition, from Condition (A.2), we have that there exist subsequences

$$\{x^{k''}\}_{k'' \in K' \subset K}, \text{ and } \{r^{k''}(x^{k''})\}_{k'' \in K' \subset K} \text{ such that}$$

$$x^{k''} \rightharpoonup \bar{x}, \ r^{k''}(x^{k''}) \rightharpoonup \bar{r}, \ \bar{r} \in F(\bar{x}), \text{ and}$$

$$\lim_{k'' \to \infty} \langle r^{k''}(x^{k''}), x^{k''} - \bar{x} \rangle + \varphi(x^{k''}) - \varphi(\bar{x}) \geq 0.$$

This implies that

$$0 = \lim_{k'' \to \infty} \langle r^{k''}(x^{k''}), x^{k''} - x^* \rangle + \varphi(x^{k''}) - \varphi(x^*) \geq \langle \bar{r}, \bar{x} - x^* \rangle + \varphi(\bar{x}) - \varphi(x^*).$$

Since $F$ is paramonotone over dom $\varphi$, it follows from Lemma 2.2 that $\bar{x}$ solves problem $(P)$. ∎

This result provides informations on the limit behavior of the generated sequence under very weak restrictions on $F$. Now, to obtain that each weak limit point of $\{x^k\}_{k \in I\!N}$ is a solution of problem $(P)$, we have to require that, for some solution $x^*$ of problem $(P)$, the mapping $F$ satisfies the following condition:

**Condition ($H_{x^*}$):**
*There exist a constant $\alpha > 0$ and a function $l$ defined on $H$ such that,*

$$\langle r(y), y - x^* \rangle + \varphi(y) - \varphi(x^*) \geq \alpha(l(y) - l(x^*)), \ \forall y \in H, \ \forall r(y) \in F(y);$$
$$l(y) = l(x^*), \text{ if } y \text{ is a solution of problem } (P),$$
$$l(y) > l(x^*), \text{ otherwise.}$$

Observe that the function $l$ can be considered as a merit function for problem $(P)$ since solving $(P)$ amounts to find a global minimum point for $l$. If $\varphi$ is the indicator function of a closed convex set and $l$ is supposed to be Lipschitz continuous on this subset, then Condition $(H_{x^*})$ is just Property (C) of [17]. It is easy to see that property $(H_{x^*})$ is satisfied for any solution of problem $(P)$ when

**(1)** $F$ is a compact-valued and paramonotone operator, by taking

$$l(y) = \inf_{r(y) \in F(y)} \langle r(y), y - x^* \rangle + \varphi(y) - \varphi(x^*), \ \forall y \in \text{dom}\varphi, \text{ and } \alpha = 1;$$

**(2)** $F$ is a strongly monotone operator with modulus $\bar{\alpha}$, by choosing

$$l(y) = \|y - x^*\|^2, \ \forall y \in H, \text{ and } 0 < \alpha \leq \bar{\alpha};$$

**(3)** $F$ is the subdifferential of a finite-valued convex continuous function $f$, if

$$l(y) = f(y) + \varphi(y), \ \forall y \in H, \text{ and } \alpha = 1.$$

Moreover, we have to impose the following condition on the sequence of approximate functions $\{\varphi^k\}_{k \in I\!N}$:

**Condition (B):**

$$\forall k \in I\!N, \varphi \leq \varphi^{k+1} \leq \varphi^k, \text{ and for any bounded subset } B \text{ of dom } \varphi,$$

$$\exists c > 0 : \ \forall k \in I\!N, \forall x, y \in B \cap \text{dom } \varphi^k, \ |\varphi^k(x) - \varphi^k(y)| \leq c \ \|x - y\|.$$

The last part of this condition is satisfied for example if $\exists c_1, c_2 > 0$ : $\forall k \in I\!N, \forall x \in \text{dom } \varphi^k, \ \forall r \in \partial\varphi^k(x) : \ \|r\| \leq c_1\|x\| + c_2$ (see [6]). When $\varphi$ is the indicator function of a closed convex subset $C$ of $H$ and $\varphi^k$ is the indicator function of a closed convex subset $C^k$ of $C$ such that $C^k \subset C^{k+1} \subset C$ for all $k \in I\!N$, then Condition (B) is satisfied.

**Theorem 2.3** *Let us suppose that all assumptions of Lemma 2.1 hold. If, in addition, $\sum_{k=0}^{+\infty} \lambda_k^2 < +\infty$, $\sum_{k=0}^{+\infty} \lambda_k = +\infty$, $F$ satisfies Conditions*

*(A.1) and $(H_{x^*})$, $l$ is Lipschitz continuous on any bounded subset of dom $\varphi$ and $\{\varphi^k\}_{k\in\mathbb{N}}, \varphi$ satisfy Condition (B), then $\lim_{k\to\infty} l(x^k) = l(x^*)$. Moreover, if $l$ is weakly lower semicontinuous on dom $\varphi$, then each weak limit point of the sequence $\{x^k\}_{k\in\mathbb{N}}$ is a solution of problem $(P)$.*

**Proof.** From the definition of the sequence $\{x^k\}_{k\in\mathbb{N}}$ with $x = x^k$, we can write

$$\langle \nabla K(x^{k+1}) - \nabla K(x^k), x^{k+1} - x^k\rangle$$
$$\leq (\lambda_k/\eta_k)[\langle r(x^k), x^k - x^{k+1}\rangle + \varphi^k(x^k) - \varphi^k(x^{k+1})].$$

Using the strong convexity of $K$, the definition of $\eta_k$, Condition (B) and boundedness of $\{x^k\}_{k\in\mathbb{N}}$, we deduce that

$$\beta\|x^{k+1} - x^k\|^2 \leq \lambda_k(1+c)\|x^{k+1} - x^k\|. \tag{33.15}$$

Moreover, since $r(x^k) \in F(x^k)$ and $F$ satisfies Condition $(H_{x^*})$, we have that, for each $k$,

$$\langle r(x^k), x^k - x^*\rangle + \varphi(x^k) - \varphi(x^*) \geq \alpha(l(x^k) - l(x^*)). \tag{33.16}$$

Since the sequence $\{x^k\}_{k\in\mathbb{N}}$ is bounded, we have by the same reasoning as before that there exists a constant $\bar{\eta} > 1$ such that $\eta_k \leq \bar{\eta}$ for all $k$. Recall also that $\eta_k \geq 1$ for all $k$. With these observations and inequality (33.16), we derive from (33.3) that

$$(\alpha/\bar{\eta})\lambda_k[l(x^k) - l(x^*)] \leq \Gamma^k(x^*, x^k) - \Gamma^{k+1}(x^*, x^{k+1}) + T^k + \lambda_k^2 u.$$

Summing up, this gives for all $N \geq 0$:

$$(\alpha/\bar{\eta})\sum_{k=0}^{N} \lambda_k[l(x^k) - l(x^*)] \leq \Gamma^0(x^*, x^0) + \sum_{k=0}^{N}(T^k + \lambda_k^2 u).$$

If we take the limit on $N$ and if we use the fact that the series $\sum_{k=0}^{+\infty}(T^k + \lambda_k^2 u)$ is convergent, we obtain that

$$\sum_{k=0}^{+\infty} \lambda_k[l(x^k) - l(x^*)] < +\infty. \tag{33.17}$$

Since the function $l$ is Lipschitz continuous on the bounded sequence $\{x^k\}_{k\in\mathbb{N}}$, since $\sum_{k=0}^{+\infty}\lambda_k = +\infty$ and we have (33.15) and (33.17), we can apply Lemma 4 of [6] to conclude that $\lim_{k\to\infty} l(x^k) = l(x^*)$.
It is then easy to prove that any weak limit point of $\{x^k\}_{k\in\mathbb{N}}$ is a solution of problem $(P)$. Indeed, let $\bar{x}$ be a weak limit point of $\{x^k\}_{k\in\mathbb{N}}$ such that

the subsequence $\{x^{k'}\}_{k'\in K\subset I\!N}$ weakly converges to $\bar{x}$. Since $l$ is weakly lower semicontinuous and the sequence $\{l(x^k)\}_{k\in I\!N}$ converges to $l(x^*)$, we have that

$$l(\bar{x}) \leq \underline{\lim}_{k'\to\infty} l(x^{k'}) = \lim_{k\to\infty} l(x^k) = l(x^*).$$

It follows that $l(\bar{x}) = l(x^*)$ which means that $\bar{x}$ solves problem $(P)$. ∎

Note that if the function $l$ is convex and bounded above on an open subset, it is locally Lipschitz on the interior of its domain (see [7], ch.I, corollary 2.4). In finite dimension, this implies that $l$ is Lipschitz on any compact subset in the interior of its domain. Observe also that any weakly lower semicontinuous function is strongly lower semicontinuous. And the converse is true in the convex case ( see [2]).

The additional conditions that ensure the weak convergence of the whole sequence are displayed in the following theorem.

**Theorem 2.4** *Let us suppose that all assumptions of Theorem 2.3 are satisfied. If, in addition, condition* (iv) *of Lemma 2.1 is satisfied for each solution of problem* $(P)$, *and* $\nabla K$ *is weakly continuous on dom* $\varphi$ *(i.e. if* $z^k \rightharpoonup z$, $z^k \in$ *dom* $\varphi$, *then* $\nabla K(z^k) \rightharpoonup \nabla K(z)$), *then, the sequence* $\{x^k\}_{k\in I\!N}$ *weakly converges to some solution of problem* $(P)$.

**Proof.** For contradiction, let us suppose that the sequence $\{x^k\}_{k\in I\!N}$ has two different weak limit points $x^1$ and $x^2$ such that $\{x^{m(k)}\}_{k\in I\!N}$ is a subsequence of $\{x^k\}_{k\in I\!N}$ weakly converging to $x^1$ and $\{x^{n(k)}\}_{k\in I\!N}$ is a subsequence weakly converging to $x^2$. We have just shown in Theorem 2.3 that $x^1$ and $x^2$ are solutions of problem $(P)$ so that, by the proof of Theorem 2.1, there exist $l_1, l_2 \in I\!R$ such that $\lim_{k\to\infty} \Gamma^k(x^1, x^k) = l_1$ and $\lim_{k\to\infty} \Gamma^k(x^2, x^k) = l_2$. By definition of the Lyapunov function, we have

$$\begin{aligned}&\Gamma^{n(k)}(x^1, x^{n(k)}) - \Gamma^{n(k)}(x^2, x^{n(k)})\\&= K(x^1) - K(x^2) - \langle \nabla K(x^{n(k)}), x^1 - x^2\rangle\\&+(\lambda_{n(k)}/\eta_{n(k)})[\langle r(x^1), x^{n(k)} - x^1\rangle - \langle r(x^2), x^{n(k)} - x^2\rangle + \varphi(x^2) - \varphi(x^1)].\end{aligned}$$

Since $\nabla K$ is weakly continuous, if we take the limit on $k$ in the last inequality, we obtain

$$l_1 - l_2 = K(x^1) - K(x^2) - \langle \nabla K(x^2), x^1 - x^2\rangle. \tag{33.18}$$

Since the role of $x^1$ and $x^2$ is symmetric, we also have that

$$l_1 - l_2 = K(x^1) - K(x^2) - \langle \nabla K(x^1), x^1 - x^2\rangle. \tag{33.19}$$

Comparing (33.18) and (33.19), we obtain $\langle \nabla K(x^1) - \nabla K(x^2), x^1 - x^2 \rangle = 0$. Since $\nabla K$ is strongly monotone, this inequality implies that $x^1 = x^2$, and the proof is complete. ∎

For example, if $K(x) = (1/2)x^T x$, for all $x \in H$, then $\nabla K$ is weakly continuous. Note that assumption (ii) of Lemma 2.1 ensures the continuity of $\nabla K$ in the strong topology. So, in a finite dimensional space, the convergence of the whole sequence toward a solution is established without any further assumption.

The strong convergence is presented in the following theorem.

**Theorem 2.5** *Let us suppose that all assumptions of Theorem 2.3 hold. If, in addition, $l$ is strongly convex on dom $\varphi$, then the sequence $\{x^k\}_{k \in \mathbb{N}}$ strongly converges to the solution of problem $(P)$.*

**Proof.** If $l$ is strongly convex with constant $s$ on dom $\varphi$, then $x^*$ is the unique solution of problem $(P)$, and for any $e^* \in \partial l(x^*)$,

$$l(x^k) - l(x^*) - \langle e^*, x^k - x^* \rangle \geq (s/2)\|x^k - x^*\|^2, \tag{33.20}$$

From Theorem 2.3, $\lim_{k \to \infty} l(x^k) = l(x^*)$, and the sequence $\{x^k\}_{k \in \mathbb{N}}$ weakly converges to $x^*$. Then, passing to the limit in (33.20) gives that $\lim_{k \to \infty} \|x^k - x^*\|^2 = 0$. This completes the proof. ∎

We have seen that, when $F$ is strongly monotone, we can take $l(y) = \|y - x^*\|^2$, for all $y \in H$, with $x^*$ denoting the unique solution of the problem. In that case, the function $l$ is obviously Lipschitz continuous on any bounded subset of $H$ and strongly convex.

Our analysis can be applied directly to the optimization problem $(OP)$ by taking for $F$ the subdifferential of any finite-valued convex and continuous function $f$. In that case, $F$ is paramonotone and $(H_{x^*})$ is satisfied. Moreover, Condition $(A)$ holds in $\mathbb{R}^n$. If, for all $x \in H$, $K(x) = (1/2)x^T x$, and for all $k \in \mathbb{N}$, $\varphi^k = \varphi = \Psi_C$, where $\Psi_C$ denotes the indicator function of a closed convex subset $C$ of $H$, then the method defined by problems $(P^k)$, $k \in \mathbb{N}$, reduces to the projected subgradient process and our results generalize well-known ones (see for example [1]).

## References

[1] Alber, Y., Iusem, A., and Solodov, M. (1998). On the Projected Subgradient Method for Nonsmooth Convex Optimization in a Hilbert Space. *Mathematical Programming*, 81:23–35.

[2] Aubin, J. and Ekeland, I., editors (1984). *Applied Nonlinear Analysis.* Wiley, New York, New York.

[3] Burachik, R. and Iusem, A. (1998). A Generalized Proximal Point Algorithm for the Variational Inequality Problem in a Hilbert Space. *SIAM Journal on Optimization*, 8:197–216.

[4] Censor, Y., Iusem, A., and Zenios, S. (1998). An Interior Point Method with Bregman Functions for the Variational Inequality Problem with Paramonotone Operators. *Mathematical Programming*, 81:373–400.

[5] Cohen, G. (1988). Auxiliary Problem Principle Extended to Variational Inequalities. *Journal of Optimization Theory and Applications*, 59:325–333.

[6] Cohen, G. and Zhu, D. (1984). Decomposition Coordination Methods in Large Scale Optimization Problems: The Nondifferentiable case and the Use of Augmented Lagrangians. In Cruz, J., editor, *Advances in Large Scale Systems Theory and Applications 1*, pages 203–266. JAI Press, Greenwich, Connecticut, USA.

[7] Ekeland, I. and Temam, R., editors (1976). *Convex Analysis and Variational Inequalities.* North–Holland, Amsterdam.

[8] Iusem, A. (1998). On some Properties of Paramonotone Operators. *Journal of Convex Analysis*, 5:269–278.

[9] Makler-Scheimberg, S., Nguyen, V., and Strodiot, J.-J. (1996). Family of Perturbation Methods for Variational Inequalities. *Journal of Optimization Theory and Applications*, 89:423–452.

[10] Mosco, U. (1969). Convergence of Convex Sets and of Solutions of Variational Inequalities. *Advances in Mathematics*, 3:510–585.

[11] Moudafi, A. and Noor, M. (2000). New Convergence Results of Iterative Methods for Set-Valued Mixed Variational Inequalities. *Mathematical Inequalities and Applications*, 3:295–303.

[12] Panagiotopoulos, P. and Stavroulakis, G. (1994). New Types of Variational Principles Based on the Notion of Quasidifferentiability. *Acta Mechanic*, 94:171–194.

[13] Rockafellar, R. (1969). Local Boundedness of Nonlinear, Monotone Operators. *The Michigan Mathematical Journal*, 16:397–407.

[14] Rockafellar, R., editor (1970). *Convex Analysis.* Princeton University Press, Princeton, New Jersey.

[15] Salmon, G., Nguyen, V., and Strodiot, J.-J. (2000a). A Perturbed and Inexact Version of the Auxiliary Problem Method for solving General Variational Inequalities with a Multivalued Operator. In

Nguyen, V., Strodiot, J.-J., and Tossings, P., editors, *Optimization, Lecture Notes in Economics and Mathematical Systems, 481*, pages 396–418. Springer–Verlag, Berlin.

[16] Salmon, G., Nguyen, V., and Strodiot, J.-J. (2000b). Coupling the Auxiliary Problem Principle and the Epiconvergence Theory to Solve General Variational Inequalities. *Journal of Optimization Theory and Applications*, 104:629–657.

[17] Zhu, D. (1997). The Decomposition Method for the Variational Inequalities with Multivalued Mapping. Private communication.

Chapter 34

# A NOTE ON RANDOM VARIATIONAL INEQUALITIES AND SIMPLE RANDOM UNILATERAL BOUNDARY VALUE PROBLEMS: WELL-POSEDNESS AND STABILITY RESULTS

Joachim Gwinner
*Institut für Mathematik*
*Fakultät für Luft- und Raumfahrttechnik*
*Universität der Bundeswehr München*
gwin@rzls3.LRT.UniBW-Muenchen.de

**Abstract** This note treats a class of random variational inequalities. In addition to existence and uniqueness results under coercivity assumptions, a stability result is presented for perturbations in the given real-valued random variables and also for perturbations in the convex closed subset with respect to Mosco convergence. This stability result is applied to random elliptic boundary value problems with unilateral Signorini boundary conditions, where randomness enters in the coefficient of the elliptic operator and in the right hand side of the partial differential equation.

**Keywords:** Random variational inequality, Coercivity, Stability, Mosco convergence, Random elliptic boundary value problem, Unilateral Signorini boundary condition

## 1 INTRODUCTION

In this note we present a symbiosis of results and methods from Calculus of Variations that now evolved to Variational Analysis, from Measure Theory, and Theory of Partial Differential Equations - fields of analysis, where C. Caratheodory contributed famous achievements.

More specifically, we deal with a class of random variational inequalities. Here the variational inequality itself is given by three mathematical



*N. Hadjisavvas and P.M. Pardalos (eds.), Advances in Convex Analysis and Global Optimization,* 531–543.

objects: a convex closed subset in a Hilbert space, a nonnegative bilinear form, and a linear form, where randomness enters in the latter two objects. We present existence and uniqueness results under coercivity assumptions. In addition, we investigate a more special class of random variational inequalities, where randomness appears separately in the form of two real-valued random variables. For that latter class we give a stability result for perturbations in the two real-valued random variables and also for perturbations in the convex closed subset with respect to Mosco convergence. This stability result is applied to random elliptic boundary value problems with unilateral Signorini boundary conditions, where randomness enters in the coefficient of the elliptic operator and in the right hand side of the partial differential equation.

For motivation to study random unilateral boundary value problems we refer to [6, 9].

## 2 RANDOM VARIATIONAL INEQUALITIES - EXISTENCE

In this section we investigate a class of random variational inequalities in separable Hilbert spaces. Here we show the existence of a unique measurable solution which, moreover, lies in an appropriate Bochner - Lebesgue space under reasonable assumptions on the data. We concentrate on variational inequalities that are defined by bilinear forms; the method of proof, however, extends to more general random variational inequalities using monotonicity methods (see e.g. [3, 10] for the deterministic case).

In the following we use some concepts of measure theory, in particular measurability of set-valued maps (correspondences) what can be found with [2, chapter 8].

Let $(\Omega, \mathcal{A}, \mu)$ be a complete $\sigma$-finite measure space and $(\mathcal{X}, (\cdot, \cdot), \|\cdot\|)$ be a separable Hilbert space. Let $\mathcal{K}$ be a closed convex nonvoid subset of $\mathcal{X}$. We are given Carathéodory functions $\lambda : \Omega \times \mathcal{X} \to \mathbb{R}$ and $\beta : \Omega \times (\mathcal{X} \times \mathcal{X}) \to \mathbb{R}$, that is, for every $y, z \in \mathcal{X}, \lambda(\cdot, y)$ and $\beta(\cdot, y, z)$ are measurable and for every $\omega \in \Omega, \lambda(\omega, \cdot)$ and $\beta(\omega, \cdot, \cdot)$ are continuous. Moreover, for every $\omega \in \Omega, \lambda(\omega, \cdot)$ and $\beta(\omega, \cdot, \cdot)$ is assumed to be a linear form, respectively a bilinear form. Thus shortly $\lambda(\omega) := \lambda(\omega, \cdot) \in \mathcal{L}(\mathcal{X}, \mathbb{R}) = \mathcal{X}^*$. Via

$$\beta(\omega, y, z) = \langle \beta(\omega) y, z \rangle_{\mathcal{X}^* \times \mathcal{X}} \quad (\forall y, z \in \mathcal{X})$$

we find that $\beta(\omega) := \beta(\omega, \cdot, \cdot) \in \mathcal{L}(\mathcal{X} \times \mathcal{X}, \mathbb{R}) \cong \mathcal{L}(\mathcal{X}, \mathcal{X}^*)$. Furthermore we assume that, for every $\omega \in \Omega$, $\beta(\omega, \cdot, \cdot)$ is nonnegative, that is, that for every $y \in \mathcal{X}$, $\beta(\omega, y, y) \geq 0$ holds.

In this setting we study the following random variational inequality: Find for every $\omega \in \Omega$, an element $\hat{y} \in \mathcal{K}$ depending on $\omega \in \Omega$ such that

$$\beta(\omega, \hat{y}, z - \hat{y}) \geq \lambda(\omega, z - \hat{y}) \quad \text{for all } z \in \mathcal{K}. \tag{34.1}$$

Our first result concerns the measurability of the solution (set-valued) mapping $\Sigma : \Omega \rightsquigarrow \mathcal{X}$ given by

$$\Omega \ni \omega \rightsquigarrow \Sigma(\omega) = \{\hat{y} \in \mathcal{K} : \hat{y} \text{ solves } (34.1)\}$$

with respect to the $\sigma$-algebra $\mathcal{B}(\mathcal{X})$ of the Borel subsets of $\mathcal{X}$. Here (and later on) we use Minty's lemma (see e.g. [3, Theorem 7.14]) which states in our context that $\hat{y} \in \Sigma(\omega)$, if and only if

$$\hat{y} \in \mathcal{K},\ \beta(\omega, z, \hat{y} - z) \leq \lambda(\omega, \hat{y} - z) \quad \text{for all } z \in \mathcal{K}.$$

PROPOSITION 2.1 *Suppose that for every $\omega \in \Omega$, (34.1) possesses a solution. Then the set-valued map $\Sigma$ is measurable.*

PROOF. With $\mathcal{X}$ separable, the metric subspace $\mathcal{K}$ is separable, too. Let $\{z_\nu\}_{\nu \in \mathbb{N}}$ be dense in $\mathcal{K}$. Then by Minty's lemma and by continuity, $\hat{y} \in \Sigma(\omega)$, if and only if

$$\hat{y} \in \mathcal{K},\ \beta(\omega, z_\nu, \hat{y} - z_\nu) \leq \lambda(\omega, \hat{y} - z_\nu) \quad \text{for all } \nu \in \mathbb{N}.$$

Therefore $\Sigma = \bigcap_{\nu \in \mathbb{N}} \Sigma_\nu$, where for any $\nu \in \mathbb{N}$, $\Sigma_\nu : \Omega \rightsquigarrow \mathcal{X}$ is given by

$$\omega \rightsquigarrow \Sigma_\nu(\omega) := \{\hat{y} \in \mathcal{K} : \beta(\omega, z_\nu, \hat{y}) - \lambda(\omega, \hat{y}) \leq \beta(\omega, z_\nu, z_\nu) - \lambda(\omega, z_\nu)\}.$$

Then $\Sigma_\nu(\omega)$ is closed by continuity, and by $\mathcal{A} \otimes \mathcal{B}(\mathcal{X}) - \mathcal{B}(\mathbb{R})$ measurability of Carathéodory functions (see e.g. [2, Lemma 8.2.6] ) the graph of the set valued map $\Sigma_\nu$ belongs to $\mathcal{A} \otimes \mathcal{B}(\mathcal{X})$. Since

$$\text{Graph } \Sigma = \bigcap_{\nu \in \mathbb{N}} \text{Graph } \Sigma_\nu \in \mathcal{A} \otimes \mathcal{B}(\mathcal{X}),$$

by the Castaing characterization theorem (see [2, Theorem 8.1.4]) the claimed measurability of $\Sigma$ follows. □

REMARKS. - If for any $\omega \in \Omega, \beta(\omega, \cdot, \cdot)$ is a symmetric bilinear form, then it is well-known that $\hat{y} \in \mathcal{K}$ solves (34.1) if and only if, $\hat{y} \in \mathcal{K}$ minimizes the convex quadratic function

$$\frac{1}{2}\beta(\omega, y, y) - \lambda(\omega, y), \quad y \in \mathcal{K}.$$

In this case, the result above follows directly from the measurability of the marginal map in stochastic optimization (see e.g. [2, Theorem 8.2.1]). - With $\{x_\nu\}_{\nu\in\mathbb{N}}$ a given dense sequence in $\mathcal{X}$, the sequence $\{z_\nu\}_{\nu\in\mathbb{N}} := \{\mathrm{proj}_{\mathcal{K}} x_\nu\}_{\nu\in\mathbb{N}}$ is easily constructed as a dense subset of $\mathcal{K}$, since the projection operator $\mathrm{proj}_{\mathcal{K}} : \mathcal{X} \to \mathcal{K}$ is nonexpansive.

The measurability result above can be combined with any existence result for deterministic variational inequalities (see e.g. [3, 10]). In this paper we concentrate on the coercive case, where we do not have only existence, but also uniqueness. To this end we extend the notion of coercivity in the deterministic theory.

DEFINITION. - We call $\beta$ *coercive*, if for any $\omega \in \Omega$ there is some constant $c_0 > 0$ such that

$$c_0 \|y\|^2 \le \beta(\omega, y, y) \quad (\forall y \in \mathcal{X}). \tag{34.2}$$

Further we call $\beta$ *uniformly coercive*, if this constant $c_0$ is independent of $\omega \in \Omega$, i.e. if there holds

$$c_0 \|y\|^2 \le \beta(\omega, y, y) \quad (\forall \omega \in \Omega, y \in \mathcal{X}). \tag{34.3}$$

Under the assumption of coercivity, the Lions-Stampacchia theorem (see e.g. [3, Theorem 3.1],[10, Theorem 2.1]) guarantees the unique solvability of (34.1) for every $\omega \in \Omega$. Moreover uniform coercivity implies the following a priori estimate for the solution $\hat{u}(\omega) := \hat{y}\,, \omega \in \Omega$. We have

$$c_0 \|\hat{u}(\omega)\|^2 \le \|\beta(\omega)\|_{\mathcal{L}(\mathcal{X},\mathcal{X}^*)} \|\hat{u}(\omega)\| \; \|z_0\| + \|\lambda(\omega)\|_{\mathcal{X}^*} \left(\|\hat{u}(\omega)\| + \|z_0\|\right)$$

with some arbitrary fixed $z_0 \in \mathcal{K}$, hence

$$\|\hat{u}(\omega)\| \le \tilde{c}\left(1 + \|\lambda(\omega)\|_{\mathcal{X}^*} + \|\beta(\omega)\|_{\mathcal{L}(\mathcal{X},\mathcal{X}^*)}\right) \tag{34.4}$$

where $\tilde{c} > 0$ only depends on $c_0$ and $\|z_0\|$. Thus the solution $\hat{u}$ shows the same $\omega$ – regularity as the data $\beta$ and $\lambda$.

Note that in the separable space $\mathcal{X}$, the measurability of $\hat{u}$ with respect to $\mathcal{B}(\mathcal{X})$ implies the measurability of the real valued function $\|\hat{u}(\cdot)\|$ (see e.g. [5, Lemma 1.5]). However, in contrary to $\mathcal{X}^* \cong \mathcal{X}$ by the Riesz isomorphism, the space $\mathcal{L}(\mathcal{X} \times \mathcal{X}, \mathbb{R}) \cong \mathcal{L}(\mathcal{X}, \mathcal{X})$ does not need to be separable (see the counterexample in [5, Section 1.2]).

Therefore in the following we introduce assumptions on the functions $\omega \mapsto \|\beta(\omega)\|, \omega \mapsto \|\lambda(\omega)\|$ involved in (34.4), instead on the mappings $\omega \mapsto \beta(\omega), \omega \mapsto \lambda(\omega)$. Then we can exploit (34.4) and can conclude that the solution $\hat{u}$ belongs to the Bochner-Lebesgue space $L^q(\Omega, \mu, \mathcal{X})$, the Banach space of (classes of) measurable maps $v : \Omega \to \mathcal{X}$ such that $\int_\Omega \|v(\omega)\|^q \, d\mu(\omega) < \infty$ (for $1 \leq q < \infty$), respectively in the Banach space of (classes of) measurable, $\mu$-essentially bounded maps $v : \Omega \to \mathcal{X}$ (for $q = \infty$). To sum up, we arrive at the following result.

PROPOSITION 2.2 *Let $(\Omega, \mathcal{A}, \mu)$ be a complete $\sigma$-finite measure space, $\mathcal{X}$ a separable Hilbert space and $\beta$ a coercive bilinear form. Then the random variational inequality (34.1) admits a unique solution $\hat{u} : \omega \in \Omega \mapsto \hat{u}(\omega) \in \mathcal{K}$.*
*Suppose in addition, that $\mu$ is a finite measure and $\beta$ is uniformly coercive. Moreover suppose, that the real valued functions $\omega \mapsto \|\beta(\omega)\|_{\mathcal{L}(\mathcal{X}, \mathcal{X}^*)}$ and $\omega \mapsto \|\lambda(\omega, )\|_{\mathcal{X}^*}$ belong to $L^q(\Omega, \mu)$ for some $1 \leq q \leq \infty$. Then we have $\hat{u} \in L^q(\Omega, \mu, \mathcal{X})$.*

In the following sections we concentrate on the mean square case $q = 2$ and not only give existence results, but also well-posedness and stability results.

## 3 ANOTHER FORMULATION AND A WELL-POSEDNESS RESULT

Let us concentrate on the mean square case $q = 2$ and start from a probability space $(\Omega, \mathcal{A}, P)$. Then we can consider the above random variational problem in the Hilbert space $X =: L^2(\Omega, P, \mathcal{X})$ of all $\mathcal{X}$ - valued $P$ - measurable random variables $V$ such that

$$E^P \|V\|^2 = \int_\Omega \|V(\omega)\|^2 \, dP(\omega) < \infty .$$

Clearly the set

$$K =: L^2(\Omega, P, \mathcal{K}) = \{V \in L^2(\Omega, P, \mathcal{X}) : V \in \mathcal{K} \text{ } P\text{-almost everywhere}\} .$$

is convex. We have also

LEMMA 3.1 *The set $K$ is closed in $L^2(\Omega, P, \mathcal{X})$.*

PROOF. Without any restriction of generality we can assume that $0 \in \mathcal{K}$, since addition with the constant random variable $\omega \in \Omega \mapsto z_0$ with some fixed $z_0 \in \mathcal{K}$ is a topological isomorphism in $L^2(\Omega, P, \mathcal{X})$. Thus the bipolar theorem applies, and we have

$$\mathcal{K} = \{z \in \mathcal{X} : (z, \zeta) \leq 1, \ \forall \, \zeta \in \mathcal{K}^0\},$$

where

$$\mathcal{K}^0 = \{\zeta \in \mathcal{X}^* : (z, \zeta) \le 1, \ \forall z \in \mathcal{K}\}.$$

Since $\mathcal{K}^0$ as a metric subspace of the separable Hilbert space $\mathcal{X}^* \cong \mathcal{X}$ is separable, too, we find a sequence $\{\zeta_\nu\}_{\nu \in \mathbb{N}} \subset \mathcal{K}^0$ such that

$$\mathcal{K} = \{z \in \mathcal{X} : (z, \zeta_\nu) \le 1, \ \forall \, \nu \in \mathbb{N}\}. \tag{34.5}$$

Now let $\{V_n\}_{n \in \mathbb{N}} \subset K$ such that $V_n \to V \, (n \to \infty)$ strongly in $L^2(\Omega, P, \mathcal{X})$. This implies $\left(V_n(\cdot), \zeta_\nu\right) \le 1$ a.s. and $\zeta_\nu \circ V_n \to \zeta_\nu \circ V \, (n \to \infty)$ strongly in $L^2(\Omega, P)$ for any $\nu \in \mathbb{N}$. Therefore by well-known results in measure theory (see e.g. [4, Propositions 3.1.4, 3.1.2]), for any $\nu \in \mathbb{N}$, there exists a subsequence converging a.s. on $\Omega$ and hence $\left(V \, (\cdot), \zeta_\nu\right) \le 1$ a.s.. By the representation (34.5) we conclude that $V(\cdot) \in \mathcal{K}$ a.s. and the closedness of $K$ follows. □

Moreover, the given Carathéodory functions $\lambda : \Omega \times \mathcal{X} \to \mathbf{R}$ and $\beta : \Omega \times (\mathcal{X} \times \mathcal{X}) \to \mathbf{R}$, give rise under appropriate growthness assumptions to Nemitskii operators $V \mapsto \lambda(\cdot, V(\cdot)) \in L^1(\Omega, P, \mathbf{R}), (V, W) \mapsto \beta(\cdot, V(\cdot), W(\cdot)) \in L^1(\Omega, P, \mathbf{R})$ such that

$$\begin{aligned} L : V &\mapsto E^P\{\lambda(\cdot, V(\cdot))\} \\ B : (V, W) &\mapsto E^P\{\beta(\cdot, V(\cdot), W(\cdot))\} \end{aligned}$$

is a continuous linear form in $X^*$, respectively a coercive bilinear form in $\mathcal{L}(X \times X, \mathbf{R}) \cong \mathcal{L}(X, X)$. Therefore immediately by the Lions-Stampacchia theorem, the following variational inequality in integrated form: Find $U \in L^2(\Omega, P, \mathcal{K})$ such that for all $V \in L^2(\Omega, P, \mathcal{K})$

$$E^P\{\beta(U, V - U)\} \ge E^P\{\lambda(V - U)\} \tag{34.6}$$

has a unique solution $\hat{U}$.

By uniqueness, both formulations (34.1) and (34.6) are equivalent.

In this setting that abstracts from the special structure of $X$ and $K$, well-posedness results are easily obtained. Let us here only discuss variation with respect to the set $\mathcal{K}$. Thus in addition, consider a given sequence $\{\mathcal{K}^{(\nu)}\}_{\nu \in \mathbb{N}}$ of closed convex nonempty subsets of $\mathcal{X}$. Thus we are led to the perturbed variational inequality: Find $U^{(\nu)} \in L^2(\Omega, P, \mathcal{K}^{(\nu)})$ such that for all $V \in L^2(\Omega, P, \mathcal{K}^{(\nu)})$

$$E^P\{\beta(U^{(\nu)}, V - U^{(\nu)})\} \ge E^P\{\lambda(V - U^{(\nu)})\}. \tag{34.7}$$

Using the notions of the limes superior (limsup) and the limes inferior (liminf) of a sequence of sets Mosco [12, Def.1.1, section 1.3] defined the sequence $\{K^{(\nu)}\}$ to converge to $K$, if and only if with respect to the sequential weak topology, $w - \limsup_{\nu\to\infty} K^{(\nu)} \subset K$ and with respect to the strong topology, $K \subset s - \liminf_{\nu\to\infty} K^{(\nu)}$. This convergence is now called *Mosco-convergence*, see e.g. [1, 3.3.2].

Then from [12], [7] (in the more general semicoercive case) one obtains the following stability result.

COROLLARY 3.2 *Let $B, L, K$, and $K^{(\nu)} = L^2(\Omega, P, \mathcal{K}^{(\nu)})$ be given as above. Let $B$ be coercive. Suppose that $\{K^{(\nu)}\}_{\nu\in\mathbb{N}}$ Mosco-converges to $K$. Then $\lim_{\nu\to\infty}\|U^{(\nu)} - \hat{U}\|_X = 0$ holds.*

Similar stabilty results can easily be given with respect to perturbations of the continuous linear form $L$ in $X^*$, respectively of the coercive bilinear form $B$ in $\mathcal{L}(X \times X, \mathbb{R})$. In the next section we consider a more specialized class of random variational inequalities, for which we can study perturbations separately of the deterministic data and of the random data directly.

# 4 SIMPLE RANDOM VARIATIONAL INEQUALITIES

## 4.1 THE PROBLEM - BASIC FORMULATIONS

From now on we study a more specialized class of random variational inequalities. Here the data of the problem decompose in deterministic data (bilinear forms and a linear form, convex subset) in the separable Hilbert space $\mathcal{X}$ and given real-valued random variables on a probability space $(\Omega, \mathcal{A}, P)$. Then in its first formulation the problem considered now reads: Find a $\mathcal{X}$-valued random variable $\hat{U}$ such that $P$-almost surely $\hat{U} \in \mathcal{K}$ and for $P$-almost every elementary event $\omega \in \Omega$, there holds for all $z \in \mathcal{K}$

$$S(\omega)\, a\Big(\hat{U}(\omega), z - \hat{U}(\omega)\Big) \quad + \quad b\Big(\hat{U}(\omega), z - \hat{U}(\omega)\Big) \geq R(\omega)\, l(z - \hat{U}(\omega))\,. \tag{34.8}$$

Here $a$ and $b$ are continuous nonnegative bilinear forms on $\mathcal{X} \times \mathcal{X}$, $l$ is a continuous linear form on $\mathcal{X}$; $R$ and $S$ are real-valued random variables. As before, $\mathcal{K}$ stands for a closed convex nonvoid subset of $\mathcal{X}$. In accordance with the standard case of uniformly strongly elliptic operators, we assume that $P$-almost surely, $S$ is contained in a compact interval that is included in $(0, +\infty)$. Hence $S$ belongs to $L^\infty(\Omega)$. Further, we assume that the bilinear form $\beta := S\,a + b$ is uniformly coercive on $\mathcal{X} \times \mathcal{X}$. This uniform coercivity is here equivalent to the coercivity of

$\underline{S}a + b$, where $\underline{S}$ is a positive constant and $S \geq \underline{S}$ holds $P$-almost surely. For the right hand side, we only assume that $R$ is in $L^2(\Omega)$. Thus for any $V, W \in L^2(\Omega, P, \mathcal{X})$ we have the estimates

$$\left|S(\omega)\, a\Big(V(\omega), W(\omega)\Big)\right| \leq \|S\|_{L^\infty(\Omega)}\, \|a\|_{\mathcal{L}(\mathcal{X},\mathcal{X}^*)}\, \|V(\omega)\|\, \|W(\omega)\| \tag{34.9}$$

$$\left|R(\omega)\, l\Big(V(\omega)\Big)\right| \leq \left|R(\omega)\right|\, \|l\|_{\mathcal{X}^*}\, \|V(\omega)\|\,. \tag{34.10}$$

Therefore we can apply Proposition 2.2 in the case $q = 2$ and find the unique solution $\hat{U}$ in the class

$$K = L^2(\Omega, P, \mathcal{K}) = \{V \in L^2(\Omega, P, \mathcal{X}) : V \in \mathcal{K}\ P\text{-almost surely}\}\,.$$

Now instead of testing with $z \in \mathcal{K}$ in (34.7), we can plug in $z = V(\omega)$, where $V \in L^2(\Omega, P, \mathcal{K})$ arbitrarily, and can integrate on $\Omega$, since by the estimates above the integrand is a $L^1(\Omega)$ function. Thus $\hat{U}$ is a solution to the following variational inequality in integrated form: Find $U \in L^2(\Omega, P, \mathcal{K})$ such that for all $V \in L^2(\Omega, P, \mathcal{K})$

$$E^P\{S\, a(U, V - U) + b(U, V - U)\} \geq E^P\{R\, l(V - U)\}\,. \tag{34.11}$$

On the other hand, by the estimates (34.9) and (34.10) we obtain

$$\left|\int_\Omega \{S\; a(V, W) + b(V, W)\}\; dP(\omega)\right| \tag{34.12}$$

$$\leq \{\|S\|_{L^\infty(\Omega)} \|a\|_{\mathcal{L}(\mathcal{X},\mathcal{X}^*)} + \|b\|_{\mathcal{L}(\mathcal{X},\mathcal{X}^*)}\}\, \|V\|_{L^2(\Omega,P,\mathcal{X})}\, \|W\|_{L^2(\Omega,P,\mathcal{X})}\,,$$

$$\left|\int_\Omega R\, l(V)\; dP(\omega)\right| \leq \|l\|_{\mathcal{X}^*}\, \|R\|_{L^2(\Omega,P)}\, \|V\|_{L^2(\Omega,P,\mathcal{X})}\,. \tag{34.13}$$

The right hand side of (34.11), respectively the left hand side of (34.11) gives rise to a continuous linear form on $L^2(\Omega, P, \mathcal{X})$, respectively a continuous nonnegative bilinear form on $[L^2(\Omega, P, \mathcal{X})]^2$, which inherits coercivity from the coercivity of $\underline{S}a+b$. Moreover, by Lemma 3.1, the convex set $K$ is closed. Therefore appealing again to the Lions-Stampacchia theorem, there exists a unique solution in $L^2(\Omega; \mathcal{K}, P)$ to the variational inequality (34.11).

By uniqueness, both problems (34.7) and (34.11) are equivalent.

## 4.2 THE PROBLEM - PROBABILISTIC FORMULATION

Now we take the view point of probability theory (see e.g. the discussion in [11, section 10.2]). We consider the joint distribution $P_T$ of the real random vector $T = (R, S)$. Thus we can get rid of the probability space $(\Omega, \mathcal{A}, P)$ and work instead in the special probability space $(\mathbb{R}^2, \mathcal{B}(\mathbb{R}^2), P_T)$, to which all our results in the preceding sections apply.

Analogous to the first formulation (34.7) our problem reads now: Find a $\mathcal{X}$-valued random variable $\hat{u}$ defined on $(\mathbb{R}^2, \mathcal{B}(\mathbb{R}^2))$ such that $P_T -$ almost surely $\hat{u} \in \mathcal{K}$ and for $P_T$-almost all $(r, s) \in \mathbb{R}^2$, there holds for all $z \in \mathcal{K}$

$$s\, a\Big(\hat{u}(r,s), z-\hat{u}(r,s)\Big) + b\Big(\hat{u}(r,s), z-\hat{u}(r,s)\Big) \geq r\, l(z-\hat{u}(r,s)\Big) . \quad (34.14)$$

By the same arguments as in the preceding subsection we find that the variational inequality (34.14) admits a unique solution $\hat{u}$ and that $\hat{u}$ lies in the class of $\mathcal{K}$-valued $L^2$-functions,

$$K_T := L^2(\mathbb{R}^2, \mathcal{K}, P_T) = \{v \in L^2(\mathbb{R}^2, \mathcal{X}, P_T) : v \in \mathcal{K}\, P_T - \text{almost surely}\},$$

where the Hilbert space $L^2(\mathbb{R}^2, \mathcal{X}, P_T)$ is equipped with the norm

$$[E^{P_T}\|v\|^2]^{1/2} = [\int_{-\infty}^{\infty}\int_{-\infty}^{\infty} \|v(r,s)\|^2\, dP_T(r,s)]^{1/2}.$$

Further $\hat{u}$ solves the following variational inequality analogously to (34.11) in integrated form: Find $u \in K_T$ such that for all $v \in K_T$

$$E^{P_T}\{s\, a(u, v-u) + b(u, v-u)\} \geq E^{P_T}\{r\, l(v-u)\}. \quad (34.15)$$

We can summarize as follows.

COROLLARY 4.1 *Both problems (34.14) and (34.15) are equivalent. They admit a unique solution $\hat{u} \in L^2(\mathbb{R}^2, \mathcal{K}, P_T)$ and the composition $\hat{U} = \hat{u} \circ T = \hat{u}(T) \in L^2(\Omega, \mathcal{K}, P)$ is the unique solution of the equivalent problems (34.7) and (34.11).*

## 4.3 SOME PERTURBED PROBLEMS

Consider a sequence $\{\mathcal{K}^{(\nu)}\}_{\nu \in \mathbb{N}}$ of closed convex nonempty subsets of $\mathcal{X}$. Suppose that the original data $\{\mathcal{K}^{(\nu)}\}_{\nu \in \mathbb{N}}$ Mosco-converges to $\mathcal{K}$. Further let sequences $\{S_n\}_{n \in \mathbb{N}}$, $\{R_n\}_{n \in \mathbb{N}}$ of real-valued random variables on the probability space $(\Omega, \mathcal{A}, P)$ be given. Suppose that $S_n \to S$ in $L^\infty(\Omega)$ and that $R_n \to R$ in $L^2(\Omega)$ for $n \to \infty$.

Thus we are led to the perturbed problem in its integrated formulation analogous to (34.11): Find $U_n^{(\nu)} \in L^2(\Omega, P, \mathcal{K}^{(\nu)})$ such that for all $V \in L^2(\Omega, P, \mathcal{K}^{(\nu)})$

$$E^P\{S_n\, a(U_n^{(\nu)}, V - U_n^{(\nu)}) + b(U_n^{(\nu)}, V - U_n^{(\nu)})\} \geq E^P\{R_n\, l(V - U_n^{(\nu)})\}\,. \tag{34.16}$$

Let us consider the joint distribution $P_{T_n}$ of the real random vector $T_n = (R_n, S_n)$ and define

$$\begin{aligned} K_{T_n}^{(\nu)} &:= L^2(\mathbb{R}^2, \mathcal{K}^{(\nu)}, P_{T_n}) \\ &= \{v \in L^2(\mathbb{R}^2, \mathcal{X}, P_{T_n}) : v \in \mathcal{K}^{(\nu)}\ P_{T_n} - \text{almost surely}\}\,, \end{aligned}$$

Then we arrive to the perturbed problem in its integrated formulation analogous to (34.15): Find $u_n^{(\nu)} \in K_{T_n}^{(\nu)}$ such that for all $v \in K_{T_n}^{(\nu)}$

$$E^{P_{T_n}}\{s\ a(u_n^{(\nu)}, v - u_n^{(\nu)}) + b(u_n^{(\nu)}, v - u_n^{(\nu)})\} \geq E^{P_{T_n}}\{r\ l(v - u_n^{(\nu)})\}\,. \tag{34.17}$$

Analogously to Corollary 4.1 we have the following result.

COROLLARY 4.2 *Both perturbed problems (34.15) and (34.17) admit unique solutions* $U_n^{(\nu)}, u_n^{(\nu)}$, *respectively, which are related by* $U_n^{(\nu)} = u_n^{(\nu)} \circ T_n$.

## 4.4 A STABILITY RESULT

Before we can establish our stability result, we need some preliminary lemmata. Firstly, one can prove similarly as Lemma 2.6 in [8] the following density result.

LEMMA 4.3 *For any given* $v \in K_T$, *there exist a subsequence* $\{T_{n_\nu}\}_{\nu\in\mathbb{N}} =: \{T^\nu\}_{\nu\in\mathbb{N}}$ *and a sequence* $\{v_\nu\}_{\nu\in\mathbb{N}}$ *such that* $v_\nu \in K_{T^\nu}^{(\nu)}$ *for* $\nu \in \mathbb{N}$ *and such that* $\|v_\nu \circ T^\nu - v \circ T\|_X \to 0$ *as* $\nu \to \infty$.

Then we need to investigate whether the sequence $\{U_n^{(\nu)}\}_{n,\nu\in\mathbb{N}}$ admits limit points that are feasible, i.e. that belong to $K$. We endeavour to obtain such a feasibility result already with weak convergence in $L^2(\Omega, P, \mathcal{X})$. Therefore we have to be somewhat more precise about the unilateral constraint.

Let us suppose that $\mathcal{K} = g + \mathcal{C}, \mathcal{K}_\nu = g_\nu + \mathcal{C}_\nu\ (\nu \in \mathbb{N})$, where $\mathcal{X} \ni g_\nu \to g$ in $\mathcal{X}\ (\nu \to \infty)$ and $\mathcal{C}, \mathcal{C}_\nu\ (\nu \in \mathbb{N})$ are convex closed *cones* with their vertices at zero in $\mathcal{X}$. Then $\mathcal{K}_\nu$ Mosco-converges to $\mathcal{K}$, if and only if $\mathcal{C}_\nu$ Mosco-converges to $\mathcal{C}$. Further we apply the equivalence of Mosco-convergence of convex lower semicontinuous functions to Mosco-convergence of their Fenchel transforms (see [1, Theorem 3.18]). Here we consider the indicator function $i_\mathcal{C} : \mathcal{X} \to \mathbb{R} \cup \{+\infty\}$ of the cone $\mathcal{C}$, given by

$$i_\mathcal{C}(x) = \begin{cases} 0 & \text{if } x \in \mathcal{C}; \\ +\infty & \text{elsewhere,} \end{cases}$$

respectively the indicator functions $i_{\mathcal{C}_\nu}$ of the cones $\mathcal{C}_\nu$. We find that the Fenchel transform $i_\mathcal{C}^*$ is simply given by

$$i_\mathcal{C}^*(\xi) = \sup\{\langle \xi, x\rangle_{\mathcal{X}^* \times \mathcal{X}} : x \in \mathcal{C}\} = i_{\mathcal{C}^-}(\xi),$$

where

$$\mathcal{C}^- = \{\xi \in \mathcal{X}^* : \langle \xi, x\rangle \leq 0 \text{ for all } x \in \mathcal{C}\}$$

denotes the *dual cone* of $\mathcal{C}$. Thus we conclude that $\mathcal{C}_\nu$ Mosco-converges to $\mathcal{C}$, if and only if in the dual space, $\mathcal{C}_\nu^-$ Mosco-converges to $\mathcal{C}^-$ $(\nu \to \infty)$. Based on this equivalence we obtain similarly as in [8, Lemma 2.7] the following feasibility result.

LEMMA 4.4 *Suppose that $\mathcal{K}_\nu$ Mosco-converges to $\mathcal{K}$ as $\nu \to \infty$, where $\mathcal{K} = g + \mathcal{C}$, $\mathcal{K}_\nu = g_\nu + \mathcal{C}_\nu$ $(\nu \in \mathbb{N})$; $g_\nu \to g$ in $\mathcal{X}$; $\mathcal{C}, \mathcal{C}_\nu$ are convex closed cones in $\mathcal{X}$. Then any limit point of the sequence $\{U_n^{(\nu)}\}_{n,\nu \in \mathbb{N}}$ with respect to weak convergence in $L^2(\Omega, P, \mathcal{X})$ belongs to $K$.*

Based on the two preceding lemmata and on Minty's lemma we can similarly as in [8, Theorem 2.8] conclude our analysis by the following result.

THEOREM 4.5 *Suppose that the bilinear form $\beta$ is uniformly coercive. Further suppose that $\mathcal{K}_\nu$ Mosco-converges to $\mathcal{K}$, where $\mathcal{K} = g + \mathcal{C}$, $\mathcal{K}_\nu = g_\nu + \mathcal{C}_\nu$ $(\nu \in \mathbb{N})$; $g_\nu \to g$ in $\mathcal{X}$; $\mathcal{C}, \mathcal{C}_\nu$ are convex closed cones in $\mathcal{X}$. Then $\{U_n^{(\nu)}\}_{n,\nu \in \mathbb{N}}$ converges strongly to the unique solution $\hat{U}$ of the equivalent problems (34.7) and (34.11) in $L^2(\Omega, P, \mathcal{X})$.*

# 5 AN APPLICATION TO A RANDOM UNILATERAL PROBLEM

In this section we treat a simple model problem to give a concrete application of our more theoretical results of the previous sections. We consider the following random unilateral boundary problem: Find a function

$\psi$ in a bounded domain $D \subset \mathbb{R}^2$ with its boundary $\partial D$ such that

$$\begin{cases} -\nabla \cdot (S\,\nabla\psi) + \psi = R\,f & \text{in } D\,. \\ q_\nu := (-S\nabla\psi)\cdot\nu & \text{on } \partial D \\ \psi \le g, q_\nu \ge 0, (\psi - g)\,q_\nu = 0 & \text{on } \partial D\,. \end{cases}$$

Here $f$ and $g$ are given data in the domain, respectively on its boundary. Randomness enters in the problem only by the given real-valued random variables $R$ and $S$ that satisfy the assumptions of section 4. Since the outward normal denoted by $\nu$, hence also the boundary flux $q_\nu$ is defined only a.e., we do not investigate the boundary value problem in its classical, i. e. pointwise formulation. Instead, with the data $f \in L^2(D), g \in L^2(\partial D)$ we turn to its variational formulation, which is obtained in the same way as for deterministic unilateral boundary value problems (see e.g. [3]). This leads to a simple random variational inequality of the form (34.7) or equivalently of the form (34.11). Indeed, let

$$\begin{aligned} \mathcal{X} &= H^1(D)\,, \\ \mathcal{K} &= \{z \in \mathcal{X} : z|_{\partial D} \le g \text{ on } \partial D \text{ a.e.}\}\,, \\ l(z) &= \int_D f(x)z(x)\,dx\,, \quad z \in \mathcal{X}\,, \\ b(z,z') &= \int_D z(x)z'(x)\,dx\,, \quad z,z' \in \mathcal{X}\,, \\ a(z,z') &= \int_D \nabla z(x).\nabla z'(x)\,dx\,, \quad z,z' \in \mathcal{X}\,. \end{aligned}$$

Thus we are in the framework of section 4. In particular, the bilinear form $\beta = Sa + b$ is uniformly coercive on $H^1(D) \times H^1(D)$, since

$$\beta(z,z) \ge \min\{\underline{S}, 1\}\,\|z\|^2_{H^1(D)}\,.$$

As before in section 4, we can shift to the probabilistic formulation of the problem. Here the class $K_T$ of feasible functions is specified by

$$K_T = \{v \in L^2(\mathbb{R}^2, P_T, H^1(D)) \,:\, v|_{\partial D} \le g \text{ a.e. and } P_T\text{-almost surely}\}\,,$$

and we arrive at a random variational problem of the form (34.14) or equivalently of the form (34.15).

Also the stability result of section 4, Theorem 4.5 readily applies and yields.

COROLLARY 5.1 *Let $f \in L^2(D))$. Let as before sequences $\{S_n\}_{n\in\mathbb{N}}, \{R_n\}_{n\in\mathbb{N}}$ of real-valued random variables be given such that $S_n \to S$ in $L^\infty(\Omega)$ and that $R_n \to R$ in $L^2(\Omega)$ for $n \to \infty$. In addition let a sequence $\{g^{(\nu)}\}_{\nu\in\mathbb{N}}$ be given such that $g^{(\nu)} \to g$ in $H^{\frac{1}{2}}(\partial D)$.*

*Then the unique solution* $\{U_n^{(\nu)}\}_{n,\nu\in\mathbb{N}}$ *of the perturbed problem (34.15) converges strongly to the unique solution* $\hat{U}$ *of the problem (34.11) in* $L^2(\Omega, P, H^1(D))$.

## References

[1] H. Attouch, *Variational Convergence for Functions and Operators*, Pitman, Boston, 1984.

[2] J.P. Aubin and H. Frankowska, *Set-valued analysis*, Birkhäuser, Boston, 1990.

[3] C. Baiocchi and A. Capelo, *Variational and Quasivariational Inequalities.* Wiley, Chichester, New York, 1984.

[4] D.L. Cohn, *Measure Theory*, Birkhäuser, Boston, 1980.

[5] G. Da Prato and J. Zabczyk, *Stochastic Equations in Infinite Dimensions*, Cambridge University, Cambridge, 1992.

[6] G. Duvaut and J.L. Lions, *Inequalities in Mechanics and Physics*, Springer, Berlin, 1976.

[7] J. Gwinner, Discretization of semicoercive variational inequalities, Aequationes Math. **42** (1991) 72–79.

[8] J. Gwinner, A class of random variational inequalities and simple random unilateral boundary value problems – existence, discretization, finite element convergence, Stochastic Anal. Appl. (to appear).

[9] H. Holden, B. Øksendal, J. Ubøe and T.-S. Zhang, *Stochastic partial differential equations*, Birkhäuser, Boston 1996.

[10] D. Kinderlehrer, G. Stampacchia, *An Introduction to Variational Inequalities and Their Applications*, Academic Press, New York, 1984.

[11] M. Loève, *Probability Theory I*, Springer, New York etc, 1977.

[12] U. Mosco, Convergence of convex sets and of solutions of variational inequalities, Advances in Math. **3**, 1969, 510 – 585.

Chapter 35

# A COMPARISON PRINCIPLE AND THE LIPSCHITZ CONTINUITY FOR MINIMIZERS

Carlo Mariconda
*Dipartimento di Matematica pura e applicata*
*Università di Padova*
*via Belzoni 7, I-35131 Padova*
*Italy*
maricond@math.unipd.it

Giulia Treu
*Dipartimento di Matematica pura e applicata*
*Università di Padova*
*via Belzoni 7, I-35131 Padova*
*Italy*
treu@math.unipd.it

**Abstract** We state some recent results on the existence, uniqueness and Lipschitz regularity of the minima of the integral functional $I(u) = \int_\Omega L(x, u, \nabla u)\,dx$ on $\bar{u} + W_0^{1,q}(\Omega)$ $(1 \leq q \leq +\infty)$ for a class of integrands $L(x, z, p) = f(p) + g(x, z)$ that are convex in $(z, p)$ and for boundary data satisfying some barrier conditions. We do not impose regularity or growth assumption on $L$.

**Keywords:** Lipschitz, Regularity, Euler equation, Barrier, Bounded slope condition



*N. Hadjisavvas and P.M. Pardalos (eds.), Advances in Convex Analysis and Global Optimization,* 545–551.

# 1 INTRODUCTION

We present here some recent results on the Lipschitz regularity of the minima of an integral functional with Lagrangian

$$L : \Omega \times \mathbf{R} \times \mathbf{R}^n \longrightarrow \mathbf{R} \cup \{+\infty\} \tag{35.1}$$
$$(x, z, p) \longmapsto L(x, z, p) \tag{35.2}$$

of the form

$$I(u) = \int_\Omega L(x, u, \nabla u)\, dx \qquad (\Omega \subset \mathbf{R}^n)$$

on a Sobolev space $\bar{u} + W_0^{1,q}(\Omega)$ $(q \geq 1)$. We first recall some of the main results on the subject. The standard Tonelli's assumptions to obtain the existence of a minimum of $I$ are

$$p \longmapsto L(x, z, p) \text{ is convex} \tag{35.3}$$

$$L(x, z, p) \geq \alpha |p|^q + \beta(x) \ (\alpha > 0, \beta \in L^1(\Omega) \text{ for } q > 1). \tag{35.4}$$

If one assumes that the minimum of $I$ is Lipschitz or that $L$ and its derivatives satisfy the growth assumptions

$$(|L| + |L_z| + |L_p|)(x, z, p) \leq C(1 + |p|^q) \tag{35.5}$$

then the minimum fulfills the integral form of the Euler equation

$$\forall \varphi \in C_c^\infty(\Omega) \quad \int_\Omega L_p(x, u, \nabla u) \cdot \nabla \varphi + L_z(x, u, \nabla u) \varphi\, dx = 0. \tag{35.6}$$

We recall that (35.6) is the quasi–linear equation in divergence form

$$\text{div } L_p(x, u, \nabla u) - L_z(x, u, \nabla u) = 0.$$

The classical approach to study the regularity and the behaviour of the minima is to impose that the equation is elliptic (the Hessian $L_{pp}$ of $L$ is positive definite) and then to apply the results on elliptic PDE's (De Giorgi–Nash's Theorem, maximum principles,...). Some natural questions arise:
1) does (35.6) still hold under (just) Tonelli's assumptions?
2) are the minima of $I$ Lipschitz in any case?
Both questions have in general a negative answer as it is shown, for instance, in [1]. There are however some positive results in this field. Namely for one dimensional problems Clarke and Vinter proved in [4] that the minimum of $I$ is Lipschitz under the assumption (35.4) and either $L(x, z, p)$ does not depend on $x$ or $L$ is convex in $(z, p)$; Giaquinta and Giusti [5] studied the Hölder regularity of the minima in

the multi–dimensional case without making use of the Euler equation for Lagrangians $L(x,z,p)$ that behave like $|p|^q$ for $|p| \to +\infty$.
We present here the main results we established in [7, 8, 9] where we showed that, without assuming (35.4) and (35.5), some suitable conditions on the boundary datum $\bar{u}$ yield the existence of minimum of $I$ that is actually Lipschitz. A first result in this direction was given by Cellina in [2] for Lagrangians of the form $L(x,z,p) = f(p)$ and for $\bar{u}$ satisfying the Bounded Slope Condition.

## 2 COMPARISON AND MAXIMUM PRINCIPLES FOR MINIMA

We assume that

$$L \text{ is a Carathéodory function, convex in } (z,p).$$

We point out that the minimum of $I$ may not exist or may not satisfy the Euler equation. In spite of this we are able to state some results for the minima that are analogous to some classical statements concerning the solutions to elliptic PDEs. We first need to recall the well known notion of sub/super–solutions to a differential equation, generalizing that of solution. We assume for simplicity that $L$ is smooth; a more general definition can be given by using the notion of subgradient of a convex function.

**Definition 1** *A function $u \in W^{1,1}(\Omega)$ is said to be a sub–solution (resp. super–solution) of (35.6) if for every $\varphi \in C_c^\infty(\Omega)$, $\varphi \geq 0$,*

$$\int_\Omega L_p(x,u,\nabla u)\cdot\nabla\varphi + L_z(x,u,\nabla u)\varphi\,dx \leq 0\,(\textit{resp.}\ \geq 0).$$

For instance, for $L(x,z,p) = |p|^2$, the sub–solutions of (35.6) are the subharmonic functions.
The following theorem allow us to compare two minima of the functional from their behaviour on the boundary of $\Omega$.

**Theorem 1** (COMPARISON PRINCIPLE FOR MINIMA)
*Assume that $I$ is strictly convex. Let $u$ be either a minimum for $I$ in $u + W_0^{1,q}(\Omega)$ or a sub–solution of (35.6) and $v$ be either a minimum for $I$ in $v + W_0^{1,q}(\Omega)$ or a super–solution of (35.6). If $u \leq v$ on $\partial\Omega$ (in the sense of Sobolev functions) then $u \leq v$ a.e. on $\Omega$.*

We point out that this result extends the classical Comparison Principle for PDEs. As a consequence we obtain the following estimate of the variations of the minima along a fixed direction.

**Theorem 2** (GRADIENT MAXIMUM PRINCIPLE FOR MINIMA)
*Let $L(x,z,p) = f(p) + g(z)$ where either $f$ or $g$ is strictly convex. Let $w$ be a minimum for $I$ in $w + W_0^{1,q}(\Omega)$ and $\tau \in \mathbf{R^n}$. Then*

$$\|w(x+\tau)-w(x)\|_{L^\infty(\Omega_\tau)} = \sup_{\partial\Omega_\tau} |w(x+\tau)-w(x)| \qquad (\Omega_\tau = (-\tau+\Omega)\cap\Omega)$$

*where the above supremum is intended in the sense of Sobolev functions. Moreover if $w \in C^1(\overline{\Omega})$ then*

$$\|\nabla w\|_{L^\infty(\Omega)} = \|\nabla w\|_{L^\infty(\partial\Omega)}.$$

The next example shows that the Gradient maximum Principle does not hold in general if $g$ depends on $x$.

**Example 1** *Let $n = 1, q = 1, \Omega = ]-1,1[$, $L(x,z,p) = f(p) + g(x,z)$ where $f(p) = p^2$ and $g(x,z) = 2\cosh(1)xz + z^2$. The function $w(x) = \sinh(x) - x\cosh(1)$ is a smooth solution of the Euler equation*

$$u'' - u = \cosh(1)x$$

*associated to the (strictly convex) functional*

$$I(u) = \int_{-1}^{1} L(x,u,u')\,dx.$$

*It follows by convexity that $w$ is a minimum for $I$ in the set of the absolutely continuous functions that coincide with $w$ in $-1$ and $1$. However $\|w'\|_{L^\infty(-1,1)} > 0 = \max\{w'(-1), w'(1)\}$.*

The corresponding Gradient Maximum Principle for the solutions of class $C^2$ of elliptic PDEs is proved under the assumption that $L$ is of class $C^3$ by differentiating the original equation. Instead we use here just the fact that $w$ is a minimum without any smoothness assumption on $L$ or on $w$.

## 3 EXISTENCE AND LIPSCHITZ REGULARITY

In what follows we work under the following assumptions on the Lagrangian.

**Definition 2** (ASSUMPTION (H))
*$L(x,z,p) = f(p) + g(x,z)$ is convex in $(z,p)$, $L(x,z,p) \geq \alpha|p| + \beta(x)$ ($\alpha \neq 0, \beta \in L^1(\Omega)$) and one of the following conditions hold:*
**(H1)** *The function $g$ is of class $C^2$ in $\overline{\Omega} \times \mathbf{R}$ and*

$$\inf\{g_{zz}(x,z) : (x,z) \in \Omega \times \mathbf{R}\} > \mathbf{0};$$

*or*

**(H2)** *The function $g$ does not depend on $x$ and either $f$ or $g$ is strictly convex.*

First we state a result on Lipschitz regularity of minima of $I$ lying between two Lipschitz functions.

**Theorem 3** (REGULARITY FOR CONSTRAINED MINIMA) *Assume that $L$ satisfies Assumption (H), $\bar{u} \in \mathrm{Lip}(\Omega)$, $\ell^1, \ell^2$ be Lipschitz and coincide with $\bar{u}$ on $\partial\Omega$. Let $w$ be a minimum for $I$ in $\bar{u} + W_0^{1,q}(\Omega)$ such that $\ell^1 \leq w \leq \ell^2$ a.e. on $\Omega$. Then $w$ is Lipschitz, with a Lipschitz constant bounded by a constant depending only on $g$, $\|\nabla \ell^1\|_{L^\infty}$ and $\|\nabla \ell^2\|_{L^\infty}$.*

The proof of the above results relies on a consequence of a theorem of [12] under (H1) and on the Gradient Maximum Principle for minima under (H2). For $q = +\infty$ the estimate on the Lipschitz constant turns out to be essential to prove the existence of minimizers of $I$ among the Lipschitz functions, under a suitable assumption on the boundary datum. We state now a condition on $(I, \bar{u})$ that ensure that the minima of $I$ are constrained.

**Definition 3** (GENERALIZED BOUNDED SLOPE CONDITION)
*Let $K > 0$, we say that the pair $(I, \bar{u})$ satisfies the Generalized Bounded Slope Condition $(GBSC)_K$ if $\bar{u}$ is Lipschitz with constant less than $K$ and for every $x_0$ in $\partial\Omega$ there exist a sub–solution $\ell^1_{x_0} \in \mathrm{Lip}(\Omega)$ and a super–solution $\ell^2_{x_0} \in \mathrm{Lip}(\Omega)$ for (35.6), both of Lipschitz constant less than $K$, such that*

$$\forall x \in \partial\Omega \qquad \ell^1_{x_0}(x) \leq \bar{u}(x) \leq \ell^2_{x_0}(x) \text{ and } \ell^1_{x_0}(x_0) = \bar{u}(x_0) = \ell^2_{x_0}(x_0).$$

In the case where the functions $\ell^i_{x_0}$ are affine the above definition is the Bounded Slope Condition. If $L(x, z, p)$ is smooth and depends only on $p$ there are some well known conditions on $\Omega$ and the eigenvalues of the Hessian of $L$ with respect to $p$ that ensure that $(I, \bar{u})$ fulfills the $(GBSC)$ for every Lipschitz boundary datum $\bar{u}$ (see for instance [6]). The $(GBSC)$ is a classical tool in PDEs to give an estimate of the gradient of regular solutions. We exploit it here in a different way to prove the existence of a smooth minimum of $I$. We state now our main result.

**Theorem 4** (EXISTENCE AND LIPSCHITZ REGULARITY FOR MINIMA) *Let $L$ be real valued and satisfy Assumption (H). Let $\bar{u} \in \mathrm{Lip}(\Omega)$ and assume that the pair $(I, \bar{u})$ satisfies the $(GBSC)_K$ for some $K > 0$. Then, for every $q \in [1, +\infty]$, the functional $I$ has a unique minimum $w$ in $\bar{u} + W_0^{1,q}(\Omega)$. Moreover $w$ is Lipschitz with Lipschitz constant bounded by an explicit constant depending only on $g$ and $K$.*

We underline the most important facts contained in the previous claim. We obtain existence of a minimum of $I$ without growth assumtions and regularity of the minimum; the validity of the Euler equation folows without growth assumptions on $L$ or on its subdifferentials. In particular when $L(x,z,p) = f(p)$, $q = +\infty$ and $\bar{u}$ satisfies the Bounded Slope Condition we obtain both Miranda's existence result [11] and Cellina's regularity theorem [2].

**Sketch of the proof.** The estimate of the Lipschitz constant given by Theorem 3 yields the existence of a minimum $w$ among the Lipschitz functions that coincide with $\bar{u}$ on the boundary of $\Omega$. Then Clarke's non-smooth version of the Euler equation [3] implies that $w$ is a minimum in $\bar{u} + W_0^{1,q}(\Omega)$. If $I$ is strictly convex the uniqueness of $w$ is straightforward. Otherwise if $\tilde{w}$ is another minimum of $I$ then $w \le \tilde{w} \le w$ on $\partial\Omega$ and $w$ is both a sub–solution and a super–solution to the Euler equation: The *Comparison Principle* yields the conclusion.

**Remark.** In [7] we proved a Comparison Principle for functionals whose Lagrangian is of the form $L(x,z,p) = f(x,p) + g(x,z)$ without requiring the strict convexity of $I$. Its application actually yields the claim of the previous theorem under (H2) and the assumption, instead of strict convexity of $f$, that the (vector) projections of the faces of the epigraph of $f$ are contained in a vector subspace of $\mathbf{R^n}$ of dimension less than $n-1$. In this case we still obtain the uniqueness of the minimum without the standard strict convexity assumption of the functional.

## References

[1] J.M. Ball, V.J. Mizel, *One-dimensional variational problems whose minimizers do not satisfy the Euler–Lagrange equation*, Arch. Rational Mech. Anal **90** (1985), 325–388.

[2] A. Cellina, *On the Bounded Slope Condition and the validity of the Euler equation*, SIAM Control Opt. (2000), to appear.

[3] F.H. Clarke, *Multiple integrals of Lipschitz functions in the calculus of variations*, Proc. Amer. Math. Soc. **64** (1977), 260–264.

[4] F. H. Clarke, R. B. Vinter *Regularity properties of solutions to the basic problem in the calculus of variations*, Trans. Amer. Math. Soc.**289** (1985), 73–98.

[5] M. Giaquinta, E. Giusti, *Differentiability of minima of nondifferentiable functionals*, Invent. Math. **72** (1983), 285–298.

[6] D.Gilbarg, N.S. Trudinger, *Elliptic Partial Differential equations of second order*, Grundenlehren der Matematischen Wissenschaften, vol. 224, Springer–Verlag, Berlin– Heidelberg–New York, 1998.

[7] C. Mariconda, G. Treu, *A Comparison Principle and the Lipschitz continuity for minimizers*, preprint, Dipartimento di Matematica pura e applicata dell' Università di Padova **27** (1999).

[8] C. Mariconda, G. Treu, *A gradient maximum principle for minima*, preprint, Dipartimento di Matematica pura e applicata dell' Università di Padova **14** (2000).

[9] C. Mariconda, G. Treu, *Existence and Lipschitz regularity for minima*, preprint, Dipartimento di Matematica pura e applicata dell' Università di Padova **15** (2000).

[10] C. Mariconda, G. Treu, *A Comparison Principle for minimizers*, Comptes Rendus de l'Academie des Sciences, Série I **330** (2000), 681–686.

[11] M. Miranda, *Un teorema di esistenza e unicità per il problema dell'area minima in n variabili*, Ann. Sc. Normale di Pisa **19** (1965), 233–249.

[12] G. Treu, M. Vornicescu, *On the equivalence of two variational problems*, Calculus of Variations and Partial Differential Equations **11** (2000), 307–319.

Chapter 36

# TUNNELING AND GENETIC ALGORITHMS FOR GLOBAL OPTIMIZATION

Susana Gómez *
*IIMAS, UNAM, Mexico*
susanag@servidor.unam.mx

Julio Solano
*IIMAS, UNAM, Mexico*
julio@uxdea4.iimas.unam.mx

Longina Castellanos†
*ICIMAF, Cuba*
longina@cidet.icmf.inf.cu

Marcos I. Quintana
*IIMAS, UNAM, Mexico*
qhmi@uxdea4.iimas.unam.mx

**Abstract** The purpose of this paper is to compare the performance of two global optimization methods very different in nature. Genetic Algorithms (GA) are stochastic global search procedures that only require objective function values. The Tunneling Algorithms (TA), classical and exponential, are deterministic, need gradient information requiring $f$ to be smooth, $f \in C^2$, and depend only on one initial point. The results obtained here show that although GA converges very fast to lower levels of the objective function, it is extremely slow to finally get the desired global solution specially for larger dimensions. TA converge to the global

*Corresponding author.
†This author was supported in part by UNAM-CRAY SC-010899 Project and TWAS Research Grant No. 97-043 RG/MATHS/LA.



*N. Hadjisavvas and P.M. Pardalos (eds.), Advances in Convex Analysis and Global Optimization,* 553–567.

solution, even for problems with a large number of local minima near the global level and larger dimensions, if enough time is given, for every initial point of the sample. Scaling the problem, showed to be extremely beneficial for TA, specially for larger dimensions, as the complexity of the objective function can be disminuished.

**Keywords:** Global optimization, Genetic Algorithms, Tunneling Algorithms.

# 1 INTRODUCTION

We want to find the best solution of the general problem

$$\min f(x), \quad f : \Re^n \to \Re, \quad x \in B \subset \Re^n \tag{36.1}$$

where the feasible region in this case is defined by bounds on the variables

$$B = \{x : l \leq x \leq u\}$$

The best or global solution $x_G^*$, implies that $f(x_G^*) \leq f(x), \forall x \in B$.

The purpose of this work, is to compare the performance of two global optimization methods very different in nature, which have shown to be effective to solve many practical problems [7, 8, 9, 13, 14, 17].

In general terms, Genetic Algorithms (GA) are stochastic global search procedures based on the mechanics of natural evolution involving a structured information exchange. These methods only require objective function values. The implementation of the genetic algorithm (GA), incorporates real–valued operators such as line recombination [12] and dynamic mutation [16], to improve the diversity of the population and therefore improve the convertion rate. It also exploits the inherent parallelism associated to the GA process through the use of subpopulations, giving as a result not only a better convergence rate but offering the possibility of implementing the algorithm in multiple processors in a transparent manner [17].

The tunneling algorithms, classical [10, 11] and exponential [2], need gradient information requiring $f$ to be smooth, $f \in C^2$, and depend only on one initial point. In order to make a fare comparison the TA run for 100 and 1000 initial random points whereas the GA have run from 100 initial random populations of 100 individuals each. To show their performance for increased dimensionality, the test problems were taken with $n = 10$ and $n = 100$.

In section 2, the basic ideas behind the TA and a scaling procedure are presented, a description of the GA used here is given in section 3, academic examples are in section 4, and the description of the experiments and the discussion on the numerical results will finally be presented in section 5.

# 2 THE TUNNELING METHODS

They are deterministic in the sense that the aim is to find a monotonic decreasing sequence of local minima

$$\{x_i^*\} \text{ such that } f(x_{i+1}^*) \leq f(x_i^*) \tag{36.2}$$

using a local optimization method and a deflation technique, to be able to pass from the valley of the last local minimum found, to another which is a better candidate for the global solution, while avoiding valleys above.

They also have an stochastic element as the search for the point in another valley is started from random directions in the neighbourhood of the last local minimum found. If this search is computationally exhaustive, the global solution will eventually be found. The algorithm basic ideas follow and a detail description can be found in [3]:

- **Minimization Phase**: it looks for a local minimum $x^*$ within the bounds:

$$x^* = \quad arg(\min f(x)) \qquad \text{with } f^* = f(x^*), \quad x \in B \tag{36.3}$$

- **Tunneling Phase**: it finds (if it exists) an $x_{tu}^* \in B$ in another valley such that $f(x_{tu}^*) \leq f(x^*)$ , and $x_{tu}^* \neq x^*$, by means of an iterative decrease of the tunneling function

$$T(x, x^*, f^*, \lambda^*, \lambda_m, x_m) \qquad \text{and stops whenever } T \leq 0.$$

$T$ can be chosen to be one of the following functions:

1. Classical tunneling function [1] (see [11])

$$T^c(x, x^*, f^*, \lambda^*, \lambda_m, x_m) = \frac{f(x) - f^*}{\prod_{i=1}^{t} \|x - x_i^*\|^{\lambda_i^*}} \cdot \frac{1}{\|x - x_m\|^{\lambda_m}} \tag{36.4}$$

2. Exponential tunneling function (see [2])

$$T^e(x, x^*, f^*, \lambda^*, \lambda_m, x_m) = (f(x) - f^*)\, exp\left(\frac{\lambda_m}{\|x - x_m\|}\right) \prod_{i=1}^{t} exp\left(\frac{\lambda_i^*}{\|x - x_i^*\|}\right) \quad (36.5)$$

In both functions, $x^*$ is the last local minimum obtained, where a pole is set to destroy it and whose strength is $\lambda^*$, and $x_m$ is the position of another pole, called the mobile pole, whose strength is $\lambda_m$, which is used when needed, to smooth or destroy the critical points of $T$ along the iterations. Both functions use the factor to keep the information of the minima at the same level $f^*$ already found, in order to avoid going back to them, i.e., where $f^* = f(x_i^*)$, with $i = 1, .., t$ and $t$ is the total number of local minima with the same function value found so far, $x_i^*$ is the location of the ith-local minima and $\lambda_i^*$ its strength. Once a minimum $x^*$ with lower function value than $f^*$ is found, $t$ is set to one.

The tunneling phase begins at a random $x_{tu}^0$ generated in a neighbourhood of the last local minimum $x^*$. The points, $x^*$ obtained in the minimization phase and $x_{tu}^*$ obtained in the tunneling phase, are found using any local optimization method. For this implementation we are using a Limited Memory Quasi-Newton routine called L-BFGS-B [19]. In the tunneling phase this routine is used to generate a sequence of points with decreasing values of the Tunneling function.

Once a point $x_{tu}^*$ is found such that $T(x_{tu}^*) \leq 0$, the minimization phase starts again from $x_{tu}^*$ to find a new local minimum $x^*$, and the general two phase procedure stops whenever a stopping criteria is satisfied (see [3]). When the method finds the global solution, the tunneling phase checks that there is not a point in another valley, and the algorithm stops.

## 2.1 SCALING BY VARIABLE TRANSFORMATION

Scaling by variable transformation converts the variables from units that typically reflect the physical nature of the problem to units that display certain desirable properties for the minimization process.

There is an important distinction between transforming variables to improve the behaviour of an optimization method and transforming the variables to change the problem category. The first type of transformation is the one we are going to use here, (see [5]).

The scaling aim is to achieve that the variables of the scaled problem should be of similar magnitude and of order unity in the feasible region.

Within optimization routines, convergence tolerances and other criteria are necessarily based upon an implicit definition of "*small*" and "*large*", and thus variables with widely varying orders of magnitude may cause dificulties for some algorithms. If typical values of all the variables are known, a problem can be transformed so that the variables are all of the same order of magnitude. If the minimization problem has simple bounds on the variables, as is the case here, say $x_i \in [l_i, u_i]$ for $i = 1, .., n$, then a new variable $y_i \in [l_i^n, u_i^n]$ can be defined as:

$$y_i = \left( \frac{u_i^n - l_i^n}{u_i - l_i} \right) x_i + \frac{u_i l_i^n - l_i u_i^n}{u_i - l_i} \tag{36.6}$$

The transformation (36.6) can be written in matrix form as

$$x = Dy + c \tag{36.7}$$

where $D$ is a diagonal matrix with i-th diagonal element $\frac{u_i - l_i}{u_i^n - l_i^n}$ and $c$ is a vector with i-th element $\frac{l_i u_i^n - u_i l_i^n}{u_i^n - l_i^n}$.

When the variables are scaled by a linear transformation of the form (36.7), the derivatives of the objective function, say $f(x)$, are also scaled. Let $g_y$ and $G_y$ denote the gradient vector and the Hessian matrix of the transformed problem $h(y) = f(x)$, and $g$ and $G$ denote the corresponding for the original one. These derivatives are then related by

$$g_y = Dg, \qquad G_y = DGD \tag{36.8}$$

Hence, even a "mild" scaling such as $x_i = 10y_i$ may have a substantial effect on the Hessian, and this in turn may significantly alter the convergence rate of the optimization algorithms, which use gradient and curvature information as the TA. If the new interval is larger, the effect is to flatten the function, and with a smaller interval the valleys become sharper.

## 3 GENETIC ALGORITHM APPROACH

This section introduces the basic concepts mechanics, representation, reproductive scheme and parameters involved in the genetic algorithm approach used in this work. In addition, some alternative operators (e.g.the dynamic mutation) and parallel schemes have also been used.

A genetic algorithm is composed of a reproductive plan for selecting successful individuals to be used in creating the offspring of the new generation, and a set of genetic operators. Let P be the initial population with np individuals where P may involve a knowledge base of np m-bit-string structures generated at random in the feasible region.

$$P = \{x_1, x_2, \ldots x_{np}\}, \qquad\qquad x_i \in R^n$$

In a simplified reproductive plan, each structure $x_i$ is evaluated according to the problem domain specific criteria and is assigned a measure generally called *fitness* (a measure of profit, utility or goodness). A probability of selection is calculated for each $x_i$ which is proportional to that fitness value. A new generation of structures is then created by selecting pairs of individuals (from the pool generated by the reproductive plan) according to their probabilities in the current generation and applying to them a set of genetic operators. This new generation replaces the current generation and its new individuals are evaluated and again assigned selection probabilities according to their fitness value. The cycle repeats until a stop criterion is fulfilled or a specified number of generations is reached.

The genetic operators can be divided into two main categories: *recombination* which involves the exchange of genetic material between individuals (usually two), and *mutation* which produces changes in the information contents of the individuals according to some probabilistic rule.

## 3.1 REPRESENTATION, GA PARAMETERS AND REPRODUCTIVE SCHEME

The most commonly used representation of individuals in a GA is that of a single-level binary string [6, 8]. Alternative encoding strategies such as integer and real-valued representations have also been suggested. In [18], it is stated that the use of real-valued representations offer a number of advantages over binary encodings: the efficiency of the GA is increased as there is no need to decode the chromosomes for each function evaluation; the use of the processor is more efficient as floating-point internal computer representations can be used; there is no loss in precision caused by discretisation; and there is greater freedom in the use of different genetic operators. This representation has been used during the development of this work. The initial population is created by generating *np*n* random real numbers, where: *np* is the population size and $n$ is the number of variables of the function under consideration. A Ranking selection scheme has been used as well as the genetic operators (i.e. extended line recombination and mutation) based on those proposed in [1], [12].

The use of ranking allows the algorithm to control the allocation of the number of offspring to the members of the population and, as a result, the loss of diversity which usually leads to premature convergence can be substantially reduced.

The extended line recombination is a method used to produce offspring based on the real values of the parent phenotypes. These new phenotypes are produced according to the following expressions.

$$\overline{x}_1 = x_1 + \alpha(x_2 - x_1), \qquad \overline{x}_2 = x_2 + \alpha(x_2 - x_1)$$

where: $\overline{x_1}$ y $\overline{x}_2$ are the new phenotypes generated from the parents $x_1$ and $x_2$, and the scaling factor ($\alpha$) is chosen randomly over an interval [-0.25, 1.25].

The mutation operator is commonly used to guarantee that the probability of searching any given individual will never be zero as well as to recover useful information sometimes present in the worst elements which would be lost otherwise. The mutation operator for a real-valued encoding representation, generates a new phenotype according to the expression,

$$\overline{x}_i = x_i + range\ \delta$$

where $i$ is the index of the member of the population to be mutated. The offspring is produced by adding to the original element $x_i$ a value taken from the interval $[-range, range]$ of the variable to be mutated . The sign is chosen with probability 0.5 and $\delta$ is calculated as

$$\delta = \sum_{i=0}^{n} \alpha_i 2^{-i}$$

if $\alpha = 0$ the mutation does not take place. Otherwise, it is mutated with probability $\delta = \frac{1}{n}$.

## 3.2 DYNAMIC MUTATION

It is well known that one of the main problems regarding GA's efficiency is its premature convergence due to the loss of diversity in the population. Several methods have been devised in order to avoid such premature convergence. Here, a description of a so called dynamic mutation is given . It is based on the similarities between the selected parents which will undertake a mutation process, and determines dynamically the probability of being mutated. Parents genetically similar will have a high probability whereas parents genetically different will be assigned a small probability. These similarities between parents are produced by the use of a distance measure which uses a least square criterion. It considers a relative distance based on the bounds of each variable of the individual. The expressions are included here for completeness [16].

Consider two individuals, $A = (a_1,...,a_n)$ and $B = (b_1,...,b_n)$ and $Min_i$ and $Max_i$ as the lower and upper limits of the $i^{th}$ digit respectively, then

$$\begin{aligned} dist(A,B) &= \frac{1}{n}\sum_{i=1}^{n}\left|\frac{a_i - b_i}{Max_i - Min_i}\right|^d \\ M(A,B) &= 1 - dist(A,B) \end{aligned}$$

where: $d$ is a distance parameter, and $M(A,B)$ is the mutation rate associated with that distance. The mating of two identical individuals will give as a result $M(A,A) = 1$, which means that there will be a compulsory mutation for each variable of the offspring. This is clearly a high mutation rate and therefore a parameter $M_m$ (maximum mutation rate) needs to be introduced. Then, the effective mutation rate can be defined as follows:

$$P_r(A,B) = M(A,B) \cdot M_m = (1 - dist(A,B)) \cdot M_m$$

In this implementation $M_m$ and $d$ are fixed ($i.e. M_m = 0.9$ and $d = 0.2$)

## 3.3 GA USING SUB-POPULATIONS

It is important to mention that the GA approach also offers the feasibility of exploiting its parallel nature by the use of sub-populations. This characteristic will produce not only a speed-up in terms of processing time, but also the generation of better solutions. In general terms the Parallel Genetic Algorithms (PGA's), as they are also commonly known, cluster around two main approaches: the island model and the fine-grained model [4]. The island model, used in this work, comprises a set of nodal GA's [16, 15], each one residing on a processing node within a multiprocessor system. Each node operates on part of the complete population which is related to the number of processors in the system and can execute a GA either independently or interacting with other sub-populations.

Three main variations of this model may be considered in any application: Farming, Migration and Diffusion. We have included the implementation of the Migration model in this work once proved that both Migration a Diffusion models have similar performance and that they outperform substantially the Farming model which presents a high communication [17]. For this model, on each processor a local sub-population is resident on which a sequential GA (nodal GA) is performed. In arbitrary intervals, single individuals might move from/to one sub-population to another (emigration or immigration).

# 4 PROBLEM FUNCTIONS

**Problem 1: Rosenbrock Function**

$$f(x) = \sum_{i=1}^{n-1} 100\left(x_{i+1} - x_i^2\right)^2 + (1 - x_i)^2 - 2.048 \leq x_i \leq 2.048, i = 1, 10$$

with $f^* = 0$ at $x^* = (1, 1, ..., 1)$

This function has only one minimum at the base of a banana shaped valley. This is a very hard problem for local optimization methods, which may fail to reach the solution if relative tolerances are not used properly.

**Problem 2: Rastrigin Function**

$$f(x) = 10n \sum_{i=1}^{n} x_i^2 - 10\cos(2\pi x_i) \qquad -5.12 \leq x_i \leq 5.12, \qquad i = 1, 10$$

with $f^* = 0$ at $x^* = (0, 0, ..., 0)$

This function has many local solutions along flat areas and thus stretching the interval makes the valleys "sharper". The scaled problem has the interval $[-1, 1]$ for the Classical and $[-0.1, 0.1]$ for the Exponential Tunneling method.

**Problem 3: Griewangk Function**

$$f(x) = \frac{\sum_{i=1}^{n} x_i^2}{4000} - \prod_{i=1}^{n} \cos\left(\frac{x_i}{\sqrt{i}}\right) + 1 \qquad -600 \leq x_i \leq 600, \qquad i = 1, 10$$

with $f^* = 0$ at $x^* = (0, 0, ..., 0)$.The scaled version of this problem has bounds in $[-100, 100]$.

**Problem 4: Schubert Function**

$$f(x_1, x_2) = \left\{\sum_{i=1}^{5} i\cos\left[(i+1)\,x_1 + i\right]\right\}\left\{\sum_{i=1}^{5} i\cos\left[(i+1)\,x_2 + i\right]\right\}$$

$$-10 \leq x_i \leq 10, \quad i = 1, 2$$

This function exhibits 760 local minima in the region considered, eighteen of which are also global minima. The objective function value at these global minima is $f^* = -186.73019$ and their coordinates can be found in [10]. Although TA is able to find all global solutions, in this work only one is found.

# 5 NUMERICAL RESULTS

In order to perform a comparative study between these methods so different in nature, and to learn about their behaviour for increasing dimensionality, we have solved four academic problems with very different characteristics.

Both methods GA and TA, use as the general stopping conditions, a maximum number of function evaluations of 50,000. This implies 500 generations for the GA. A function value of less than $10^{-3}$ is also a general stopping condition.

TA were run from 100 random initial points, and for dimension $n = 100$ a set of 1000 random initial points were also considered. GA started from 100 different initial populations each one of 100 individuals (random points).

The number of populations was taken to favour the diversity in the GA which is a crucial aspect for a good behaviour of evolutionary methods. Alternatively, for the TA where every run depends on only one initial point, their performance becomes better if sufficient computer time (number of function evaluations) is allowed. We also report results for a larger number of function evaluations (500,000). Two variants of the GA are explored:

- Simple GA with dynamic mutation (G1).- This implementation has used the standard routines contained in the MATLAB Genetic Algorithms Toolbox V. 1.2 , developed in the University of Sheffield, U.K. 100 different initial populations of 100 individuals have been used for $n = 10$ and $n = 100$. A line recombination operator has also been used in the algorithm with real encoding and probability of 0.85. Dynamic mutation has also been used here. A Ranking selection method has been considered with a GGAP=0.9 in order to keep a 10% of the best individuals in each generation.

- GA using sub-populations (G2).- As stated previously the GA approach also offers the feasibility of exploiting its parallel nature. This characteristic will produce in theory not only a speed-up in terms of processing time, but also in the generation of better solutions, even if performed in only one cpu as implemented in this work. For this application, 2, 4, 5 and 10 sub-populations have been considered to evolve the GA. A Migration scheme is used with the total population of 100 individuals, GGAP=0.9, INSR (Insertion rate)=1, and crossover and mutation probabilities of 0.9 and 0.08 respectively.

In Table I, a comparison test is made for $n = 10$. for the Tunneling Algorithms,we report the percentage of success for the sample of 100 initial random points when 50000 (%1) and 500000 (%2) function evaluations are allowed. Also the most successful run (Best) obtained from one of the initial random points of the sample. All the runs were made with our code TUNNEL described in [3]. For the GA, the percentage of success (%), refers to the success of 100 initial populations of 100 individuals each. The best values are referred, as in the case of TA, to the number of function evaluations performed by the best population. Increasing the number of allowed generations (500), does not make a significant improvement in the convergence of the GA and thus is not reported here. This behaviour can be seen in Fig 1. The result NC implies that there was not convergence for ANY of the initial points (TA), or the initial populations (GA).

| Prob | Tun Cla | | | Tun Exp | | | Gen1 | | Gen2 | | | |
|---|---|---|---|---|---|---|---|---|---|---|---|---|
| | | | | | | | | | Sp=2 | | Sp=10 | |
| Ros | %1 | %2 | Best | %1 | %2 | Best | % | Best | % | Best | % | Best |
| | 100 | 100 | 81 | 100 | 100 | 81 | 0 | NC | 1 | 41270 | 0 | NC |
| Rast | %1 | %2 | Best | %1 | %2 | Best | | | | | | |
| SCA | 100 | 100 | 34 | 100 | 100 | 64 | | | | | | |
| Rast | %1 | %2 | Best | %1 | %2 | Best | % | Best | % | Best | % | Best |
| NSCA | 1 | 43 | 12397 | 2 | 2 | 2718 | 69 | 6120 | 93 | 6800 | 94 | 17200 |
| Griew | % | % | Best | % | % | Best | | | | | | |
| SCA | 100 | 100 | 23 | 100 | 100 | 23 | | | | | | |
| Griew | %1 | %2 | Best | % | % | Best | % | Best | % | Best | % | Best |
| NSCA | 95 | 100 | 31 | 100 | 100 | 31 | 60 | 4860 | 92 | 2750 | 76 | 7480 |

*Table I.- Results for $n = 10$*

In Table II a comparison test is made for $n = 100$. For TA, we also report the percentage of success for the sample of 100 (np1) and 1000 (np2) initial random points.

In Table III we report the Schubert function with dimension $n = 2$. TA was run from the sample of 100 initial random points, and GA with

| Prob | | Tun Cla | | | Tun Exp | | | Gen1 | | Gen2 Sp=2 | | Gen2 Sp=10 | |
|---|---|---|---|---|---|---|---|---|---|---|---|---|---|
| | | %1 | %2 | Best | %1 | %2 | Best | % | Best | % | Best | % | Best |
| Ros | np1 | 100 | 100 | 108 | 100 | 100 | 108 | 0 | NC | 0 | NC | 0 | NC |
| | np2 | 100 | 100 | 85 | 100 | 100 | 108 | | | | | | |
| Rast SCA | np1 | 100 | 100 | 105 | 100 | 100 | 7877 | 0 | NC | | | | |
| | np2 | 100 | 100 | 80 | 100 | 100 | 7272 | | | | | | |
| Rast NSCA | np1 | 22 | 99 | 16688 | 100 | 100 | 7724 | 0 | NC | 0 | NC | 0 | NC |
| | np2 | 99.9 | 99.9 | 3657 | 100 | 100 | 6613 | | | | | | |
| Griew SCA | np1 | 100 | 100 | 3 | 100 | 100 | 3 | 0 | NC | | | | |
| | np2 | 100 | 100 | 3 | 100 | 100 | 3 | | | | | | |
| Griew NSCA | np1 | 100 | 100 | 3 | 100 | 100 | 3 | 0 | NC | 0 | NC | 0 | NC |
| | np2 | 100 | 100 | 3 | 100 | 100 | 3 | | | | | | |

*Table II. Results for $n = 100$*

100 initital populations. Only one global minimum was obtained here, because GA can not guarantee not to find several times the same global solution. Although TA avoids going back to the minima already found as explained in section 2 we decided to stop the runs at the first global to make a fair comparison.

| Prob | Tun Cla | | Tun Exp | | Gen1 | |
|---|---|---|---|---|---|---|
| | % | Best | % | Best | % | Best |
| Schub | 100 | 9 | 100 | 9 | 100 | 1080 |

*Table III Results for Schubert function where $n = 2$*

The need to scale the problems with high density of local minima, when these minima are almost at the same level of the objective function, and close to each other is clear. The performance seems to be radically better between the scaled and the non-scaled problems. It becomes evident here, that the TA percentage of success, depends more on the maximum number of function evaluations allowed, than on the position of the initial point. These methods seem to converge from every initial guess if enough computer time is allowed. When dimensionality increases for this set of problems, TA became extremely fast.

For the GA, it can be observed, that the convergence for all functions improves when sub-populations are used. Nevertheless, when dimensionality increases ($n = 100$), convergence was not attained for any initial population.

In Fig 1. we show the typical behaviour of the performance of GA in all the examples tested here. The graph Average relates to the 100 initial populations and the graph Best, shows the most successful population. From these results it becomes clear the ability of these methods to get a low value of the objective function in few generations, but due to the lack of diversity, in spite of the use of dynamic mutation, they show great dificulty in achieving the right precision.

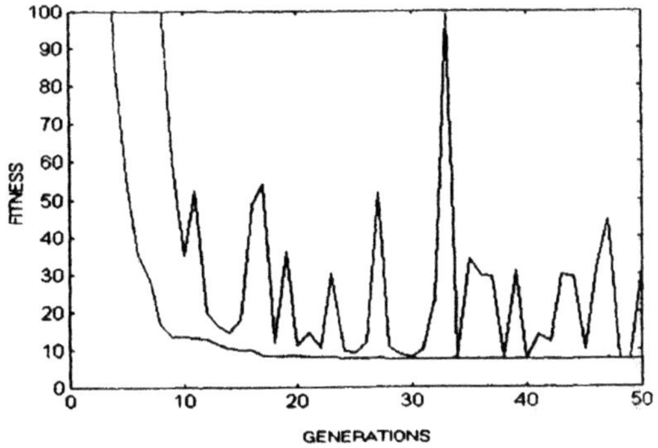


*Figure 36.1* Convergence behaviour of the GA for the Griewangk function

## 6 CONCLUSIONS

GA converge very fast to lower levels of the objective function as shown in Fig 1. but is extremely slow to finally get the desired global solution accuracy. For problems with several local minima at lower values of the objective function, as problems 2, 3 and 4, there is no guarantee of locating the global valley in a small number of generations, and thus the use of a local method at these points may not converge to the global solution.

TA converge to the global solution, even for problems with a large number of local minima near the global level, if enough time is given, for every initial point of the sample. Scaling the problem, is extremely beneficial for the TA, specially for larger dimensions, as the complexity of the objective function can be disminuished.

It becomes clear from the results presented in this work, that a hybrid method that takes advantage of the fast initial decrease of the objective function achieved with GA, and the ability of the TA to find the global solution starting from initial points which may not be in the global valley, could result in a fast and robust global method. We are now implementing a hybrid GA-TA algorithm, that produces GA-generations until it is automatically detected that the objective function does not decrease significantly, and continue from the best individual with TA iterations. This work will be reported in the near future.

## Notes

1. Throughout this paper, $||.||$ is the squared 2-norm.

## References

[1] Bäck, T., D.B. Fogel and Z. Michalewicz, eds., *Handbook in Evolutionary Computation*, Oxford University Press, 1997.

[2] Barron C., Gomez S., *The Exponential tunneling method*, Technical Report, IIMAS–UNAM, No.3, pp. 1–23, 1991

[3] Castellanos L., Gomez S.,*A New Implementation of the Tunneling Methods for Bound Constrained Global Optimization*, Sub. to ACM TOMS, 2000.

[4] Dorigo M., Maniezzo V., *Parallel Genetic Algorithms: Introduction and Overview of Current Research* in Stender J. Ed., Parallel Genetic Algorithms: Theory and Appl., 105 Press, pp. 5–41, 1993.

[5] Gill P.E., Murray W., Wright M.H., *Practical Optimization*, Acad. Press, 1981.

[6] Goldberg, D., *Genetic Algorithms in Search, Optimization and Machine Learning*, Addison-Wesley, 1989.

[7] S. Gomez, O. Gosselin y J. Barker, *Gradient–based History–Matching with a global optimization method.* Proc. of SPE Annual Tech. Conference, Vol. Reservoir Engineering Part 2, pp. 415–427, 1999.

[8] S. Gomez y D. Romero, *Two Global Methods for Molecular Geometry Optimization.* Progress in Mathematics Birkhauser, Vol. 121, pp. 503–509, 1994.

[9] B. Kepping, S. Gomez y C. Pantelides, *Using the Exponential Tunneling method for the time Optimal Control of High Dimensional systems*, manuscript, 1998.

[10] Levy A.V., Gomez S., *The tunneling method applied to global optimization*, en P.T. Boggs, R.H. Byrd and R.B. Schnabel (eds), Numerical Optimization SIAM, pp. 213–244, 1985.

[11] Levy A.V., Montalvo A., *The tunneling algorithm for the global minimization of functions*, SIAM J. Sci. Stat. Comp., Vol.6, No.1, pp. 15–29, 1985.

[12] Mühlenbein, H. and Schlierkamp-Voosen, D., *Predictive Models for the Breeder Genetic Algorithm*, Evolutionary Comp., Vol. 1, No. 1 25-49, 1993.

[13] Nichita D.V., Gomez S. and Luna E., *Multiphase Equilibria Calculations Using a Global Optimization Method.* Manuscript, 2000.

[14] Nichita D.V., Gómez S. and Luna E., *Multiphase Equilibria Calculations by Direct Minimization of Gibbs Free Energy Using the Tunneling Global Optimization Method.* Submitted to SPE Eastern Regional Meeting, 2000.

[15] Pettey C.B., Lenze MR, *A Theoretical Investigation of Parallel Genetic Algorithms*, in Proc. of the Third Int. Conf. on Genetic Algorithms, George Mason University, Los Altos, CA: Morgan Kaufmann, 1989 pp. 398–405.

[16] Sefrioui, M., Périaux, J. and Ganascia, J-C., *Fast Convergence thanks to diversity*, Evolutionary Computing, EC'96, San Diego, CA, 1996.

[17] Solano, J., Rodríguez, K. and García, D., *Model–based spectral estimation of Doppler signals using parallel genetic algorithms*, Artificial Intelligence in Medicine, No. 19, (2000), 75–89.

[18] Wright, A.H., *Genetic Algorithms for Real Parameter Optimization*, in Foundations of Genetic Algorithms, Rawlins G., ed., 205-218, 1991.

[19] Zhu C., Byrd R., Nocedal J., Lu. P., *Algorithm 778: L-BFGS-B-FORTRAN subroutines for Large Scale Bound Constrained Optimization*, TOMS 23(4): 550-560, 1997.

# Chapter 37

# CONVEXITY AND MONOTONICITY IN GLOBAL OPTIMIZATION

Hoang Tuy
*Institute of Mathematics*
*P.O. Box 631, Bo Ho, Hanoi, Vietnam*
htuy@hn.vnn.vn

**Abstract** Convexity and monotonicity are two properties of crucial importance in the deterministic approaches to global optimization. An overwhelming majority of deterministic global optimization methods developed over the last three decades are based on exploiting convexity in some form or another . On the other hand, a recently initiated theory of monotonic optimization is based on exploiting monotonicity solely. By drawing a parallel between the two approaches: d.c. (difference-convex) optimization and d.m. (difference-monotonic) optimization, this paper focuses on aspects which make the d.m. approach particularly attractive from a numerical point of view, at least in some important cases of interest. An improved form of an earlier developed basic algorithm for d.m. optimization is presented and applied to polynomial programming to illustrate the wide applicability of the new approach.

**Keywords:** Difference of convex (d.c.) functions, Difference of monotonic (d.m.) functions, Normal sets, Canonical monotonic optimization problem, Polyblock algorithm, Polynomial programming.

## 1 INTRODUCTION

Convexity plays an essential role in modern optimization theory. Since the linear space generated by convex functions consists of functions of the form $u(x) - v(x)$ where $u(x), v(x)$ are convex functions, it is natural that the d.c. (difference-convex) structure underlies a wide variety of nonconvex problems.

A theory of d.c. optimization has been developed over the last three decades to deal with optimization problems formulated by means of d.c.



*N. Hadjisavvas and P.M. Pardalos (eds.), Advances in Convex Analysis and Global Optimization,* 569–594.

functions, i.e. optimization problems of the form

$$\min\{f(x)|\ g_i(x) \leq 0 \quad i = 1, \ldots, m\} \tag{37.1}$$

where $f(x), g_1(x), \ldots, g_m(x)$ are d.c. functions. For a review of d.c. optimization see [12]. More recent advances can be found in [5] and [14], [16], [1].

Due to the lattice structure of the linear space of d.c. functions with respect to the operations of upper and lower envelope:

$$(f_1 \vee f_2)(x) = \max\{f_1(x), f_2(x)\}, \qquad (f_1 \wedge f_2)(x) = \min\{f_1(x), f_2(x)\},$$

any d.c. programming problem can be reduced to the canonical form

$$\min\{f(x)|\quad g(x) \leq 0 \leq h(x)\} \tag{37.2}$$

with all functions $f, g, h$ being convex (this problem differs from a convex program only by the presence of the additional reverse convex constraint $h(x) \geq 0$).

Despite its elegance and usefulness for studying a wide class of problems of interest, the theory of d.c. optimization has limitations. As a matter of fact, within its framework only properties pertinent to convexity or reverse convexity can be exploited, whereas many functions encountered in various applications are not d.c. and many other, known theoretically to be d.c., do not admit an easily computable effective d.c. representation. On the other hand, quite a few functions which are not d.c. or for which we do not know an effective d.c. representation, are monotonic, i.e. increasing or decreasing, in the sense that $f(x) \leq f(x')$ ($f(x) \geq f(x')$, resp.) whenever $0 \leq x \leq x'$.

A function which can be represented as the difference of two increasing functions, will be called a *d.m.* (difference-monotonic) *function.* Like d.c. functions, d.m. functions form a vector space which is also a lattice with respect to the upper and lower envelope operations. Therefore, for the formulation and study of nonconvex optimization problems, d.m. functions are as convenient as d.c. functions and perhaps even more convenient, since in most cases a d.m. representation of a function is much easier to find than a d.c. representation.

Motivated by these facts, a theory of monotonic optimization [8], [15], [17], has been recently developed to provide a general mathematical framework for studying optimization problems of the form (37.1), where each of the functions $f(x), g_i(x)$, is a d.m. function. Using the lattice structure of the class of d.m. functions, one can convert any such optimization problem to a canonical form (37.2) in which all functions $f, g, h$ are increasing on $R^n_+$. This provides an unified approach to global optimization by reducing the study of every global optimization problem of an extremely large class to the study of a canonical d.m. optimization problem, which becomes thus the core of the whole theory.

A geometric concept associated with monotonicity is that of *normal set*, introduced in mathematical economics [7] more than twenty years ago, which is defined as any set $G \subset R^n_+$ such that $x \in G$ whenever $0 \leq x \leq x'$, and $x' \in G$. Just as convex sets are essentially lower level sets of quasiconvex functions, normal sets are essentially lower level sets of increasing functions.

There is a striking parallel between monotonicity and convexity. Just as the most important elementary convex sets are polytopes which can be defined as convex hulls of finite sets, the most important elementary normal sets are *polyblocks*, defined as normal hulls of finite sets (the normal hull of a set being the smallest normal set containing it). As is well known, from the classical separation property of compact convex sets it follows that any compact convex set can be approximated by a sequence of enclosing polytopes. Likewise, a specific separation property of normal compact sets can be established from which it follows that any normal compact set can be approximated by a sequence of enclosing polyblocks. Furthermore, just as the separation property of compact convex sets gives the foundation for the polyhedral outer approximation method for maximizing convex functions over compact convex sets, this specific separation property of compact normal sets gives rise to a polyblock outer approximation method for maximizing increasing functions over compact normal sets.

The purpose of the present paper is to discuss monotonic optimization as developed in [15], [17], from a parallel perspective with d.c. optimization. We shall show how starting from a simple separation property it is possible to build up a general optimization theory based solely on exploiting monotonicity in much the same way as d.c. optimization is based on exploiting convexity. At the same time, we shall focus on some particular aspects which make the monotonic approach attractive from the numerical point of view, at least in some important cases. It seems that with monotonic analysis and monotonic optimization a new research field is emerging which sheds new lights on many difficult problems so far resistant to existing deterministic methods, including d.c. optimization methods. Successful applications to several important classes of nonconvex optimization problems such as: multiplicative programming, generalized multiplicative programming, linear fractional programming, optimization over the efficient set, etc., [8], [18], [19], [6], have confirmed the wide applicability of the new approach.

The paper is divided into 8 sections. In Section 2 we shall briefly recall the key role of the separation property of convex sets in modern optimization theory. In Section 3, the essentials of d.c. optimization theory will be reviewed to show how this separation property is behind the construction of almost every basic algorithm for d.c. optimization. In Section 4, we shall introduce and study monotonic functions and normal sets which are essentially level sets of increasing functions. By focussing on geometrical properties of normal sets pertinent to optimization, we shall show how these sets behave like convex sets in many respects. Section 5 will be devoted to the d.m. structure generated by functions which are differences of monotonic functions. Section 6 will be devoted to the basic algorithms of d.m. optimization, and Section 7 to some improved variants of these basic algorithms. Finally, in Section 8, to illustrate the usefulness of d.m. optimization we shall present a new and easily implementable approach to linearly constrained monotonic optimization, including polynomial and fractional programming.

As a matter of notation, for any two vectors $x', x \in R^n$ we write $x' \geq x$ and say that $x'$ *dominates* $x$ if $x'_i \geq x_i$ $\forall i = 1, \ldots, n$; we write $x' > x$ and say that $x'$ *strictly dominates* $x$ if $x'_i > x_i$ $\forall i = 1, \ldots, n$. Let $R^n_+ = \{x \in R^n | \ x \geq 0\}$ and $R^n_{++} = \{x \in R^n | \ x > 0\}$. For $x \in R^n_+$ let $I(x) = \{i | \ x_i = 0\}$ and define

$$K_x = \{x' \in R^n_+ | \ x'_i > x_i \ \forall i \notin I(x)\}, \qquad \text{cl}K_x = \{x' \in R^n_+ | \ x' \geq x\}. \tag{37.3}$$

If $a \leq b$ then $[a, b]$ denotes the box formed by all $x$ such that $a \leq x \leq b$. Also $(a, b] := \{x | \ a < x \leq b\}, [a, b) := \{x | \ a \leq x < b\}$. As usual $e$ is the vector of all ones and $e^i$ the $i$-th unit vector of the space under consideration, i.e. the vector whose $i$-th component equals 1, and all other components equal zero.

## 2 SEPARATION PROPERTY AS THE CORNERSTONE OF OPTIMIZATION THEORY

Whereas the fundamental role of convexity in modern optimization is well known, it is much less obvious which key property of convexity is responsible for this role.

Close scrutiny shows that the single property that lies at the foundation of almost all algorithmic and theoretical developments of convex and local optimization is the *separation* property of convex sets, namely:

Given a closed convex set $C \subset R^n$ and any point $x \in R^n \setminus C$ there exists a closed halfspace $L$ in $R^n$ such that

$$C \subset L, \quad \text{while } x \notin L. \tag{37.4}$$

It is this property that is used, in one or another of its equivalent formulations, in such constructions as:

- Subdifferential of convex functions
- Linearization (approximation of convex functions by affine functions).
- Cutting plane (Outer Approximation)
- Duality
- Lagrange multipliers
  etc.

If a function is differentiable in the neighbourhood of a point then locally at this point it can be approximated by an affine function. More generally, a function may have a convex minorant in the neigbourhood of a point. Therefore, convexity also plays a fundamental role in nonconvex local optimization.

Nevertheless, many functions and sets are nonconvex, and in many problems we are interested in global rather than local optimality. One may ask whether convexity can play any role in nonconvex global optimization ?

The development during these last decades has shown that not only is convexity a useful tool, it also plays an essential role even in nonconvex optimization. This is no wonder since, as can easily be observed, a lot of convexity is present in a very broad class of problems, even if it is sometimes hidden and not easy to exploit for numerical purposes.

## 3 D.C. STRUCTURE AND THE CANONICAL D.C. PROGRAMMING PROBLEM

A striking property of euclidean spaces is that for any closed set $X \subset R^n$ one can find a continuous convex function $g(x)$ on $R^n$ such that

$$x \in S \Leftrightarrow g(x) - \|x\|^2 \leq 0, \quad x \in \text{int}S \Leftrightarrow g(x) - \|x\|^2 < 0$$

(see e.g. [5]). In particular $S = \{x \in R^n | \; g(x) - \|x\|^2 \leq 0\}$. Due to this property, any continuous constraint such as $h(x) \leq 0$ with $h(x)$ continuous can be converted into a constraint of the form $g(x) - \|x\|^2 \leq 0$ with $g(x)$ continuous convex. Also any continuous optimization problem of the form

$$\min\{f(x)| \quad x \in X\}, \quad f(x) \text{ continuous, } X \text{ closed } \subset R^n$$

can in principle be reduced to minimizing a linear function under constraints of the form

$$g_i(x) \leq 0 \quad i = 1, \ldots, m$$

where each function $g_i(x)$ is a *d.c. function* (difference of two convex functions).

From a practical point of view a host of optimization problems encountered in various applications are or can be formulated in terms of d.c. functions (see e.g. [12]). In fact, it can be proved that any $C^2$-function, in particular any polynomial, is d.c. Hence, the class DC($X$) of d.c. functions on a compact convex set $X$ is dense in the Banach space of continuous functions on $X$ with the usual supnorm topology. Furthermore, the class of d.c. functions turned out to be most favourable for optimization theory, due to its stability under linear operations and the operations of lower and upper envelope of frequent use in optimization. More precisely, (see e.g. [14]):

**Proposition 1** *If $f, g \in$ DC($X$) then the following functions also belong to* DC($X$) :
*1)* $\alpha f + \beta g$, $(\alpha, \beta \in R)$;
*2)* $\min\{f, g\}$, $\max\{f, g\}$.
*If* $f(x) \in$ DC($X$) *then* $|f(x)| \in$ DC($X$).

As consequences of the above properties it follows that (see e.g. [12], or [14])

(i) A conjunctive system of d.c. inequalities

$$g_i(x) \leq 0 \ \forall i = 1, \ldots, m$$

is equivalent to a single d.c. inequality $\max_{i=1,\ldots,m} g_i(x) \leq 0$.

(ii) A disjunctive system of d.c. inequalities

$$g_i(x) \leq 0 \text{ for at least one } i = 1, \ldots, m$$

is equivalent to a single d.c. inequality $\min_{i=1,\ldots,m} g_i(x) \leq 0$.

(iii) Any problem of minimizing or maximizing a d.c. function $f$ : $[0, b] \to R$, under d.c. constraints (briefly, any d.c. optimization problem) can be cast into the *canonical form*

$$\min\{f(x) | \ g(x) \leq 0 \leq h(x)\} \qquad \text{(CDC)}$$

where $f(x), g(x), h(x)$ are all convex.

Thus the fundamental problem of d.c. optimization is (CDC) in which all the nonconvexity is concentrated on a unique reverse convex constraint $h(x) \geq 0$. Several methods are currently available for solving

(CDC) (see e.g. [2], [5], [14]). One method in which the separation property is used in the most direct way is the outer approximation method which proceeds basically as follows. Let $\gamma$ be any real number. Then solving (CDC) can be reduced to solving a sequence of *d.c. feasibility problems* of the following form

*Find a feasible solution to (CDC) such that* $f(x) < \gamma$ *if there is one.*

Under mild conditions, this amounts to checking the nonemptiness of the set $G_\gamma = \{x|\ g(x) \leq 0 \leq h(x),\ f(x) \leq \gamma\}$. Since $g(x), f(x)$ are convex, the two convex constraints $f(x) \leq \gamma,\ g(x) \leq 0$ can be incorporated into a single one. Then, by changing the notation the d.c. feasibility problem can be put in the form

*Find a point* $x \in R^n$ *such that* $g(x) \leq 0 \leq h(x)$.

Setting $G = \{x|\ g(x) \leq 0\}$ this in turn reduces to solving the *convex maximization* problem

$$\max\{h(x)|\ x \in G\} \qquad \text{(CM)}$$

where $h(x)$ is a convex function and $G$ is a closed convex set (assumed to be compact for simplicity). If $P$ is a polytope containing $G$ and $v$ a vertex of $P$ such that $v \notin G$ then by the separation property there exists a halfspace such that the polytope $P' = L \cap P$ still contains $G$ but excludes $v$. Using this observation, one can solve (CM) by an outer approximation (OA) method which consists of adaptively constructing a nested sequence of polytopes enclosing $G$: $P_1 \supset P_2 \supset \ldots \supset G$ in such a way that

$$\max\{h)x)|\ x \in P_k\} \searrow \max\{h(x)|\ x \in G\}.$$

Thus in principle any d.c. optimization problem can be solved by the OA method. In practice, especially for large scale problems, other methods such as inner approximation, branch and bound, dualization, or combined methods, may be more efficient by taking account of the specific problem's structure (see [2], [5], [14], [9], [1]). In each case, however, a close look into the algorithm and its foundation inevitably reveals some variant of the separation property of convex sets, either for linear or convex relaxation to compute bounds over the partition sets, or for dualization to handle non convex constraints, or for decomposition and dimension reduction, etc.

The question arises as to whether it is possible to construct a deterministic approach to global optimization without convexity.

We shall show in this paper that a positive and satisfactory answer to that question is furnished by monotonic optimization. This is an approach based on exploiting the monotonic structure underlying an extremely large variety of mathematical programming problems. The key role in this approach will be given to a specific separation property of normal sets, similar to but not quite the same as the separation property of convex sets.

# 4 MONOTONIC FUNCTIONS

A function $f(x) : R^n_+ \to R$ is said to be *increasing* if

$$0 \leq x \leq x' \in R^n_+ \quad \Rightarrow \quad f(x) \leq f(x')$$

It is said to be increasing on a box $[0, b] \subset R^n_+$ if the above property is restricted to all $x, x'$ in this box. A function is decreasing if $0 \leq x \leq x' \in R^n_+ \quad \Rightarrow \quad f(x') \leq f(x)$. Of course, if $f(x)$ is decreasing then $-f(x)$ is increasing. A function which is either increasing or decreasing is said to be monotonic.

Examples of increasing functions on $R^n_+$ abound in mathematics: polynomials with nonnegative coefficients, or more generally, posynomials, etc. In real life (engineering, economics), we often encounter cost, utility, efficiency functions, etc. which are neither convex not concave, not even d.c., but simply increasing.

A geometric concept closely connected with monotonicity is that of *normal set* which is defined as any set $G \subset R^n_+$ such that

$$0 \leq x \leq x' \in G \quad \Rightarrow \quad x \in G$$

If a function $g(x) : R^n_+ \to R$ is increasing then for any $\gamma \in R$ the level set $G = \{x \in R^n_+ | \; g(x) \leq \gamma\}$ is normal. Conversely, it can easily be proved that for any normal set $G \subset R^n_+$ there exists an increasing function $g(x)$ such that $\text{cl}G = \{x \in R^n_+ | \; g(x) \leq 0\}$.

**Proposition 2** *The union or intersection of any family of normal sets is normal.*

*Proof* Straightforward. □

Let $G \subset [0, b]$ be a closed normal set. For every $z \in [0, b]$ define

$$\pi_G(z) = \lambda z, \quad \text{for} \quad \lambda = \max\{\alpha | \; \alpha z \in G\} \tag{37.5}$$

The set $\partial^+ G := \{y = \pi_G(z) | \; z \in R^n_+\}$ is called the *upper boundary* of $G$.

**Proposition 3** *An increasing function $f(x)$ achieves its maximum over a closed normal set $G$ on $\partial^+ G$.*

*Proof* For any $z \in G$ the point $x = \pi_G(z)$ satisfies $x \geq z$ hence $f(x) \geq f(z)$. □

Clearly, the singleton $\{0\}$, the whole orthant $R^n_+$ and the set $R^n_+ \setminus R^n_{++}$ are trivial normal sets. A box $[0, z]$ with $0 \leq z \leq b$ is also a normal set. An elementary normal set which is the union of finitely many boxes is called a *polyblock*. In other words, a polyblock is a set of the form $P = \cup_{z \in T}[0, z]$ where $T$ is a finite subset of $R^n_+$. Each point $z \in T$ is called a *vertex* of the polyblock and we write $T = \text{vert}(P)$. A vertex $z$ is said to be *proper* if there is no $z' \in T \setminus \{z\}$ such that $z \leq z'$. Of course a polyblock is fully determined by its proper vertex set, denoted by $\text{pvert}(P)$.

Given a set $A \subset R^n_+$ the smallest normal set containing $A$ is called the *normal hull* of $A$ and is denoted by $A^{\rceil}$. Clearly

$$A^{\rceil} = \cup_{z \in A}[0, z]$$

so a polyblock is the normal hull of its proper vertex set.

**Proposition 4** *The union or intersection of finitely many polyblocks is a polyblock.*

*Proof* We need only show that the intersection of finitely many polyblocks is a polyblock. But $(\cup_{z \in T}[0, z]) \cap (\cup_{y \in V}[0, y]) = \cup_{z \in T, y \in V}([0, z] \cap [0, y])$ and $[0, z] \cap [0, y] = [0, u]$ with $u_i = \min\{z_i, y_i\}$. □

We next show a separation property of normal sets whose key role for monotonic optimization is similar to that of the classical separation property of convex sets for d.c. optimization.

**Proposition 5** *Let $G \subset [0, b]$ be a normal closed set. For any $z \in [0, b] \setminus G$ the cone $K_y$ with $y = \pi_G(z)$ separates $z$ strictly from $G$ (i.e. contains $z$ but is disjoint from $G$).*

*Proof* Recall that $K_y := \{x \in R^n_+ |\ y_i < x_i\ \forall i \notin I(y)\}$, where $I(y) = \{i |\ y_i = 0\}$. Since $z \notin G$, it is clear that $y_i < z_i\ \forall i \notin I(y)$, hence $z \in K_y$. On the other hand, if $x \in G \cap K_y$ then $y_i < x_i\ \forall i \in I(y) = I(z)$, hence for sufficiently small $\varepsilon > 0$ we have $y + \varepsilon z \in [0, x] \subset G$, contradicting the definition of $y = \pi_G(z)$. Therefore $K_y$ is disjoint from $G$. □

**Proposition 6** *If $0 \leq x \leq z \leq b$ then $P = [0, z] \setminus K_x$ is a polyblock in $[0, b]$ with vertex set $V = \{z^{*i} |\ i \notin I(x)\} \subset R^n_{++}$, where*

$$z^{*i} = z - (z_i - x_i)e^i.$$

*Proof* Let $K_i = \{y \in R^n_+ |\ x_i < y_i\}$. Since $K_x = \cap_{i \notin I(x)} K_i$, we have $P = [0, z] \setminus K_x = \cup_{i \notin I(x)}([0, z] \setminus K_i)$. But

$$[0, z] \setminus K_i = \{y |\ 0 \leq y_i \leq x_i,\ 0 \leq y_j \leq z_j\ \forall j \neq i\} = [0, z^{*i}]$$

where $z^{*i}$ denotes the vector such that $z^{*i}_j = z_j\ \forall j \neq i,\ z^{*i}_i = x_i$, i.e. $z^{*i} = z - (z_i - x_i)e^i$. To prove that $V \subset R^n_{++}$, consider any $z^{*i}$ with

$i \notin I(x)$. Then for every $j \neq i$ we have $z_j^{*i} = z_j > 0$, while $z_i^{*i} = x_i > 0$. □

**Proposition 7** *Let $G$ be a compact set contained in a box $[0, b] \subset R_+^n$. Then the following assertions are equivalent:*
*(i) $G$ is normal;*
*(ii) $G$ is the intersection of a family of polyblocks in $[0, b]$.*

*Proof* (i)⇒ (ii). The box $[0, b]$ is a polyblock containing $G$. Let $E$ be the intersection of all polyblocks containing $G$. Clearly $G \subset E$. If $z \in E \setminus G$ then by Proposition 5 there exists $y \in \partial^+ G$ such that $z \in K_y$ but $K_y \cap G = \emptyset$, i.e. $z \notin P := [0, b] \setminus K_y$, while $P$ is a polyblock containing $G$, contradicting $z \in E \subset P$.

(ii)⇒ (i) Obvious because any polyblock is normal and the intersection of a family of normal sets is normal. □

A set $H \subset R_+^n$ is called *reverse normal* if $0 \leq x \leq x' \notin H \Rightarrow x \notin H$. It is said to be *reverse normal* in a box $[0, b]$ if $0 \leq x \leq x' \in [0, b] \setminus H \Rightarrow x \notin H$. Clearly:

(i) A set $H$ is reverse normal if and only if the set $H^b = R_+^n \setminus H$ is normal.

(ii) If $h(x)$ is an increasing function on $[0, b]$ the set $H = \{x \in [0, b] | \ h(x) \geq 0\}$ is reverse normal.

Let $H$ be a closed reverse normal set in $[0, b]$. For every $z \in [0, b]$ define

$$\rho_H(z) = b - \mu(b - z), \quad \text{for} \quad \mu = \max\{\alpha | \ b - \alpha(b - z) \in H\} \qquad (37.6)$$

The set $\partial^- G := \{y = \rho_H(z) | \ z \in [0, b]$ is called the *lower boundary* of $H$.

**Proposition 8** *An increasing function $f(x)$ achieves its minimum over a closed reverse normal set $H$ on $\partial^- H$.*

*Proof* For any $z \in H$ the point $y = \rho_H(z)$ satisfies $y \leq z$ hence $f(y) \leq f(z)$. □

# 5 THE D.M. STRUCTURE

Many functions which may not be monotonic can however be represented as difference of two increasing functions. For example, if $S$ is a closed subset of $R_+^n$ then one can show that

$$S = \{x \in R_+^n | \ g(x) - \|x\|^2\}$$

where $g(x) = \sup\{\langle x, y\rangle + \|y\|^2 | \ y \in R_+^n \setminus S\}$ and $\|x\|^2$ are both increasing functions on $R_+^n$.

A function $f : R_+^n \rightarrow R$ which is the difference of two increasing functions: $f(x) = f_1(x) - f_2(x)$, where each function $f_i(x), i = 1, 2$, is increasing, is called a *d.m.* (difference of monotonic) or *d.i.* (difference of increasing) *function*. Denote by DM[0, b] the class of d.m. functions on [0, b]. In fact this class enjoys several properties of remarkable interest for optimization theory.

**Proposition 9** (i) DM[0, b] *is dense in the Banach space* C[0, b] *of continuous functions on* [0, b] *with the supnorm topology.*
(ii) DM[0, b] *is a vector lattice*

*Proof* (i) It is easy to see that every polynomial is d.m. on any box $[0, b] \subset R^n_+$. Indeed, any polynomial $P(x)$ can be written as $P(x) = P^+(x) - P^-(x)$ where $P^+(x)$ (sum of all terms with nonnegative coefficients) and $P^-(x)$ (sum of all terms with nonpositive coeffcients) are both increasing. Since the set of polynomials is dense in C[0, b] it follows that DM[0, b] is dense in C[0, b].
(ii) That DM[0, b] is a vector space is plain. To see that it is a lattice we write

$$\max_{i=1,\dots,m}\{p_i - q_i\} = \sum_{k=1,\dots,m} p_k - \min\{\sum_{k\neq i} p_k + q_i\},$$

$$\min_{i=1,\dots,m}\{p_i - q_i\} = \sum_{k=1,\dots,m} p_k - \max_{i=1q,\dots,m}\{\sum_{k\neq i} p_k + q_i\}$$

and note that the sum, the upper and the lower envelopes of finitely many increasing functions are increasing. □
Just as with DC($X$), the lattice structure of DM[0, b] immediately implies that:
(i) A conjunctive system of d.m. inequalities is equivalent to a single one:

$$g_i(x) \le 0 \ \forall i = 1, \dots, m \quad \Leftrightarrow \quad \max_{i=1,\dots,m} g_i(x) \le 0.$$

(ii) A disjunctive system of d.m. inequalities is equivalent to a single one:

$$g_i(x) \le 0 \text{ for at least one } i = 1, \dots, m \quad \Leftrightarrow \quad \min_{i=1,\dots,m} g_i(x) \le 0.$$

Consider now an arbitrary d.m. optimization problem, i.e. an optimization problem of the form

$$\min\{f_1(t) - f_2(t) | \ g_{i,1}(t) - g_{i,2}(t) \le 0 \ (i = 1, \dots, m), \ \ 0 \le t \le c\} \quad \text{(DM)}$$

where $f_1, f_2, g_{i,1}, g_{i,2}$ are increasing functions on $[0, c] \subset R^p_+$.

**Proposition 10** *Any problem (DM) can be cast into the* canonical form

$$\min\{f(x) | \ g(x) \le 0 \le h(x)\} \quad \text{(CDM)}$$

*where $f(x), g(x), h(x)$ are increasing functions on a box $[0,b] \subset R^n_+$ ($n = p+2$).*

*Proof* Since $f_2(t)$ is increasing on $[0,c]$, we have $d_0 := f_2(c) - f_2(0) \geq 0$, so by introducing a new variable $u_0 = f_2(c) - f_2(t)$ we can write (DM) as

$$\begin{gathered} \min\{f_1(t) + u_0 - f_2(c)| \\ u_0 + f_2(t) - f_2(c) \geq 0,\ g_{i,2}(t) - g_{i,1}(t) \geq 0\ (i = 1, \dots, m) \\ 0 \leq t \leq c,\ 0 \leq u_0 \leq d_0\} \end{gathered} \tag{37.7}$$

where $f_1(t) + u_0$ and $u_0 + f_2(t)$ are increasing on $[0,c] \times [0,d_0]$. Next the system of d.m. constraints in (37.7) can be replaced by a single one

$$w_1(t, u_0) - w_2(t, u_0) \geq 0.$$

where $w_1, w_2$ are increasing. Setting $d_1 = w_2(c, d_0) - w_2(0,0)$ and introducing a further additional variable $u_1 = w_2(c, d_0) - w_2(t, u_0)$ we can split this constraint into two ones:

$$w_1(t, u_0) + u_1 \geq w_2(c, d_0),\ w_2(t, u_0) + u_1 \leq w_2(c, d_0), \quad 0 \leq u_1 \leq d_1.$$

Finally, the problem (DM) is equivalent to

$$\begin{gathered} \min\{f_1(t) + u_0 - f_2(t)| \\ w_1(t, u_0) + u_1 \geq w_2(c, d_0) \geq w_2(t, u_0) + u_1, \\ (t, u_0, u_1) \in [0,c] \times [0, d_0] \times [0, d_1]\}, \end{gathered}$$

which is a problem of the form (CDM) where $x = (t, u_0, u_1) \in R^n_+$, $n = p+2$, $b = (c, d_0, d_1)$. □

It is worth noticing that the canonical d.m. optimization problem (CDM) has exactly the same form as (CDC), except that all the functions $f(x), g(x), h(x)$ in (CDM) are increasing rather than convex as in (CDC).

## 6 THE BASIC ALGORITHM OF D.M. OPTIMIZATION

We now use the separation property of normal sets to construct an efficient algorithm for solving the canonical d.c. optimization problem (CDM).

Setting

$$G = \{x \in R^n_+ |\ g(x) \leq 0\}, \quad H = \{x|\ h(x) \geq 0\} \tag{37.8}$$

we can rewrite (CDM) as

$$\min\{f(x)|\ x \in G \cap H\} \tag{37.9}$$

where $G$ is a normal set, $H$ a reverse normal set. We shall assume that $G$ and $H$ are closed, $G \cap H \subset [0, b]$ and $f(x)$ is lower semicontinuous on $G$. By setting $x = b - y$ and

$$\tilde{f}(y) = -f(b - y), \quad \tilde{G} = b - G, \quad \tilde{H} = b - H$$

we convert this problem to a maximization problem, viz:

$$\max\{\tilde{f}(y) | \quad y \in \tilde{G} \cap \tilde{H}\}$$

where $\tilde{H}$ is a closed normal set, $\tilde{G}$ is a closed reverse normal set, $\tilde{G} \cap \tilde{H} \subset [0, b]$ and $\tilde{f}(y)$ is increasing on $[0, b]$.

Thus, by changing the notation we can consider, instead of a minimization, the maximization problem

$$\max\{f(x) | \, x \in G \cap H\} \tag{CDM}$$

where $G$ is a closed normal set, $H$ a closed reverse normal set and

$$G \cap H \subset [a, b], \quad x > a \;\forall x \in H, \tag{37.10}$$

with $0 < a < b$. The additional assumption $x > a \; \forall x \in H$, which is innocuous and can always be made to hold by a translation, is necessary to ensure theoretical convergence of the polyblock approximation algorithm we are going to describe.

The basic idea of the algorithm [17] is simple: inductively build up a decreasing sequence of polyblocks $P_1, P_2, \ldots, P_k, \ldots$, enclosing at least an optimal solution, such that:

$$\max\{f(x) | \, x \in P_k\} \searrow \max\{f(x) | \, x \in G \cap H\}. \tag{37.11}$$

Observe that if a vertex $z$ of $P_k$ does not belong to $H$ then, by the reverse normality of $H$, the whole box $[0, z]$ is disjoint from $H$ and hence cannot contain any feasible solution. Therefore every vertex $z$ of $P_k$ that does not belong to $H$ can be removed. With this in mind, the construction of the sequence $P_1 \supset P_2 \ldots \supset P_k \supset \ldots$ can be carried out as follows: Start with $P_1 = [0, b]$. Suppose $P_k$ has been already constructed and $T_k \subset H$ is its proper vertex set. Then the maximum of $f(x)$ is achieved at some $z^k \in T_k$ which can be computed as a maximizer of $f(x)$ over $T_k$. Since $z^k \in H$ if $z^k \in G$ then it is feasible, hence optimal because $f(z^k) \geq f(x) \; \forall x \in P_k$ and $P_k$ contains at least an optimal solution. If $z^k \notin G$ then we can separate $z^k$ from $G$ by a cone $K_{y^k} = y^k + R^n_{++}$ where $y^k = \pi_G(z^k)$. Using Proposition 6, the vertex set $T'_k$ of the polyblock $P'_k = P_k \setminus K_{y^k}$ can be computed and after removing the improper elements and also all elements that do not belong to $H$, the proper vertex set $T_{k+1}$ of $P_{k+1}$

is obtained. If $T_{k+1} = \emptyset$, the problem is infeasible. Otherwise, since $T_{k+1} \subset H$ and $P_{k+1}$ still contains at least an optimal solution (because no optimal solution has been lost by the removal of $K_{y^k}$ and all the vertices which do not belong to $H$), the procedure can be continued.

In that way, either the procedure terminates at some iteration $k$, (giving then an optimal solution or the evidence that the problem is infeasible), or it continues endlessly. We shall prove that the latter case (37.11) holds.

In practice a tolerance $\varepsilon$ is given, so at each iteration $k$ we can update the current best feasible solution $\bar{x}^k$ and remove all $z \in T'_k$ such that $f(z) \leq f(\bar{x}^k) + \varepsilon$. If we do that, then the situation $T_{k+1} = \emptyset$ means that no better feasible solution than $\bar{x}^k$ can be found, hence, if $\bar{x}^k$ exists then it is an $\varepsilon$-optimal solution (i.e. a feasible solution such that $f(\bar{x}^k) \geq f(x) - \varepsilon$ for all feasible solutions $x$); on the other hand the problem is infeasible if $\bar{x}^k$ does not exist. Also, if $y^k \in H$ and $f(z^k) - f(y^k) \leq \varepsilon$ then the procedure can be terminated with $y^k$ accepted as an $\varepsilon$-optimal solution. We can thus state

**Algorithm 1** (for a maximization problem)
*Initialization.* If $a \notin G$ or $b \notin H$ the problem is infeasible (because then $G \cap H = \emptyset$). Otherwise, let $T_1 = \{b\}$. Let $\bar{x}^1$ be the best feasible solution available, $CBV = f(\bar{x}^1)$. If no feasible solution is available, let $CBV = -\infty$. Set $k = 1$.
*Step 1.* Select $z^k \in \operatorname{argmax}\{f(z) | \quad z \in T_k\}$.
Compute $y^k = \pi_G(z^k)$ If $f(z^k) - f(y^k) \leq \varepsilon$, and $y^k \in H$, then terminate: $y^k$ is an $\varepsilon$-optimal solution. Otherwise, determine the new current best value $CBV$ and the new current best feasible solution $\bar{x}^{k+1}$.
*Step 2.* Determine $T'_k = (T_k \setminus \{z^k\}) \cup \{z^{k,1}, \ldots, z^{k,n}\}$ where

$$z^{k,i} = z^k - (z_i^k - y_i^k)e^i, \; i = 1, \ldots, n. \tag{37.12}$$

From $T'_k$ remove all improper elements, all $z \notin H$ and all $z$ such that $f(z) \leq CBV + \varepsilon$ and let $T_{k+1}$ be the resulting set.
*Step 3.* If $T_{k+1} = \emptyset$, terminate: if $CBV = -\infty$, the problem is infeasible; if $CBV > -\infty$, the current best feasible solution $\bar{x}^k$ is an $\varepsilon$-optimal solution.
If $T_{k+1} \neq \emptyset$, set $k \leftarrow k + 1$ and return to Step 1.

**Remark.** The improper members of $T'_k$ can be removed by the following rule which can be justified easily (see [17]):
For every $z \in T_k \setminus \{z^k\}$ such that $z \geq y^k$ do: if $z_j \geq z_j^k \; \forall j \neq i$ then remove $z^{k,i}$.

**Theorem 1** *Algorithm 1 is finite unless $\varepsilon = 0$, in which case (37.11) holds, and each of the generated sequences $\{z^k\}, \{y^k\}$ contains a subsequence converging to an optimal solution.*

*Proof* Let us agree to call any vector $z^{k,i}$ obtained from $z^k$ by formula (37.12) a child of $z^k$. If the Algorithm is infinite, it generates at least one infinite sequence $z^{l_1}, z^{l_2}, \ldots, z^{l_h}, \ldots$ contained in $H$ such that $z^{l_{h+1}}$ is a child of $z^{l_h}$, i.e.

$$z^{l_{h+1}} = z^{l_h} - (z^{l_h}_{i_h} - y^{l_h}_{i_h})e^{i_h}. \tag{37.13}$$

where $y^{l_h} = \pi_G(z^{l_h})$ and $i_h \in \{1, \ldots, n\}$. (Such a sequence can be obtained by taking $z^{l_1} = b$, and for $h \geq 1$, selecting $z^{l_{h+1}}$ among the children of $z^{l_h}$ that have infinitely many descendants.)

We show that any sequence satisfying (37.13) converges to an optimal solution. Clearly $z^{l_1} \geq z^{l_2} \geq \ldots \geq z^{l_k} \geq \ldots \geq 0$. Let $\tilde{z}$ be an accumulation point of this bounded sequence ($z^k \in [0, b]\ \forall k$). Since $\|z^{l_1} - \tilde{z}\| \geq \|z^{l_2} - \tilde{z}\| \geq \ldots$ and a subsequence of this decreasing sequence tends to zero, it follows that $\lim_{h\to+\infty} \|z^{l_h} - \tilde{z}\| = 0$, i.e. $\tilde{z} = \lim_{h\to+\infty} z^{l_h}$. This implies $z^{l_h} - z^{l_{h+1}} \to 0$, and hence, since $z^{l_{h+1}}_{i_h} = y^{l_h}_{i_h}$,

$$z^{l_h}_{i_h} - y^{l_h}_{i_h} = |z^{l_h}_{i_h} - z^{l_{h+1}}_{i_h}| \leq \|z^{l_h} - z^{l_{h+1}}\| \to 0 \quad (h \to +\infty).$$

But $z^{l_h} - y^{l_h} = \lambda_h z^{l_h}$, so $z^{l_h}_{i_h} - y^{l_h}_{i_h} = \lambda_h z^{l_h}_{i_h} \to 0$. On the other hand, in view of the fact $z^{l_h} \in H$, we have $z^{l_h}_i \geq \min_{j=1,\ldots,n} a_j > 0\ \forall i$, hence $\lambda_h = (z^{l_h}_{i_h} - y^{l_h}_{i_h})/z^{l_h}_{i_h} \to 0$. Therefore, $z^{l_h} - y^{l_h} \to 0$. and consequently, $\tilde{z} = \lim_{h\to+\infty} x^{l_h} = \lim_{h\to+\infty} z^{l_h}$. Since $y^{l_h} \in G\ \forall h$ while $z^{l_h} \in H\ \forall h$, we can conclude $\tilde{z} \in G \cap H$. Furthermore, since $f(z^{l_h}) \geq f(x)\ \forall x \in G \cap H$ and $\{z^{l_h}\} \subset H$, it follows from the upper semicontinuity of $f(x)$ on $H$ that $f(\tilde{z}) \geq f(x)\ \forall x \in G \cap H$, i.e. $\tilde{z}$ is a global maximizer of $f(x)$ over $G \cap H$. If $\varepsilon > 0$, we would have had, for $h$ sufficiently large, $f(z^{l_h}) - f(y^{l_h}) \leq \varepsilon$ and the algorithm would have stopped at iteration $l_h$. Therefore, the algorithm can be infinite only if $\varepsilon = 0$. □

## Agorithm for Minimization Problems

Although the problem (37.9) could be solved by transforming it into a maximization problem and applying the above algorithm, it is more convenient to apply a direct algorithm obtained by expressing Algorithm 1 in terms of the original data of (37.9) (in particular interchanging the roles of $G$ and $H$). For the minimization problem (37.9) the assumption (37.10) is now replaced by

$$G \cap H \subset [0, b], \quad x < b' < b\ \forall x \in G. \tag{37.14}$$

**Algorithm 2** (for a minimization problem)
*Initialization.* If $a \notin G$ or $b \notin H$ the problem is infeasible (because then $G \cap H = \emptyset$). Otherwise, let $T_1 = \{0\}$. Let $\bar{x}^1$ be the best feasible solution available, $CBV = f(\bar{x}^1)$. If no feasible solution is available, let $CBV = +\infty$. Set $k = 1$.
*Step 1.* Select $z^k \in \text{argmin}\{f(z)| \quad z \in T_k\}$. Compute $y^k = \rho_G(z^k)$ If $f(y^k) - f(z^k) \leq \varepsilon$ and $y^k \in G$, then terminate: $y^k$ is an $\varepsilon$-optimal solution. Otherwise, determine the new current best value $CBV$ and the new current best feasible solution $\bar{x}^{k+1}$.
*Step 2.* Determine $T'_k = (T_k \setminus \{z^k\}) \cup \{z^{k,1}, \ldots, z^{k,n}\}$ where

$$z^{k,i} = z^k + (y_i^k - z_i^k)e^i, \; i = 1, \ldots, n. \tag{37.15}$$

From $T'_k$ remove all improper elements, all $z \notin G$ and all $z$ such that $f(z) \geq CBV - \varepsilon$ and let $T_{k+1}$ be the resulting set.
*Step 3.* If $T_{k+1} = \emptyset$, terminate: if $CBV = +\infty$, the problem is infeasible; if $CBV < +\infty$, the current best feasible solution $\bar{x}^k$ is accepted as an $\varepsilon$-optimal solution.
If $T_{k+1} \neq \emptyset$, set $k \leftarrow k + 1$ and return to Step 1.

Here an element $z \in T \subset [0, b]$ is said to be improper if $z \leq z'$ for some $z' \in T \setminus \{z\}$.

# 7 IMPROVED VARIANTS OF THE BASIC ALGORITHM

I has been observed in some computational experiments that, as it stands, Algorithm 1 tends to be slow when the objective function $f(x)$ does not differ much from the function $g(x)$. To get over the difficulty in such cases, and enhance flexibility as well as efficiency of the polyblock method, we propose two alternative approaches: *successive improvement* through a sequence of feasibility problems and *branch and bound* method with Algorithm 1 as bounding subroutine.

## 7.1 SUCCESSIVE IMPROVEMENT

Given a tolerance $\varepsilon > 0$, a point $x \in [0, b]$ is called an *$\varepsilon$-approximate feasible solution* of (CDM) if $g(x) \leq \varepsilon, \; h(x) \geq 0$. An $\varepsilon$-approximate feasible solution $\bar{x}$ such that $f(\bar{x}) + \varepsilon \geq f(x) \; \forall x \in G \cap H$ is called an *$\varepsilon$-approximate optimal solution.* For any $r \in [f(0), f(b)]$ consider the following feasibility problem:

FP($r$): *Find an $\varepsilon$-approximate feasible solution $x$ of* (CDM) *such that $f(x) \geq r + \varepsilon$, if there is one.*

If an efficient procedure is available for solving FP($r$) for any given $r \in [f(a), f(b)]$, we can proceed as follows to find an $\varepsilon$-approximate optimal solution of (CDM) :

0. Start with $r = f(a)$.

1. Solve FP($r$). If FP($r$) is feasible, i.e. an $\varepsilon$-approximate feasible solution $x^{[r]}$ is found such that $f(x^{[r]}) \geq r + \varepsilon$ then set CBS $= x^{[r]}$ and go to Step 2. If FP($r$) is infeasible with $r = f(a)$ the problem (CDM) is infeasible. If FP($r$) is infeasible with $r > f(a)$, then CBS is an $\varepsilon$-approximate optimal solution of (CDM).

2. Reset $r \leftarrow f(x^{[r]})$ and go back to Step 1.

By this approach the feasibility problem FP($r$) constitutes the core of monotonic optimization. Its rationale is that since the feasibility problem FP($r$) does not involve the objective function, it can be solved faster by a suitable modification of the basic algorithm. Intuitively, if $f(x)$ does not differ much from $g(x)$ then the criterion for selecting $z^k$ in Step 1 as prescribed in Algorithm 1 means that $z^k$ must be the farthest point from $\partial^+ G$, thus forcing the procedure to go through many iterations to reach $\partial^+ G$. Therefore, to solve FP($r$) fast, the idea is to select $z^k$ in Step 1 to be the vertex the nearest to $G$, i.e. corresponding to the smallest value of $g(x)$. With this modification in Step 1, Algorithm 1 will bring rapidly the point $z^k$ close to $G$. We are thus led to the following version of Algorithm 1 for solving (CDM):

**Algorithm 1A** (assuming $a \in G$, $b \in H$, otherwise the problem (CDM) is infeasible).

*Initialization.* Set $r = f(a)$, $T_1 = \{b\}$, $k = 1$.

*Step 1.* Select $z^k \in \text{argmin}\{g(z) | \ z \in T_k\}$.
Compute $y^k = \pi_G(z^k)$.

*Step 2.* If $y^k \in H$ and $f(y^k) \geq r + \varepsilon$ then reset $r \leftarrow f(y^k)$, $y^k \leftarrow x^{[r]}$, $T_k \leftarrow T_k \setminus \{z | \ f(z) < f(y^k) + \varepsilon\}$ and go back to Step 1.
If $g(z^k) \leq \varepsilon$, then reset $r \leftarrow f(z^k)$, $x^{[r]} \leftarrow z^k, T_k \leftarrow T_k \setminus \{z | \ f(z) < f(z^k) + \varepsilon\}$ and go back to Step 1.
Otherwise, go to Step 3.

*Step 3.* Determine $T'_k = (T_k \setminus \{z^k\}) \cup \{z^{k,1}, \ldots, z^{k,n}\}$ where

$$z^{k,i} = z^k - (z^k_i - y^k_i)e^i, \ i = 1, \ldots, n. \tag{37.16}$$

From $T'_k$ remove all improper elements, all $z \notin H$ and all $z$ such that $f(z) < r + \varepsilon$ and let $T_{k+1}$ be the resulting set.

*Step 4.* If $T_{k+1} = \emptyset$, terminate: if $x^{[r]}$ is defined, it is an $\varepsilon$-approximate optimal solution of (CDM); otherwise, (CDM) is infeasible.
If $T_{k+1} \neq \emptyset$, set $k \leftarrow k + 1$ and return to Step 1.

**Theorem 1A** *Algorithm 1A (with $\varepsilon > 0$) terminates after finitely many steps, yielding an $\varepsilon$-optimal solution or establishing the infeasibility of the problem.*

*Proof* If the algorithm is infinite then $r$ must remain constant after a number of iterations. Using then the same argument as in the proof of Theorem 1, we can show the existence of an infinite sequence $\{l_k\} \subset \{1, 2, \ldots\}$ such that $z^{l_k}$, $y^{l_k}$ converge to a common limit $\tilde{z}$. Since $z^{l_k} \in H$, $f(z^{l_k}) \geq r + \varepsilon$, $y^{l_k} \in G$ for every $k$, it follows that $\tilde{z} \in G \cap H$, hence for sufficiently large $k$ we must have $z^{l_k} \in H$, $g(z^{l_k})) \leq \varepsilon$, $f(z^{l_k}) \geq r + \varepsilon$, so that the algorithm must stop at this iteration $l_k$. This contradiction proves the finiteness of the algorithm. □

## 7.2 BRANCH AND BOUND

Another method for solving (CDM) is to use the branch and bound strategy. At any iteration of Algorithm 1 the value $f(z^k) = \max\{f(x) | x \in T_k\}$ gives an upper bound of $f(x)$ over all the feasible solutions. Therefore, to solve (CDM) one can use a rectangular branch and bound method, in which branching is performed by rectangular subdivision, and an upper bound over any subrectangle $[p, q]$ of $[0, b]$ is computed by applying a number of iterations of Algorithm 1 to the problem $\max\{f(x) | x \in G \cap H \cap [p, q]\}$. More specifically, an upper bound $\mu(M)$ over the feasible portion in a partition set $M = [p, q]$ and the vectors $z(M), y(M)$ such that $f(z(M)) = \mu(M), y(M) = \pi_G(z(M))$ can be computed by the following subroutine (where CBV denotes the current best value for (CDM), $\varepsilon > 0$ is the tolerance).

**Bounding Subroutine**

Select $L$ (user supplied) = maximal number of iterations to be executed.

If $g(p) > 0$ ($p \notin G$) or $h(q) < 0$ ($q \notin H$) then $[p, q] \cap G \cap H = \emptyset$ and we can set $\mu(M) = -\infty$. So suppose $p \in G, q \in H$.

If $g(q) \leq 0$ then take $\mu(M) = f(q)$, $z(M) = y(M) = q$ and terminate the subroutine ($f(q)$ is then the exact maximum of $f(z)$ over the feasible solutions in $[p, q]$).

*Step 0.* Compute

$$\beta_i = \sup\{\beta > 0 | \ p + \beta e^i \in G, \ p_i + \beta \leq q_i\}, \quad q' = p + \sum_{i=1}^n \beta_i e^i \quad (37.17)$$

$$\alpha_i = \sup\{\alpha > 0 | \ q' - \alpha e^i \in H, \ p_i \leq q'_i - \alpha\}, \quad p' = q' - \sum_{i=1}^n \alpha_i e^i \quad (37.18)$$

If $g(p') > 0$ or $h(q') < 0$ set $\mu(M) = -\infty$. If $g(q') \leq 0, h(q') \geq 0$, set $\mu(M) = f(q')$, $z(M) = y(M) = q'$ and terminate the subroutine ($f(q')$ is then the exact maximum of $f(z)$ over the feasible solutions in $[p, q]$).

If $g(p') \leq 0,\ h(q') \geq 0$, set $[p,q] \leftarrow [p',q'],\ z^1 = q, T_1 = \{z^1\}$. Set $\nu = 1$ ($\nu$: iteration counter for the subroutine).

*Step 1.* Compute $y^\nu$ = intersection of the surface $g(x) = 0$ with the segment between $p$ and $z^\nu$ (this can be done because $g(p) \leq 0,\ g(z^\nu) > 0$). Compute

$$z^{\nu,i} = z^\nu - (z_i^\nu - y_i^\nu)e^i,\ i = 1, \ldots, n.$$

Let $T'_\nu = (T_\nu \setminus \{z^\nu\}) \cup \{z^{\nu,1}, \ldots, z^{\nu,n}\}$. If $h(y^\nu) \geq 0$ (i.e. $y^\nu$ is feasible) then use it to update CBV.

*Step 2.* Let $T_{\nu+1}$ be the set that remains from $T'_\nu$ after removing all improper elements and all $z$ such that: $z \notin H$, or $f(z) \geq CBV + \varepsilon$. If $T_{\nu+1} = \emptyset$, set $\mu(M) = CBV + \varepsilon$. Otherwise, compute

$$z^{\nu+1} \in \operatorname{argmax}\{f(z) |\ z \in T_{\nu+1}\}$$

If $\nu = L$ set $\mu(M) = f(z^{\nu+1}),\ z(M) = z^{\nu+1},\ y(M) = y^{\nu+1}$. Otherwise, set $\nu \leftarrow \nu + 1$ and return to Step 1.

As for the subdivision rule the best is to use a normal subdivision rule as described e.g. in [14]. For large scale problems whose dimension cannot be reduced, this branch and bound approach may give better results than a direct application of Algorithm 1.

**7.2** HEURISTIC AND STOCHASTIC METHODS.

A point $x \in G \cap H$ (where $G$ and $H$ are defined by (37.8)) is called an *upper basic solution* (ubs for short) of the monotonic system

$$g(x) \leq 0 \leq h(x) \tag{37.19}$$

if $x \leq x' \in G \cap H$ implies $x = x'$. Clearly an ubs $x$ must belong to $\partial^+ G$ (upper boundary of $G$) because if $x \notin \partial^+ G$ then there is $y \in K_x \cap G$ and since $H$ is reverse normal and $x \in H$, one must have $y \in H$, i.e. $y \in G \cap H$, but $y \neq x$ (because $y \in K_x$), conflicting with $x$ being an ubs. Furthermore, the search for an optimal solution of (CDM) can be restricted to the set of ubs of (37.19). From Algorithm 1 it is easy to derive the following representation theorem

**Theorem 2** (i) *Every ubs $x$ of a monotonic system (37.19) is characterized by a sequence $\{i_0, i_1, \ldots, i_k, \ldots\}$, where $i_k \in \{1, 2, \ldots, n\}$, such that $x = \lim_{k\to\infty} z^k$ where $z^0 \geq z^1 \geq z^2 \geq \ldots$, $z^k \in H$ for all $k$ and $z^k$ is defined by*

$$z^0 = b,\ z^{k+1} = z^k - (z_{i_k}^k - y_{i_k}^k)e^{i_k} \quad y^k = \pi_G(z^k) \tag{37.20}$$

By providing a way of encoding the set of feasible solutions, this Theorem may facilitate the encoding problem in the genetic approach to (CDM).

Also, if we write $x = [i_0, i_1, \ldots, i_k]$ to denote the point $x = z^k$ defined by (37.20) then each point $x^i = [i_0, i_1, \ldots, i_k, i_{k+1}]$ with $i_{k+1} = i \in \{1, \ldots, n\}$ can be considered a neighbour of $x$. This fact could be used to define the moves when applying the simulated annealing or the tabu search approach to (CDM).

**Remark.** For the canonical monotonic minimization problem (37.9), a successive improvement version, and a branch and bound version of Algorithm 2 can be constructed in the same manner as above. Noting that the minimum of $f(x)$ over $G \cap H$ is achieved on $\partial^- H$, one can define a lower basic solution (lbs) of (37.19) to be any solution $x$ of this system such that $x \geq x' \in G \cap H$ implies $x = x'$. The corresponding representation theorem states that any lbs $x$ of (37.19) is characterized by a sequence $\{i_0, i_1, \ldots, i_k, \ldots\}$, where $i_k \in \{1, 2, \ldots, n\}$, such that $x = \lim_{k\to\infty} z^k$ where $z^0 \leq z^1 \leq z^2 \leq \ldots,\ z^k \in G$ for all $k$ and $z^k$ is defined by

$$z^0 = 0,\ z^{k+1} = z^k + (y^k_{i_k} - z^k_{i_k})e^{i_k} \quad y^k = \rho_H(z^k).$$

# 8 D.M. OPTIMIZATION UNDER LINEAR CONSTRAINTS

As an illustrative application, in this section we consider the important problem of linearly constrained d.m. optimization:

$$\max\{h(x) - g(x)|\ \langle c^i, x\rangle \leq d_i\ (i = 1, \ldots, m),\ 0 \leq x \leq b\} \qquad (LDM)$$

where $c_i \in R^n$, $b \in R^n_{++}$, and $h(x), g(x)$ are increasing functions on $[0, b]$. This problem includes as special case the *linearly constrained polynomial programming problem* (in particular, the *linearly constrained quadratic optimization problem*), which has begun to be extensively studied in recent years [10], [11], [9], [1]. In fact, any polynomial $P(x)$ can be written as $P(x) = P^+(x) - P^-(x)$, where $P^+(x), P^-(x)$ are polynomials grouping all terms of $P(x)$ with positive and negative coefficients, respectively. By adding if necessary the constant $-g(0)$ to the objective function, we may, without loss of generality, assume that $g(0) = 0$.

## 8.1 CONVERSION TO THE CANONICAL FORM

Let $c_i = c_i^+ - c_i^-$ where $c_i^+ = (c_{i1}^+, \ldots, c_{in}^+), c_i^- = (c_{i1}^-, \ldots, c_{in}^-)$ with $c_{ij}^+ = \max\{c_{ij}, 0\}, c_i^- = \max\{-c_{ij}, 0\}$. Consider first the system of linear constraints

$$\langle c_i, x\rangle \leq d_i\ (i = 1, \ldots, m), \quad 0 \leq x \leq b \qquad (37.21)$$

**Lemma 1** *Let* $\alpha = \sum_{i=1}^{m} \langle c_i^+, b\rangle$. *The system (37.21) is equivalent to*

$$(x,t) \in G_0 \cap H_0 \quad \textit{where} \tag{37.22}$$

$$G_0 = \{(x,t)|\ t + \sum_{i=1}^{m} \langle c_i^+, x\rangle \le \alpha, \quad (x,t) \in [0,(b,\alpha)]\} \tag{37.23}$$

$$H_0 = \{(x,t)|\ t + \min_{i=1,\dots,m} (\sum_{k\neq i} \langle c_k^+, x\rangle + \langle c_i^-, x\rangle + d_i) \ge \alpha\} \tag{37.24}$$

*Proof* We can write the system (37.21) as

$$\max_{i=1,\dots,m} \{\langle c_i^+, x\rangle - \langle c_i^-, x\rangle - d_i\} \le 0$$

or equivalently as

$$p(x) - q(x) \le 0 \tag{37.25}$$

where

$$p(x) = \sum_{i=1}^{m} \langle c_i^+, x\rangle,\ q(x) = \min_{i=1,\dots,m} (\sum_{k\neq i} \langle c_k^+, x\rangle + \langle c_i^-, x\rangle + d_i) \tag{37.26}$$

Since $p(x)$ is increasing, we have $p(x) \le p(b) = \alpha$ for $0 \le x \le b$, so if (37.25) holds then setting $t = \alpha - p(x)$ yields

$$p(x) + t \le \alpha \le q(x) + t, \quad 0 \le x \le b,\ 0 \le t \le \alpha. \tag{37.27}$$

Conversely, (37.27) obviously implies (37.25). So (37.21) is equivalent to (37.27) with $p(x), q(x)$ defined from (37.26). Hence the result. □

Now let $\beta = g(b)$ and write $h(x) - g(x) = h(x) - \beta + [\beta - g(x)]$.

**Proposition 11** *Let* $\alpha = \sum_{i=1}^{m} \langle c_i^+, b\rangle, \beta = g(b)$. *The problem (LDM) is equivalent to*

$$\max\{h(x) + u - \beta|\ (x,t,u) \in G \cap H\} \quad \textit{where} \tag{MP}$$

$$G = \{(x,t,u) \in [0,(b,\alpha,\beta)] \mid$$
$$t + \sum_{i=1}^{m} \langle c_i^+, x\rangle \le \alpha,\ u + g(x) \le \beta\} \tag{37.28}$$

$$H = \{(x,t,u)|\ t + \min_{i=1,\dots,m} (\sum_{k\neq i} \langle c_k^+, x\rangle + \langle c_i^-, x\rangle + d_i) \ge \alpha\} \tag{37.29}$$

*Proof* Using the pevious lemma we can write the problem as

$$\max\{h(x)+u-\beta|\ u+g(x)\le\beta,\ p(x)+t\le\alpha\le q(x)+t,\\ 0\le x\le b, 0\le t\le\alpha, 0\le u\le\beta\}$$

where $\beta = g(b)$. Hence the result. □

**Lemma 2** *For every $z=(x,t,u)\in[0,(b,\alpha,\beta)]\backslash G$ define $\lambda_z=\max\{\theta|\ \theta(x,t,u$ $G\}$. Then $\lambda_z=\min\{\lambda_0,\lambda_1\}$ where $\lambda_0$ and $\lambda_1$ are determined from the equations*

$$\lambda_0=\frac{\alpha}{t+\sum_{i=1}^{m}\langle c_i^+,x\rangle},\qquad \lambda_1 u+g(\lambda_1 x)=\beta \qquad (37.30)$$

*Proof* By definition $\lambda_z=\max\{\lambda|\ \lambda t+\lambda\sum_{i=1}^{m}\langle c_i^+,x\rangle\le\alpha,\ \lambda u+g(\lambda x)\le\beta\}$, hence the result. □

**Remark.** Computing $\lambda_1$ according to (37.30) is generally an easy task. In fact, if $g(x)$ is quadratic then $\lambda_1$ is merely a root of the quadratic equation in $\lambda$ : $\lambda u+g(\lambda x)=\beta$. In any case, the function $\varphi(\lambda):=\lambda u+g(\lambda x)$ is monotone increasing in the interval $0\le\lambda\le 1$, so if no better method is available, the following binary search procedure can be used (noting that $\varphi(0)<\beta,\ \varphi(1)>\beta$) :

*Computing* $\lambda_1$.

0. Set $l=0, L=1$.

1. Compute $\varphi(L)$. If $\varphi(L)=\beta$ terminate: $\lambda_1=L$. Otherwise go to Step 2.

2. If $\varphi(L)>\beta$, reset $l\leftarrow(l+L)/2$, and go back to Step 1. Otherwise go to Step 3

3. If $L-l\le\varepsilon$ (tolerance), terminate and accept $\lambda_1=l$. Otherwise, reset $L\leftarrow(l+L)/2$ and go back to Step 1.

## 8.2 POLYBLOCK ALGORITHM FOR (LDM)

For convenience we use the notation $z=(x,t,u)$, so that $z_i=x_i$ for $i=1,\dots,n$ and $z_{n+1}=t, z_{n+2}=u$. Let $w^0=(b,\alpha,\beta)$.

By translating if necessary we can assume that

$$x>a,\ t>\theta,\ u>\sigma\quad \forall z=(x,t,u)\in G\cap H. \qquad (37.31)$$

for some $(a,\theta,\sigma)\in R^n\times R\times R$ such that $0<(a,\theta,\sigma)<(b,\alpha,\beta)$. Then (LDM) has the canonical form

$$\max\{f(z)|\ U(z)\le 0\le V(z),\ z\in[0,(b,\alpha,\beta)]\}$$

where for $z = (x, t, u)$ :

$$f(z) = h(x) + u - \beta,$$

$$U(z) = \max\{t + \sum_{i=1}^{m} \langle c_i^+, x\rangle - \alpha,\ u + g(x) - \beta\}$$

$$V(z) = t + \min_{i=1,\dots,m} (\sum_{k\neq i} \langle c_k^+, x\rangle + \langle c_i^-, x\rangle + d_i - \alpha$$

and the condition (37.10) is satisfied . Furthermore, assume

$$(a, \theta, \sigma) \in G, \quad (b, \alpha, \beta) \in H,$$

otherwise (LDM) is clearly infeasible.

**Algorithm 3** (for (LDM).
*Initialization.* Set $r = h(a) + \sigma - \beta,\ z^1 = (b, \alpha, \beta),\ T_1 = \{z^1\},\ k = 1.$
*Step 1.* Select $z^k = (x^k, t^k, u^k) \in \operatorname{argmin}\{U(z)|\ z \in T_k\}$.
Compute $\lambda_k = \lambda_{z^k}$ (see Lemma 2), $\tilde{z}^k = \lambda_k z^k$.
*Step 2.* If $\tilde{z}^k \in H$ and $f(\tilde{z}^k) = h(\lambda_k x^k) + \lambda_k u^k - \beta \geq r + \varepsilon$ then reset $r \leftarrow f(\tilde{z}^k),\ \tilde{z}^k \leftarrow z^{[r]},\ T_k \leftarrow T_k \setminus \{z|\ f(z) < f(\tilde{z}^k) + \varepsilon\}$ and go back to Step 1.
If $U(z^k) \leq \varepsilon$, then reset $r \leftarrow f(z^k),\ z^{[r]} \leftarrow z^k, T_k \leftarrow T_k \setminus \{z|\ f(z) < f(z^k) + \varepsilon\}$ and go back to Step 1.
Otherwise, go to Step 3.
*Step 3.* Determine $T_k' = (T_k \setminus \{z^k\}) \cup \{z^{k,1} \dots, z^{k,n}\}$ where

$$z^{k,i} = z^k - (z_i^k - \tilde{z}_i^k)e^i,\ i = 1, \dots, n. \tag{37.32}$$

From $T_k'$ remove all improper elements, all $z \notin H$ and all $z$ such that $f(z) < r + \varepsilon$ and let $T_{k+1}$ be the resulting set.
*Step 4.* If $T_{k+1} = \emptyset$, terminate: if $z^{[r]}$ is defined, it is an $\varepsilon$-approximate optimal solution of (LDM); otherwise, (LDM) is infeasible.
If $T_{k+1} \neq \emptyset$, set $k \leftarrow k + 1$ and return to Step 1.

**Remarks.** (i) The above algorithm reduces (LDM), in particular any linearly constrained polynomial program, to a sequence of univariate problems of computing the zero of a monotone increasing function of one variable.

(ii) The minimization of a difference of two increasing functions over a polytope $D$ can be reduced to the above problem since $\min\{h(x) - g(x)|\ x \in D\} = -\max\{g(x) - h(x)|\ x \in D\}$.

## 8.3 EXTENSION: FRACTIONAL PROGRAMMING

The above method can be extended to solve fractional programming problems of the form

$$\max\{\frac{h(x)}{g(x)} | \langle c_i, x\rangle \leq d_i \ (i = 1, \ldots, m), \ a \leq x \leq b\} \tag{FP}$$

where $h(x), g(x)$ are continuous increasing on $[0, b]$ and $g(x) > 0 \ \forall x \in [0, b]$.

By replacing the objective function with $\frac{h(x)}{g(x)} + M$ where

$$M > \max\{\frac{|h(x)|}{g(x)} | 0 \leq x \leq b\}$$

we transform the objective function into $\frac{h(x) + Mg(x)}{g(x)}$ with $h(x) + Mg(x) > 0 \ \forall x \in [0, b]$. Therefore, without loss of generality we can assume $h(x) > 0 \ \forall x \in [0, b]$.

Under these conditions $\frac{h(x)}{g(x)}$ is positive on $[0, b]$, so instead of maximizing this quotient we can maximize $\log(\frac{h(x)}{g(x)}) = \tilde{h}(x) - \tilde{g}(x)$. Here $\tilde{h}(x) = \log h(x), \tilde{g}(x) = \log g(x)$ are increasing functions, therefore the above algorithm applies.

Let $\beta = 1/g(0)$. In terms of the functions $h(x), g(x)$ rather than their logarithms, it is easily seen that the problem equivalent to (FP) is

$$\max\{uh(x) | \ (x, t, u) \in G \cap H\} \quad \text{where} \tag{MFP}$$

$$G = \{(x, t, u) \in [0, (b, \alpha, \beta)] \mid t + \sum_{i=1}^{m} \langle c_i^+, x\rangle \leq \alpha, \ ug(x) \leq \beta\} \tag{37.33}$$

$$H = \{(x, t, u) | \ t + \min_{i=1,\ldots,m} (\sum_{k \neq i} \langle c_k^+, x\rangle + \langle c_i^-, x\rangle + d_i) \geq \alpha, \ x > a\} \tag{37.34}$$

So the previous method applies, except that instead of $u + h(x)$ and $u + g(x)$ we now have $uh(x)$ and $ug(x)$.

## References

[1] C.A. Floudas, *Deterministic Global Optimization*, Kluwer Academic Publishers, 2000.

[2] R. Horst and H. Tuy, *Global Optimization (Deterministic Approaches)*, third edition, Springer-Verlag, 1996.

[3] H. Konno and T. Kuno, Multiplicative Programming Problems, in R. Horst and P.M. Pardalos (eds.), *Handbook on Global Optimization*, 1995, Kluwer Academic Publishers, pp. 369-406.

[4] H. Konno, Y. Yajima and T. Matsui, Parametric simplex algorithms for solving a special class of nonconvex minimization problems, *Journal of Global Optimization*, **1**(1991) 65-81.

[5] H. Konno, P.T. Thach and H. Tuy, *Optimization on Low Rank Nonconvex Structures*, Kluwer Academic Publishers, 1997.

[6] L.T. Luc, Monotonic Approach to Optimization over the Weakly Efficient and Efficient Sets. *Acta Mathematica Vietnamica*, to appear.

[7] V.L. Makarov and AS.M. Rubinov, *Mathematical Theory of Economic Dynamic and Equilibria*, Springer 1977.

[8] A. Rubinov, H. Tuy and H. Mays, Algorithm for a monotonic global optimization problem, *Optimization*, to appear.

[9] H.D. Sherali and W.P. Adams, *A reformualtion-Linearization Technique for Solving Discrete and Continuous Nonconvex Problems*, Kluwer Academic Publishers, 1998.

[10] H.D. Sherali and C.H. Tuncbilek, A reformualtion-convexification approach to solving nonconvex quadratic programming problems, *Journal of Global Optimization*, **7**(1995), 1-31.

[11] N.Z. Shor, *Nondifferentiable Optimization and Polynomial Problems*, Kluwer Academic Publishers, 1998.

[12] H. Tuy, The complementary convex structure in global optimization, *Journal of Global Optimization*, **2**(1992) 21-40.

[13] H. Tuy, D.C. Optimization: Theory, Methods and Algorithms, in R. Horst and P.M. Pardalos (eds.), *Handbook on Global Optimization*, 1995, Kluwer Academic Publishers, pp. 149-216.

[14] H. Tuy, *Convex Analysis and Global Optimization*, Kluwer Academic Publishers, 1998.

[15] H. Tuy, Normal sets, polyblocks and monotonic optimization, *Vietnam Journal of Mathematics*, 1999, 277-300.

[16] H. Tuy, On Some Recent Advances and Applications of D.C. Optimization, in V.H. Nguyen, J-J. Strodiot, P. Tossings (eds), *Optimization*, Lecture Notes in Economics and Mathematical Systems, 481, Springer-Verlag 2000, pp.473-497.

[17] H. Tuy, Monotonic optimization: Problems and solution approaches. Preprint, Institute of Mathematics, Hanoi 1999. To appear in SIOPT 2000.

[18] H. Tuy and Le Tu Luc, A new approach to optimization under monotonic constraint, *Journal of Global Optimization*, to appear.

[19] H. Tuy and Ng.T. Hoai-Phuong: A Unified Monotonic Approach to Generalized Linear Fractional Programming, Preprint, Institute of Mathematics, Hanoi 2000.

# Nonconvex Optimization and Its Applications

1. D.-Z. Du and J. Sun (eds.): *Advances in Optimization and Approximation.* 1994 ISBN 0-7923-2785-3
2. R. Horst and P.M. Pardalos (eds.): *Handbook of Global Optimization.* 1995 ISBN 0-7923-3120-6
3. R. Horst, P.M. Pardalos and N.V. Thoai: *Introduction to Global Optimization* 1995 ISBN 0-7923-3556-2; Pb 0-7923-3557-0
4. D.-Z. Du and P.M. Pardalos (eds.): *Minimax and Applications.* 1995 ISBN 0-7923-3615-1
5. P.M. Pardalos, Y. Siskos and C. Zopounidis (eds.): *Advances in Multicriteria Analysis.* 1995 ISBN 0-7923-3671-2
6. J.D. Pintér: *Global Optimization in Action.* Continuous and Lipschitz Optimization: Algorithms, Implementations and Applications. 1996 ISBN 0-7923-3757-3
7. C.A. Floudas and P.M. Pardalos (eds.): *State of the Art in Global Optimization.* Computational Methods and Applications. 1996 ISBN 0-7923-3838-3
8. J.L. Higle and S. Sen: *Stochastic Decomposition.* A Statistical Method for Large Scale Stochastic Linear Programming. 1996 ISBN 0-7923-3840-5
9. I.E. Grossmann (ed.): *Global Optimization in Engineering Design.* 1996 ISBN 0-7923-3881-2
10. V.F. Dem'yanov, G.E. Stavroulakis, L.N. Polyakova and P.D. Panagiotopoulos: *Quasidifferentiability and Nonsmooth Modelling in Mechanics, Engineering and Economics.* 1996 ISBN 0-7923-4093-0
11. B. Mirkin: *Mathematical Classification and Clustering.* 1996 ISBN 0-7923-4159-7
12. B. Roy: *Multicriteria Methodology for Decision Aiding.* 1996 ISBN 0-7923-4166-X
13. R.B. Kearfott: *Rigorous Global Search: Continuous Problems.* 1996 ISBN 0-7923-4238-0
14. P. Kouvelis and G. Yu: *Robust Discrete Optimization and Its Applications.* 1997 ISBN 0-7923-4291-7
15. H. Konno, P.T. Thach and H. Tuy: *Optimization on Low Rank Nonconvex Structures.* 1997 ISBN 0-7923-4308-5
16. M. Hajdu: *Network Scheduling Techniques for Construction Project Management.* 1997 ISBN 0-7923-4309-3
17. J. Mockus, W. Eddy, A. Mockus, L. Mockus and G. Reklaitis: *Bayesian Heuristic Approach to Discrete and Global Optimization.* Algorithms, Visualization, Software, and Applications. 1997 ISBN 0-7923-4327-1
18. I.M. Bomze, T. Csendes, R. Horst and P.M. Pardalos (eds.): *Developments in Global Optimization.* 1997 ISBN 0-7923-4351-4
19. T. Rapcsák: Smooth Nonlinear Optimization in $R^n$. 1997 ISBN 0-7923-4680-7
20. A. Migdalas, P.M. Pardalos and P. Värbrand (eds.): *Multilevel Optimization: Algorithms and Applications.* 1998 ISBN 0-7923-4693-9
21. E.S. Mistakidis and G.E. Stavroulakis: *Nonconvex Optimization in Mechanics.* Algorithms, Heuristics and Engineering Applications by the F.E.M. 1998 ISBN 0-7923-4812-5

# Nonconvex Optimization and Its Applications

22. H. Tuy: *Convex Analysis and Global Optimization.* 1998 ISBN 0-7923-4818-4
23. D. Cieslik: *Steiner Minimal Trees.* 1998 ISBN 0-7923-4983-0
24. N.Z. Shor: *Nondifferentiable Optimization and Polynomial Problems.* 1998 ISBN 0-7923-4997-0
25. R. Reemtsen and J.-J. Rückmann (eds.): *Semi-Infinite Programming.* 1998 ISBN 0-7923-5054-5
26. B. Ricceri and S. Simons (eds.): *Minimax Theory and Applications.* 1998 ISBN 0-7923-5064-2
27. J.-P. Crouzeix, J.-E. Martinez-Legaz and M. Volle (eds.): *Generalized Convexitiy, Generalized Monotonicity: Recent Results.* 1998 ISBN 0-7923-5088-X
28. J. Outrata, M. Kočvara and J. Zowe: *Nonsmooth Approach to Optimization Problems with Equilibrium Constraints.* 1998 ISBN 0-7923-5170-3
29. D. Motreanu and P.D. Panagiotopoulos: *Minimax Theorems and Qualitative Properties of the Solutions of Hemivariational Inequalities.* 1999 ISBN 0-7923-5456-7
30. J.F. Bard: *Practical Bilevel Optimization.* Algorithms and Applications. 1999 ISBN 0-7923-5458-3
31. H.D. Sherali and W.P. Adams: *A Reformulation-Linearization Technique for Solving Discrete and Continuous Nonconvex Problems.* 1999 ISBN 0-7923-5487-7
32. F. Forgó, J. Szép and F. Szidarovszky: *Introduction to the Theory of Games.* Concepts, Methods, Applications. 1999 ISBN 0-7923-5775-2
33. C.A. Floudas and P.M. Pardalos (eds.): *Handbook of Test Problems in Local and Global Optimization.* 1999 ISBN 0-7923-5801-5
34. T. Stoilov and K. Stoilova: *Noniterative Coordination in Multilevel Systems.* 1999 ISBN 0-7923-5879-1
35. J. Haslinger, M. Miettinen and P.D. Panagiotopoulos: *Finite Element Method for Hemivariational Inequalities.* Theory, Methods and Applications. 1999 ISBN 0-7923-5951-8
36. V. Korotkich: *A Mathematical Structure of Emergent Computation.* 1999 ISBN 0-7923-6010-9
37. C.A. Floudas: *Deterministic Global Optimization: Theory, Methods and Applications.* 2000 ISBN 0-7923-6014-1
38. F. Giannessi (ed.): *Vector Variational Inequalities and Vector Equilibria.* Mathematical Theories. 1999 ISBN 0-7923-6026-5
39. D.Y. Gao: *Duality Principles in Nonconvex Systems.* Theory, Methods and Applications. 2000 ISBN 0-7923-6145-3
40. C.A. Floudas and P.M. Pardalos (eds.): *Optimization in Computational Chemistry and Molecular Biology.* Local and Global Approaches. 2000 ISBN 0-7923-6155-5
41. G. Isac: *Topological Methods in Complementarity Theory.* 2000 ISBN 0-7923-6274-8
42. P.M. Pardalos (ed.): *Approximation and Complexity in Numerical Optimization: Concrete and Discrete Problems.* 2000 ISBN 0-7923-6275-6
43. V. Demyanov and A. Rubinov (eds.): *Quasidifferentiability and Related Topics.* 2000 ISBN 0-7923-6284-5

# Nonconvex Optimization and Its Applications

44. A. Rubinov: *Abstract Convexity and Global Optimization.* 2000 ISBN 0-7923-6323-X
45. R.G. Strongin and Y.D. Sergeyev: *Global Optimization with Non-Convex Constraints.* 2000 ISBN 0-7923-6490-2
46. X.-S. Zhang: *Neural Networks in Optimization.* 2000 ISBN 0-7923-6515-1
47. H. Jongen, P. Jonker and F. Twilt: *Nonlinear Optimization in Finite Dimensions.* Morse Theory, Chebyshev Approximation, Transversability, Flows, Parametric Aspects. 2000 ISBN 0-7923-6561-5
48. R. Horst, P.M. Pardalos and N.V. Thoai: *Introduction to Global Optimization.* 2nd Edition. 2000 ISBN 0-7923-6574-7
49. S.P. Uryasev (ed.): *Probabilistic Constrained Optimization.* Methodology and Applications. 2000 ISBN 0-7923-6644-1
50. D.Y. Gao, R.W. Ogden and G.E. Stavroulakis (eds.): *Nonsmooth/Nonconvex Mechanics.* Modeling, Analysis and Numerical Methods. 2001 ISBN 0-7923-6786-3
51. A. Atkinson, B. Bogacka and A. Zhigljavsky (eds.): *Optimum Design 2000.* 2001 ISBN 0-7923-6798-7
52. M. do Rosário Grossinho and S.A. Tersian: *An Introduction to Minimax Theorems and Their Applications to Differential Equations.* 2001 ISBN 0-7923-6832-0
53. A. Migdalas, P.M. Pardalos and P. Värbrand (eds.): *From Local to Global Optimization.* 2001 ISBN 0-7923-6883-5
54. N. Hadjisavvas and P.M. Pardalos (eds.): *Advances in Convex Analysis and Global Optimization.* Honoring the Memory of C. Caratheodory (1873-1950). 2001 ISBN 0-7923-6942-4

KLUWER ACADEMIC PUBLISHERS – DORDRECHT / BOSTON / LONDON

GPSR Compliance

*The European Union's (EU) General Product Safety Regulation (GPSR) is a set of rules that requires consumer products to be safe and our obligations to ensure this.*

*If you have any concerns about our products, you can contact us on ProductSafety@springernature.com*

In case Publisher is established outside the EU, the EU authorized representative is:

Springer Nature Customer Service Center GmbH
Europaplatz 3
69115 Heidelberg, Germany

**Batch number: 10755248**

Printed by Printforce, the Netherlands